Preventing, Identifying and Treating Diseases in Dairy Cattle

Preventing, Identifying and Treating Diseases in Dairy Cattle

Edited by Bethany Crawford

MURPHY & MOORE
www.murphy-moorepublishing.com

Published by Murphy & Moore Publishing,
1 Rockefeller Plaza,
New York City, NY 10020, USA
www.murphy-moorepublishing.com

Preventing, Identifying and Treating Diseases in Dairy Cattle
Edited by Bethany Crawford

International Standard Book Number: 978-1-63987-448-4 (Hardback)

Cataloging-in-Publication Data

Preventing, identifying and treating diseases in dairy cattle / edited by Bethany Crawford.
p. cm.
Includes bibliographical references and index.
ISBN 978-1-63987-448-4
1. Dairy cattle. 2. Dairy cattle--Diseases. 3. Dairy cattle--Diseases--Prevention. 4. Veterinary medicine.
I. Crawford, Bethany.
SF961 .P74 2021
636.208 96--dc23

TABLE OF CONTENTS

PREFACE

Dairy cattle are the cattle cows bred in order to produce large quantities of milk in order to produce various dairy products. Veterinary diseases are of prime concern as they pose a risk to public health and cause damage to economy of the state. Several factors like poor sanitation and improper management are cited for the transmission and advancement of these diseases. Some of the diseases associated with dairy cattle are bacteria spongiform encephalopathy (mad cow disease), digital dermatitis, anthrax and pneumonia. Proper hygiene and use of vaccinations are some of the preventive measures taken to ensure efficient production and improve the general health of dairy cattle. The book aims to shed light on some of the unexplored aspects of diseases in dairy cattle and the recent researches in this field. Such selected concepts that redefine prevention, identification and treatment of diseases in dairy cattle have been presented herein. The extensive content of this book provides the readers with a thorough understanding of the subject.

After months of intensive research and writing, this book is the end result of all who devoted their time and efforts in the initiation and progress of this book. It will surely be a source of reference in enhancing the required knowledge of the new developments in the area. During the course of developing this book, certain measures such as accuracy, authenticity and research focused analytical studies were given preference in order to produce a comprehensive book in the area of study.

This book would not have been possible without the efforts of the authors and the publisher. I extend my sincere thanks to them. Secondly, I express my gratitude to my family and well-wishers. And most importantly, I thank my students for constantly expressing their willingness and curiosity in enhancing their knowledge in the field, which encourages me to take up further research projects for the advancement of the area.

Editor

1

Assessment of the main plasma parameters included in a metabolic profile of dairy cow based on Fourier Transform mid-infrared spectroscopy

Luigi Calamari*, Annarita Ferrari, Andrea Minuti and Erminio Trevisi

Abstract

Background: Although a metabolic profile represents a valid tool utilized in dairy herds to determine abnormalities in blood chemistry related to an increased risk of production diseases, there are no studies on application of Fourier Transform mid-infrared (FT-MIR) spectroscopy. This study assesses the potential application of FT-MIR to analyze the main blood biochemical parameters included in the metabolic profile of dairy cows. Infrared transmission spectra were acquired for 35 plasma samples (two replicates on each sample) of Italian Friesian dairy cows (14 primiparous and 21 pluriparous), all without clinical events, and at different stages of lactation, although mainly in the transition phase. Each sample was also analyzed independently using accepted reference clinical chemical methods and these results were used as calibrating values to perform predictive models by PLS method using cross validation.

Results: Measured blood parameters concentrations were all within the reference ranges reported for healthy dairy cows. The number of extracted factors with the PLS procedure for each prediction model ranged between 3 and 7. The coefficient of determination (R^2) of the prediction models ranged between 0.1 to values close to 1. R^2 values greater than 0.9 were observed for the prediction models of total cholesterol, total protein, globulin, and albumin; values between 0.75 and 0.9 were observed for urea, NEFA, and total bilirubin, while values of R^2 lower than 0.6 were observed for all minerals and for enzyme activity. The range error ratio (RER) and prediction to deviation (RPD) ranged from 5.1 to 43.8 and from 1 to 13.8 for RER and RPD, respectively. Values of RPD greater than 5 were observed for total cholesterol, total protein, albumin, and globulin. RPD ranged between 2 and 5 for the prediction models of urea, NEFA, and total bilirubin, while RPD and RER were low for minerals and enzyme activities.

Conclusions: Although the results of this study require further validation, the use of FT-MIR spectroscopy was possible and provides fairly accurate measurement of various parameters of great importance in the evaluation of the metabolic and inflammatory status in dairy cows.

Keywords: Dairy cows, Metabolic profile, FT-MIR spectroscopy, Infrared model prediction

Background

Blood biochemistry is mainly used in diagnostic investigation to provide supportive evidence of a suspected diagnosis, as a prognostic indicator, or for monitoring the progress of disease in animals that are being treated. Payne et al. [1] proposed that blood tests could be included in a metabolic profile of dairy herds to indicate an inability to maintain homeostasis, with the aim to determine abnormalities in some blood parameters and, the signaling of an increase in the risk of production diseases. Recently, Clarke and Haselden [2] extended the meaning of metabolic profiling and include the measurement of any parameter in animal fluids that is able to reflect a dynamic response to genetic modification, and physiological, pathophysiological and developmental stimuli. Although substrate and analytical methods are similar in evaluating disease diagnosis and metabolic profiling, there are different approaches to

* Correspondence: luigi.calamari@unicatt.it
Istituto di Zootecnica, Facoltà di Scienze Agrarie, Alimentari e Ambientali, Università Cattolica del Sacro Cuore, Piacenza 29122, Italy

sampling and interpretation. To diagnose disease, a small population of clinically affected animals are selected for the blood analysis. Diagnosis of disease is based on changes in one or more blood analytes. Conversely, for a proper assessment of the metabolic profile blood, samples are collected from clinically healthy animals within certain well-defined physiological phases [3]. The metabolic profile is then used to evaluate disease risk in contrast to disease diagnosis. When used in association with animal, diet and management assessments, a metabolic profile can be a useful tool for prediction of periparturient problems and infertility, to diagnose metabolic diseases, to assess nutritional status [4–6], stress conditions [7, 8], and to determine welfare condition [9, 10].

Because a metabolic profile considers clinically healthy animals, the variability of analytes is lower than the variability in blood biochemistry for diagnosis of diseases. As such, a metabolic profile requires high precision and an accurate analytical method.

The blood tests included in a metabolic profile are in generally analyzed by using an automated or semi-automated analyzer. However, some critical aspects can restrict its use, including the cost of analytical testing and in some cases the long waiting time between the sampling and return of analysis results. Human studies have demonstrated the potential of infrared spectroscopic methods for reliable clinical chemistry [11, 12], as well as clinical applications and diseases diagnosis [13].

Applications of infrared spectroscopy as an analytical tool in different areas of animal production are receiving growing interest and attention. This technology is used to assess feed composition and digestibility, traceability assessment, and for determination of the main composition of animal products (meat, milk, fish, cheese, eggs) [14]. For more information see the review paper by Berzaghi and Riovanto [14]. Infrared spectroscopy has several advantages over other analytical techniques: rapidity of analysis, no use of chemicals, minimal or no sample preparation, and easy applicability in different work environments (on/in/at line applications) [11, 15].

To the best of our knowledge no reports have investigated the possibility of using infrared spectroscopy to analyze a metabolic profile in dairy cows. Considering the results obtained in human studies that demonstrate the potential application of infrared spectroscopy in this field, the objective of the present research was to test the feasibility of developing prediction models of the main blood biochemical parameters included in a metabolic profile for dairy cows using MIR spectrometry.

Methods

Animal and management conditions

The research protocol and the animal care were in accordance with the Directive 2010/63/EU (European Union, 2010) on the protection of animals used for scientific purposes. The cows recruited in this trial were involved in Research Projects authorized by Italian Ministry of Health and approved by the ethics committee (Commissione per la valutazione etica di sperimentazioni animali e di correttezza della gestione dell'animal care) prot. N. 25906/13 of 22 Nov 2013.

The Italian Friesian dairy cows involved in this study were raised in the experimental free stall barns of the Università Cattolica del Sacro Cuore (Piacenza, Italy). Dry cows and lactating cows were kept in two adjoining pens. The cows were moved from a far-off pen to a late pregnancy pen one month before the expected calving day and the heifers were moved two months before their expected calving day. After calving the cows were housed in the pen for lactating cows. Fresh potable water was available *ad libitum* within each pen. Cows were fed TMR once daily for *ad libitum* DMI, at 0830 h and 1030 h for lactating and dry cows, respectively. Diets were formulated to meet requirements according to NRC system (2001).

The trial involved 35 Italian Friesian dairy cows (14 primiparous and 21 pluriparous) of good genetic merit, in the last three months of pregnancy and till 280 days in milk. The 40 % of cows (14 animals) were in the transition period (from 21 days before calving till 21 days after calving).

Blood sampling and analyses with reference methods

Blood samples were collected from all cows all at one time. The samples were collected in the morning, before the feed distribution, by venipuncture from the jugular vein, using 10-mL Li-heparin treated tubes (Vacuette, containing 18 IU of Li-heparin mL^{-1}, Kremsmünster, Austria). Samples were immediately cooled in an ice-water bath after collection.

The blood was centrifuged (3500 × g for 16 min at 4 °C) and the plasma obtained was separated into two aliquots: the first fraction was immediately used to collect the infrared spectra; the second one was stocked at −20 °C until metabolites analysis and the results were used as calibration values.

Plasma metabolites used as calibrating values were analyzed at 37 °C by an automated clinical analyzer (ILAB 600, Instrumentation Laboratory, Lexington, MA), using the methodology showed in Table 1. Commercial kits were used to measure glucose, total cholesterol, urea, calcium, inorganic phosphorus, magnesium, total protein, albumin, total bilirubin, and creatinine (Instrumentation Laboratory SpA, Werfen, Monza, Milan, Italy), NEFA and zinc (Wako, Chemicals GmbH, Neuss, Germany), and β-OH-butyric acid (BHBA, kit Ranbut, Randox Laboratories Limited, Crumlin, County Antrim, United Kingdom Randox, UK). A Trinder end point [Glucose oxidase (GOD)/Peroxidase (POD)] was used to measure

Table 1 Methodologies used to measure the plasma parameters with reference methods

Parameter	Methodology	Wavelength (nm)	CV[a]
Glucose	Endpoint	510	1.50
Total cholesterol	Endpoint	510	2.10
NEFA[b]	Endpoint	546	1.50
BHBA[c]	Endpoint	340	5.25
Urea	Endpoint	340	1.20
Creatinine	Endpoint	510	5.40
Ca	Endpoint	570	1.40
Inorganig P	Endpoint	340	2.00
Mg	Rate	340	1.40
Na	ISE device[g]		0.90
K	ISE device[g]		1.30
Cl	ISE device[g]		1.50
Zn	Endpoint	546	
Ceruloplasmin	Endpoint	546	3.48
Total protein	Endpoint	546	1.20
Albumin	Endpoint	600	1.80
Total bilirubin	Endpoint	546	6.70
Haptoglobin	Endpoint	450	13.54
AST[d]	Rate	340	2.10
GGT[e]	Rate	405	3.72
AP[f]	Rate	405	1.70

[a]Calculared on the results obtained between runs according to the National Committee for Clinical Laboratory Standards (Document EP3-T: Guidelines for Manufacturers for Establishing
Performance Claims for Clinical Chemistry Methods, Replication Experiment Evaluation", Villanova, PA, 1982.)
[b]Non esterified fatty acids;
[c]β-OH-butyric acid;
[d]Aspartate amino transferase
[e]γ-glutamyl transferase
[f]Alkaline phosphatase
[g]Ion selective electrodes

glucose. Total cholesterol (cholesterol and cholesterol esters) was also measured using Trinder end point [Cholesterol oxidase (CHOD)/Peroxidase (POD)], after a hydrolysis of cholesterol esters to free cholesterol. Urea was measured with end point method using the couple urease glutamate dehydrogenase (GLDH) enzyme system. Colorimetric methodology based on the reaction of calcium with o-cresolphthalein complexone was used to measure Ca with end point method. Inorganic phosphorus was measured with end point UV method, based on the reaction between phosphate ions in an acidic medium with ammonium molybdate to form a phosphomolybdate complex. The magnesium determination was based on the reaction of magnesium with Xylidyl Blue (as chelator) at alkaline pH, which yields a purple colored complex. Total protein were measured with the modified biuret methodology, based on the reaction of peptide bonds with Cu^{++} ions in alkaline solution to form a colored complex. Albumin was measured with an end point colorimetric method, based on the binding between albumin with green bromocresol resulting in a spectral change of the dye from yellow to green. Total bilirubin was determined with an end point analysis using modified Jendrassik-Grof method, based on the reaction between total bilirubin with diazotized sulfanilic acid in presence of lithium dodecylsulfate to form azobilirubin. Creatinine was analyzed with an end point colorimetric method, based on the reaction of creatinine with picric acid under alkaline conditions. NEFA were determined with an Trinder end point [Acyl coenzyme A oxidase (ACOD)/ Peroxidase (POD)] assay, after the acylation of coenzyme A by NEFA contained in the sample. BHBA was measured with a kinetic UV method, based on the oxidation od D-3 hydroxybutyrate to acetoacetate by 3-Hydroxybutyrate dehydrogenase. Electrolytes, Na, K, and Cl, were measured using a potentiometric system, with specific electrodes. Kinetic analysis was adopted to determine the activity of enzymes: alkaline phosphatase (AP, EC 3.1.3.1), aspartate aminotransferase (AST, EC 2.6.1.1), γ-glutamyltransferase (GGT, EC 2.3.2.2) using Instrumentation Laboratory kits (Instrumentation Laboratory SpA, Werfen, Monza, Milan, Italy). Ceruloplasmin was measured using the method described by Sunderman and Nomoto [16], adapted to ILAB 600 condition. The method is based on measurement of p-phenylenediamine dihydrochloride oxidation by the oxidase activity of ceruloplasmin. Finally, haptoglobin was measured using the method described by Skinner et al. [17] and Owen et al. [18] and adapted to ILAB 600 condition. The method is based on peroxidase activity of methaemoglobin-haptoglobin complex measured by the rate of oxidation of guaiacol (hydrogen donor) in presence of hydrogen peroxide (oxidizing substrate).

FT-MIR spectroscopy

FT-MIR measurements were performed with a MilkoScan FT 120 (Foss Electric, Hillerød, Denmark). FT-MIR spectra were collected (two replicates on each sample of the calibration datasets, and one replicate on each sample of the validation dataset) as % transmittance in the range of 5012–926 cm^{-1}. The transmittance spectrum measured on each sample was standardized using the FT-MIR Equalizer sample (Foss Electric, Hillerød, Denmark), in order to use the same calibration for several instruments. The instrument was standardized with the equalizer sample before reading the spectra of the samples collected from each farm.

In order to not use areas of the spectra that introduce noise to the calibration, only the following areas were used to develop the prediction model: 2971.7 to 2432.4 cm^{-1}, 2272.4 to 1716.8 cm^{-1}, and 1543.2 to 964.5 cm^{-1}. The two areas 3626.5 to 2970.7 cm^{-1} and 1716.8 to 1543.2 cm^{-1}, corresponding to water absorption areas, were not used to calibrate plasma parameters. The 5012.0 to 3626.5 cm^{-1} area was excluded as this contains very little interesting information (Foss Electric, 2002).

Data processing

All analyses were performed using the statistical software package SAS 9.2 (SAS Inst. Inc., Cary, NC). Data were tested for non-normality by the Shapiro test. In case of non-normality, parameters were normalized by log or exponential transformation. Transformations were performed for total bilirubin, haptoglobin, AST, GGT, AP, NEFA, and BHBA.

The infra-red data were processed using the SAS statistical package. In order to suppress side effects from water absorption [19], before statistical analysis the transmittance data of each area was referred to a reference value, dividing the %T of each area by the %T of the reference area (1492 to 1481 cm^{-1}). The absorbance was then calculated (−log10 of the corrected %T) for each area.

Partial least-squares (PLS) analysis was used to process the data by SAS. The selection of the optimal number of the extracted PLS factors for the prediction model of each plasma variable was obtained using cross validation to prevent overfitting. The predictors and the responses were centered and scaled to have a mean of 0 and a standard deviation of 1. Centering the predictors and the response ensures that the criterion for choosing successive factors is based on how much variation they explain, in either the predictors or the responses or both. Scaling serves to place all predictors and responses on an equal footing relative to their variation in the data. The accuracy of the predictive ability of the model is assessed by cross-validation, by means of the root mean square error of cross validation (RMSECV). All prediction residuals were then combined to compute the predicted residual of sum of squares (PRESS) [20]. A statistical model comparison [21] was applied to test whether the differences between the cross-validated residuals from models with different number of extracted factors were significant. The chosen prediction model was based on the results obtained with the cross validation, and the number of factors chosen was the fewest with residuals that were not significantly larger than the residuals of a model with minimum PRESS.

The accuracy of the calibration was evaluated considering the coefficients of determination R^2 for predicted versus measured values in cross-validation; the ratio of the range of each variable in the data was set to its standard deviation of prediction errors; and the ratio between the standard deviation of a variable in the data set (SD) by its standard deviation of prediction errors. Values for R^2 between 0.66 and 0.81 indicate approximate quantitative predictions, whereas, a value for R^2 between 0.82 and 0.90 is adequate for a good prediction. Calibration models having a value for R^2 above 0.91 are considered to be excellent [22]. The practical utility of the calibration was also evaluated by using the range error ratio (RER). This ratio was calculated by dividing the range of each variable by its standard deviation of prediction errors [23]. Values of RER below 3 indicate that a model has little practical utility; RER values between 3 and 10 are limited to good practical utility and RER values above 10 indicate models of high utility value [23]. Because RER could be susceptible to the presence of extreme values at both ends of the range, the ratio of the SD to the standard deviation of prediction errors, called the ratio of prediction to deviation (RPD), was also considered. Based on the RPD value, five levels of prediction accuracy according to Saeys et al. [24] were considered. A value for the RPD below 1.5 indicates that the calibration is not usable. Values for RPD between 1.5 and 2.0 reveal a possibility to distinguish between low and high values, while a value between 2.0 and 2.5 allows for approximate quantitative predictions. For values between 2.5 and 3.0, and above 3.0, the prediction is classified as good and excellent, respectively.

Repeatability (S_r) was obtained as the standard deviation of the results obtained with the prediction models applied to the two replicates and infrared spectra of each sample. Repeatability relative standard deviation (RSD_r) was also calculated.

Results

Table 2 presents descriptive statistics. The blood parameters with the lowest variability were the electrolytes. CV was 2.4, 4.9, and 2.3 % for Na, K, and Cl, respectively. Plasma Ca was characterized by low variability (CV of 4.7 %). Greater variability was found for Mg (9.8 %), inorganic P (19.34 %) and Zn (23.7 %).

The greatest variability was observed for haptoglobin (CV of 108.0 %), total bilirubin (CV of 116.3 %) and NEFA (87.9 %). These variables were characterized by an asymmetrical distribution, with high skewness values, indicating that the tail on the right side was longer or fatter than the left side. These variables were also processed after natural logarithm transformation, with only a slight reduction of their variability.

The plasma parameters of energy metabolism, apart from the NEFA, were characterized by high variability with high CV and/or high range. The CV of glucose was not high (16.7 %), but the range was very wide (range of 4.54 mmol L^{-1}), with a high kurtosis value (5.6),

Table 2 Descriptive statistic (35 plasma samples)

Item	Mean	SD	Min	Max	Median	SK[a]	KU[b]
Glucose, mmol L^{-1}	4.02	0.67	1.66	6.20	3.95	−0.13	5.64
Total cholesterol, mmol L^{-1}	3.36	1.59	0.80	5.87	3.18	0.05	−1.57
NEFA[c], mmol L^{-1}	0.33	0.29	0.06	1.18	0.20	1.06	0.30
NEFA[c], ln(mmol L^{-1})	−1.52	0.94	−2.83	0.16	−1.61	0.17	−1.49
BHBA[d], mmol L^{-1}	0.49	0.13	0.28	0.85	0.47	1.00	0.78
BHBA[d], ln(mmol L^{-1})	−0.74	0.25	−1.28	−0.16	−0.76	0.34	0.06
Urea, mmol L^{-1}	4.21	1.42	1.50	7.52	4.02	0.21	−0.61
Creatinine, μmol L^{-1}	99.54	10.67	83.62	125.89	97.57	0.47	−0.51
Ca, mmol L^{-1}	2.58	0.12	2.36	2.86	2.61	0.33	−0.16
Inorganig P, mmol L^{-1}	1.81	0.35	1.19	2.89	1.74	1.09	1.36
Mg, mmol L^{-1}	1.02	0.10	0.57	1.12	1.04	−2.97	9.99
Na, mmol L^{-1}	144.20	3.44	138.78	151.78	143.81	0.38	−0.68
K, mmol L^{-1}	4.07	0.20	3.59	4.57	4.08	−0.09	0.20
Cl, mmol L^{-1}	107.51	2.48	102.27	112.98	107.46	−0.09	−0.19
Zn, μmol L^{-1}	11.43	2.71	5.17	16.19	11.40	−0.48	−0.19
Ceruloplasmin, μmol L^{-1}	3.24	1.02	1.99	6.63	3.03	1.48	2.41
Total protein, g L^{-1}	76.07	6.27	63.57	86.89	77.10	−0.35	−0.67
Albumin, g L^{-1}	35.90	2.65	25.79	39.55	36.26	−2.07	5.68
Globulin, g L^{-1}	40.05	6.33	27.99	52.08	40.61	−0.23	−0.78
Total bilirubin, μmol L^{-1}	3.98	4.63	0.29	22.48	2.14	2.57	6.98
Total bilirubin, ln(μmol L^{-1})	0.94	0.91	−1.24	3.11	0.76	0.35	0.24
Haptoglobin, g L^{-1}	0.25	0.27	0.03	0.91	0.10	1.46	0.69
Haptoglobin, ln(g L^{-1})	−1.91	0.98	−3.51	−0.09	−2.30	0.60	−0.87
AST[e], U L^{-1}	106.67	69.78	49.98	394.07	89.35	3.15	9.75
AST[e], ln(U L^{-1})	4.56	0.42	3.91	5.98	4.49	1.90	4.32
GGT[f], U L^{-1}	28.78	9.66	16.27	63.94	26.63	1.50	3.64
GGT[f], ln(U L^{-1})	3.32	0.31	2.79	4.16	3.28	0.40	0.20
AP[g], U L^{-1}	50.85	22.75	21.44	103.73	42.84	0.87	−0.21
AP[g], ln(U L^{-1})	3.84	0.43	3.07	4.64	3.76	0.23	−0.88

[a]Skewness
[b]Kurtosis
[c]Non esterified fatty acids
[d]β-OH-butyric acid
[e]Aspartate amino transferase
[f]γ-glutamyl transferase
[g]Alkaline phosphatase

indicating that more of the variance is the result of infrequent extreme deviations (leptokurtic distributions). Conversely, total cholesterol was characterized by a high CV (47.3 %), in tune with its wide range (range of 5.1 mmol L^{-1}) and a slight negative kurtosis (−1.57). BHBA was characterized by an intermediate CV compared with those of glucose and total cholesterol. For this ketone body the maximum value observed cannot be considered among the high values.

Plasma total protein and albumin were characterized by a low CV (8.2 and 7.4 %, respectively). The variability of globulin (CV of 15.8 %) was only slightly greater. Besides the low variability of these three variables, their range could be considered wide. For albumin only, the wide range was observed together with a high kurtosis (5.68), indicating that more of the variance is the result of infrequent extreme deviations.

Enzymes activities were characterized by high variability, with positive skewness. The natural logarithm transformation decreased both the skewness and their variability.

Table 3 shows the statistics of the prediction models based on FT-MIR spectroscopy for each plasma variable. The number of extracted factors with the PLS procedure

Table 3 Calibration curves and cross validation

Item	Extracted factors[a]	R^2	RMSECV[b]	RER[c]	RPD[d]	S_r^e	RSD_r^f
Glucose, mmol L^{-1}	5	0.66	0.3893	11.7	1.7	0.1550	3.86
Total cholesterol, mmol L^{-1}	7	0.99	0.1156	43.8	13.8	0.0690	2.06
NEFA[g], mmol L^{-1}	6	0.86	0.1094	10.2	2.7	0.0325	9.80
NEFA[g], ln(mmol L^{-1})	5	0.78	0.4414	6. 8	2.1	0.1214	7.99
BHBA[h], mmol L^{-1}	5	0.38	0.1012	5.6	1.3	0.0245	4.98
BHBA[h], ln(mmol L^{-1})	7	0.60	0.1572	7.1	1.6	0.0669	9.03
Urea, mmol L^{-1}	6	0.90	0.4623	13.0	3.1	0.2454	5.83
Creatinine, µmol L^{-1}	4	0.60	6.7608	6.3	1.6	0.3080	0.31
Ca, mmol L^{-1}	3	0.60	0.0791	6.3	1.6	0.0048	0.19
Inorganig P, mmol L^{-1}	5	0.55	0.2336	7.3	1.5	0.0545	3.00
Mg, mmol L^{-1}	4	0.55	0.0701	7.8	1.5	0.0063	0.62
Na, mmol L^{-1}	4	0.57	2.2673	5.7	1.5	0.3289	0.23
K, mmol L^{-1}	2	0.10	0.1920	5.1	1.0	0.0043	0.11
Cl, mmol L^{-1}	6	0.70	1.3219	8.1	1.9	0.3198	0.30
Zn, µmol L^{-1}	4	0.49	1.9577	5.6	1.4	0.2240	1.96
Ceruloplasmin, µmol L^{-1}	3	0.59	0.6575	7.1	1.5	0.0705	2.17
Total protein, g L^{-1}	5	0.99	0.7452	31.3	8.4	0.1561	0.21
Albumin, g L^{-1}	6	0.96	0.5082	27.1	5.2	0.1362	0.38
Globulin, g L^{-1}	4	0.98	0.8128	29.6	7.8	0.2665	0.67
Total bilirubin, µmol L^{-1}	5	0.75	2.2967	9.7	2.0	0.6654	16.71
Total bilirubin, ln(µmol L^{-1})	3	0.66	0.5320	8. 2	1.7	0.0225	2.40
Haptoglobin, g L^{-1}	4	0.62	0.1678	5.2	1.6	0.0148	5.97
Haptoglobin, ln(g L^{-1})	4	0.66	0.5734	6.0	1.7	0.0494	2.59
AST[i], U L^{-1}	6	0.69	39.5398	8.7	1.8	9.1112	8.54
AST[i], ln(U L^{-1})	4	0.54	0.2856	7.2	1.5	0.0282	0.62
GGT[l], U L^{-1}	4	0.40	7.5093	6.3	1.3	0.3565	1.27
GGT[l], ln(U L^{-1})	3	0.42	0.2344	5.8	1.3	0.0154	0.46
AP[m], U L^{-1}	7	0.66	13.2962	6.2	1.7	8.3024	16.33
AP[m], ln(U L^{-1})	7	0.64	0.2614	6.02	1.6	0.1696	4.42

[a]Number of extracted factors with the PLS procedure
[b]Root mean square error of cross validation
[c]Range error ratio, obtained dividing the range of each variable by its standard deviation of prediction errors
[d]Ratio of prediction to deviation, obtained dividing the standard deviation of each variable with its standard deviation of prediction errors
[e]Repeatability of the standard deviation
[f]Repeatability relative standard deviation
[g]Non esterified fatty acids
[h]β-OH-butyric acid
[i]Aspartate amino transferase
[l]γ-glutamyl transferase
[m]Alkaline phosphatase

for each prediction model ranged between 3 and 7. A low number of extracted factors was observed for some minerals (electrolytes, Ca, and Zn).

The coefficient of determination (R^2) of the prediction models ranged between 0.1 to values close to 1. Values greater than 0.9 were observed, in decreasing order, for the prediction models of total cholesterol (Fig. 1), total protein (Fig. 2), globulin, and albumin. The R^2 values ranged between 0.75 and 0.9 for the prediction model of urea, NEFA, and total bilirubin. Values of R^2 lower than 0.6 were observed for all minerals, with a slightly greater value only for Cl. The prediction models of enzymes activity were also characterized by low R^2 values, ranging between 0.39 and 0.66.

The range error ratio (RER) and ratio to deviation (RPD) ranged between 5.1 and 43.8 for the former, and between 1 and 13.8 for the latter. Values of RPD greater than 5 were observed for total cholesterol, total protein,

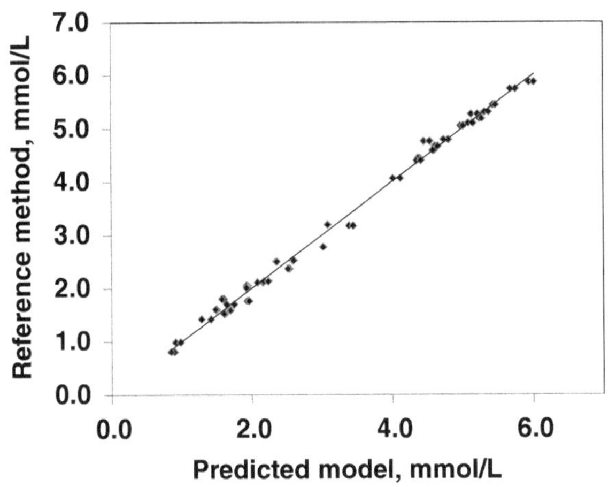

Fig. 1 Linear regression plots for total cholesterol. Relationship between total cholesterol concentration (mmol L^{-1}) measured by the reference method and by FT-MIR spectrometry in plasma using mid-infrared spectra (70 observations). (R2 = 0.995; root mean square error of cross validation = 0.1156; RER = 43.8; RER is the range error ratio, and is obtained dividing the range of total cholesterol with its standard deviation of prediction errors; RPD = 13.8, where RPD is the ratio of prediction to deviation, and is obtained dividing the standard deviation of total cholesterol with its standard deviation of prediction errors)

albumin, and globulin. RPD ranged between 2 and 5 for the prediction models of urea, NEFA, and total bilirubin. In general RPD and RER were low for minerals and enzymes activities.

The repeatability (S_r) between the two values obtained with the two replicates of infrared spectra was good for almost all parameters considered in this study. Poor repeatability was observed for NEFA, BHBA, total bilirubin, AST and AP.

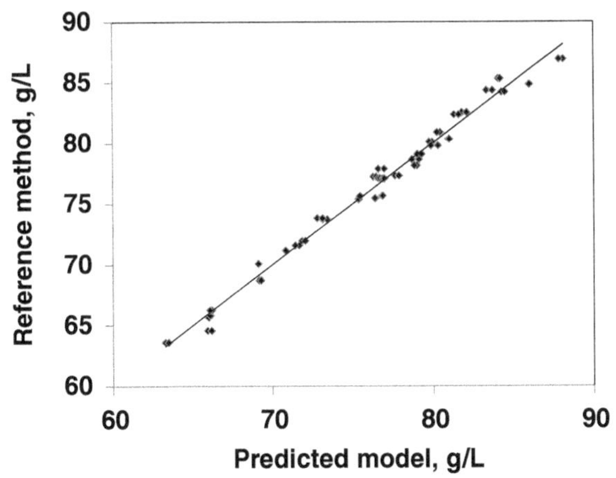

Fig. 2 Linear regression plots for total protein. Relationship between total protein concentration (g L^{-1}) measured by the reference method and by FT-MIR spectrometry in plasma using mid-infrared spectra (70 observations). (R2 = 0.985; root mean square error of cross validation = 0.7452; RER = 31.3; RER is the range error ratio, and is obtained dividing the range of total protein by its standard deviation of prediction errors; RPD = 8.4; RPD is the ratio of prediction to deviation, and is obtained dividing the standard deviation of total protein by its standard deviation of prediction errors)

Discussion

This study presented results concerning the use FT-MIR spectroscopy to predict plasma parameters in the metabolic profile of farm animals. In order to build a prediction model that accurately predicts unknown samples, the calibration data set should contain samples that represent all the possible sources of variation encountered when the prediction model will be used on an unknown sample. For this reason, the selection of samples is very important to provide the largest range of information for the building of the calibration data set [14]. The dairy cows selected for this study were all clinically healthy, consequently the variability observed through the data is representative of a normal physiological condition. Although the cows were not affected by clinical disease, the subjects of a population are often characterized by the presence of high variability for many blood biomarkers, and also for the occurrence of sub-clinical conditions [25]. In agreement with this consideration, all the concentrations of blood parameters measured in the present study were within the reference ranges [6, 26] reported for dairy cows without clinical signs of disease. To better represent the variability of blood biomarkers, we took blood samples of various dairy cow physiological stages, and reserved the transition period as having most relevant importance (14 of 35 samples). We made this choice because the peripartum period is the most critical phase for dairy cows. Important physiological, metabolic, and nutritional changes occur during transition [27, 28], and consequently, blood indices are subjected to marked changes [29–31]. In particular, parameters of energy and protein metabolism are strongly affected, as are glucose, NEFA, BHBA, urea and creatinine, as well as parameters related to an inflammatory response, such as haptoglobin, ceruloplasmin, total bilirubin, albumin, paraoxonase and cholesterol [25, 32, 33].

This study utilized a low number of blood samples, and the sampling was mainly focused around calving and limited to only one herd. The low number of samples used in this research suggested use of internal validation involving validation of the calibration models using the same sample set as that used for calibration development. Cross validation is a very reliable method of validation; it seeks to validate the calibration model on an independent test data set, but it does not use samples for testing only. These findings need to be confirmed with a further study on a wider population belonging to more herds, and with external validation with an independent and representative set of test objects, in order to give relevant and reliable estimates on the prediction ability of the predictive models.

The accuracy of infrared analysis is affected by the quality of the reference assays used [34]. In our study the reference assays used to analyze some blood parameters (i.e., BHBA, creatinine, ceruloplasmin, total bilirubin, haptoglobin, and GGT) were characterized by not having optimal repeatability between runs, with a CV greater than 3 %. For these blood parameters the developed calibration curves where characterized by a RPD lower than 2, which is considered to be the minimal threshold for acceptable approximate quantitative predictions. Among these blood variables without optimal repeatability in the reference chemistry, only the calibration curve of total bilirubin could be considered near to acceptability. Our results suggest that an improvement in the reference assays for all these blood parameters without optimal repeatability could improve the predictive ability of the calibration curve based on FT-MIR spectroscopy.

Poor prediction ability characterized the calibration curves of all minerals analyzed in this study. A similar result has been described in milk, mainly for electrolytes. In particular, Mg, Na and K showed poor prediction ability in FT-MIR spectroscopy [35]. Conversely, Soyeurt et al. [35] and Toffanin et al. [36] have suggested the potential of FT-MIR spectroscopy to predict Ca and P content in milk. Because the correlations between Ca and P concentrations estimated by the FT-MIR predictions and the known milk components were inferior to the correlation calculated based on the cross-validation, Soyeurt et al. [35] concluded that these equations of calibration were obtained from a real spectral absorbance. On the whole, these results confirm the difficulty in predicting mineral content in plasma with FT-MIR spectroscopy, in particular for minerals that are not included in organic compounds. Nevertheless, a similar condition also occurs for milk Ca, mainly included in casein under an organic (20 %) or inorganic form (46 % as tri-calcium phosphate) [37]. The difficulty observed in our study to predict mineral content in blood is likely due to two main reasons. First, among the minerals measured in the blood, a large proportion is in ionized form and not included in organic compounds. About 50 % of total Ca in plasma is in the ionized form and about 45 % is linked to protein, whereas, the quota of Mg ionized is about the 70 % [19]. Second, the variability "captured" for the mineral from the samples in this dataset was lower compared to variability measured for other blood parameters. A low variability, in particular for electrolytes, was observed. A better assay of minerals requires a dataset containing samples with a greater variability, which allows for an improvement in the predictive ability of FT-MIR spectroscopy. To increase the variability of these indices, it would be useful to include some animals with clinical diseases in the population.

A poor predictive ability characterized the calibration curves of all enzymes activities measured in this study. The inadequate estimation was probably due to the poor variability of these indices and also the non-normal distribution of the data, with a very low proportion of high values in this dataset. The log transformation did not provide an improvement in the predictive ability of the calibration curve. For a better assessment of enzymes activities, it is necessary to obtain a wider distribution of the data, in particular a representative number of samples with high values could improve the prediction ability of FT-MIR spectroscopy. An improvement is probable, considering that with FT-MIR spectroscopy, some authors were able to measure plasminogen concentration in milk and the distinction between plasmin and plasminogen [38], as well as secondary and tertiary structural changes in bovine plasminogen [39].

An excellent prediction ability was obtained for several of the parameters, mainly regarding energy, protein metabolism and inflammatory status. These results are very promising, but further studies are necessary to confirm the results and require validation on a larger population, data on different breeds, and herds raised in different environments. Among energy parameters, an excellent calibration curve was developed for total cholesterol. Interestingly, the determination of cholesterol in dairy products by using FT-MIR spectroscopy also showed good results [40].

Among the parameters of protein metabolism excellent calibration curves were developed for total protein, albumin and globulin. From this point of view, the FT-MIR technique is a widely used tool in many different fields and is even used to evaluate parameters with a "complex etiology" like secondary and tertiary protein structure studies [41, 42], differentiation of plasmin and plasminogen in milk [38], and determination of secondary and tertiary structural changes in bovine plasminogen [39] and casein [40].

For inflammatory status, it appears possible to predict with FT-MIR spectroscopy the blood parameters used to calculate the liver functionality index (LFI) according to Bertoni and Trevisi [4]. LFI measures the variation of some negative acute-phase proteins, which are reduced during inflammation, or related parameters (albumin, cholesterol, and total bilirubin) to help evaluate changes in liver function caused by inflammatory events. LFI takes into account changes in albumin, lipoproteins (measured as total cholesterol) and total bilirubin (its secretory enzymes are synthesized by the liver) occurring between 3 and 28 days in milk [2]. The lower the value of LFI, the more severe are the consequences of the inflammatory events that occur in the transition period and the inability of cows to adapt their metabolism to these challenges [31], causing a worsening of health status (clinical or subclinical problems). Our results may open an interesting perspective for an easier and more cost-effective approach to monitoring farm animals during critical periods, such as in the transition phase of dairy cows.

Conclusions

Although the results of this study require further validation, they are very promising. We have shown that FT-MIR spectroscopy offers fairly accurate measurement of various plasma biomarkers of great importance for the evaluation of the metabolism and inflammatory status of dairy cows. All the animals selected in this investigation were healthy (i.e., without clinical events of disease) and the variability of the range of the checked parameters was restricted to a normal physiological condition. In these circumstances the accuracy required for the estimation of the biomarkers is very high and was reached for only some parameters. As such, FT-MIR technology may have even more success for the study of animals with pathological conditions (clinical and subclinical), especially when blood is used as main substrate for disease diagnosis. In this context, pathological changes occur for many markers and, in these conditions, it is certainly acceptable to have a greater range of error from an indirect assessment method with IR.

Abbreviations

RPD: ratio to deviation; RER: ranger error ratio; NEFA: non esterified fatty acids; BHBA: β-OH-butyric acid; AST: aspartate amino transferase; GGT: γ-glutamyl transferase; AP: alkaline phosphatase.

Competing interests

None of the authors of this paper has a financial or personal relationship with other people or organizations that could inappropriately influence or bias the content of the paper.

Authors' contributions

LC conceived with ET the study, participated in its design, and carried out infrared analysis and data processing, and drafted the manuscript. AF participated in study design and data analysis using reference methods. AM participated in its design, participated to the sample collection and data processing. ET conceived the study together LC, participated in study design and its coordination at farm level and helped to draft the manuscript. All authors read and approved the final manuscript.

Acknowledgements

This study was financially co-supported by the Italian Ministry of Health (Project IZSLER PRF 2006201) and Università Cattolica del Sacro Cuore (Cod. Res. 2842290).

References

1. Payne JM, Dew SM, Manston R, Faulks M. The use of a metabolic profile test in dairy herds. Vet Rec. 1970;87:150–8.
2. Clarke CJ, Haselden JN. Metabolic Profiling as a Tool for Understanding Mechanisms of Toxicity. Toxicol Pathol. 2008;36:140–7.
3. Friedrichs KR, Harr KE, Freeman KP, Szladovits B, Walton RM, Barnhart KF, et al. ASVCP reference interval guidelines: determination of de novo reference intervals in veterinary species and other related topics. Vet Clin Pathol. 2012;41:441–53.
4. Bertoni G, Trevisi E. Use of the liver activity index and other metabolic variables in the assessment of metabolic health in dairy herds. Vet Clin North Am Food Anim Pract. 2013;29:413–31.
5. Ingraham RH, Kappel LC. Metabolic profile testing. Vet Clin North Am Food Anim Pract. 1988;4:391–411.
6. Calamari L, Bertoni G, Maianti M, Cappa V. Sulla utilità di nuovi parametri ematochimici nella valutazione del profilo metabolico delle lattifere. Zootecnica e Nutruzione Animale. 1989;15:191–210.
7. Abeni F, Calamari L, Stefanini L. Metabolic conditions of lactating Friesian cows during the hot season in the Po valley. 1. Blood indicators of heat stress. Int J Biometeorol. 2007;52:87–96.
8. Calamari L, Petrera F, Stefanini L, Abeni F. Effects of different feeding time and frequency on metabolic conditions and milk production in heat-stressed dairy cows. Int J Biometeorol. 2013;57:785–96.
9. Abeni F, Calamari L, Calza F, Speroni M, Bertoni G, Pirlo G. Welfare assessment based on metabolic and endocrine aspects in primiparous cows milked in a parlor or with an automatic milking system. J Dairy Sci. 2005;88:3542–52.
10. Calamari L, Bertoni G: Model to evaluate welfare in dairy cow farms. Ita. J. Anim. Sci. 2009; 8:301–23.
11. Shaw RA, Kotowich S, Leroux M, Mantsch HH. Multianalyte serum analysis using mid-infrared spectroscopy. Ann Clin Biochem. 1998;35(Pt 5):624–32.
12. Shaw RA, Mantsch HH: Infrared Spectroscopy in Clinical and Diagnostic Analysis. In: Mayers RA, editor. Encyclopedia of Analytical Chemistry Chichester: John Wiley & Sons Ltd.; 2000. p. 83–102.
13. Hahn H, Pallua JD, Pezzei C, Huck-Pezzei V, Bonn GK, Huck CW. Infrared-spectroscopy: a non-invasive tool for medical diagnostics and drug analysis. Curr Med Chem. 2010;17:2956–66.
14. Berzaghi P, Riovanto R. Near infrared spectroscopy in animal science production: principles and applications. Ital J Anim Sci. 2010;8:39.
15. Kruse-Jarres JD, Janatsch G, Gless U, Marbach R, Heise HM. Glucose and other constituents of blood determined by ATR-FTIR-spectroscopy. Clin Chem. 1990;36:401–2.
16. Sunderman FW, Nomoto S. Measurement of human serum ceruloplasmin by its p-phenylenediamine oxidase activity. Clin Chem. 1970;16:903–10.
17. Skinner JG, Brown RA, Roberts L. Bovine haptoglobin response in clinically defined field conditions. Vet Rec. 1991;128:147–9.
18. Owen JA, Better FC, Hoban J. A simple method for the determination of serum haptoglobins. J Clin Pathol. 1960;13:163–4.
19. Eletrics F: MilkoScan FT120. Type 71200. Reference manual. Issue 9 GB. 2002.
20. Geladi P, Kowalski BR. Partial least-squares regression: a tutorial. Anal Chim Acta. 1986;185:1–17.
21. Van der Voet H. Comparing the predictive accuracy of models using a simple randomization test. Chemom Intell Lab Syst. 1994;25:313–23.
22. Williams PC. Near-Infrared Technology Getting the Best out of Light. A Short Course in the Practical Implementation of Near Infrared Spectroscopy for the User. 11th ed. Nanaimo: PDK Projects Inc; 2003.
23. Williams PC. Implementation of near-infrared technology. In: Williams PC, Norris K, editors. Near-Infrared Technology in the Agricultural and Food Industries. 2nd ed. St. Paul: American Association of Cereal Chemists; 1987. p. 143–67.
24. Saeys W, Mouazen AM, Ramon H. Potential for Onsite and Online Analysis of Pig Manure using Visible and Near Infrared Reflectance Spectroscopy. Biosyst Eng. 2005;91:393–402.
25. Bionaz M, Trevisi E, Calamari L, Librandi F, Ferrari A, Bertoni G. Plasma paraoxonase, health, inflammatory conditions, and liver function in transition dairy cows. J Dairy Sci. 2007;90:1740–50.
26. Bertoni G, Calamari L, Trevisi E. Nuovi criteri per l'individuazione dei valori di riferimento di taluni parametri ematici in bovine da latte. La Selezione Vet. 2000;41:261–8.
27. Drackley JK. Biology of Dairy Cows During the Transition Period: the Final Frontier? J Dairy Sci. 1999;82:2259–73.
28. LeBlanc S. Monitoring metabolic health of dairy cattle in the transition period. J Reprod Dev. 2010;56:S29–35.
29. Calamari L, Librandi F, Trevisi E, Bertoni G. Transition period in dairy cows: immune system, inflammatory conditions and liver activity. J Dairy Sci. 2002;85:985.
30. Soriani N, Trevisi E, Calamari L. Relationships between rumination time, metabolic conditions, and health status in dairy cows during the transition period. J Anim Sci. 2012;90:4544–54.
31. Loor JJ, Bertoni G, Hosseini A, Roche JR, Trevisi E. Functional welfare – using biochemical and molecular technologies to understand better the welfare state of peripartal dairy cattle. Anim Prod Sci. 2013;53:931–53.

32. Calamari L, Soriani N, Panella G, Petrera F, Minuti A, Trevisi E. Rumination time around calving: an early signal to detect cows at greater risk of disease. J Dairy Sci. 2014;97:3635–47.
33. Trevisi E, Amadori M, Cogrossi S, Razzuoli E, Bertoni G. Metabolic stress and inflammatory response in high-yielding, periparturient dairy cows. Res Vet Sci. 2012;93:695–704.
34. Barbano DM, Clark JL. Infrared Milk Analysis — Challenges for the Future. J Dairy Sci. 1989;72:1627–36.
35. Soyeurt H, Bruwier D, Romnee J-M, Gengler N, Bertozzi C, Veselko D, et al. Potential estimation of major mineral contents in cow milk using mid-infrared spectrometry. J Dairy Sci. 2009;92:2444–54.
36. Toffanin V, De Marchi M, Lopez-Villalobos N, Cassandro M. Effectiveness of mid-infrared spectroscopy for prediction of the contents of calcium and phosphorus, and titratable acidity of milk and their relationship with milk quality and coagulation properties. Int Dairy J. 2015;41:68–73.
37. Corradini C. Chimica E Tecnologia Del Latte. Milano: Tecniche N; 1995.
38. Ozen BF, Hayes KD, Mauer LJ. Measurement of plasminogen concentration and differentiation of plasmin and plasminogen using Fourier-transform infrared spectroscopy. Int Dairy J. 2003;13:441–6.
39. Hayes KD, Ozen BF, Nielsen SS, Mauer LJ. FTIR determination of ligand-induced secondary and tertiary structural changes in bovine plasminogen. J Dairy Res. 2003;70:461–6.
40. Curley DM, Kumosinski TF, Unruh JJ, Farrell HM. Changes in the secondary structure of bovine casein by Fourier transform infrared spectroscopy: effects of calcium and temperature. J Dairy Sci. 1998;81:3154–62.
41. Van Der Ven C, Muresan S, Gruppen H, De Bont DBA, Merck KB, Voragen AGJ. FTIR spectra of whey and casein hydrolysates in relation to their functional properties. J Agric Food Chem. 2002;50:6943–50.
42. Qi PX, Wickham ED, Farrell HM. Thermal and alkaline denaturation of bovine beta-casein. Protein J. 2004;23:389–402.

2

The effect of sub-clinical infection with *Mycobacterium avium* subsp. *paratuberculosis* on milk production in a New Zealand dairy herd

Andrew Bates[1*], Rory O'Brien[2], Simon Liggett[2] and Frank Griffin[3]

Abstract

Background: Johne's disease is a major production limiting disease of dairy cows. The disease is chronic, progressive, contagious and widespread; there is no treatment and there is no cure. Economic losses arise from decreased productivity through reduced growth, milk yield and fertility and capital losses due to premature culling or death. This study attempts to address the effect of subclinical JD on milk production under New Zealand pastoral dairy farming conditions using a new testing approach. Blood samples were taken from all lactating animals from a single seasonally calving New Zealand dairy herd in the autumn of 2013 and 2014. Samples were subject to serological assay for antibodies to *Mycobacterium avium* subsp. *paratuberculosis* using a combination of four ELISA tests in parallel followed by selective quantitative fecal PCR to confirm the fecal shedding characteristics of ELISA positive cows. ELISA status was classified as Not-Detected, Low, Moderate or High and fecal PCR status as Not-Detected, Moderate or High.

Results: A mixed generalized regression model indicated that, compared to cows where MAP was not detected, daily milk solids production was 4% less for high ELISA positive cows ($p = 0.004$), 6% less for moderate fPCR cows ($p = 0.036$) and 12% less for high fPCR cows ($p < 0.001$).

Conclusions: This study confirms that sub-clinical JD can have a significant impact on milk production and that the testing methodology used stratified the animals in this herd on their likely impact on production and disease spread. This allowed the farmer to prioritize removal of heavily shedding, less-productive animals and so reduce the risk of infection of young stock. This is the first longitudinal study based in New Zealand looking at the effect of Johne's infection status on daily milk production allowing for intermediary and confounding factors.

Keywords: Johne's disease, Milk production, ELISA, Quantitative PCR

Background

Johne's disease (JD) is a chronic disease of ruminants caused by intestinal infection with *Mycobacterium avium* subsp. *paratuberculosis* (MAP). Johne's infection is predominantly subclinical in most dairy cows with farmers becoming aware of the disease when the clinical signs of infection such as diarrhoea and wasting become apparent [1]. Whitlock and Buergelt [2] suggested a bovine JD "iceberg effect" whereby, for every clinically affected animal born on the farm, a minimum of 25 other animals are likely to be infected.

Many studies have found MAP infection to be associated with a statistically significant reduction in milk production [3–6], although this depends on the cow's age [3], farm system [7], the genetics of the cow [8] and strain of MAP [9]. However, few studies have examined the effect of MAP infection on milk production under seasonal, pastoral systems in NZ. Comparing differences in milk production between MAP positive and negative cows is not straight forward because of differences in quantification of milk yield, trial design, diagnostic tests,

* Correspondence: andrew.bates@vetlife.co.nz
[1]Vetlife, Centre for Dairy Excellence, Wilson Street, Geraldine, New Zealand
Full list of author information is available at the end of the article

farm system and the presence of confounding and intermediary factors [6, 10].

Farmer perception has a major impact on the uptake of disease control schemes and engagement at the individual farm level [11]. Many farmers without direct experience of clinical JD in their herds consider that JD is not a major problem for their farm. Norton et al. [12] found only 10% (22/225) of NZ farmers that had not observed clinical cases in their herd considered the disease a serious problem. Similar findings were reported by Sorge et al. [13] in a Canadian survey where most dairy farmers did not consider JD a serious problem for their operation. Without clear evidence that MAP infection is linked to decreased milk production, the low clinical incidence of JD in most herds allows farmers to relegate MAP infection to the realm of natural losses.

The specificity of ELISA tests may be compromised by common antigens shared between MAP, *Mycobacterium avium* and other saprophytic environmental mycobacteria. The sensitivity of ELISA tests, particularly for subclinically infected animals in the early stages of JD, is also influenced by the dynamics of antibody production [14]. While detection of the organism via fecal culture on Herrold's egg yolk medium has been a definitive test for MAP infection this requires prolonged incubation periods of up to 16 weeks and may be compromised by overgrowth by contaminating gut organisms [15–17]. Internationally, the rapid, direct and quantitative measurement of MAP shedding in feces of infected and affected animals by quantitative PCR is rapidly becoming a standard and widely used method for JD diagnostic testing [18–20].

As no single diagnostic test satisfies all criteria in terms of sensitivity, specificity, speed of turnaround, cost and convenience, combinations of tests are used to achieve optimal diagnosis [21, 22]. In NZ, a herd testing protocol based on an initial herd screening with a serological ELISA for multiple MAP antigens (Paralisa™) [23] coupled with a quantitative fecal PCR (fPCR) test to confirm the status of ELISA positive animals [24] has been developed. This allows farmers and their advisers to stratify shedders according to disease status and environmental risk. However, the impact of subclinical infection, as defined by this test, has not yet been characterized for milk solids production under NZ pastoral systems.

The present longitudinal study attempted to quantify the effect of MAP infection on individual lactation test day milk production on a single, seasonally calving, NZ pastoral dairy herd allowing for potential confounders.

Methods

Study animals

A spring calving, pasture based, Friesian dairy herd (1250 cows at peak milk) in the South Canterbury region of NZ was selected for the study. In the 5 years preceding this study, the herd had culled 3–5% of the milking herd annually from suspected clinical JD based on clinical signs observed by the owner. In 2009, MAP had been isolated, and JD confirmed histopathologically, from gut and mesenteric lymph node samples from each of 4 cull animals suspected of clinical JD. One year later, all milking cows over 2 years old were subject to a single serologically based ELISA (Paralisa™) with fPCR performed on a small subset of the ELISA positive animals. At this test, 97/1086 (8.93%) were ELISA positive and approximately 20% of these ELISA positive cows were shedding high levels (exceeding $\geq 1 \times 10^4$ genomes/mL) of MAP as determined by fPCR. Considering the high prevalence of ELISA-positive animals the farmer was reluctant to cull all seropositives, most of which appeared healthy and productive. Persistent losses (> 3% pa) of clinically affected animals continued from 2010 to 2013. In 2014 a decision was made to rescreen the herd using serial ELISA and fPCR testing to identify animals which were shedding high levels of MAP, for future culling.

Sample acquisition and treatment

In the autumn of 2014 and again in the autumn of 2015, a coccygeal tail vein blood sample was collected into a plain blood tube from all milking cows in the enrolled herd. Samples were transported to DRL (Disease Research Ltd, Mosgiel, NZ) and assayed for circulating antibody to MAP by serum ELISA using a combination of two ELISA tests, Paralisa™ (DRL, Mosgiel, NZ) and IDEXX Paratuberculosis Screening Ab Test (IDEXX Laboratories, Inc., Westbrook, ME, USA). The Paralisa™ methodology was based on previously published procedures for ELISA immunoassays used to diagnose immune reactions to MAP infection in farmed red deer [23, 25]. In addition to the IgG_1 antibody responses to a denatured antigen in the form of Purified Protein Derivative J (Johnin) and a native protein in the form of Protoplasmic Antigen (PPA), in this study an additional MAP-specific recombinant protein antigen, Ag_1Del_1, was incorporated into the Paralisa™ test protocol. Final test results were arrived at by considering the antibody level to the IDEXX test and the three Paralisa™ test antigens in parallel. IDEXX ELISA assays were performed and interpreted according to the instructions supplied by the kit manufacturer. Results were classified as follows; for the Paralisa™ serological assays, a classification of Not Detected was returned for results of < 50 ELISA Units (EU) for Johnin, PPA, and Ag_1Del_1 antigens, readings of 50–100 EU in any one test were classified as Low, readings of 101–150 EU as Moderate, and readings of > 150 EU as High. For the IDEXX tests, results were classified as Not Detected ($S/P < 55$), Low ($S/P \geq 55$), Moderate ($S/P \geq 100$) or High ($S/P \geq 150$) based on the response relative to the kit positive control.

Seven days later, a single fecal sample (10 g approx.) was collected from each cow testing Low, Moderate or High to any of the ELISA tests and forwarded to DRL for quantitation of MAP shedding in feces by fPCR [24, 26]. Quantitation of MAP DNA titer in fecal samples was accomplished using a standard curve comprising DNA dilution standards spanning 7 serial log dilutions of MAP genomic DNA prepared from MAP laboratory strain 316f and results extrapolated and reported as 'MAP genome copies/mL' equivalents. DNA standards ranged from 16.5 μg/mL to 1.65×10^{-5} μg/mL; 3uL of DNA standard was utilized in each (20uL) PCR reaction such that, given a MAP genome size of 4.8Mbp [27], these values equated to a topmost standard of 1×10^7 genomes/20uL (5×10^8 genomes/mL) down to a lowermost standard of 10 genomes/20uL reaction (or 500 genomes/mL).

Fecal sample data were stratified into shedding categories with MAP shedding scores of $\geq1\times10^3$ to $<1\times10^4$ genomes/mL classified as Moderate and counts exceeding $\geq1\times10^4$ genomes/mL as High. In this study, fecal samples which returned shedding scores of $<1\times10^3$ genomes/mL feces were conservatively classified as Not Detected. Classification of MAP status by ELISA and fPCR results is summarized in Table 1.

Milk production data

For each cow, milk production was measured at 4 herd test dates spread every 60–70 days throughout the lactation. At each herd test, individual cow production in kg milk solids (kgMS) and individual somatic cell count (ISCC) was recorded as cells/mL for the test day. Third-party electronic access to individual cow age, breed, calving date, milk quality and production data were granted.

Table 1 Classification scheme of MAP status from ELISA and fPCR results in a study looking at the effect of MAP infection on milk production in a South Canterbury dairy herd

Test				MAP status
ELISA				
Paralisa™			IDEXX[a]	
Johnin[a]	PPA[a]	$Ag_1Del_1^a$		
< 50 EU	< 50 EU	< 50 EU	Not Detected	Not Detected
50–100 EU in any test			Low	Low
101–150 EU in any test			Moderate	Moderate
> 150 EU in any test			High	High
fPCR				
$<1\times10^3$ genomes/mL				Not Detected
$\geq1\times10^3$ - $<1\times10^4$ genomes/mL				Moderate
$\geq1\times10^4$ genomes/mL				High

[a]ELISA results were interpreted in parallel
[b]fPCR results were interpreted in series with ELISA results

Statistical analysis

The outcome variable was level of kg milk solids, (continuous) recorded at each herd test day for the 2013–2014 season and for the 2014–2015 season. In the NZ seasonal dairying context, season refers to the period in milk from calving in the spring (July–August) to dry off in the autumn (April–May) and typically lasts 280 days. The predictor variables included how many days in milk from calving at test day (continuous), parity at sampling date, breed (categorized as > 75%, 50–75%, < 50% proportion of Friesian genetics), milking season (categorical; 2013–14 or 2014–15) and MAP infection status. Test day somatic cell linear score (LS, indicating the $\log_2$ of the SCC transformed as ($\log_2$(ISCC/100,000)) + 3 [28] was also included as a predictor variable. A separate time variable was created for milk test-day number indicating the number of milk test days for a given animal since the beginning of the lactation. Models were constructed with parity at sampling date treated as a continuous variable and also divided into primiparous and multiparous to account for differences in lactation shape [4, 5]. The model with the lowest value for Akaike Information Criteria (AIC) was selected.

ELISA test and fPCR fecal status as categorized above were used to describe MAP status modifying the methodology described by Smith et al. [4]. Five stages of MAP infection were identified: Undetected, Moderate Latent, High Latent, Moderate Shedding and High Shedding. The aim of the present study was to assess the relationship between test result and milk production in the current lactation. Given that cows were tested 60 days from the end of lactation and the greater sensitivity of the ELISA tests used, cows were assigned a MAP status based on the test results for the current lactation. Cows with a negative ELISA and fPCR result were classified as undetected. Cows with at least one positive test result were considered infected. Animals with a positive fPCR result were classified as shedding for all milk test days of the current lactation; moderate shedding if the fPCR $\geq1\times10^3$ to $<1\times10^4$ genomes/mL and high shedding when fPCR $\geq1\times10^4$ genomes/mL. Cows with a positive ELISA result in the absence of a positive fPCR result were classified as latent infection. For all milk test days prior to a positive ELISA result cows were classified latent as it was assumed that cows were infected in calf-hood.

Database summaries and plots were used to explore the data. All variables were assessed for correlation using a correlation matrix and where a correlation > 0.2 was found, a variance inflation factor to assess collinearity was calculated using auxiliary regressions of one of the correlated variables on the remaining explanatory variables in the model. When the variance inflation factor was > 10, or if on rerunning the model without the variable the remaining coefficients changed in value by more than 20%, the collinear variables were assessed for biological plausibility. In this situation, the least useful variable was discarded from the final model using a Likelihood ratio test.

The effect of the input variables was investigated using pair-wise combinations in an ANOVA with a Bonferroni adjustment for multiple comparisons. Individual variables and their two-way interaction terms were carried forward to a multivariable linear regression model if ANOVA indicated the level of significance was < 0.1. Given that there were only a small number of variables a hand built model was constructed for the dependent variable (kg milk solids at each test day). Model structure was based on that of Smith et al. [4]. Age, MAP status, breed, days in milk and LS were all modelled as fixed effects. Given that there were only 2 seasons and the average number of lactations per cow was 1.51, milking season was modelled as a fixed effect with cow modelled as a random effect. To allow for correlation of test day results within individual cows, first order autoregression using test day number of a cow within lactation to identify the time lag between milk test day observations in individual cows was used. The model structure was:

$$\begin{aligned}\text{kgMS}_{\text{ilt}} = \beta_0 &+ \beta_{1,\text{p}}\text{parity} + \beta_2\text{DIM}_{\text{ilt}} + \beta_3 \exp[-0.1 \times \text{DIM}_{\text{ilt}}] + \beta_{4,\text{b}}\text{breed}\\ &+\delta_{1,\text{p}}\text{DIM}_{\text{ilt}\ \text{x}}\text{parity} + \delta_{2,\text{p}} \exp[-0.1\\ &\times\text{DIM}_{\text{ilt}}] \times \text{parity} + \beta_{5,\text{y}}\text{season}\\ &+\beta_{6,\text{n}}\text{MAP}_{\text{ilt}} + (\rho_{\text{i}}\varepsilon_{\text{i,t}-1} + \mu_{\text{iltpmn}})\end{aligned}$$

In this model, the outcome is daily milk production (in kgMS/day), i indicates cow, l indicates the present lactation, and t indicates milk test day; $\beta_{1,p}$ is the fixed effect of the dichotomized pth parity ($p = 1, > 1$); β_2 is the effect of Days in Milk (DIM); β_3 is the effect of Wilmink's correction; $\beta_{4,\text{b}}$ is the effect of breed b = > 75%, 50–75%, < 50% Friesian genetics; $\beta_{5,y}$ is the fixed effect of the yth season (y = 2013–14; 2014–15); and $\beta_{6,n}$ is the fixed effect of the nth JD status (n = Undetected, Moderate-Latent, High-Latent, Low-Shedding, High-Shedding). The interaction coefficient $\delta_{1,p}$ is the effect of the interaction between DIM and the pth parity, and the interaction coefficient $\delta_{2,p}$ is the effect of the interaction between Wilmink's correction and the pth parity [29]. The term $\rho_i\varepsilon_{i,t-1}$ provides the first-order autoregression between milk test days in individual cows; and u_{iltpmn} is the error term for each test date [4].

A dependent variable such as milk production may be associated with many input variables which may in turn be associated with each other [30]. A causal web was used to understand the likely nature of the relationship between potential input variables and to allow for confounding and intermediary variables [6]. A confounding variable has to be associated with the exposure variable and the dependent variable, the latter causally. This distorts the relationship between the exposure and dependent variable unless taken into account by retaining them in the regression model [6]. Conversely, intermediate variables are on the causal pathway between exposure and dependent variable and the effect of the dependent variable is mediated in part or completely through the intermediate variable. Leaving intermediate variables within a regression model distorts the relationship between the exposure and the dependent variable towards the null [31]. In the model, confounding variables were retained if the adjusted estimate of the effect of MAP status differed by more than 10% from the crude estimate excluding the confounder. Where the causal web suggested that variables may be acting as partial intermediaries between MAP status and milk production, the direct effect of MAP status on milk production was also estimated by excluding the intermediary variables from the model [32, 33]. In addition, variables initially excluded with $p < 0.1$ under ANOVA pairwise comparison were reintroduced to check if they were confounders within the model. All regression models were tested using standard diagnostic techniques for homoscedasticity, normality of residuals, linearity of predictor-outcome association and the effect of outliers. Statistical analysis was performed using the statistical programme R v3.3.2 [34].

Results

Effect of JD status on milk production

Records from 259 cows were discarded (128 from 2013 to 14 and 131 from 2014 to 15) due to missing herd test data. Analysis of the missing data using the R package "MissMech" indicated that the missing data were not normally distributed (Hawkins p value < 0.001) but there was no evidence the missing data were heteroscedastic (non parametric test of homoscedasticity 0.427). Consequently, we concluded that the missing data were not normally distributed but there was no evidence they were not missing completely at random [35]. Subsequently, these missing data were excluded from analysis.

Across both seasons complete production records were available from 1122 cows in 2013–14 and 1069 cows in 2014–15. At the end of the 2013–14 season (May 2014), 388 cows were culled (114 with an elevated ELISA and/or fPCR status) and 335 heifers joined the herd in July 2014. Thus, of the 2191 lactation records with a known ELISA status, 1468 were from cows present in both years, 388 from cows present only in 2013–14 and 335 from cows present only in 2014–15. Of these ELISA positive animals, 405 had a known fPCR status. Test results and herd descriptive data are presented in Table 2, while descriptive data by category of MAP infection are presented in Table 3. Centiles and

Table 2 Milk solids production, age and days in milk for cows which underwent screening for MAP infection

Season	2013–14			2014–15			*p*-value
Variable (Centile)	10th	50th	90th	10th	50th	90th	
Age (Years)	3	5	8	3	5	8	0.465[a]
Days in Milk	215	258	268	219	255	267	< 0.001[a]
Milk Solids (kg/cow)	307	433	552	337	480	590	< 0.001[a]
ISCC (cells/mL)	31 k	87 k	400 k	25 k	53 k	197 k	< 0.001[a]
Proportion Friesian	0.50	0.75	1.00	0.50	0.75	1.00	0.860[a]
Cows in milk	1122			1069			
ELISA Status	Number (Proportion)			Number (Proportion)			
Not Detected	825 (0.74)			961 (0.90)			< 0.001[b]
Low	157 (0.14)			39 (0.04)			< 0.001[b]
Moderate	63 (0.06)			27 (0.03)			< 0.001[b]
High	77 (0.07)			42 (0.04)			0.002[b]
Total	1122			1069			
fPCR Status of ELISA positive animals							
Not Detected	235 (0.79)			80(0.74)			0.557[b]
Moderate	32 (0.11)			11 (0.10)			0.865[b]
High	30 (0.10)			17 (0.16)			0.351[b]
Total	297			108			

[a]Wilcoxon Rank Sum
[b]χ^2 test with Holme-Bonferonni correction

non-parametric test results are presented where the data distribution was not normal.

Looking at the results for the ELISA (Paralisa™) test rather than the combination of ELISA and IDEXX tests used for classification in Table 2, in 2010–11 97/1086 (8.9%) animals were positive, compared to 365/1122 (33.0%) ELISA (Paralisa™) test positive animals in 2013–14 indicating that the prevalence of JD had continued to increase in this herd. In 2014–15 59/1069 (5.5%) were positive to the ELISA (Paralisa™) test. The change in prevalence of JD in this herd will be the subject of a separate study.

The effect of MAP infection status on milk production was confounded by age, season, days in milk and breed. However, the coefficients for the effect of MAP status changed by < 10% when somatic test score was included in the model and the impact of these changes in milk production was slight (+/− 0.01 kgMS). Consequently, LS was adjudged to be acting as an intermediary variable in this dataset and was excluded from the model [6].

The random effect of cow was significant in the final model ($p < 0.001$). The coefficients for the final mixed model illustrating the effect of MAP infection status on test day milk production for both years are presented in Table 4. The coefficients represent the effect of a unit change in the predictor variable on the predicted daily milk yield in kgMS given that all the other variables in the model do not change. There were no significant interactions between any of the variables in the model although there was a trend for the effect of MAP infection status on milk production to be greater in 2014–15 than 2013–14 ($p = 0.065$). However, the AIC value for the model with interaction were 10,205 with 20 degrees of freedom compared to 10,206 and 16 degrees of freedom for the model without interaction. Correspondingly the Bayesian Information Criterion (BIC) was 10,346 for the model with interaction and 10,319 for the simple model. The slight decrease in AIC indicated relatively little danger of underfitting in the simple model, while the larger increase in BIC suggested more danger of over fitting if the interaction was included [36]. Thus, coupled with a non-significant interaction term at the expense of increased model complexity and degrees of freedom the interaction term was not considered worthwhile for this data set.

In each season the regression model predicted that after adjusting for the effects of age, days in milk, season and breed, cows that had a latent infection status or were shedding MAP produced less milk than cows where MAP was not detected (not significant for moderate latent cows). Tests of the significance of the effects of differing MAP infection status on milk production with Tukey contrasts for multiple comparisons are shown in Table 5.

Generally, all infection statuses were associated with reduced milk production compared to cows where MAP

Table 3 Descriptive data by MAP infection status

MAP infection status	Average DIM (SD)	Average daily milk production in kgMS (SD)	Cows in category (proportion)	Average age (SD)	Total milk test results	First lactation results
Undetected	283 (19)	1.78 (0.58)	1779 (0.81)	5.0 (2.0)	7067	1828
Moderate latent	278 (22)	1.77 (0.59)	192 (0.09)	5.7 (1.7)	761	52
High latent	282 (18)	1.75 (0.49)	130 (0.06)	5.7 (1.9)	518	40
Moderate shedding	282 (18)	1.66 (0.53)	43 (0.02)	4.7 (1.5)	172	32
High shedding	282 (18)	1.59 (0.58)	47 (0.02)	5.5 (1.7)	188	24
Total	283 (19)	1.77 (0.58)	2191	5.2 (2.0)	8706	1976

was not detected (not significant for moderate latent cows). There was no difference within the categories of shedding nor within the categories of latency. Moderate and high shedders produced less milk than all other categories but this effect was only significant for high shedding cows. The model predicted milk solid production at the average number of days in milk (141 days) by MAP status together with pairwise comparison with Tukey adjusted p-values is shown in Table 6.

A smoothed plot of the predicted milk solids compared with the observed average milk solids and the 95% confidence interval for the range of observed milk solids

Table 4 Results of a linear mixed model predicting daily kgMS production

Input variable	Coefficient	SE	p-value
Intercept	1.77	0.03	< 0.001
Parity			
Primiparous	Ref		
Multiparous	0.73	0.03	< 0.001
Days in Milk	−0.002	0.0001	< 0.001
Wilmink's correction[a]			
exp(− 0.1 × DIM)	−0.13	0.22	0.546
Cow breed			
> 75% Friesian	Ref		
50–75% Friesian,	−0.09	0.02	< 0.001
< 50% Friesian	−0.07	0.02	< 0.001
DIM × parity			
DIM × Primiparous	Ref		
DIM × Multiparous	−0.002	0.0002	< 0.001
Wilmink's correction × parity			
exp(−0.1 × DIM × primiparous)	Ref		
exp(−0.1 × DIM × multiparous)	−0.21	0.24	0.379
Season			
2013–14	Ref		
2014–15	0.22	0.01	< 0.001
MAP status[b]			
Undetected	Ref		
Moderate latent	−0.03	0.02	0.136
High latent	−0.07	0.03	0.004
Moderate shedding	−0.07	0.04	0.036
High shedding	−0.20	0.04	< 0.001

Ref = the reference category for each categorical variable
[a]Overall significance of Wilmink's correction and its interaction in the model < 0.001
[b]Overall significance of MAP status in the model < 0.001

Table 5 Comparison tests for the effect of MAP infection status on daily milk production. *P*-values are Tukey contrasts for multiple comparison

Comparison	Difference	SE	*p*-value
Significant differences			
High latent vs negative 2–0	− 0.07	0.03	0.004
Moderate shedding vs negative 3–0	− 0.07	0.03	0.036
High shedding vs negative 4–0	− 0.20	0.04	< 0.001
Moderate latent vs high shedding 1–4	+ 0.17	0.05	0.002
High latent vs high shedding 2–4	+ 0.13	0.05	0.050
Non-significant differences			
Moderate latent vs negative 1–0	−0.03	0.02	0.136
Moderate latent vs high latent 1–2	+ 0.04	0.03	0.679
Moderate latent vs moderate shedding 1–3	+ 0.04	0.05	0.890
Moderate shedding vs high shedding 3–4	+ 0.12	0.06	0.194
High latent vs moderate shedding 2–3	−0.00	0.05	1.000

at each herd test is depicted in Fig. 1 to demonstrate that the model adequately predicted the observed lactations. Ninety-five percent confidence intervals for the observed milk yield at each of four test dates in each lactation are represented by vertical lines in each lactation.

Discussion

This study supports the hypothesis that sub-clinical MAP infection reduces milk production. To our knowledge this is the first longitudinal study based in NZ looking at the effect of MAP infection status on daily milk production while allowing for intermediary and confounding factors.

We recognize that our interpretation of these results is based on an imperfect testing methodology in that neither test had 100% sensitivity and specificity nor were all cows tested with both ELISA and fPCR. Nevertheless, in a single herd study, Aly et al. [37] identified that qPCR testing of faecal samples from ELISA positive cows in high MAP bioburden pens was the most cost effective strategy for confinement systems. Whilst pen sampling is not applicable in pasture systems, these workers also suggested that whole herd milk or single antigen ELISA serum test followed by fPCR on ELISA positive cows was the next most cost effective strategy and increased sensitivity to 68.4%.

O'Brien et al. [26] described a synergistic effect through the incorporation of a dual ELISA tests which resulted in an 18 and 17% increase in overall test sensitivity over individual tests used in isolation (IDEXX and Paralisa™, respectively) for defined fecal shedding states in a dataset comprising 1069 matched bovine fecal samples submitted for routine testing. These authors reported a sensitivity of the composite ELISA tests of 92%, with a specificity of 59% for detection of $\geq 1 \times 10^3$ MAP genomes/mL based on this matched dataset. These authors acknowledged that while it was possible that the increased sensitivity observed through the inclusion of additional antigens arose at the cost of test specificity, true specificity values could not be derived from the dataset as the samples were submitted as part of routine testing from a number of infected herds. It is vital to maximise the sensitivity of the initial screening test to identify the maximum numbers of putative shedders. While Aly et al., [37] suggest that specificity is the most important parameter in developing a cost effective strategy to control MAP infection, for serial test systems it is the specificity of the composite tests (ELISA + fPCR) rather than individual test results that are critical.

Table 6 Model predicted daily milk production (kgMS/day) at different MAP infection status. *P*-values are Tukey contrasts for multiple comparison

infection status	Predicted daily milk production and 95% CI (kgMS/day)	Significance of Tukey adjusted pairwise comparison of MAP status[a].
Undetected	1.74 (1.71–1.76)	b
Moderate Latent	1.71 (1.66–1.77)	b
High Latent	1.67 (1.60–1.73)	b
Moderate Shedding	1.64 (1.56–1.72)	ab
High Shedding	1.53 (1.42–1.63)	a

[a]Values with different letters are significantly different at $p < 0.05$

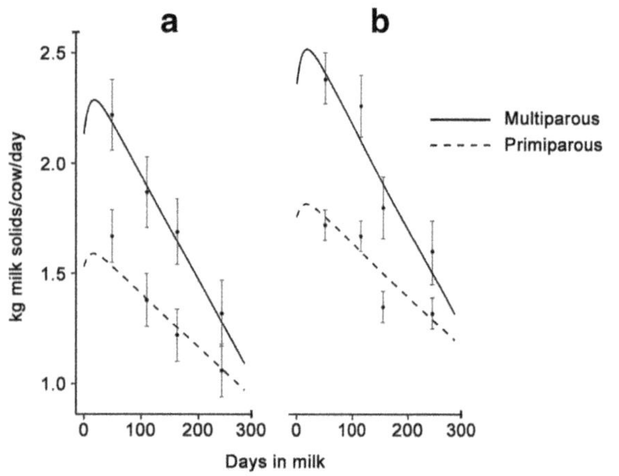

Fig. 1 Predicted milk yield for multiparous and primiparous cows in 2013–14 (**a**) and 2014–15 (**b**) from a model investigating the effect of MAP infection status on milk yield. Ninety-five percent confidence intervals for the observed milk yield at each of four test dates in each lactation are represented by vertical lines in each lactation

The ELISA screening test was designed to have maximal sensitivity to select a group of animals at risk of shedding, from which a subgroup of high shedders were identified with the confirmatory fPCR test which has a specificity of virtually 100% [37]. Using these figures, a priori, we believed our combined ELISA test had a sensitivity of 0.92 and a specificity of 0.81 which would suggest that 21% would test positive (true and false positives). In total we tested 2191 cows and found 19% were test positive.

Although the results are derived from one herd over a two-year period and so cannot be readily extrapolated to other NZ dairy herds, they are in line with the effect of MAP infection outlined by others although differences in testing methodology make direct comparison difficult. Using a single serum ELISA test and fecal culture where > 30 MAP cfu/g feces was classified as high shedding, Smith et al. [4] reported a reduction in daily milk production equivalent to around 0.2–0.3 kgMS. Using a similar testing methodology, Aly et al. [6] reported that fecal positive cows produced the equivalent of 0.2 kgMS less per day compared to cows where MAP was not detected. Smith et al. [4] reported that latent infected animals produced an equivalent 0.2 kgMS/day more milk than negative cows and they postulated a positive genetic association between susceptibility to MAP infection and milk production. There was no positive association between latent status and milk production in the current study; moderate latent cows had no significant difference in production compared to cows where MAP was not detected and high latent cows produced 0.07 kgMS/day less. Smith et al. [4] found that, although latent cows produced more milk than cows where MAP was not detected, the difference decreased with time spent in the latent infection state. In the present study, cows were tested once near the end of lactation and so the tested population may represent cows more advanced in the infection process than when testing is quarterly.

Other studies have also identified a positive or null association between MAP infection and milk production [38, 39]. Johnson et al. [40] found no effect on milk production of subclinical JD infection (as diagnosed by serum ELISA and fecal radiometric culture). The authors of these studies suggested that the inconsistency reported between different studies in the effect of MAP status on milk production is because the effect of MAP status depends upon the parity of the cow [40] or the production potential [41] and that younger, more productive cows showed a positive association between MAP status and milk production whereas older cows only showed a negative association between MAP status and milk production in the lactation before culling. However, the herd age distribution in these studies was considerably younger than in the current herd with Johnson et al. [40] reporting 59% of the study herds in their first or second lactation and the average age at 3.67 years. In the present study, 45% of animals were in their first or second lactation and the average age was 5.2 years. These workers also found no difference in the prevalence of MAP positive cows with age whereas in this NZ study, prevalence of MAP infection increased with age.

In the present study, latent and shedder status were subdivided to see if differences in the cow's response to MAP infection (from whatever cause) was reflected in a measurable production response. The strength of the ELISA response has been predictive of the decrease in milk production in other studies [42] and has been recommended as a factor to guide culling decisions [43].

In both their early study and a later follow up, Smith et al. [4, 5] were able to look at the effect of changes in MAP infection status as cows were sampled multiple times during the lactation. These authors found that some infected animals went on to become high shedders (> 50 cfu/g) while others remained low shedders for the duration of the study (7 years). Although both groups suffered a drop-in production, this partially recovered in some low shedding cows.

Smith et al. [5] found that the number of fecal positive tests was a good indicator of the effect on milk production; cows with multiple fecal positive test results were more likely to have a progressing infection and ongoing decreased milk production. As these authors acknowledge, multiple tests are likely to prove cost prohibitive on commercial dairy farms. However, the quantitative fPCR used in the present study may offer an advantage in allowing classification and stratification of shedding animals. The present study cannot determine how the screening test results used relate to the infection path described by Smith et al. [5] but they indicate that, based on the differences in predicted daily milk solids production over a 282 day lactation, high shedding animals have a more dramatic reduction in milk production (−59.22 kgMS over 282 days lactation) than low shedders (−28.20 kgMS over 282 days lactation). Knowledge of the average, predicted drop in milk solids can help farmers to prioritize animals for culling although a thorough evaluation of the most appropriate control strategy must consider the number of cows in each MAP class as well as their fecal MAP burden.

Fecal shedding in infected animals had a consistently more negative effect on production than latent infection as has been found in other studies [3]. However, as these workers point out this may be due to differences in test specificity rather than to differences in pathology. With a specificity of 59–81% for the pooled ELISA and close to 100% for the fPCR [37], more false positive cows will be identified by the ELISA test than the fPCR. These MAP negative cows would decrease the apparent effect of a positive ELISA status on milk production.

The present results may have been biased by the loss of animals with incomplete records, especially if these were low producing animals, culled before the autumn MAP test. However, the loss of these cows would have biased the results towards non-significance [4]. Similarly, lack of sensitivity in any of the single ELISA tests would have had the same effect by failing to detect MAP infected animals but this was offset by using multiple ELISA tests in parallel. In the current study, fecal shedders were classified as any cow that had a positive fPCR whereas Smith et al. [4], with a single ELISA classification, defined all ELISA positives as fecal shedders. In the present study, with the reduced specificity from ELISA tests in parallel, classifying all ELISA positive cows as fecal shedders would have led to false positives. We believe that our parallel ELISA testing maximises sensitivity and so reduces the number of false negatives that go forward to fPCR testing. The greater sensitivity of fPCR compared to fecal culture [19] then supports the contention that the risk of failing to identify shedders in this group was low. Although fecal contamination during collection of fecal samples could lead to false positive fecal results the quantitative nature of fPCR means that it is unlikely that this would be sufficient to change the fPCR shedding classification of the cow. Conversely, while tissue and for many practical purposes, fecal culture on solid medium remains the gold standard for specificity, the long incubation period required means that this test may be compromised with overgrowth of contaminating gut bacteria [24]. Sensitivity of fecal culture in high prevalence herds has been estimated at 53% for sub-clinical cows [44].

We found no interaction between age and MAP infection status ($p > 0.1$) although numerically suggestive of an increase in effect with age as described by others [6, 22, 45]. The small sample size in the current study and the relative lack of MAP positive 2 year olds may have meant there was insufficient power to detect such a relationship at a statistically significant level. In pastoral systems, matching of feed supply to cow demand is highly dependent on seasonal factors [46] and so the ability of cows to milk to their potential is variable year to year. This is consistent with differences in the degree to which MAP status impacts on milk production suggested by the statistical trend for an interaction between season and MAP status in this model. This contrasts with models from confined all year round calving systems [4, 5] where no such interaction was present but where feed supply-cow demand is much less influenced by seasonal climate. However, others have suggested that interaction terms should be interpreted more conservatively ($p < 0.001$) to avoid Type 1 errors and so we elected to use the simpler model without interaction [47].

In this study we describe the application of DNA standards to facilitate quantitative judgements of MAP shedding. These standards spanned the range of MAP shedding observed in clinical samples by the testing laboratory using this approach and were linear in the assay over the 7 logs (typically $r^2 = 0.999$). Although detection of MAP DNA at lower titers is readily achievable, sample replicates may become poor because of stochastic variation [20] and, in JD affected herds, extremes of diagnostic sensitivity for MAP bacilli in feces may be of dubious clinical relevance or may be otherwise attributable to pass-through shedding in a contaminated environment; for these reasons a conservative lower detection cut point was favored. While there is surprisingly little agreement on standardization of fPCR data for quantitative reporting of MAP fecal shedding [47] in this study DNA standards were favored for quantitation as they are quick and inexpensive to prepare and to quantify accurately and also stable in storage, allowing a set of quantitation standards to be prepared that are highly reproducible from day to day and which facilitate objective comparisons between samples. More importantly they are constant, comparable and reproducible across laboratories, geography and time.

While recommendations for optimal diagnostic strategies can be made for NZ informed by international best practice and studies performed overseas, they must also be considered in the light of cost and local availability of diagnostic services. Due to the extreme demands placed on diagnostic tests by the dynamics of MAP infection and the persistent, chronic nature of JD progression in cattle, single, non-quantitative tests used in isolation may be insufficiently exact to maximally inform management decisions particularly in the early, subclinical stages of disease when bioindicators are absent. Diagnostic tests based on multiple ELISA and fPCR are quick, inexpensive and quantitative while also amenable to the incorporation of additional antigens to broaden their diagnostic repertoire at little additional cost.

Conclusion

In this herd we identified a significant effect of subclinical MAP infection on milk solids production. The testing methodology used in the present study allowed the farmer to identify cows that were infectious and our model suggests they are likely to under produce compared to cows where MAP is not detected. Although our findings are from a single herd study and cannot be extrapolated to the wider farming environment, they are in line with similar studies involving small numbers of herds [4, 37].

Greater knowledge allows better decisions to be made about which animals to cull to both reduce the infectious pressure within the herd and the production losses associated with infection. Early culling of sub-clinically affected animals can be justified both on epidemiological grounds and because of the associated deficits in production. While there is contention as to the impact and costs associated with subclinical MAP infection in dairy cows, the deficits in milk production seen in the current study suggest that composite diagnostic testing to identify and cull fPCR shedders and high latent infected cows may be justified.

Abbreviations

DIM: Days in milk; DRL: Disease Research Ltd; EU: ELISA units; fPCR: Fecal PCR; ISCC: Individual somatic cell count; JD: Johne's disease; LS: Somatic cell linear score; MAP: *Mycobacterium avium* subsp. *paratuberculosis*; NZ: New Zealand; PPA: Protoplasmic antigen

Acknowledgements

The authors would like to gratefully acknowledge the support of the farmer and farm staff, veterinary colleagues, Vetlife and DRL. We acknowledge and thank Dr Douwe Bakker for providing Ag_1Del_1 antigen. The impact of cited references on the design and analysis of this work is also acknowledged.

Funding

This work was carried out without external funding.

Authors' contributions

AB, RO'B, SL and FG designed the study. RO'B and SL prepared fPCR and immunoassay data, respectively. AB analyzed the data and performed statistical analysis. AB, RO'B, SL and FG interpreted the data. All authors took part in writing the manuscript and have approved the final version.

Consent for publication

Not applicable.

Competing interests

The authors declare that they have no competing interests.

Author details

[1]Vetlife, Centre for Dairy Excellence, Wilson Street, Geraldine, New Zealand. [2]Disease Research Limited, Invermay Agricultural Centre, Mosgiel, New Zealand. [3]Department of Microbiology & Immunology, University of Otago, Dunedin, New Zealand.

References

1. Hunnam J. Johne's control is possible: does anyone have a MAP? In: Proceedings of The Society of Dairy Cattle Veterinarians of the NZVA Annual Conference. NZVA; 2014. p. 133–50.
2. Whitlock RH, Buergelt C. Preclinical and clinical manifestations of paratuberculosis (including pathology). Vet Clin North Am Food Anim Pract. 1996;12(2):345–56.
3. Gonda MG, Chang YM, Shook GE, Collins MT, Kirkpatrick BW. Effect of *Mycobacterium paratuberculosis* infection on production, reproduction, and health traits in US Holsteins. Prev Vet Med. 2007;80(2–3):103–19.
4. Smith RL, Grohn YT, Pradhan AK, Whitlock RH, Van Kessel JS, Smith JM, Wolfgang DR, Schukken YH. A longitudinal study on the impact of Johne's disease status on milk production in individual cows. J Dairy Sci. 2009;92(6):2653–61.
5. Smith RL, Grohn YT, Pradhan AK, Whitlock RH, Van Kessel JS, Smith JM, Wolfgang DR, Schukken YH. The effects of progressing and nonprogressing *Mycobacterium avium* ssp. *paratuberculosis* infection on milk production in dairy cows. J Dairy Sci. 2016;99(2):1383–90.
6. Aly SS, Anderson RJ, Adaska JM, Jiang J, Gardner IA. Association between *Mycobacterium avium* subspecies *paratuberculosis* infection and milk production in two California dairies. J Dairy Sci. 2010;93(3):1030–40.
7. Küpper J, Brandt H, Donat K, Erhardt G. Associations between paratuberculosis status and milk production traits in Holstein cattle under consideration of interaction effects between test result and farm and lactation number. Archiv Tierzucht. 2013;56(94):924–33.
8. Mortier RA, Barkema HW, De Buck J. Susceptibility to and diagnosis of *Mycobacterium avium* subspecies *paratuberculosis* infection in dairy calves: a review. Prev Vet Med. 2015;121(3–4):189–98.
9. Stevenson K. Genetic diversity of *Mycobacterium avium* subspecies *paratuberculosis* and the influence of strain type on infection and pathogenesis: a review. Vet Res. 2015;46:64.
10. Raizman EA, Wells SJ, Godden SM, Fetrow J, Oakes JM. The associations between culling due to clinical Johne's disease or the detection of *Mycobacterium avium* subsp. *paratuberculosis* fecal shedding and the diagnosis of clinical or subclinical diseases in two dairy herds in Minnesota, USA. Prev Vet Med. 2007;80(2–3):166–78.
11. Garforth C. Livestock keepers' reasons for doing and not doing things which governments, vets and scientists would like them to do. Zoonoses Public Health. 2015;62:29–38.
12. Norton S, Heuer C, Jackson R. A questionnaire-based crosssectional study of clinical Johne's disease on dairy farms in New Zealand. NZ Vet J. 2009;57:34–43.
13. Sorge U, Kelton D, Lissemore K, Godkin A, Hendrick S, Wells S. Attitudes of Canadian dairy farmers toward a voluntary Johne's disease control program. J Dairy Sci. 2010;93(4):1491–9.
14. Gardner IA, Nielsen SS, Whittington RJ, Collins MT, Bakker D, Harris B, Sreevatsan S, Lombard JE, Sweeney R, Smith DR, et al. Consensus-based reporting standards for diagnostic test accuracy studies for paratuberculosis in ruminants. Prev Vet Med. 2011;101(1–2):18–34.
15. Bogli-Stuber K, Kohler C, Seitert G, Glanemann B, Antognoli MC, Salman MD, Wittenbrink MM, Wittwer M, Wassenaar T, Jemmi T, et al. Detection of *Mycobacterium avium* subspecies *paratuberculosis* in Swiss dairy cattle by real-time PCR and culture: a comparison of the two assays. J Appl Microbiol. 2005;99(3):587–97.
16. Scott HM, Fosgate GT, Libal MC, Sneed LW, Erol E, Angulo AB, Jordan ER. Field testing of an enhanced direct-fecal polymerase chain reaction procedure, bacterial culture of feces, and a serum enzyme-linked immunosorbent assay for detecting *Mycobacterium avium* subsp *paratuberculosis* infection in adult dairy cattle. Am J Vet Res. 2007;68(3):236–45.
17. Clark DL Jr, Koziczkowski JJ, Radcliff RP, Carlson RA, Ellingson JL. Detection of *Mycobacterium avium* subspecies *paratuberculosis*: comparing fecal culture versus serum enzyme-linked immunosorbent assay and direct fecal polymerase chain reaction. J Dairy Sci. 2008;91(7):2620–7.
18. Slana I, Kralik P. Perspective: is it time to change the gold standard in MAP detection? In: Proceedings of the 12th International Colloquium on

Paratuberculosis. Parma: Internstional Association for Paratuberculosis; 2014. p. 62. http://www.paratuberculosis.net.
19. Douarre PE, Cashman W, Buckley J, Coffey A, O'Mahony JM. Isolation and detection of *Mycobacterium avium* subsp. *paratuberculosis* (MAP) from cattle in Ireland using both traditional culture and molecular based methods. Gut Pathog. 2010;2(1):11.
20. Plain KM, Marsh IB, Waldron AM, Galea F, Whittington AM, Saunders VF, Begg DJ, de Silva K, Purdie AC, Whittington RJ. High-throughput direct fecal PCR assay for detection of *Mycobacterium avium* subsp. *paratuberculosis* in sheep and cattle. J Clin Microbiol. 2014;52(3):745–57.
21. Collins MT, Wells SJ, Petrini KR, Collins JE, Schultz RD, Whitlock RH. Evaluation of five antibody detection tests for diagnosis of bovine paratuberculosis. Clin Diagn Lab Immunol. 2005;12(6):685–92.
22. Beaudeau F, Belliard M, Joly A, Seegers H. Reduction in milk yield associated with *Mycobacterium avium* subspecies *paratuberculosis* (Map) infection in dairy cows. Vet Res. 2007;38(4):625–34.
23. Griffin JF, Spittle E, Rodgers CR, Liggett S, Cooper M, Bakker D, Bannantine JP. Immunoglobulin G1 enzyme-linked immunosorbent assay for diagnosis of Johne's disease in red deer (*Cervus elaphus*). Clin Diagn Lab Immunol. 2005;12(12):1401–9.
24. O'Brien R, Hughes A, Liggett S, Griffin F. Composite testing for ante-mortem diagnosis of Johne's disease in farmed New Zealand deer: correlations between bacteriological culture, histopathology, serological reactivity and faecal shedding as determined by quantitative PCR. BMC Vet Res. 2013;9:72.
25. Griffin JFT, Spittle E, Mackintosh CG, Rodgers CR, Liggett S, Cooper M. Serodiagnosis of Johne's disease in red deer (*Cervus elaphus*). In: Proceedings of the 8th international colloquium on Paratuberculosis. Copenhagen: International Association for Paratuberculosis, Inc.; 2005. p. 496–500.
26. O'Brien R, Liggett S, Bates A, F G. Johne's disease diagnosis in New Zealand: An update. In: Proceedings of The Society of Dairy Cattle Veterinarians of the NZVA Annual Conference; Hamilton, New Zealand. NZVA; 2016: 41–50.
27. Li L, Bannantine JP, Zhang Q, Amonsin A, May BJ, Alt D, Banerji N, Kanjilal S, Kapur V. The complete genome sequence of *Mycobacterium avium* subspecies *paratuberculosis*. Proc Natl Acad Sci U S A. 2005;102(35):12344–9.
28. Dabdoub S, Shook G. Phenotypic relations among milk yield, somatic cell count and clinical mastitis. J Dairy Sci. 1984;67:163–4.
29. Wilmink J. Adjustment of test-day milk, fat, and protein yield for age, season and stage of lactation. Livest Prod Sci. 1987;16:335–48.
30. Dohoo IR, Martin SW, Stryhn H. Veterinary epidemiologic research. Charlottetown: AVC Inc.; 2003.
31. Schisterman EF, Cole SR, Platt RW. Overadjustment bias and unnecessary adjustment in epidemiologic studies. Epidemiology. 2009;20(4):488–95.
32. Martin J. Confounder or Intermediate Variable? http://www.ctspedia.org/do/view/CTSpedia/BiasConfoundIntermed. Accessed 10 Feb 2017.
33. Taylor JM, Wang Y, Ahdieh L, Chmiel JS, Detels R, Giorgi JV, Kaslow R, Kingsley L, Margolick J. Causal pathways for CCR5 genotype and HIV progression. J Acquir Immune Defic Syndr. 2000;23(2):160–71.
34. R Development Core Team. R: A language and environment for statistical computing. R Foundation for Statistical Computing. http://www.r-project.org. Accessed 27 July 2017.
35. Jamishidian M. MissMech: an R package for testing homoscedasticity, multivariate normality, and missing completely at random (MCAR). J Stat Softw. 2014;56:1–31.
36. Dziak JJ, Coffman D, Lanza S, Runzie L. Sensitivity and specificity of information criteria. Technical Report Series 12–119. 2012. https://methodology.psu.edu/media/techreports/12-119.pdf
37. Aly SS, Anderson RJ, Whitlock RH, Fyock TL, McAdams SC, Byrem TM, Jiang J, Adaska JM, Gardner IA. Cost-effectiveness of diagnostic strategies to identify *Mycobacterium avium* subspecies *paratuberculosis* super-shedder cows in a large dairy herd using antibody enzyme-linked immunosorbent assays, quantitative real-time polymerase chain reaction, and bacterial culture. J Vet Diagn Investig. 2012;24(5):821–32.
38. McNab W, Meek AH, Martin SW, Duncan JR. Associations between dairy production indices and lipoarabinomannan enzyme-immunoassay results for paratuberculosis. Can J Vet Res. 1991;55:356–61.
39. Wilson DJ, Rossiter C, Han HR, Sears PM. Association of *Mycobacterium paratuberculosis* infection with reduced mastitis, but with decreased milk production and increased cull rate in clinically normal dairy cows. Am J Vet Res. 1993;54(11):1851–7.
40. Johnson YJ, Kaneene JB, Gardiner JC, Lloyd JW, Sprecher DJ, Coe PH. The effect of subclinical *Mycobacterium paratuberculosis* infection on milk production in Michigan dairy cows. J Dairy Sci. 2001;84(10):2188–94.
41. Benedictus G, Dijkhuizen AA, Stelwagen J. Economic losses due to paratuberculosis in dairy cattle. Vet Rec. 1987;121(7):142–6.
42. Lombard JE, Garry FB, McCluskey BJ, Wagner BA. Risk of removal and effects on milk production associated with paratuberculosis status in dairy cows. J Am Vet Med Assoc. 2005;227(12):1975–81.
43. Dorshorst NC, Collins MT, Lombard JE. Decision analysis model for paratuberculosis control in commercial dairy herds. Prev Vet Med. 2006; 75(1–2):92–122.
44. Sockett DC, Carr DJ, Collins MT. Evaluation of conventional and radiometric fecal culture and a commercial DNA probe for diagnosis of *Mycobacterium paratuberculosis* infections in cattle. Can J Vet Res. 1992;56(2):148–53.
45. McKenna SL, Keefe GP, Tiwari A, VanLeeuwen J, Barkema HW. Johne's disease in Canada part II: disease impacts, risk factors, and control programs for dairy producers. Can Vet J. 2006;47(11):1089–99.
46. Roche J, Burke C, Meier S, Walker C. Nutrition x reproduction interaction in pasture-based systems: is nutrition a factor in reproductive failure? Anim Prod Sci. 2011;51:1045–66.
47. Marsh I, Plain K, Whittington R. Molecular diagnostic tests for Johne's disease; time to standardise. In: Proceedings of the 12th international colloquium on Paratuberculosis. Parma: Internstional Association for Paratuberculosis; 2014. p. 61. http://www.paratuberculosis.net.

3

Assessment of genetic variation for pathogen-specific mastitis resistance in Valle del Belice dairy sheep

Marco Tolone[1*†], Cristian Larrondo[2†], José M. Yáñez[2], Scott Newman[3], Maria Teresa Sardina[1] and Baldassare Portolano[1]

Abstract

Background: Mastitis resistance is a complex and multifactorial trait, and its expression depends on both genetic and environmental factors, including infection pressure. The objective of this research was to determine the genetic basis of mastitis resistance to specific pathogens using a repeatability threshold probit animal model.

Results: The most prevalent isolated pathogens were coagulase-negative staphylococci (CNS); 39 % of records and 77 % of the animals infected at least one time in the whole period of study. There was significant genetic variation only for *Streptococci* (STR). In addition, there was a positive genetic correlation between STR and all pathogens together (ALL) (0.36 ± 0.22), and CNS and ALL (0.92 ± 0.04).

Conclusion: The results of our study support the presence of significant genetic variation for mastitis caused by *Streptococci* and suggest the importance of discriminating between different pathogens causing mastitis due to the fact that they most likely influence different genetic traits. Low heritabilities for pathogen specific-mastitis resistance may be considered when including bacteriological status as a measure of mastitis presence to implement breeding strategies for improving udder health in dairy ewes.

Keywords: Ewes, Mastitis, Pathogen, Resistance

Background

Mastitis is one of the most common diseases affecting dairy sheep. Mastitis leads to major economic losses, mainly due to discarded milk, reduced milk production and quality, alteration of cheese-making properties, early culling, and increased health care costs [1–8]. The alterations or reductions of the dry matter of milk and its composition, have a substantial effect on the economic and industrial values of the milk, considering that almost all is processed into fermented products and cheeses [4, 9, 10]. Mastitis resistance is a complex and multifactorial trait, and its expression depends on both genetic and environmental factors, including infection pressure. In the broadest sense, resistance could be defined as the ability to avoid any infection and/or the quick recovery from an infection [11, 12], and involves different factors such as to avoid entry of the pathogen into the mammary gland, to induce an immune response capable of limiting pathogen development in the udder and to recover from the infection, as well as controlling the pathogenic effects of the infection, such as tissue damage [13].

Over 100 different micro-organisms can cause mastitis, in particular coliform bacteria, staphylococci and streptococci [14]. In dairy sheep the most important agents involved in clinical mastitis are the bacterial infections, and the most frequently isolated pathogens are coagulase-negative staphylococci (CNS); that are present on and around the udder skin [9] with a different pathogenicity causing clinical and subclinical mastitis [15–18]. The bacterial pathogens responsible for infection of the mammary gland may be grouped into two main categories: major and minor pathogens. Major pathogen infection generally results in clinical illness or strong inflammatory

* Correspondence: marco.tolone@unipa.it
†Equal contributors
1Dipartimento Scienze Agrarie e Forestali, Università degli Studi di Palermo, Viale delle Scienze, Palermo 90128, Italy
Full list of author information is available at the end of the article

responses and reduced milk yields, whereas minor pathogen infection is usually subclinical [19].

Selection for genetic resistance to mastitis can be done directly or indirectly. Direct selection corresponds to the diagnosis of the infection: the actual trait [i.e., bacteriological examination of milk and/or observation of clinical cases of mastitis] is measured on the animal or its relatives. Indirect selection corresponds to a prediction of the bacteriological status of the udder based on traits related to the infection [e.g., inflammatory parameters]: an indicator trait for mastitis is measured on the animal itself or its relatives [20]. Simple and indirect methods have been widely applied based on the evaluation of the degree of inflammation or of internal mammary lesions [21]. Their accuracy was established by bacteriological analysis as a reference method [17]. Among the indirect methods, the most frequently used to detect mastitis are milk somatic cell count (SCC). SCC is considered as a good measure to indirectly select for mastitis resistance in cattle, especially when a direct measure of clinical mastitis incidence is not available [18, 22]. In cattle, values of SCC between 250 and 300 × 10^3 cells/mL are recommended as satisfactory discrimination thresholds to distinguish between healthy and infected udders. In sheep there is no widely accepted threshold [15, 23] but some studies suggested a critical limit of 500 × 10^3 cells/mL [24]. There are few studies concerning genetic variation of mastitis in sheep according to bacteriological status [22, 25]. A genetic selection approach could be one of the strategies for controlling mastitis and has been shown to be a valid option, together with management, to prevent mastitis cases [1, 26]. Studies have reported genetic variation accounting for resistance to mastitis in Valle del Belice dairy sheep [22, 24]. These authors have defined mastitis as a binary trait distinguishing between ewes with at least one case of mastitis (1) and ewes without (0) in a defined period of lactation and was analyzed using a linear model approach. This definition excluded alternative definitions, for example multiple cases of mastitis within lactation, and ignored the etiology of intra-mammary infections. The purpose of this study was to determine the genetic bases of pathogen-specific resistance to mastitis in Valle del Belice dairy sheep using a threshold repeated model.

Methods

Data

Data were collected between 2006 and 2011 in five Valle del Belice flocks, with a total of 2350 ewes and 5856 animals in the pedigree. Observations for this study included 1795 primiparous, 1285 secondiparous and 2225 multiparous dairy ewes. All ewes were milked twice daily (morning and evening), and records for milk yield (MY), bacteriological status (infected or not infected), and SCC were collected at approximately monthly intervals, following an A4 recording scheme (monthly records with two daily milking) which is defined by the International Committee for Animal Recording [27]. The milk samples were collected during routine milking so avoiding any harmful process to individuals. The consent for sample collection was obtained by the animals' owners. Moreover, Sample collection, animal management and cares were in agreement with the Directive 2010/63/EU. The observed bacteriological colonies were identified as: *Escherichia coli* (ESCCL), *Staphylococcus aureus* (STHAU), *Streptococcus dysgalactiae* (STPDG), *Streptococcus uberis* (STPUB), *Streptococcus agalactiae* (STPAG) and *Bacillus spp.* (BACIL), *Corynebacterium spp.* (CORLT), *Pasteurella spp.* (PASCL), *Pseudomonas spp.* (PSELT), coagulase negative staphylococci (CNS) and *Streptococcus spp.* (STR). Ewes were considered infected if at least one record with positive bacteriological test during lactation period was recorded, while they were considered healthy if the bacteriological test did not show a positive result. Ewes were measured more than one time during the same lactation. Thus, the repeatability of records is across and within lactations. Table 1 shows average number of records per ewe within lactations. Moreover, ewes were considered infected if more than five colony forming units (CFU) per 10 μl of milk of one species of bacteria were isolated. The response variable used in the model corresponds to the binary disease status, coded as 0 or 1 to represent uninfected or infected individuals, respectively.

Trait definition and statistical model

Phenotypic observations of infection status were defined as a repeated binary trait. The binary trait distinguished between sheep with infected udder status (1) and uninfected udder status (0) within each lactation for each particular pathogen described above. SCC was also recorded and normalized through a logarithmic transformation into somatic cell score (SCS) according to the formula of Ali and Shook [28]:

$$\text{Somatic cell score (SCS)} = \log_2(\text{SCC}/100,000) + 3$$

Table 1 Mean, standard deviation (SD), minimum (Min) and maximum (Max) number of records per ewe within lactation

Lactation	Mean	SD	Min	Max
1	3.47	1.97	1	12
2	4.13	2.25	1	20
3	4.46	2.33	1	15
4	4.75	2.4	1	12
5	7.47	5.75	1	32

The binary trait was analyzed using the following repeatability threshold probit animal model:

$$Pr(Y_{ijklmn}) = \Phi(\mu + OP_i + MY_j + FYS_l + PE_m + A_n)$$

Where y_{ijklmn} is the observation for the specific pathogen causing mastitis (CNS, STR, ESCCL, STHAU, STPDG, STPUB, STPAG, BACIL, CORLT, PASCL, PSELT and ALL); Φ is the normal cumulative density function; μ is the fixed effect of the overall mean; OP_i is the order of parity fitted as fixed effect (with 5 classes); MY_j is the milk production yield fitted as covariate; FYS_l is the flock-year-season random effect (51 classes); PE_m is the random permanent environmental effect of the individual m across lactations (2350 levels with records); and A_n is the random animal effect (5856 levels in the pedigree). The implicit residual variance on the underlying scale is 1 for the probit model (standard normal). Parameters of the univariate threshold models were estimated using ASREML version 3.0 [29].

Heritabilities

Heritabilities for resistance to different pathogens were calculated as:

$$h^2 = \frac{\sigma_a^2}{\sigma_a^2 + \sigma_{FYS}^2 + \sigma_{PE}^2 + \sigma_e^2}$$

Where σ_a^2 is the animal additive genetic variance, σ_{FYS}^2 is the variance associated with flock-year-season, σ_{PE}^2 is the variance due to permanent environment and σ_e^2 is residual variance. For all traits, the animal effect was assumed ~ $N(0, \mathbf{A}\sigma_a^2)$, where **A** is the additive genetic relationship matrix among all animals included in the pedigree (5856). Similarly, FYS and PE effects were assumed ~ $N(0, \mathbf{I}\sigma_{FYS}^2)$ and ~ $N(0, \mathbf{I}\sigma_{PE}^2)$, where **I** is an identity matrix with order equal to number of FYS and PE classes (51 and 2350) respectively.

Results

Arithmetic mean and standard deviation of MY, SCC and SCS, for infection status (infected and uninfected) of udders are shown in Table 2. Daily average MY values were 1275 ± 544 and 1338 ± 558 g, for infected and uninfected udders, respectively. Mean SCC was 2908 ± 4926 and 1155 ± 3083 (x 10^3 cells/mL), for infected and uninfected udders, respectively; whereas mean SCS was 6.21 ± 2.45 and 4.55 ± 2.13. Overall mean values for both infected and uninfected udders were 1314 ± 553 g, 1815 ± 3972 (x 10^3 cells/mL), and 5.18 ± 2.3, for MY, SCC and SCS, respectively.

There was a low prevalence of isolation due to CORLT, ESCCL, PASCL, PSELT, STPDG, STPUB, STPAG, and BACIL, in all the observations and according to infection status of animals. Models did not converge for these pathogens, most likely due to the low incidence (zero inflation).

Table 2 Arithmetic mean and SD for MY, SCC, and SCS of infected and uninfected udders

	n	MY(g)	SCC(x 10^3) [a]	SCS
		Mean ± SD	Mean ± SD	Mean ± SD
Infected	8019	1275 ± 544	2908 ± 4926	6.21 ± 2.45
Not infected	13,246	1338 ± 558	1155 ± 3083	4.55 ± 2.13
All	21,265	1314 ± 553	1815 ± 3972	5.18 ± 2.33

SD standard deviation, *MY* milk yield, *SCC* somatic cell count, *SCS* somatic cell score
n number of observations
[a] Cells/mL of milk

Absolute (AF) and relative (RF) frequency distribution according to infection status (infected or uninfected) observations within all the records data set, are shown in Table 3. The most prevalent isolated pathogens were CNS with 7951 (39 %) observations, followed by STHAU (940; 5 %) and STR (541; 3 %). Considering all pathogens (ALL), 8019 (38 %) milk samples were infected.

Table 4 shows absolute (AF) and relative (RF) frequency distributions according to udder status (infected or uninfected) of animals per pathogen. Similarly, the most prevalent isolated pathogens affecting ewes were CNS (1811; 77 %), followed by STHAU (513; 21.83 %) and STR (217; 9.23 %) when only animals were analyzed.

Table 5 shows components of variance due to Flock-Year-Season effect, permanent environment effect, additive genetic effect, phenotypic effect and heritabilities for resistance to CNS, STR and ALL. Due to the low frequency of isolation of ESCCL, STHAU, STPDG, STPUB, STPAG, BACIL, CORLT, PASCL, and PSELT in the total records and isolation of pathogens per animals (Tables 3 and 4), problems associated with convergence were found when analyzing resistance to these bacteria. In contrast, there was statistically significant genetic

Table 3 Absolute (AF) and relative (RF) frequency distribution according to udder status of observations ($n = 20,519$)

Pathogen	Status [a]	AF	RF
CNS	0	12,568	0.61
	1	7951	0.39
STHAU	0	19,579	0.95
	1	940	0.05
STR	0	19,978	0.97
	1	541	0.03
ALL	0	13,246	0.62
	1	8019	0.38

CNS coagulase-negative *staphylococci*, *STHAU Staphylococcus aureus*, *STR streptococci*, *ALL* observed bacteriological colonies as described in M&M section
[a] Udder status as binary record: 0 = not infected, 1 = infected

Table 4 Absolute (AF) and relative (RF) frequency distribution according to udder status of animals ($n = 2350$) per pathogen

Pathogen	Status [a]	AF	RF
CNS	0	539	0.23
	1	1811	0.77
STHAU	0	1837	0.78
	1	513	0.22
STR	0	2133	0.91
	1	217	0.09
ALL	0	612	0.26
	1	1738	0.74

CNS coagulase-negative *staphylococci*, *STHAU Staphylococcus aureus*, *STR streptococci*, *ALL* observed bacteriological colonies as described in M&M section
[a] Udder status as binary record: 0 = not infected, 1 = infected

variation for CNS, STR, and ALL pathogens (Table 5). Variances due to FYS random effect were 0.18, 0.09 and 0.79, whereas due to permanent environment effect were 0.39, 0.39, and 0.40 for CNS, STR and ALL, respectively. Variances of additive and phenotypic effects were 0.03, 0.15, 0.04; and 1.60, 1.62, 2.23, for CNS, STR and ALL, respectively. The heritability obtained for STR (0.09) was significant different from zero, whereas for CNS (0.02) and ALL (0.02) were not.

Table 6 shows phenotypic and genetic correlations for resistance to mastitis caused by STR, CNS and ALL estimated using a multivariate repeatability linear model. All of the estimated phenotypic and genetic correlations were significantly different from zero. The phenotypic correlation between ALL and STR was low, however, the genetic correlation between these traits was moderately high indicating that there was a direct relationship between these traits in genetic terms. On the other hand, both phenotypic and genetic correlations between ALL and CNS were high, indicating that there was a strong positive relationship, both phenotypic and genetic, between these traits. These results suggest that resistance to CNS is a similar to the resistance when it is measured as ALL. In addition, the phenotypic correlation between STR and CNS was negative and the genetic correlation between these traits was low, thus, indicating that selection for improved STR will not have an impact on CNS resistance.

Discussion

The overall infection prevalence considering frequency of infection on all the samples of records was 37.7 %, close to the value of 42 % reported in a previous investigation in the same breed [25], and higher than the values of 26.2 and 24.6 % reported by Pengov [15] and Gonzalo et al. [16], respectively. However, in the study of Pengov [15] only one milk sample ($n = 496$ samples, 251 ewes) of udder halves was considered, whereas the study of Gonzalo et al. [16] was based only on subclinical mastitis prevalence. When the prevalence was assessed on all the animals it increased to 74 %, higher than any prevalence reported in previous studies.

Probably the high SCC reported in our study are a consequence of inadequate preventive management, a lack of strict hygiene conditions and extensive management practices, generating a high number of subclinical mastitis cases due to environmental pathogens. Moreover, our results suggested that ewes have higher SCC than cows and it is therefore necessary to establish an acceptable threshold in dairy sheep considering the difference in SCC between breeds and other factors [15, 18, 22].

Leitner et al. [9] suggested categories for classification of SCC in sheep and goat related to quality of milk and infection status. These researchers suggested that infection of 25, 50 and 75 % of the udders in a given herd was associated with 4.1 to 12.2 % of milk loss in sheep and 0.8 to 2.3 % in goats [9]. Mavrogenis et al. [30] suggested that an increase of 0.5 cells/mL × 10^6 SCC above the mean resulted in reduction of mean individual daily production of milk by 18 g.

In the present study, there was a 68 g difference in mean MY between infected and non- infected ewes. For SCC, the mean value for infected animals was approximately 3-fold higher than uninfected animals, similar to values reported in a previous study [24]. However mean SCC for healthy animals were different. Mean SCC for uninfected animals was different to the value of 89 cells/mL × 10^3 reported by Pengov [15] and 311 cells/mL × 10^3 reported by Leitner et al. [4], and similar to the value of 1490 cells/mL × 10^3 reported by Kern et al. [31]. These studies focused on Domestic Highland, East Friesand, and Awassi breeds, including their crosses and the Assaf breed, respectively. Moreover, considering the whole data set, mean SCC for infected animals was

Table 5 Estimates of components of variance and their standard errors for infectious status

Resistance Trait Infection Status	σ^2_{FYS}	σ^2_{PE}	σ^2_a	σ^2_P	h^2
ALL	0.79 ± 0.17	0.40 ± 0.03	0.04 ± 0.02	2.23 ± 0.17	0.02 ± 0.01
CNS	0.18 ± 0.04	0.39 ± 0.03	0.03 ± 0.02	1.60 ± 0.05	0.02 ± 0.01
STR	0.09 ± 0.03	0.39 ± 0.07	0.15 ± 0.07	1.62 ± 0.05	0.09 ± 0.04

σ^2_{FYS} Flock-Year-Season effect, σ^2_{PE} permanent environment effect, σ^2_a) additive genetic effect, σ^2_P phenotypic effect, h^2 heritabilities
CNS coagulase-negative *staphylococci*, *STR streptococci*, *ALL* observed bacteriological colonies as described in M&M section

Table 6 Phenotypic (above diagonal) and genetic (below diagonal) correlations and standard errors for resistance to mastitis

Infection status	STR	CNS	ALL
STR	-	−0.08 ± 0.02	0.17 ± 0.01
CNS	0.24 ± 0.25	-	0.87 ± 0.01
ALL	0.36 ± 0.22	0.92 ± 0.04	-

CNS coagulase-negative *staphylococci*, *STR streptococci*, *ALL* observed bacteriological colonies as described in M&M section

similar to reported values in the literature [4, 24]. Mean SCS for uninfected and mean SCS of whole data set were similar to the values reported in Valle del Belice [24, 25], and Churra dairy sheep breeds [32]. For mean SCS of infected animals, Riggio et al. [24] reported a value of 6.42 and Leitner et al. [33] a value of 6.32 in Israel-Assaf and Awassi sheep, similar to the value of 6.21 obtained in our research. Another study reported lower values of mean SCS of infected animals using different breeds [32].

Our study confirms that CNS is the most prevalent etiological group of bacteria in the infected dairy ewes. The frequency of isolation of CNS on record (39 %) was lower than other studies ranging from 60 to 90 % [3, 24, 32, 34–36]. Moreover, a high percentage (77 %) of animals were found infected at least one or more times in the period of study, showing the importance of this group of bacteria in this population.

Most cases of CNS infection produce subclinical mastitis, although intramammary infections in its subclinical form by CNS have been described as the main single factor affecting udder health and profitability in small ruminants [9]. Besides, due to the high prevalence of CNS during the ewe's lactations, subclinical cases could persist, significantly increase SCC and consequently cause clinical mastitis. This is a possible explanation of the observed differences of SCC between infected and non-infected animals and frequency of animals infected by CNS. Moreover, considering the opportunistic nature of CNS [17], with adequate hygiene practices, correct milking routine and periodic revision of milking equipment, intramammary infections by CNS could be reduced.

In this investigation, CNS were classified as minor pathogens. Researchers have reported that some species of CNS can cause high SCC, similar to those of major pathogens [15, 32, 37] and even clinical mastitis [32, 38]. Ariznabarreta et al. [32] described that *Staphylococcus caprae* and *S. simulans* were associated with high log SCC, 6.43 and 6.35, respectively, in contrast with other CNS bacteria such as *S. chromogenes*, *S. hominis*, *S. capitis*, *S. haemolyticus* and *S. epidermidis* ranging from 5.93 to 6.09. Therefore, there was variation in the inflammatory response according to the involved CNS species and their pathogenicity in milk measured through SCC, even on average ten times higher than in dairy cattle [15]. STHAU was the second one more frequently isolated bacteria in our study (5 %) followed by STR (3 %). This order differs from other authors, which reported that CNS is the most prevalent group of bacteria followed by STR [15, 31, 35, 38]. For STHAU, ewes infected at least one time or more in the period of study were 22 % (513). These findings are different respect to what reported by Riggio et al. [24] with values of 10.47 % of milk samples and similar to other studies ranging from 2 to 5.5 % [15, 32]. Infection due to STHAU is related with subclinical to acute clinical mastitis (gangrenous mastitis) with different clinical symptoms according to the virulence of the strains and in severe cases lead towards culling of the affected sheep [2, 17]. The high percentage of animals affected by STHAU in the period of study could be related with clinical mastitis cases and culling of ewes in this population. This was in agreement with Mavrogenis et al. [30] which identified STHAU as the most prevalent bacteria in clinical mastitis cases.

In sheep a heritability estimate of 0.09 for infection status assessed by bacteriological analyses was reported by Riggio et al. [24] and Tolone et al. [25] in the Valle del Belice breed using a threshold animal model assuming a probit link function. Gonzalo et al. [39] estimated genetic parameters of SCC in Churra sheep considering the type of mammary pathogen using a multitrait repeatability animal model. They reported that the effect related to the type of pathogen accounted for 32.5 % of the total variance in SCC, a value similar to that obtained for the residual effect (34.9 %), indicating a high relative importance of the type of pathogen in the decomposition of the variance for SCC. In addition, Holmberg et al. [40] in dairy cattle reported genetic variances for different pathogens ranging from 0.024 to 0.188, similar to the values of the present research (0.03 to 0.15). These results showed the importance of differentiating between the types of mammary bacteria assessed by bacteriological analyses in genetic mastitis studies.

Variances due to permanent environment and FYS effects were high and were important factors to explain the phenotypic variance resistance against CNS, STR and ALL. The possible explanations of these results for CNS group of pathogens are their nature and their high frequency of isolation in this sheep breed. CNS group of bacteria are related with inadequate management and hygiene practices, which could be different among the flocks, through the year and among them. Therefore, due to opportunistic nature of CNS, poor flock management and inadequate milking hygiene could increase the probability of occurrence of mastitis, and flocks may act as reservoirs of some CNS species. Taking into account the predominant sheep husbandry system in Sicily based on grazing with animals kept outdoors, reductions in pasture quantity and quality through the year as in summer (peak of lambing in Valle del Belice sheep) could be a stress factor to increase the susceptibility due to ALL infection. High temperatures in summer are associated with heat stress [8], and as occurs in dairy cattle heat stress is recognized as a factor which increases susceptibility to mastitis. Gonzalo et al. [41] reported that month within flock and flocks were accounted 44.1 % of the variance on bulk tank bacteria count, whereas Portolano et al. [8] reported that flock-year of lambing effect explains 27 % of the variance of time interval between lambing and first record with mastitis.

Heritabilities for pathogen-specific mastitis were in agreement with results of De Haas [20] in dairy cattle ranging from 0.02 to 0.10. However, this study only included heritabilities of pathogens involved in clinical mastitis cases and were estimated through threshold and linear models. For genetic correlations, the one estimated between CNS and ALL (0.92) was positive and very high suggesting that both are the same traits. This could be explained for the high frequency of isolation of CNS in the records (77 %). Thus, a high percentage of ALL group is explained by CNS pathogens. Furthermore, due to the fact that phenotypic variation for CNS and ALL is determined primarily by an environmental component both type of traits (CNS and ALL) could be controlled more effectively by applying a correct management measures instead of selective breeding on these population.

In the Valle del Belice breed, where the current selection is mainly practiced on a "within farm" approach and based on own performance of ewes, it is unlikely that selection for mastitis resistance is successful, independent of the use of infection status or SCS.

Conclusions

The results of our study support the presence of significant genetic variation for resistance to one specific pathogen causing mastitis (i.e. *Streptococci*). The high genetic correlation between ALL and CNS indicate that both are almost the same trait. The opportunistic nature of CNS and the high environmental influence of CNS resistance suggest that improvement of flock management and adequate milking hygiene could reduce significantly the incidence of mastitis caused by this pathogen in Valle del Belice dairy sheep.

Abbreviations

A, animal; AF, absolute frequency; ALL, all pathogens together; BACIL, *Bacillus spp.*; CFU, five colony forming units; CNS, coagulase-negative staphylococci; CORLT, *Corynebacterium spp.*; ESCCL, *Escherichia coli*; FYS, flock year season; MY, milk yield; OP, order of parity; PASCL, *Pasteurella spp.*; PE, permanent environmental; PSELT, *Pseudomonas spp.*; RF, relative frequency; SCC, somatic cell count; SCS, somatic cell score; STHAU, *Staphylococcus aureus*; STPAG, *Streptococcus agalactiae*; STPDG, *Streptococcus dysgalactiae*; STPUB, *Streptococcus uberis*; STR, *Streptococci*; STR, *Streptococcus spp*

Acknowledgements

We acknowledge Dr. M.L Scatassa (Istituto Zooprofilattico Sperimentale della Sicilia) for bacteriological analyses.

Funding

No funding was obtained for this study.

Authors' contributions

MT and CL contributed equally to this work, they carried out the design of the study, performed statistical analysis and drafted the manuscript. JMY participated in statistical analysis and drafted the manuscript. SN helped to draft the manuscript and to interpret the results. MTS helped to collect the data and revised critically the manuscript. BP participated in the design of the study and gave the final approval of the version to be submitted. All authors read and approved the final manuscript.

Competing interests

The authors declare that they have no competing interests.

Consent for publication

Not applicable

Author details

[1]Dipartimento Scienze Agrarie e Forestali, Università degli Studi di Palermo, Viale delle Scienze, Palermo 90128, Italy. [2]Faculty of Veterinary and Animal Sciences, University of Chile, Av. Santa Rosa, La Pintana, Santiago 11735, Chile. [3]Genus plc, Hendersonville, TN 37075, USA.

References

1. Barillet F, Rupp R, Mignon-Grasteau S, Astruc JM, Jacquin M. Genetic analysis of mastitis resistance and somatic cell score in French Lacaune dairy sheep. Genet Sel Evol. 2001;33:397–415.
2. Leitner G, Chaffera M, Zamirb S, Mora T, Glickmana A, Winklera M, Weisblita M, Sarana A. Udder disease etiology, milk somatic cell counts and NAGase activity in Israeli Assaf sheep throughout lactation. Small Rumin Res. 2001; 39:107–12.
3. Bergonier D, Cremoux R, Rupp R, Lagriffoul G, Berthelot X. Mastitis of dairy small ruminants. Vet Res. 2003;34:689–716.
4. Leitner G, Chaffer M, Shamay A, Shapiro F, Merin U, Ezra E, Saran A, Silanikove N. Changes in milk composition as affected by subclinical mastitis in sheep. J Dairy Sci. 2004;87:46–52.
5. Carlén E, Schneider M, Strandberg E. Survival analysis for genetic evaluation of mastitis in Dairy Cattle: A simulation study. J Dairy Sci. 2005;88:797–803.
6. Barillet F. Genetic improvement for dairy production in sheep and goats. Small Rumin Res. 2007;70:60–75.
7. Legarra A, Ramon M, Ugarte E, Perez-Guzman MD, Arranz J. Economic weights of somatic cell score in dairy sheep. Animal. 2007;1:205–12.
8. Portolano B, Finocchiaro R, Van Kaam J, Riggio V, Maizon DO. Time-to-event analysis of mastitis at first-lactation in Valle del Belice ewes. Livest Sci. 2007; 110:273–9.
9. Leitner G, Silanikove N, Merin U. Estimate of milk and curd yield loss of sheep and goats with intramammary infection and its relation to somatic cell count. Small Rumin Res. 2008;74:221–5.
10. Riggio V, Maizon D, Portolano B, Bovenhuis H, van Arendonk J. Effect of somatic cell count level on functional longevity in Valle del Belice dairy sheep assessed using survival analysis. J Dairy Sci. 2009;92:6160–6.
11. Rupp R, Boichard D. Genetics of resistance to mastitis in dairy cattle. Vet Res. 2003;34:671–88.
12. Rupp R, Foucras G. Genetics of mastitis in dairy ruminants. In: Breeding of disease resistance in farm animals. 2010. p. 183–212.
13. Rupp R, Bergonier D, Dion S, Hygonenq MC, Aurel MR, Robert-Granie C, Foucras G. Response to somatic cell count-based selection for mastitis resistance in a divergent selection experiment in sheep. J Dairy Sci. 2009;92:1203–19.
14. Smith K, Hogan J. The world of mastitis 2nd International Symposium on mastitis and milk quality. Canada: Vancouver; 2001.

15. Pengov A. The role of coagulase-negative Staphylococcus spp and associated somatic cell counts in the ovine mammary gland. J Dairy Sci. 2001;84:572–4.
16. Gonzalo C, Ariznabarreta A, Carriedo J, San Primitivo F. Mammary pathogens and their relationship to somatic cell count and milk yield losses in dairy ewes. J Dairy Sci. 2002;85:1460–7.
17. Contreras A, Sierra D, Sánchez A, Corrales JC, Marco JC, Paape MJ, Gonzalo C. Mastitis in small ruminants. Small Rumin Res. 2007;68:145–53.
18. Riggio V, Portolano B. Genetic selection for reduced Somatic Cell Counts in sheep milk: A review. Small Rumin Res. 2015;126(1):33–42.
19. White LJ, Schukken YH, Lam TJG, Medley GF, Chappell MJ. A multispecies model for the transmission and control of mastitis in dairy cows. Epidemiol Infect. 2001;127:567–76.
20. De Haas Y. Somatic cell count patterns Improvement of udder health by genetics and management. Wageningen, NL: PhD Thesis, Wageningen University; 2003.
21. De la Cruz M, Serrano E, Montoro V, Marco J, Romeo M, Baselga R, Albizu I, Amorena B. Etiology and prevalence of subclinical mastitis in the Manchega sheep at mid-late lactation. Small Rumin Res. 1994;14:2175–180.
22. Tolone M, Riggio V, Portolano B. Estimation of genetic and phenotypic parameters for bacteriological status of the udder, somatic cell score, and milk yield in dairy sheep using a threshold animal model. Livest Sci. 2013; 151:134–9.
23. Riggio V, Pesce LL, Morreale S, Portolano B. Receiver-operating characteristic curves for somatic cell scores and California mastitis test in Valle del Belice dairy sheep. Vet J. 2013;196:528–32.
24. Maurer J, Schaeren W. Udder health and somatic cell counts in ewes. Agrarforschung. 2007;14:162–7.
25. Riggio V, Portolano B, Bovenhuis H, Bishop S. Genetic parameters for somatic cell score according to udder infection status in Valle del Belice dairy sheep and impact of imperfect diagnosis of infection. Gen Sel Evol. 2010;42:30.
26. Heringstad B, Klemetsdal G, Steine T. Selection responses for clinical mastitis and protein yield in two norwegian dairy cattle selection experiments. J Dairy Sci. 2003;86:2990–9.
27. ICAR (International Committee for Animal Recording), 2014. International agreement of recording practices. Available online: http://www.icar.org/index.php/publications-technicalmaterials/recording-guidelines/ 2014.
28. Gilmour AR, Gogel BJ, Cullis BR, Thompson R. ASReml User Guide Release 3. 0 VSN International Ltd, Hemel Hempstead, HP1 1ES, UK www.vsni.co.uk. 2009;
29. Ali A, Shook G. An optimum transformation for somatic cell count in milk. J Dairy Sci. 1980;63:487–90.
30. Mavrogenis A, Koumas A, Kakoyiannis C, Taliotis C. Use of somatic cell counts for the detection of subclinical mastitis in sheep. Small Rumin Res. 1995;17:79–84.
31. Kern G, Traulsen I, Kemper N, Krieter J. Analysis of somatic cell counts and risk factors associated with occurrence of bacteria in ewes of different primary purposes. Livest Sci. 2013;157:597–604.
32. Ariznabarreta A, Gonzalo C, San Primitivo F. Microbiological quality and somatic cell count of ewe milk with special reference to staphylococci. J Dairy Sci. 2002;85:1370–5.
33. Leitner G, Chaffer M, Caraso Y, Ezra E, Kababea D, Winkler M, Glickman A, Saran A. Udder infection and milk somatic cell count, NAGase activity and milk composition–fat, protein and lactose–in Israeli-Assaf and Awassi sheep. Small Rumin Res. 2003;49:157–64.
34. Contreras A, Corrales JC, Sierra D, Marco JC. Prevalence and aetiology of non-clinical intramammary infection in Murciano- Granadina goats. Small Rumin Res. 1995;17:71–8.
35. González-Rodríguez MC, Gonzalo C, San Primitivo F, Cármenes P. Relationship between somatic cell count and intramammary infection of the half udder in dairy ewes. J Dairy Sci. 1995;78:2753–9.
36. Las Heras A, Domínguez L, Fernández-Garayzábal J. Prevalence and aetiology of subclinical mastitis in dairy ewes of the Madrid region. Small Rumin Res. 1999;32:21–9.
37. Marco JC. Mastitis en la oveja Latxa: epidemiología, diagnóstico y control. PhD Thesis. España: Universidad de Zaragoza; 1994.
38. Fthenakis GC. Prevalence and aetiology of subclinical mastitis in ewes of Southern Greece. Small Rumin Res. 1994;13:293.
39. Gonzalo C, Ariznabarreta A, Othmane M, Carriedo J, De La Fuente L, San Primitivo F. Genetic parameters of somatic cell count in dairy sheep considering the type of mammary pathogen effect. J Anim Breed Genet. 2003;120:282–7.
40. Holmberg M, Fikse WF, Andersson-Eklund L, Artursson K, Lunde A. Genetic analyses of pathogen-specific mastitis. J Anim Breed Genet. 2012;129:129–37.
41. Gonzalo C, Carriedo J, Beneitez E, Juarez M, De La Fuente L, San Primitivo F. Short Communication: bulk tank total bacterial count in dairy sheep: factors of variation and relationship with somatic cell count. J Dairy Sci. 2006;89:549–52.

4

Diurnal variation of NMR based blood metabolites in calves fed a high plane of milk replacer

Morteza H. Ghaffari[1], Jayden A. R. MacPherson[1], Harma Berends[2] and Michael A. Steele[1*]

Abstract

Background: Blood profiles have been used to monitor herd health status, diagnose disorders, and predict the risk of diseases in cattle and calves. Characterizing plasma metabolites in dairy calves could provide further insight into daily metabolic variations and the mechanisms that lead to metabolic diseases. In addition, by understanding physiological ranges of plasma metabolites relative to meal and the time of feeding in healthy animals, veterinarians can accurately diagnose abnormalities with a blood test. For diagnostic purposes, nuclear magnetic resonance (NMR) spectroscopy shows promise as a new and reliable method to determine a large number of blood metabolites simultaneously.

Results: Results demonstrated that the concentration of specific metabolites in plasma (i.e., lysine, isoleucine, leucine, tyrosine, glutamine, creatine, and 1-methylhistidine) fluctuated around meal times, while others (i.e., glutamic acid, methanol, formic acid, and acetic acid) maintained a stable temporal concentration. In addition to temporal changes in concentration, results also characterized differences for overall plasma metabolite concentrations; for example, methionine had the lowest (38 μM) while glutamine had the highest concentration (239 μM) amongst plasma AA. This is the first report describing how the plasma metabolome changes during 24-h period in young calves fed an elevated plane of milk replacer twice daily.

Conclusions: Data from this pilot study will help to establish reference standards for future metabolic diagnostics in dairy calves. In addition, this pilot study illustrated that feeding milk replacer may influence plasma metabolite concentrations. With the rapid implementation of blood metabolomics in monitoring animal health, it is then important to consider the time of feeding during the day when interpreting metabolomics analysis results.

Keywords: Calf, Plasma, Metabolomics, NMR spectroscopy

Background

Plasma metabolites are widely used to monitor the health and metabolic status of cattle herds [1]. Nuclear magnetic resonance (NMR) spectroscopy is a method that can be applied to identify and quantify many metabolites in biological samples, including blood. The minimal sample preparation and high analytical reproducibility along with the non-destructive characteristics of NMR spectroscopy represents significant advantages for employing NMR in the field of metabolomics research compared with other methods [2]. Identification of metabolomic profiles using NMR that relate to specific outcomes such as metabolic disorders and diseases could improve the speed of diagnosis and treatments [3]. Reported metabolite research values vary in sampling times during the day as well as the physiological stage of the animal. This represents a serious knowledge gap for veterinary diagnostics as the plasma concentrations of some metabolites, such as non-esterified fatty acids, glucose, and amino acids (AA), have been shown to be affected by time of feeding in cattle [4, 5]. This is important to consider for clinical investigations to avoid misinterpretation of results due to the meal effects.

* Correspondence: masteele@ualberta.ca
[1]Department of Agricultural, Food, and Nutritional Science, University of Alberta, Edmonton T6G 2P5, Canada
Full list of author information is available at the end of the article

Blood metabolite concentrations may also fluctuate throughout the day, therefore sampling time differences between studies may be a primary reason for the discrepancies in published literature. Having temporal reference data for metabolite concentrations of dairy calves could further strengthen our understanding of physiologically healthy concentrations versus metabolic disorders (i. e., chronic diarrhea, pneumonia, and insulin sensitivity). Moreover, most of the currently available data describing metabolic responses comes from studies in adult cattle, and these findings may not represent metabolomic profiles of calves. Therefore, it is important to understand and characterize the metabolome of calves to establish physiologically healthy fluctuations in metabolites throughout the day.

In dairy calves, it has been shown that increasing meal size results in slower gastric emptying [6]. The rate at which nutrients empty from the abomasum can also influence the rate at which those nutrients are delivered to the small intestine and absorbed into the bloodstream [7]. Therefore, the objective of this report is to describe diurnal variations of plasma metabolites in dairy calves fed an elevated plane of milk replacer twice daily using NMR spectroscopy. In this pilot study, NMR-based metabolomic analysis technology was used because it allowed us to study a large number of metabolites in the blood of dairy calves. To the best of our knowledge, this is the first time that a 24-h period of plasma metabolome has been characterized by NMR spectroscopy in calves fed an elevated plane of nutrition.

Results

All calves were clinically healthy at time of sampling. The mean daily starter feed intake and milk replacer intake from day 19 to day 23 of age were 46.0 ± 21.4 g/day and 7.51 ± 0.49 L/day, respectively.

Nuclear magnetic resonance spectroscopy

Twenty-four hour mean values of the plasma essential amino acids (EAA; histidine, lysine, methionine, phenylalanine, valine, leucine, and isoleucine) and nonessential amino acids (NEAA; alanine, glutamine, glutamic acid, glycine, proline, and tyrosine) are presented in Fig. 1a. Glutamine had the highest concentration (238.6 μM) while methionine was the lowest (38 μM) amongst plasma AA. Valine and lysine were the most abundant essential plasma AA. Mean baseline and four-hour postprandial plasma concentrations of AA are presented in Fig. 1b. A four-hour postprandial sampling time was selected based on the knowledge that the soluble components of milk enter the small intestine within 2 to 3 h after a meal [8]. In the morning meal, the 4-h postprandial plasma concentrations of histidine, lysine, leucine, isoleucine, valine, methionine, phenylalanine, alanine, proline, tyrosine, and glutamine were significantly higher ($P < 0.05$) compared with baseline. In the evening meal, the 4-h postprandial plasma concentrations of isoleucine, lysine, tyrosine, and glutamine were significantly higher ($P < 0.05$), and valine ($P = 0.07$) and alanine ($P = 0.08$) tended to be higher compared with baseline. The plasma concentrations of glutamic acid and glycine did not significantly change in response to time of meal feeding. The patterns of diurnal variation of plasma EAA and NEAA are displayed in supplementary file (Additional files 1 and 2).

The plasma AA concentrations varied markedly according to the time of day. Twenty-four-hour mean plasma glucose, glycerol, citric acid, and lactic acid concentrations are shown in Fig. 2a. Mean baseline and 4-h postprandial plasma recordings of glucose, glycerol, citric acid, and lactic acid concentrations are shown in Fig. 2b. Twenty-four hour mean concentrations of other plasma metabolites such as creatine, betaine, acetic acid, formic acid, dimethyl sulfone, 1-methylhistidine, 3-hydroxybutyric acid, methanol, pyruvic acid, scyllitol, ethanol, and creatinine are presented in Fig. 3a and b. Mean baseline and 4-h postprandial plasma recordings of other metabolite concentrations are presented in Fig. 3c and d.

The 4-h postprandial plasma concentrations of betaine (in the morning meal) and citric acid (in the evening meal) were significantly higher ($P < 0.05$) compared to their baseline. Plasma concentration of glucose tended to be higher ($P = 0.10$) at 4-h postprandial compared to baseline after both meal feedings. Comparison of postprandial plasma concentrations of 1-methylhistidine ($P < 0.05$) and creatine ($P < 0.01$) after both meal feedings revealed significantly higher concentrations at 4-h postprandial compared to their baseline. The patterns of diurnal variation of other plasma metabolites are displayed in an additional file (Additional file 3). Plasma concentration of creatine significantly increased after feeding to reach a maximum level of 222.3 μM in less than 3 h, followed by a progressive decrease until the next feeding. After the morning feeding, the plasma glycerol concentration numerically increased and there were small fluctuations throughout the day. Plasma concentration of pyruvic acid numerically increased after the morning meal, but its concentration significantly decreased after the evening meal. The plasma concentrations of other metabolites such as acetic acid, formic acid, 1-methylhistidine, dimethyl sulfone, 3-hydroxybutyrate, methanol, scyllitol, ethanol, and creatinine did not significantly change in response to time of meal feeding.

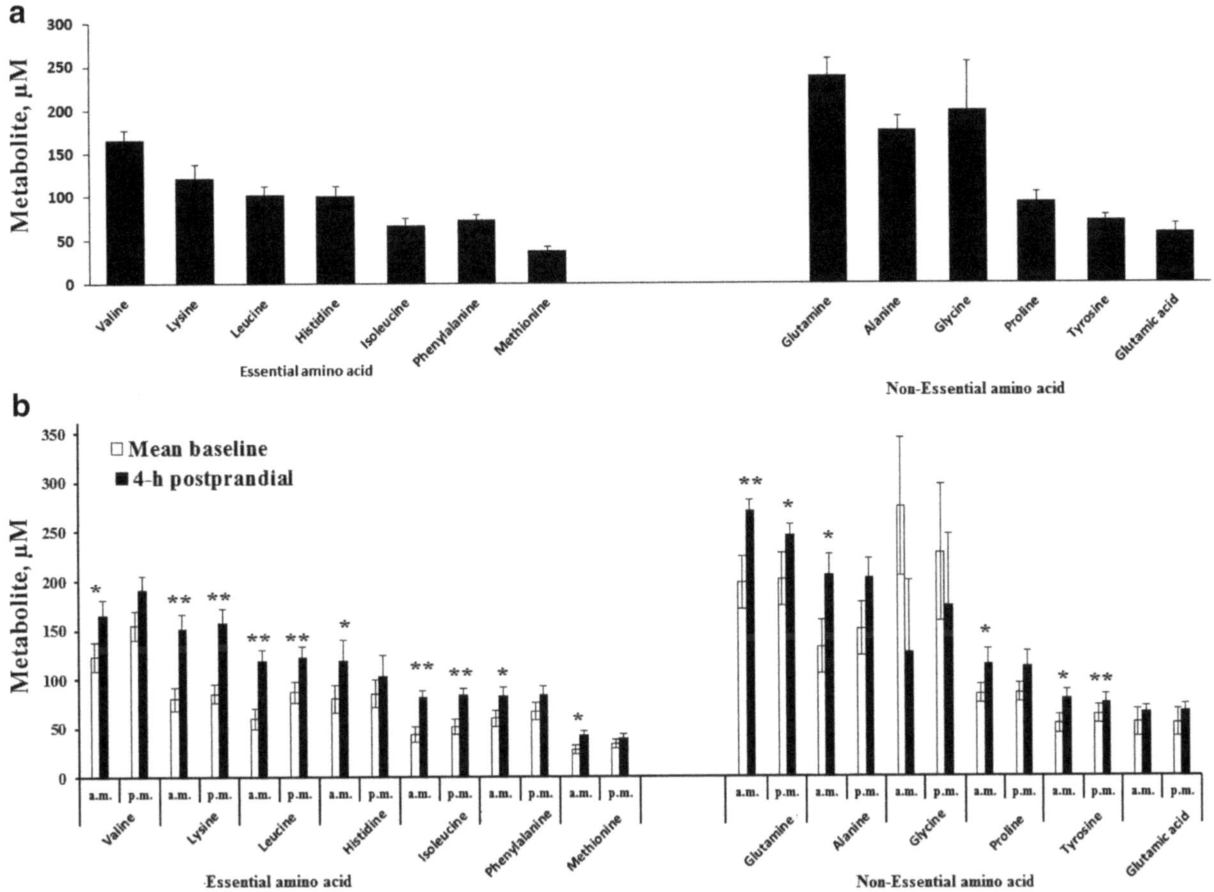

Fig. 1 Twenty-four hour mean values (**a**), mean baseline and 4-h postprandial concentrations (**b**) of blood essential and non-essential amino acids in Holstein calves fed elevated plane of nutrition twice daily. Data are expressed as means ± SE. For each time point, * = significant difference at $P < 0.05$; ** = significant difference at $P < 0.01$

Discussion

Metabolomics is becoming an important tool in clinical research, allowing for very early and accurate diagnosis of disorders and diseases [9]; however, this technique is not implemented on most of dairy farms as a tool to selectively tailor management decisions in calves and cattle. The plasma metabolite profiles in dairy calves have not been properly characterized. Thus, this pilot study specifically focused on measuring the postprandial and diurnal variation of blood metabolites using NMR in three male Holstein dairy calves fed an elevated plane of milk replacer twice daily.

Amino acids

For dairy calves, milk provides most of the nutrients, such as the AA, needed to support adequate growth for young calves before they are able to digest solid feeds [10]. The postprandial concentrations of all EAA and some NEAA in the plasma significantly increased after ingestion of milk replacer in this pilot study. The observed difference in the plasma level of EAA after milk feeding is in agreement with previously reported data, indicating that the plasma EAA concentrations increased following milk feeding in pre-weaned calves [11]. In this study, the different postprandial concentrations of plasma AA and the degree of increase after the meal can partially be explained by different AA digestibility [12]. Plasma glutamine levels increased after MR feeding in this pilot study, which is consistent with the results of McCormick and Webb [13] who reported that glutamine was the only amino acid increased after feeding in plasma of calves. In contrast, Hammon and Blum [10] reported no change in the plasma glutamine concentration in calves fed only milk replacer during the first week of life. Glutamine was the most abundant free AA in calf plasma, which is consistent with previous findings underlining the important role of glutamine in maintaining net protein synthesis and accretion [14]. Studies on dietary glutamine utilization in human indicated at least 50% of the dietary glutamate and glutamine is taken up by the GIT and is completely oxidized to CO_2 [15]. This means that the plasma

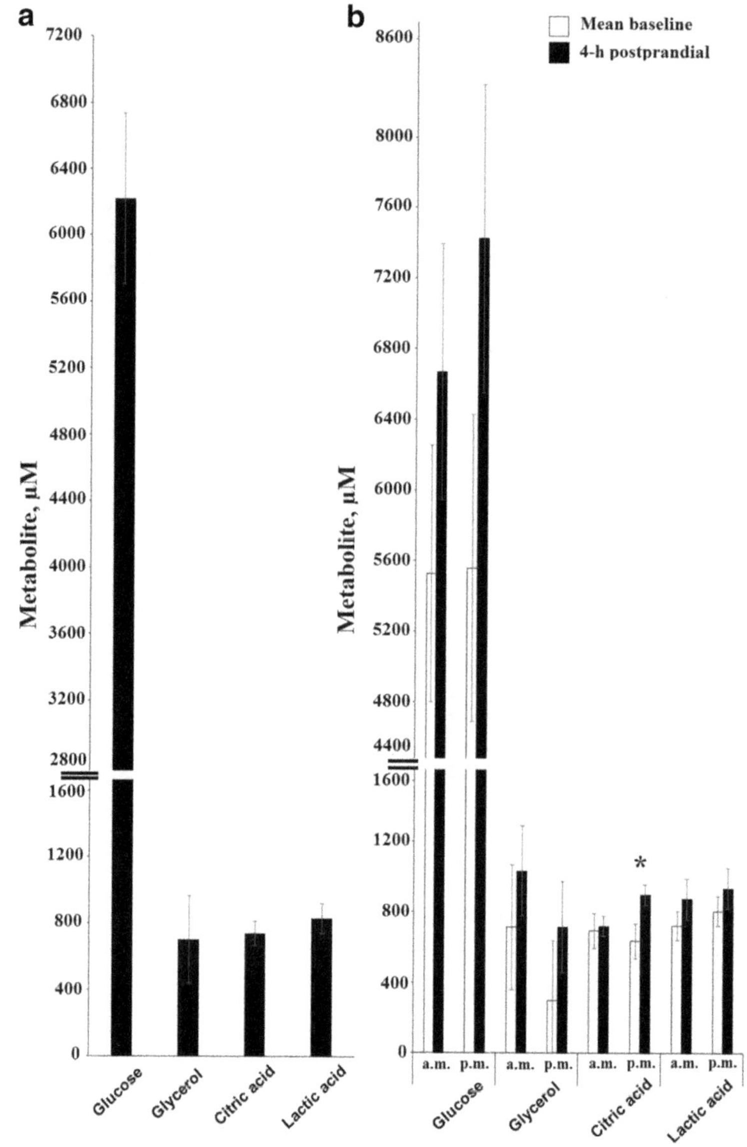

Fig. 2 Twenty-four hour mean values (**a**), mean baseline and 4-h postprandial concentrations (**b**) of blood glucose, glycerol, citric acid, and lactic acid in Holstein calves fed elevated plane of nutrition twice daily. Data are expressed as means ± SE. For each time point, * = significant difference at $P < 0.05$

glutamine concentration of calves in this study may be affected by the utilization and metabolism of glutamine by small-intestine.

In contrast to all other post-prandial results, the plasma concentration of glycine was numerically lower than the baseline level 4 h after the meal. Glycine is synthesized in the body through various complex pathways involving serine, threonine, choline, and hydroxyproline via inter-organ metabolism involving primarily the liver and kidneys [16]. The lower postprandial glycine concentration in plasma of dairy calves might be associated with low concentration of glycine in milk [17, 18]. Wang et al. [16] reported that the intake of glycine from milk provides at most only 20% of the glycine required by the 7-day old piglet; thus, 80% of the glycine needed by the piglets must be provided through endogenous synthesis.

The plasma AA concentrations in dairy calves are likely more affected by diets than mature cows. Halfpenny et al. [19] reported that plasma AA concentrations in dairy cows are relatively constant during the day. In addition, Huntington [20] showed that AA absorption in lactating cows was relatively constant until 12 h after feeding, likely because of the continuous outflow of microbial and dietary protein from the rumen, which is a key difference between pre-weaned calves and more mature weaned animals. Therefore, the results of this study indicated that the plasma AA status of young calves is highly related to milk replacer intake.

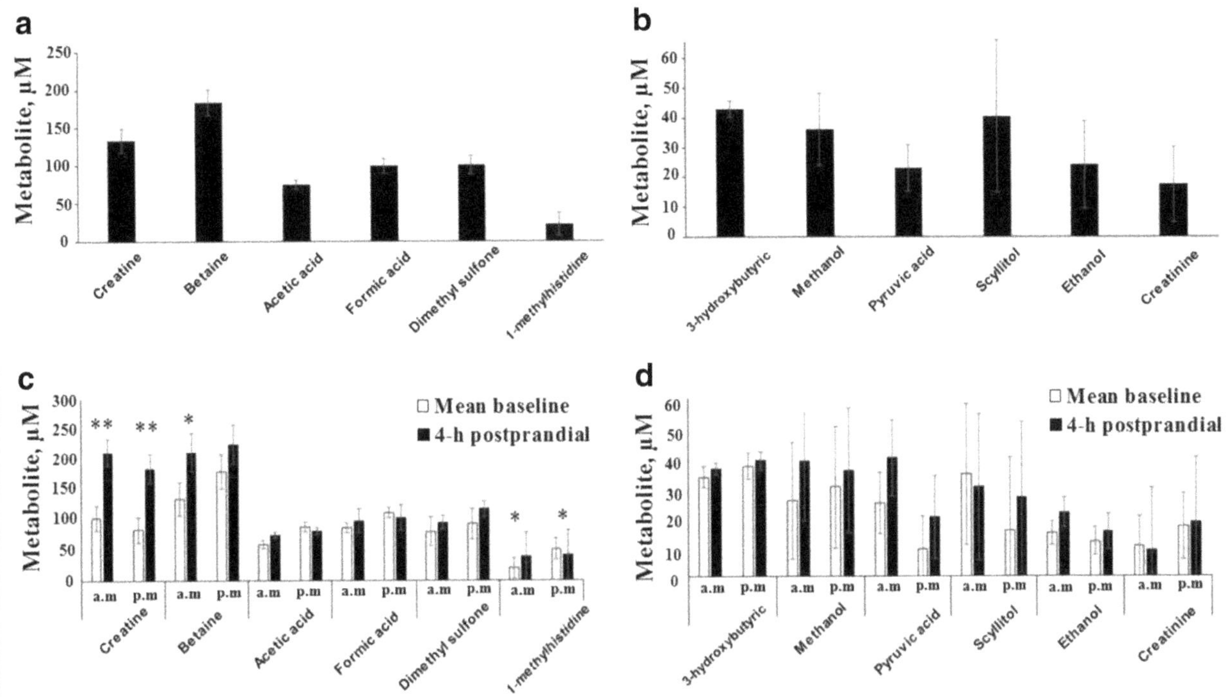

Fig. 3 Twenty-four hour mean values (**a, b**) and mean baseline and 4-h postprandial concentrations (**c, d**) of blood metabolites in Holstein calves fed elevated plane of nutrition twice daily. Data are expressed as means ± SE. For each time point, * = significant difference at $P < 0.05$; ** = significant difference at $P < 0.01$

Additional metabolites

Glucose, lactic acid, citric acid, and glycerol were the most abundant metabolites in the plasma of the dairy calves in this report. The concentration of glucose in the calves in this study were within the range suggested for plasma glucose concentrations in dairy calves fed high plane of milk replacer at 4 week of age, which vary between 6500 and 8000 µM/L [6]. The postprandial concentrations of glucose in the plasma tended to increase after ingestion of milk replacer in this pilot study, which is in line with the results of previous studies [6, 21]. Plasma glucose concentration can be affected by the level of milk feeding, milk carbohydrate sources [22], endogenous glucose production from gluconeogenesis in the liver [23], and age [24]. As calves age, the concentration of plasma glucose decreases, which is mainly due to a shift in the primary source of energy from glucose to volatile fatty acid [25]. Therefore, milk feeding can be considered an important factor for influencing the postprandial plasma glucose concentration in dairy calves and should be considered when evaluating glucose concentrations in calves.

Lactic acid was the most abundant organic acid in the plasma of these calves. A post-prandial rise of lactate concentrations in the plasma has been previously reported in milk-fed calves [26, 27], although in this pilot study the rise was not significant in the calves. The accumulation of lactic acid is an important contributory factor to metabolic acidosis in neonatal calves with diarrhea in the previous study [28]. So far, limited information is available regarding the plasma concentration of citric acid in dairy calves. Citric acid is formed from acetyl-CoA and oxaloacetic acid in a reaction catalyzed by citrate synthase [29]. In this study, the plasma citric acid level increased after the evening meal. Reasons for the increased plasma level of citric acid following evening milk feeding in calves are unclear.

In this pilot study, plasma concentrations of creatine, betaine, and 1-methylhistidine were affected by time of meal feeding in calves. The plasma creatine concentrations were found to increase after each meal. Creatine is a nitrogenous amine found in muscle tissue that is involved in ATP synthesis and muscle function [30]. The elevation of the plasma creatine concentrations following the consumption of a milk replacer meal can be mainly due to the breakdown of proteins supplied by the milk replacer. Based on NMR spectroscopy, Basoglu et al. [3] reported that the concentration of plasma creatine increased in neonatal calves with diarrhea compared with healthy calves because of enhanced intestinal barrier function. The postprandial concentration of betaine in the plasma was also found to increase after the morning meal in this pilot study. Plasma betaine concentration in human beings can be affected by absorption from the intestine, synthesis from choline, metabolism to dimethylglycine, storage in muscle, liver and kidney tissues, as well as urinary elimination [31].

However, the factors causing the postprandial level of betaine rise in plasma are unclear. Plasma concentration of 1-methylhistidine observed in dairy calves also increased and was four times higher than those in a mature cow [32]. However, in this study, the postprandial increase of 1-methylhistidine may have been a consequence of increased breakdown of milk proteins. Therefore, it can be concluded that the diurnal variations of some plasma metabolites (i.e., acetic acid, formic acid, ethanol, and creatinine) were independent of meal feeding whereas the diurnal variations in most AA, creatine, 1-methylhistidine in plasma were strongly affected by meal feeding in calves.

Limitations

This study was a preliminary investigation of the diurnal variation of blood metabolites in 20 day old calves fed milk replacer. The major limitation of this pilot study was biological replication, and the findings could be very specific to this age group and this feeding regimen. However, this pilot study will provide the groundwork for further studies investigating post-prandial responses of blood metabolites in calf feeding programs.

Conclusion

Metabolomic profiling is an emerging diagnostic tool based on the comprehensive determination of many metabolites contained in bio-fluids and in biological samples. This pilot study indicated that some plasma metabolites exhibit a definite pattern and change with time relative to milk replacer feeding. These findings will be useful in guiding the design and interpretation of future metabolite-based studies in calves. Future metabolomics studies with larger sample sizes and different ages are required to verify the findings of this pilot study, and to explore its clinical application for the diagnosis and treatment of diseased animals.

Methods

This pilot study was conducted at the Trouw Nutrition Ruminant Research Centre (Boxmeer, The Netherlands). All procedures and sampling were approved by the Animal Care and Use Committee of Utrecht University (2014.III.04.045). Three Holstein bull calves (average age 20.0 ± 1.0 days) weighing 56.0 ± 3.3 kg were housed in individual hutches (1.07 × 1.60 m) bedded with wheat straw. All calves received reheated dam colostrum from a bottle (4 L divided into two feedings at 1 and 6 h of birth). All calves were fed milk replacer (150 g/L; 24% crude protein (skimmed milk powder), 18% crude fat, and 45% lactose; %1.83 lysine, %0.55 methionine, and %0.83 cystine + methionine, Sloten B.V. Deventer, The Netherlands) from nipple buckets. Calves were stepped up to 8 L/day of milk replacer during the first week of life which consisted of a total of 2.5 L for day 2 and 3, 3.0 L for day 4 and 5, and 3.5 L for day 6 and 7. From day 8 onward calves were fed a high plane of nutrition of 8 L/day which was divided equally and offered in two daily feedings (07:00 h and 17:00 h). Calves had ad libitum access to water, pelleted calf starter (18.2% crude protein, 11.2% crude fibre, 2.2% crude fat, 3 mm pellet; AgruniekRijnvallei, Wageningen, The Netherlands), and wheat straw (3 cm chop length), provided in buckets in the front of the hutch. Jugular catheters (Intraflon 2 13G, Ecouen, France) were inserted in the morning (08:00 am) of the sampling day. Starting at 12:00 pm, blood samples were collected every hour for 24 h via the jugular catheters in EDTA (10 ml) vacutainers. Immediately after each collection, all catheters were flushed with a 2% heparinized saline solution. Blood samples were immediately centrifuged at 2772×*g* for 30 min, after which the plasma was separated, frozen at −20 °C and shipped on dry ice to The Metabolomics Innovation Centre at the University of Alberta, Canada for NMR spectrum analysis. The NMR samples were prepared as described by Psychogios et al. [33]. All ^{1}H–NMR spectra of blood samples were collected on a 500 MHz (Varian Inc., Palo Alto, CA, USA) spectrometer equipped with either a Z-gradient PFG Varian cold-probe or a 5 mm HCN Z-gradient pulsed-field gradient (PFG) room-temperature probe. The resulting ^{1}H–NMR spectra were processed and analyzed using the Chenomx NMR Suite Professional software package version 6.0 (Chenomx Inc., Edmonton, AB, Canada), as previously described [34].

Statistical analysis

Data of the plasma metabolome are presented as means ± standard error of the mean, obtained using PROC MEANS of SAS [35]. T-tests were used to determine the significance of changes between baseline and four-hour postprandial. The mean baseline is the average of the metabolite concentration 2 h before the meal (at 7:00 h and 17:00 h).

Additional files

Additional file 1: Concentrations of blood essential amino acids measured at hourly interval for 24 h from 05:00 to 04:00 h. Feeding times are marked with an arrow. Data are expressed as means ± SE.

Additional file 2: Concentrations of blood non-essential amino acids measured at hourly interval for 24 h from 05:00 to 04:00 h. Feeding times are marked with an arrow. Data are expressed as means ± SE.

Additional file 3: Concentrations of blood metabolites with response to meal feeding times of 7:00 and 17:00 measured at hourly interval for 24 h. Feeding times are marked with an arrow. Data are expressed as means ± SE.

Abbreviations

AA: Amino acids; BW: Body weight; EAA: Essential amino acids; GIT: Gastrointestinal tract; NEAA: Nonessential amino acids; NMR: Nuclear magnetic resonance; TCA: Tricarboxylic acid

Acknowledgements

The support of staff at the Trouw Nutrition Ruminant Research Center (Boxmeer, The Netherlands) is greatly appreciated.

Funding

Research reported in this publication was supported by funding provided Trouw Nutrition (Boxmeer, the Netherlands), Alberta Milk (Edmonton, AB, Canada) and The Natural Sciences and Engineering Research Council of Canada (438351).

Authors' contributions

MHG was a major contributor in analysing the data and writing the manuscript. JARM conducted research, compiled data, and participated in drafting the manuscript. HB was involved in conducting research and reviewed the manuscript. MAS directed the research, reviewed the data and manuscript, and directed revisions. All authors have read and approved the manuscript.

Consent for publication

Not applicable.

Competing interests

The authors declare that they have no competing interests.

Author details

[1]Department of Agricultural, Food, and Nutritional Science, University of Alberta, Edmonton T6G 2P5, Canada. [2]Trouw Nutrition R&D, P.O. Box 220, 5830 AE Boxmeer, The Netherlands.

References

1. Sun LW, Zhang HY, Wu L, Shu S, Xia C, Xu C, Zheng JS. 1H-Nuclear magnetic resonance-based plasma metabolic profiling of dairy cows with clinical and subclinical ketosis. J Dairy Sci. 2014;97:1552–62. http://dx.doi.org/10.3168/jds.2013-6757
2. Emwas AH. The strengths and weaknesses of NMR spectroscopy and mass spectrometry with particular focus on metabolomics research. Methods Mol Biol. 2015;1277:161–93. doi:10.1007/978-1-4939-2377-9-13.
3. Basoglu A, Baspinar N, Tenori L, Hu X, Yildiz R. NMR based metabolomics evaluation in neonatal calves with acute diarrhea and suspected sepsis: a new approach for biomarker/S. Metabolomics. 2014;4:134. http://dx.doi.org/10.4172/2153-0769.1000134
4. Ndibualonji BB, Dehareng D, Beckers F, Van Eenaeme C, Godeau JM. Continuous profiles and within-day variations of metabolites and hormones in cows fed diets varying in alimentary supplies before short-term feed deprivation. J Anim Sci. 1997;75:3262–77. doi:10.2527/1997.75123262x.
5. Vacher PY, Schmitz M, Hirni H, Blum JW. Postprandial plasma variations of 3-methylhistidine compared to those of lysine, homoarginine and xylose on the normal conditions and in malabsorption in the suckling calf. Reprod Nutr Dev. 1990;30:471–82.
6. MacPherson JAR, Berends H, Leal LN, Cant JP, Martín-Tereso J, Steele MA. Effect of plane of milk replacer intake and age on glucose and insulin kinetics and abomasal emptying in female Holstein Friesian dairy calves fed twice daily. J Dairy Sci. 2016; 99:8007–17. doi:10.3168/jds.2015-10826.
7. Bell FR, McLeay LM. The effect of duodenal infusion of milk, casein, lactose and fat on gastric emptying and acid secretion in the milk-fed calf. J Physiol. 1978;282:51–7.
8. Drackley JK. Calf nutrition from birth to breeding. Vet Clin Food Anim. 2008; 24:55–86. doi:10.1016/j.cvfa.2008.01.001.
9. Zhang A, Sun H, Yan G, Wang P, Wang X. Metabolomics for biomarker discovery: moving to the clinic. BioMed Res Inter. 2015;1-6. http://dx.doi.org/10.1155/2015/354671
10. Hammon HM, Blum JW. Free amino acids in plasma of neonatal calves are influenced by feeding colostrum for different durations or by feeding only milk replacer. J Anim Physiol Anim Nutr. 1999;82:193–204. doi:10.1046/j.1439-0396.1999.00229.x.
11. Blum JW, Hammon H. Endocrine and metabolic aspects in milk-fed calves. Domest Anim Endocrinol. 1999;17:219–30. http://dx.doi.org/10.1016/S0739-7240(99)00039-9
12. Li H, Diao YQ, Zhang NF, Fan ZY. Growth, nutrient utilization and amino acid digestibility of dairy calves fed milk replacers containing different amounts of protein in the preruminant period. Asian-Aust J Anim Sci. 2008; 21:1151–8. http://dx.doi.org/10.5713/ajas.2008.70562
13. McCormick ME, Webb KE Jr. Plasma free, erythrocyte free and plasma peptide amino acid exchange to calves in steady state and fasting metabolism. J Nutr. 1982;112:276–82.
14. Curthoys NP, Watford M. Regulation of glutaminase activity and glutamine metabolism. Annu Rev Nutr. 1995;15:133–59.
15. Haisch M, Fukagawa NK, Matthews DE. Oxidation of glutamine by the splanchnic bed in humans. Amer J Physiol. 2000;278:E593–602.
16. Wang W, Wu Z, Dai Z, Yang Y, Wang J, Wu G. Glycine metabolism in animals and humans: implications for nutrition and health. Amino Acids. 2013;45:463–77. doi:10.1007/s00726-013-1493-1.
17. Wu G, Knabe DA. Free and protein-bound amino acids in sow's colostrum and milk. J Nutr. 1994;124:415–24.
18. Davis TA, Nguyen HV, Garciaa-Bravo R. Amino acid composition of human milk is not unique. J Nutr. 1994;124:1126–32.
19. Halfpenny AF, Rook JAF, Smith GH. Variations with energy nutrition in the concentrations of amino acids of the blood plasma in the dairy cow. Br J Nutr. 1969;23:547–57. http://dx.doi.org/10.1079/BJN19690063
20. Huntington GB. Net absorption of glucose and nitrogenous compounds by lactating Holstein cows. J Dairy Sci. 1984;67:1919–27. doi:10.3168/jds.S0022-0302(84)81525-8.
21. Stahel P, Cant JP, MacPherson JAR, Berends H, Steele MA. A mechanistic model of intermittent gastric emptying and glucose-insulin dynamics following a meal containing milk components. PLos One 2016; (in press) doi:10.1371/journal.pone.0156443.
22. Klinger S, Noci B, Müller K, Breves G. Intestinal glucose absorption in calves as affected by different carbohydrate sources. J Anim Physiol Anim Nutr. 2013;97:342–52. doi:10.1111/j.1439-0396.2012.01277.x.
23. Hammon HM, Sauter SN, Reist M, Zbinden Y, Philipona C, Morel C, Blum JW. Dexamethasone and colostrum feeding affect hepatic gluconeogenic enzymes differently in neonatal calves. J Anim Sci. 2003;81:3095–106.
24. Omidi-Mirzaei H, Khorvash M, Ghorbani GR, Moshiri B, Mirzaei M, Pezeshki A, Ghaffari MH. Effects of the step-up/step-down and step-down milk feeding procedures on the performance, structural growth, and blood metabolites of Holstein dairy calves. J Dairy Sci. 2015;98: 7975–81. doi:10.3168/jds.2014-9260.
25. Hammon HM, Schiessler G, Nussbaum A, Blum JW. Feed intake patterns, growth performance, and metabolic and endocrine traits in calves fed unlimited amounts of colostrum and milk by automate, starting in the neonatal period. J Dairy Sci. 2002;85:3352–62. doi:10.3168/jds.S0022-0302(02)74423-8.
26. Hüsier BR, Blum JW. Metabolic and endocrine changes in response to endotoxin administration with or without oral arginine supplementation. J Dairy Sci. 2002;85:1927–35. doi:10.3168/jds.S0022-0302(02)74269-0.
27. Kaufhold JN, Hammon HM, Bruckmaier RM, Breier BH, Blum JW. Postprandial metabolism and endocrine status in veal calves fed at different frequencies. J Dairy Sci. 2000;83:2480–90. doi:10.3168/jds.S0022-0302(00)75140-X.
28. Omole O, Brocks O, Nappert DR, Naylor JMG, Zello GA. High-performance liquid chromatographic assay of (6)-lactic acid and its enantiomers in calf serum. J Chromatogr. 1999;727:23–9.
29. Caudarella R, Vescini F, Buffa A, Stefoni S. Citrate and mineral metabolism: kidney stones and bone disease. Front Biosci. 2003;1(8):s1084–106.
30. East DB. Biochemical pathways of creatine and creatine phosphate. University of Tennessee Honors Thesis Projects. http://trace.tennessee.edu/cgi/viewcontent.cgi?article=1536&context=utk_chanhonoproj.
31. Keaveney EM, Price RK, Hamill LL, Wallace JM, McNulty H, Ward M, Strain JJ, Ueland PM, Molloy AM, Piironen V, von Reding W, Shewry PR, Ward JL, Welch RW. Postprandial plasma betaine and other methyl

donor-related responses after consumption of minimally processed wheat bran or wheat aleurone, or wheat aleurone incorporated into bread. Br J Nutr. 2015;113:445–53.

32. Houweling M, van der Drift SGA, Jorritsma RA, Tielens GM. Technical note: Quantification of plasma 1- and 3-methylhistidine in dairy cows by high-performance liquid chromatography– tandem mass spectrometry. J Dairy Sci. 2012;95:3125–30. doi:10.3168/jds.2011-4769.
33. Psychogios N, Hau DD, Peng J, Guo AC, Mandal R, Bouatra S, Sinelnikov I, Krishnamurthy R, Eisner R, Gautam B, Young N, Xia J, Knox C, Dong E, Huang P, Hollander Z, Pedersen TL, Smith SR, Bamforth F, Greiner R, McManus B, Newman JW, Goodfriend T, Wishart DS. The human serum metabolome. PLoS One. 2011;6:e16957. doi:10.1371/journal.pone.0016957.
34. Ametaj BN, Zebeli Q, Saleem F, Psychogios N, Lewis MJ, Dunn SM, Xia J, Wishart DS. Metabolomics reveals unhealthy alterations in rumen metabolism with increased proportion of cereal grain in the diet of dairy cows. Metabolomics. 2010;6:583–94. doi:10.1007/s11306-010-0227-6.
35. SAS Institute Inc. SAS/STAT 9.1 User's Guide. Cary: SAS Institute Inc; 2004.

5

Cortisol modulates inflammatory responses in LPS-stimulated RAW264.7 cells via the NF-κB and MAPK pathways

Junsheng Dong[1,2], Jianji Li[1,2], Luying Cui[1,2], Yefan Wang[1,2], Jiaqi Lin[1,2], Yang Qu[1,2] and Heng Wang[1,2*]

Abstract

Background: The uteruses of most dairy cattle are easily infected by bacteria, especially gram-negative bacteria, following parturition. Macrophages are important cells of the immune system and play a critical role in the inflammatory response. In addition, cortisol levels become significantly increased due to the stress of parturition in dairy cattle, and cortisol is among the most widely used and effective therapies for many inflammatory diseases. In this study, we assessed the anti-inflammatory effects and potential molecular mechanisms of cortisol using a Lipopolysaccharide (LPS)-induced RAW264.7 macrophage cell line.

Results: Cortisol significantly suppressed the production of prostaglandin E_2 (PGE_2) and decreased the gene and protein expression of inducible NO synthase (iNOS) and cyclooxygenase-2 (COX-2) in a dose-dependent manner. Moreover, cortisol inhibited the mRNA expression of pro-inflammatory cytokines including tumor necrosis factor alpha (TNFα), interleukin-1β (IL-1β), and interleukin-6 (IL-6) and decreased IL-1β secretion in an LPS-treated RAW264.7 macrophage cell line. Moreover, we found that cortisol suppressed nuclear factor-kappa B (NF-κB) signaling in RAW264.7 macrophages stimulated with LPS. This suppression was mediated by the inhibition of IκBα degradation and NF-κB p65 phosphorylation. In addition, cortisol also suppressed the phosphorylation of mitogen-activated protein kinases (MAPK) such as extracellular signal-regulated kinase (ERK1/2), p38 MAPK, and c-Jun N-terminal kinase/stress-activated protein kinase (JNK).

Conclusions: These results suggest that high cortisol levels can attenuate LPS-induced inflammatory responses in the RAW264.7 macrophage cell line by regulating the NF-κB and MAPK signaling pathways.

Keywords: Macrophage, Cortisol, Anti-inflammatory, LPS, NF-κB, MAPKs

Background

Postpartum uterine infection and inflammation are the primary causes of reproductive failure in dairy cows [1, 2]. Almost all cows are susceptible to bacterial infection at the openings of anatomic barriers including the vulva, vagina, and cervix. Lipopolysaccharide (LPS) is the most common pathogenic endotoxin component in the outer membrane of gram-negative bacteria and can disturb the balance between immunity and inflammatory responses [3]. Inflammation is a major risk factor for many diseases, and macrophages are important immune cells that act as the first line of defense against invading agents (bacteria, viruses, and fungi) [4, 5]. During inflammation, macrophages produce excessive amounts of inflammatory mediators such as prostaglandin E_2 (PGE_2), inducible nitric oxide synthase (iNOS) and cyclooxygenase-2 (COX-2) and pro-inflammatory cytokines including interleukin-1β (IL-1β), IL-6, and tumor necrosis factor-alpha (TNFα) [6]. Moreover, iNOS and COX-2 are believed to be the most important inflammatory mediators [7]. Overproduction of these mediators can be harmful to animal organs.

Nuclear factor kappa-B (NF-κB) and mitogen-activated protein kinase (MAPK) are important signaling molecules in the Toll-like receptor (TLR) pathway [8, 9]. NF-κB plays an important role in regulating the inflammatory responses by increasing the expression of inflammatory mediators and pro-inflammatory cytokines such

* Correspondence: sdaulellow@163.com
[1]College of Veterinary Medicine, Yangzhou University, Yangzhou, Jiangsu 225009, China
[2]Jiangsu Co-innovation Center for the Prevention and Control of Important Animal Infectious Diseases and Zoonoses, Yangzhou, Jiangsu 225009, China

as PGE_2, iNOS, COX-2, IL-1β, IL-6 and TNFα [10]. Under unstimulated conditions, heterodimers of NF-κB components, mainly p50/p65, remain in the cytoplasm in an inactive form due to linkage to the inhibitor of κB (IκB) protein. However, when induced by LPS, NF-κB (p50/p65) is released through the phosphorylation and degradation of IκB. As a result, NF-κB p65, which is believed to play a central role in inflammation, enters the nucleus and encodes various cytokines and chemokines [11–13]. The MAPKs represent a specific class of serine/threonine kinases that respond to extracellular signals, including extracellular signal-regulated kinase 1/2 (ERK1/2), p38, and c-Jun NH2-terminal kinase (JNK). Similar to NF-κB, the MAPK signaling pathways are involved in LPS-induced iNOS and COX-2 expression in activated macrophages [14]. Even more importantly, MAPKs play essential roles in the activation of NF-κB [15]. Therefore, inhibition of the NF-κB and MAPK pathways may be a potential therapeutic approach to inflammatory injury.

Dairy cows have high levels of cortisol due to many kinds of stress during the perinatal period, such as pregnancy, labor, and lactation [16–18]. Furthermore, cortisol effectively protects immune cells from excessive inflammation [19]. However, neither the anti-inflammatory activity of cortisol on macrophages nor the mechanism of this process has been reported.

In this study, we demonstrated the anti-inflammatory properties of cortisol on LPS-induced inflammation injury in the RAW264.7 macrophage cell line. We further investigated the ability of cortisol to inhibit the activation of NF-κB and mitogen-activated protein kinases (MAPKs) to clarify the mechanism of its anti-inflammatory effects. This study may reveal a vital role for endogenous glucocorticoids and the underlying mechanism of glucocorticoid-mediated anti-inflammatory activity in the postpartum cow uterus, thus proposing a scientific basis for the prevention and treatment of endometritis in dairy cattle.

Methods

Reagents

Dulbecco's modified Eagle's medium (DMEM), fetal bovine serum (FBS), and other tissue culture reagents were purchased from Gibco BRL Co. (Grand Island, NY, USA). Cortisol (H0888) and LPS (*Escherichia coli* 0111:B4) were purchased from Sigma (St. Louis, MO, USA). The Cell-Counting Kit-8 (CCK-8) reagents were obtained from Dojindo Molecular Technologies, Inc. (Kumamoto, Japan). Enzyme-linked immunosorbent assay (ELISA) kits for PGE_2, IL-1β, IL-6, and TNFα were purchased from R & D Systems, Inc. (Minneapolis, MN, USA). β-actin, iNOS, COX-2, ERK1/2, phospho-ERK1/2, p38, phospho-p38, JNK, phospho-JNK, NF-κB p65, phospho-NF-κB p65, IκBα and phospho-IκBα antibodies were purchased from Cell Signaling Technology (Boston, MA, USA).

Cell culture and viability assays

The RAW264.7 macrophage cell line was obtained from the American Type Culture Collection (ATCC, MD, US). The cells were cultured at 37 °C in DMEM supplemented with 2 mM glutamine, 100 U/mL penicillin, 100 μg/mL streptomycin and 10% fetal bovine serum (FBS) in a 5% CO_2 environment [20]. To evaluate cell viability, RAW264.7 cells (5×10^3 cells/well) were seeded in 96-well plates and incubated for 18 h before experimental interventions. The cells were then treated with several concentrations of cortisol for 24 h. Ten microliters of the CCK-8 solution was added to each well, and the plate was incubated at 37 °C for 2 h. The optical density was then read at 450 nm using a microplate reader (Tecan, Austria).

PGE_2, IL-1β, IL-6, and TNFα assays

RAW 264.7 cells were seeded in 12-well plates (5×10^5 cells/mL) and incubated at 37 °C for 18 h. The cells were co-treated with cortisol (5, 15 and 30 ng/mL) and LPS (1 μg/mL) for 6, 12 and 24 h. Supernatant levels of PGE_2, IL-1β, IL-6, and TNFα were measured by ELISA according to the manufacturer's instructions.

RNA extraction and real-time quantitative reverse transcription PCR

RAW 264.7 macrophages were treated with 1 μg/mL LPS in the presence or absence of cortisol (0, 5, 15 and 30 ng/mL). After 6-, 12- and 24-h incubation periods, total RNA was isolated from RAW 264.7 macrophages according to the manufacturer's instructions using Trizol reagent (Invitrogen, US). The extracted RNA was quantified using a Nanodrop 2000 spectrophotometer (Thermo, USA). The RNA (900 ng) was then converted to cDNA as previously described [21]. The PCR contained 10 μL SYBR Green PCR mix, 0.5 μL each primer, and 1 μL cDNA template in a final reaction volume of 20 μL (Takara, Japan). The real-time PCR cycling conditions were 95 °C for 2 min, 40 cycles of 95 °C for 10 s, 60 °C for 30 s, and 72 °C for 30 s using a CFX connect real-time PCR system (BIO-RAD, US). The rat β-actin primers were used as the endogenous control. Relative gene expression was calculated using the comparative Ct method ($2^{-\Delta\Delta Ct}$) as previously described [22]. The primer sequences used in this study are presented in Table 1.

Western blot analysis

RAW264.7 macrophages were stimulated with LPS alone or together with cortisol as described above. The total proteins were extracted, and protein concentrations were determined using a bicinchoninic acid (BCA) protein

Table 1 Primer sequences used for qRT-PCR amplification

Gene	Forward primer	Reverse primer	Accession number	Product size(bp)
β-actin	TGCTGTCCCTGTATGCCTCT	TTTGATGTCACGCACGATTT	NM_031144.3	224
IL-1β	ACCTGTGTCTTTCCCGTGG	TCATCTCGGAGCCTGTAGTG	NM_031512.2	159
TNFα	GCCTCCCTCTCATCAGTTCTA	GGCAGCCTTGTCCCTTG	NM_012675.3	246
IL-6	AGTTGTGCAATGGCAATTCTGA	AGGACTCTGGCTTTGTCTTTCT	NM_012589.2	223
iNOS	TTCCAGAATCCCTGGACAAG	TGGTCAAACTCTTGGGGTTC	NM_012611.3	180
COX-2	AGAAGGAAATGGCTGCAGAA	GCTCGGCTTCCAGTATTGAG	NM_017232.3	194

assay kit (BioChain, US). Total proteins were separated by sodium dodecyl sulfate-polyacrylamide gel electrophoresis (SDS-PAGE) and transferred to polyvinylidene difluoride (PVDF) membranes (Millipore, Germany). The membranes were immunoblotted with primary antibodies specific for iNOS, COX-2, NF-κB p65, phospho-NF-κB p65, IκBα, phospho-IκBα, p-ERK1/2, ERK1/2, p-p38, p38, p-JNK, JNK, and β-actin at 4 °C overnight and then incubated with HRP-conjugated secondary antibodies (CST, US) at room temperature for 1 h. The blots were washed with PBS-T, and the proteins of interest were detected using a chemiluminescence (ECL) assay according to the manufacturer's instructions.

Statistical analysis

Each experiment was repeated at least three times, and all data are expressed as means ± standard error of the mean (SEM) for the number of experiments. Statistically significant differences throughout this study were calculated by one-way ANOVA followed by Dunnett's test (SPSS 17.0 software). A two-sided p-value less than 0.05 was considered statistically significant.

Results

Effect of cortisol on RAW264.7 macrophage viability

The effect of cortisol on the RAW264.7 macrophage cell line viability was assessed using a CCK-8 assay. As shown in Fig. 1, cortisol did not affect the viability of the RAW 264.7 cells at concentrations from 5 to 15 ng/mL, but it did alter cell growth at 20 to 60 ng/mL. Therefore, cortisol concentrations of 5, 15, and 30 ng/mL were selected for further investigation.

Cortisol modulation of extracellular PGE_2, TNFα, IL-1β, and IL-6 production in LPS-induced RAW264.7 macrophages

To investigate the inhibitory effects of cortisol on the extracellular production of inflammatory mediators and pro-inflammatory cytokines including PGE_2, TNFα, IL-1β, and IL-6 by LPS-induced RAW264.7 macrophages, cytokine-specific ELISAs were used to determine the levels of each molecule in RAW264.7 culture supernatants. As depicted in Fig. 2a, the PGE_2 concentration in the culture medium of the LPS-treated group was significantly ($p < 0.01$) increased compared with the control group at 12 and 24 h. However, co-incubation with cortisol significantly ($p < 0.05$) suppressed this increased production in a dose-dependent manner. The expression levels of IL-1β, IL-6, and TNFα induced by LPS were significantly upregulated at the indicated time points ($p < 0.01$). However, cortisol significantly suppressed the extracellular levels of IL-1β when compared with the LPS treated group in a dose-dependent manner (Fig. 2b). The levels of TNFα and IL-6 were not affected by cortisol treatment (Fig. 2c and d).

Effects of cortisol on the protein and mRNA expression levels of iNOS and COX-2 in LPS-stimulated RAW264.7 macrophages

Since COX-2 and iNOS are enzymes for PGE_2 and NO synthesis, we further investigated the inhibitory effects of cortisol treatment on COX-2 and iNOS expression using Western blotting and RT-PCR, respectively. As shown in Fig. 3a, the mRNA expression level of iNOS dramatically ($p < 0.01$) increased following stimulation of macrophages with LPS at 12 and 24 h. The mRNA expression levels of iNOS in the experimental groups were down-regulated by cortisol treatment at all concentrations ($p < 0.05$). Similarly, the COX-2 mRNA levels were significantly increased by stimulation of macrophages

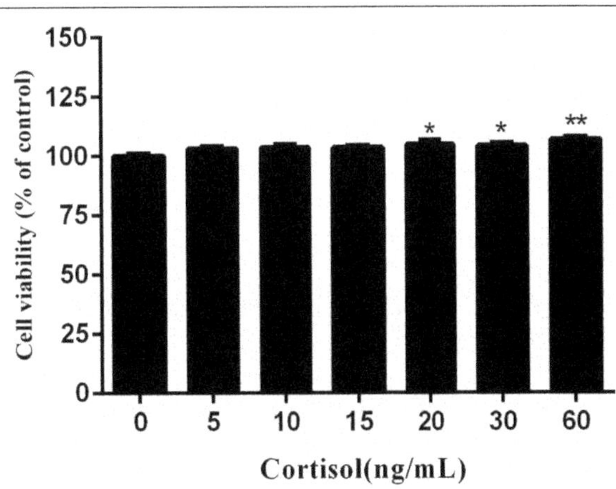

Fig. 1 Effects of different concentrations of cortisol on RAW264.7 cell viability as measured by the CCK-8 assay. The data shown are means ± SEM ($n = 6$). *$p < 0.05$ and **$p < 0.01$ vs. control group

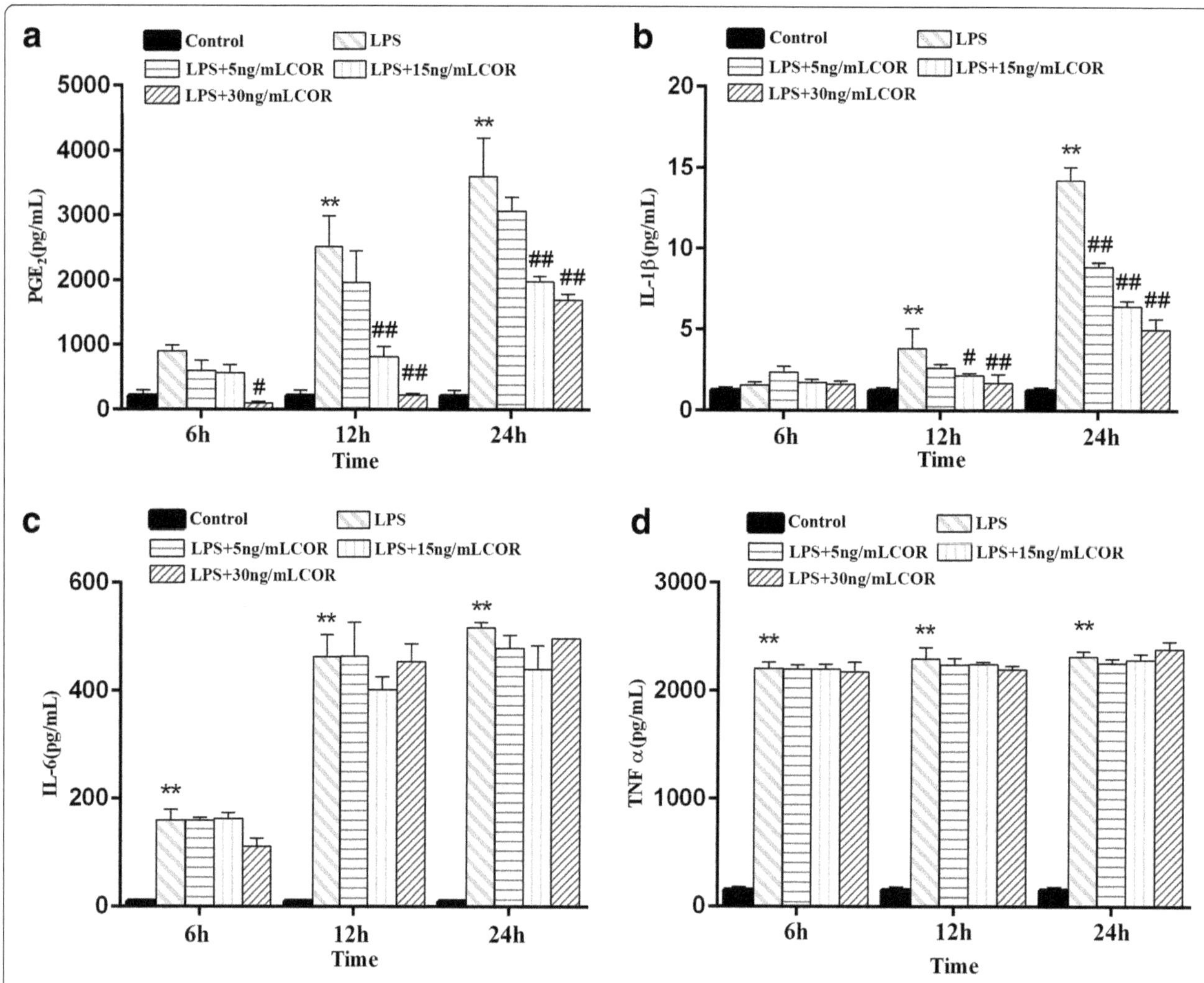

Fig. 2 Effect of cortisol on PGE_2 and cytokine production in LPS-stimulated RAW 264.7 macrophages. RAW264.7 cells were co-treated with cortisol (0, 5, 15 and 30 ng/mL) and LPS (1 μg/mL) for 0, 6, 12, and 24 h. Levels of PGE_2 (**a**), IL-1β (**b**), IL-6 (**c**), and TNFα (**d**) in culture supernatants were measured by ELISA. The data presented are the means±SEM. ** $p < 0.01$ vs. the control group; # $p < 0.05$, ## $p < 0.01$ vs. the LPS group

with LPS at 6, 12, and 24 h (Fig. 3b). In addition, COX-2 mRNA levels were also inhibited by cortisol in a dose-dependent manner.

The protein expression levels of iNOS and COX-2 were significantly ($p < 0.01$) upregulated by stimulation of macrophages with LPS at 24 h. However, these effects were markedly ($p < 0.01$) inhibited by cortisol treatment in a dose-dependent manner (Fig. 3c and d).

Inhibitory effect of cortisol on LPS-induced TNFα, IL-1β and IL-6 mRNA expression

To determine the protective effect of cortisol on RAW264.7 macrophage inflammatory responses induced by LPS, we examined the mRNA expression levels of TNFα, IL-1β and IL-6 by RT-PCR. As shown in Fig. 4, the expression of TNFα, IL-1β and IL-6 induced by LPS was significantly upregulated at the indicated time points, whereas dose-dependent reductions in LPS-stimulated TNFα, IL-1β, and IL-6 mRNA expression levels were observed in macrophages after co-incubation with cortisol ($p < 0.01$).

Effects of cortisol on NF-κB activation in LPS-stimulated RAW264.7 macrophages

NF-κB is an important transcription factor that regulates the expression of most pro-inflammatory cytokines, as well as the levels of iNOS, COX-2, and PGE_2. We investigated the critical proteins of this signaling pathway by Western blotting to determine the effect of cortisol on the NF-κB activity. As shown in Fig. 5, significant ($p < 0.01$) degradation of IκBα and increased expression of p-IκBα and p-p65 were observed in the cells following LPS exposure for 30 min, which indicated increased NF-κB activity. However, the degradation of IκBα and phosphorylation of IκBα and p65 were decreased after 45 min. Cortisol significantly inhibited the LPS-induced phosphorylation of

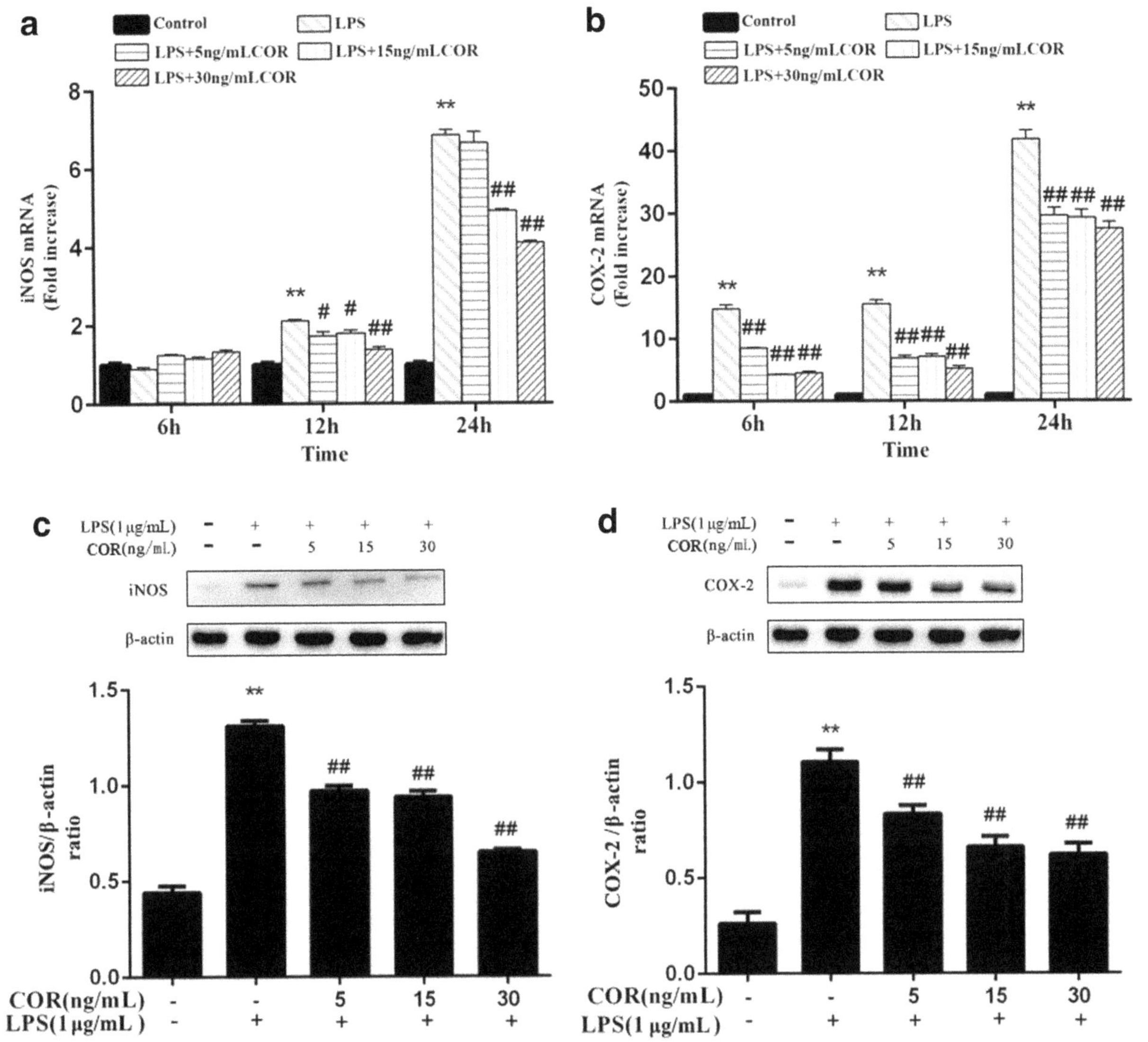

Fig. 3 Effects of cortisol on the mRNA and protein expression levels of iNOS and COX-2 in LPS-stimulated RAW264.7 cells. **a** and **b** Cells were co-treated with cortisol (5,15 and 30 ng/mL) and LPS (1 μg/mL) for 0, 6, 12, and 24 h. RNA was isolated and analyzed by RT-PCR. **c** and **d** Cells were co-treated with cortisol (5, 15 and 30 ng/mL) and LPS (1 μg/mL) for 24 h. Total proteins were isolated and analyzed by Western blot. The data presented are the means±SEM. ** $p < 0.01$ vs. the control group; # $p < 0.05$, ## $p < 0.01$ vs. the LPS group

p65, as well as phosphorylation and degradation of IκBα, in a dose-dependent manner. The data showed that the NF-κB activity in RAW264.7 macrophages induced by LPS was significantly ($p < 0.01$) inhibited by cortisol.

Effects of cortisol on the phosphorylation of MAPKs in LPS-stimulated RAW264.7 macrophages

MAPKs play important roles in the regulation of various physiological processes [23]. To determine the effect of cortisol on the MAPK pathway, we investigated the critical proteins of this signaling pathway by Western blot. The phosphorylation levels of ERK1/2, JNK, and p38 MAPK were significantly ($p < 0.01$) increased after the cells were treated with LPS for 30 min (Fig. 6a). However, the levels of phosphorylation were decreased after 45 min. Cortisol significantly ($p < 0.01$) inhibited the LPS-induced phosphorylation of ERK1/2, JNK, and p38 MAPK in a dose-dependent manner (Fig. 6b).

Discussion

In this study, we examined the anti-inflammatory activities of cortisol in LPS-induced RAW264.7 cells. Cortisol significantly inhibited the expression levels of inflammatory mediators and pro-inflammatory cytokines (Figs. 3 and 4). Moreover, the NF-κB and MAPK activities in LPS-induced RAW264.7 macrophages were obviously alleviated by cortisol (Figs. 5 and 6).

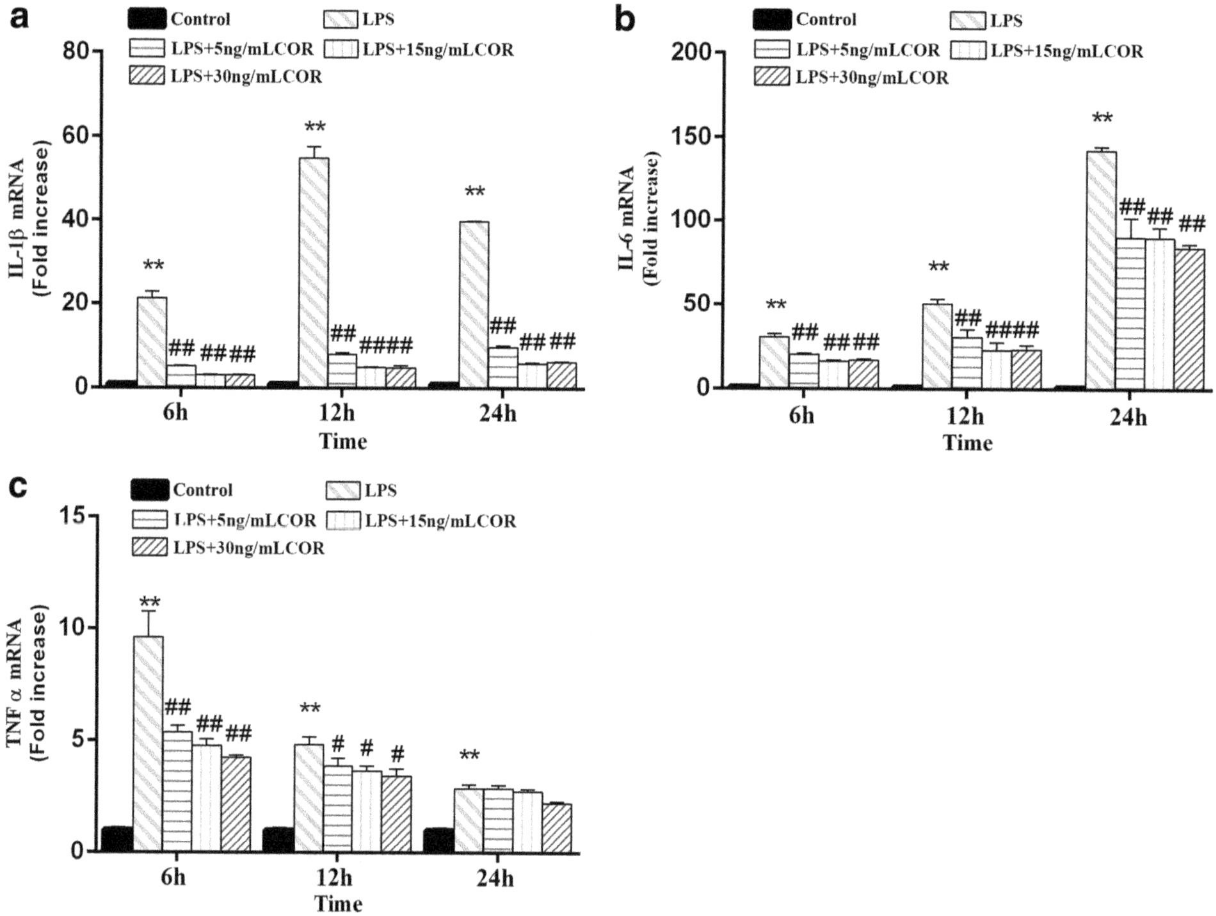

Fig. 4 Effects of cortisol on IL-1β (**a**), IL-6 (**b**) and TNFα (**c**) mRNA expression in LPS-stimulated RAW264.7 cells. RAW264.7 cells were co-treated with cortisol (5, 15 and 30 ng/mL) and LPS (1 μg/mL) for 0, 6, 12, and 24 h. RNA was isolated and analyzed by RT-PCR. The data presented are the means±SEM. $^{**}p < 0.01$ vs. the control group; # $p < 0.05$, ## $p < 0.01$ vs. the LPS group

After parturition, dairy cows are more susceptible to endometritis, which is the primary cause of reproductive failure [2]. If not treated in a timely manner, the inflammatory response generates more serious consequences that lead to endometritis and even purulent uterine inflammation. Cortisol is a major regulator of inflammation and may play a role in preventing inflammation in the body [24]. Perinatal stress triggers the release of corticotropin-releasing hormone (CRH) from the hypothalamus, which acts on the pituitary to release adrenocorticotropin hormone (ACTH) and subsequently on the adrenal glands to release cortisol into blood circulation. In addition, cortisol production peaks due to the stress of parturition [16, 25].

A rapid inflammatory response is produced in LPS-stimulated RAW264.7 macrophages, which could release a large number of pro-inflammatory cytokines (TNFα, IL-1β, and IL-6) and inflammatory mediators (PGE_2, COX-2 and iNOS) [6]. This is beneficial to attract circulating immune effector cells, such as neutrophils, to fight infection [26], but excessive inflammatory responses can injure tissues and organs. Therefore, the expression of inflammatory mediators and pro-inflammatory cytokines needs to be tightly regulated during an inflammatory response [27, 28]. In this study, we demonstrated the protective effect of cortisol against LPS-induced inflammation injury in the RAW264.7 macrophage cell line. The results showed that the gene expression and production of TNFα, IL-1β, and IL-6 were significantly increased in RAW264.7 macrophages stimulated with LPS, which induced a drastic inflammatory response. As expected, cortisol effectively inhibited the mRNA expression levels of IL-1β, IL-6, and TNFα in a dose-dependent manner, which protected macrophages from LPS-induced inflammation injury. Interestingly, there were inconsistencies between the mRNA expression and secretory protein levels of IL-6 and TNFα, which may be related to the translational regulation or the cytoplasm storage of these molecules. However, the specific mechanism behind this phenomenon requires further investigation.

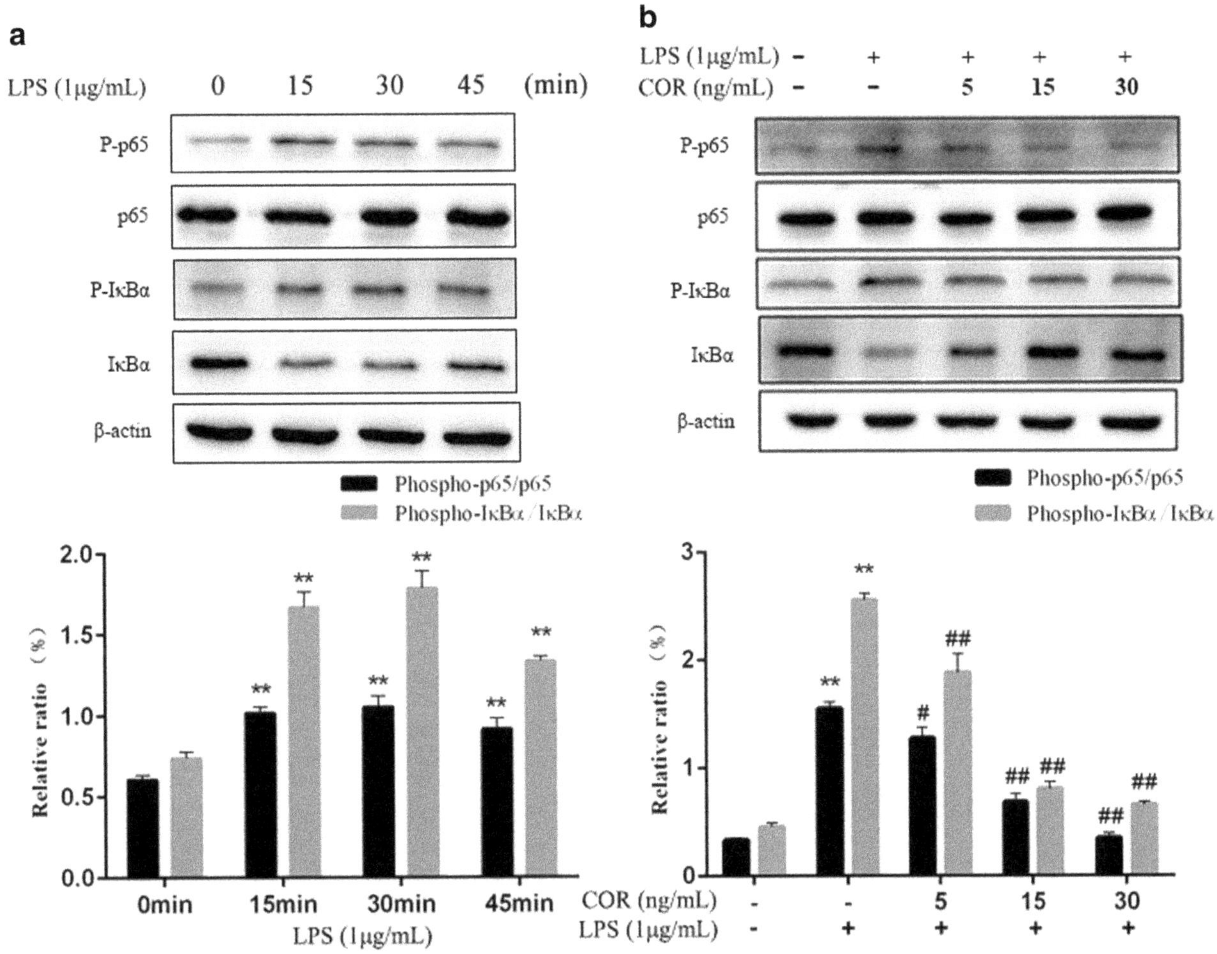

Fig. 5 Inhibitory effects of cortisol on LPS-stimulated NF-κB p65 and IκBα phosphorylation in RAW 264.7 cells. **a** Cells were stimulated with LPS (1 μg/mL) alone for 0, 15, 30 and 45 min. **b** Cells were co-treated with cortisol (5, 15 and 30 ng/mL) and LPS (1 μg/mL) for 30 min. Total proteins were isolated and subjected to Western blotting. The data presented are the means±SEM. ** $p < 0.01$ vs. the control group; # $p < 0.05$, ## $p < 0.01$ vs. the LPS group

NO and PGE_2 are important inflammatory mediators that result in serious inflammatory diseases. iNOS catalyzes the oxidative deamination of L-arginine and ultimately leads to significant nitric oxide (NO) production. Similarly, COX-2 is a key enzyme involved in the biosynthesis of prostaglandin E_2 (PGE_2). Thus, reducing the levels of iNOS and COX-2 would be an effective strategy for suppressing inflammatory responses. Our study demonstrated that cortisol inhibited extracellular production of PGE_2 in a dose-dependent manner at all time points. Moreover, cortisol inhibited LPS-induced iNOS and COX-2 mRNA and protein levels in a dose-dependent manner. These results indicated that cortisol could effectively inhibit the LPS-induced inflammatory response.

NF-κB is an important regulatory transcription factor that plays a critical role in regulating the expression of iNOS, COX-2 and pro-inflammatory cytokines such as TNFα, IL-1β and IL-6. Once activated by LPS, phosphorylation of IκBα is strongly enhanced (IκBα is the principal inhibitory protein of NF-κB), which leads to the rapid proteasomal degradation [29, 30] of IκBα. Nuclear factor κB dimers (p50:p65) are released and phosphorylated, which quickly enter the nucleus and bind specifically to defined DNA sequences to promote target gene expression [11, 12, 31]. Thus, we investigated the inhibitory effect of cortisol on LPS-induced NF-κB activation. The present study showed that the degradation of IκBα and the phosphorylation of IκBα and p65 were significantly increased after stimulation with LPS for 30 min, which suggested obviously increased NF-κB activation. However, the degradation of IκBα and the phosphorylation of IκBα and p65 were reduced after 45 min due to the synthesis and secretion of IκBα, which was consistent with the results of previous studies [32, 33]. Our results suggested that cortisol could reduce the degradation of IκBα and phosphorylation of IκB α and p65 at

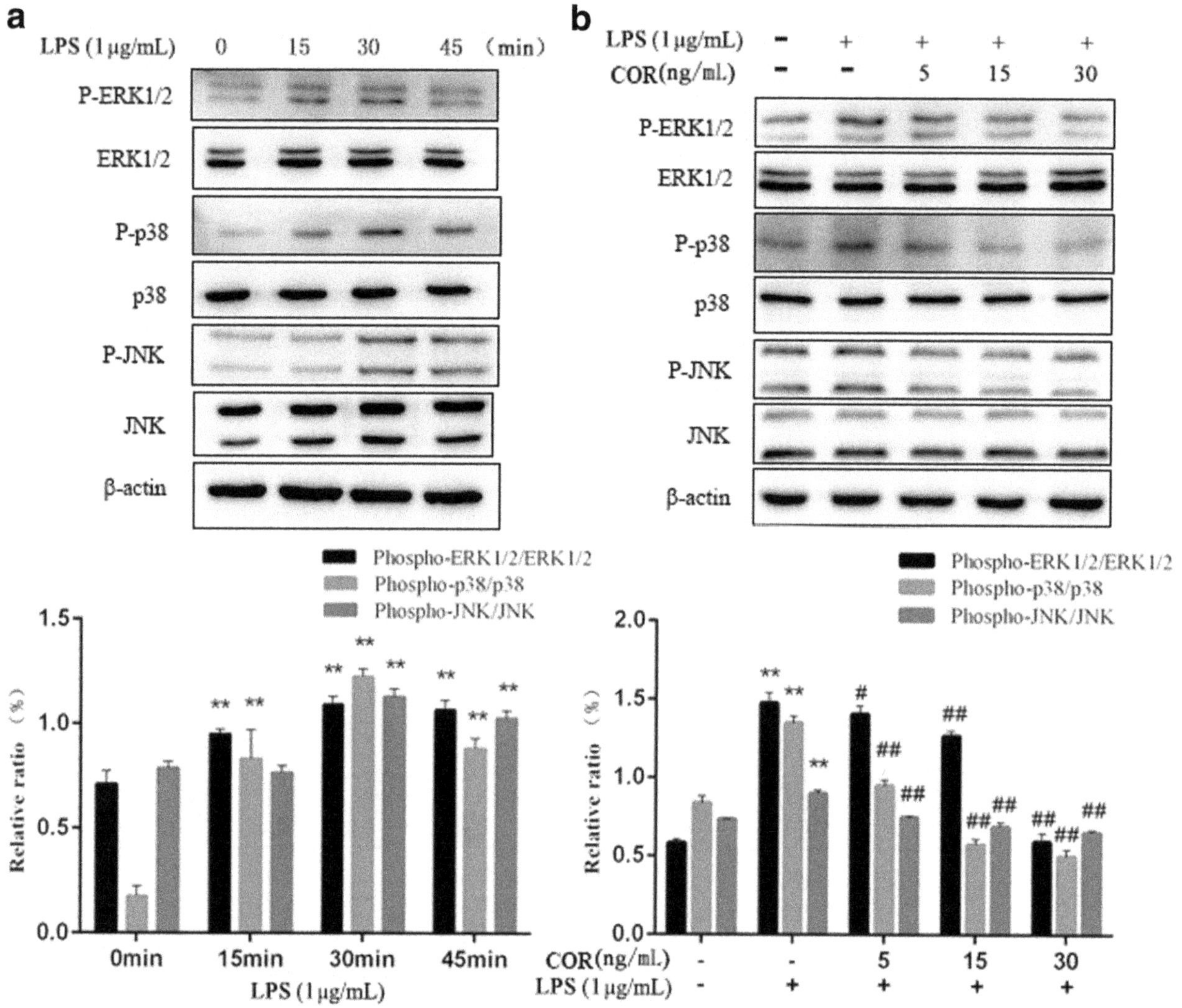

Fig. 6 Inhibitory effects of cortisol on MAPK phosphorylation in RAW264.7 macrophages. **a** Cells were stimulated with LPS (1 μg/mL) alone for 0, 15, 30 and 45 min. **b** Cells were co-treated with cortisol (5, 15 and 30 ng/mL) and LPS (1 μg/mL) for 30 min. Total proteins were isolated and subjected to Western blotting. The data presented are the means±SEM. ** $p < 0.01$ vs. the control group; # $p < 0.05$, ## $p < 0.01$ vs. the LPS group

30 min in a dose-dependent manner, demonstrating a significant inhibitory effect on NF-κB activity. Thus, cortisol could inhibit the inflammatory mediator and pro-inflammatory cytokine expression by downregulating the NF-κB pathway. Ultimately, inflammatory injury in the LPS-induced RAW264.7 macrophage cell line was significantly weakened.

In addition to NF-κB, much evidence has shown that the MAPK pathway also plays an important role in the inflammatory response. The MAPK family includes ERK1/2, JNK, and p38, which play a critical role in the transcriptional regulation of the LPS-induced expression of iNOS and COX-2 [34]. Moreover, MAPKs are known as upstream activators of NF-κB [35], as demonstrated by the inhibition of NF-κB transcriptional activation by specific MAPK inhibitors [36]. In the present study, the phosphorylation of ERK1/2, JNK, and p38 was significantly increased after LPS stimulation of RAW 264.7 macrophages, indicating that LPS activated the MAPK signaling pathway in the RAW264.7 macrophage cell line. Cortisol treatment obviously inhibited the phosphorylation of ERK1/2, JNK, and p38. These results suggested that the anti-inflammatory effects of cortisol are related to inhibition of MAPK phosphorylation in LPS-induced RAW 264.7 cells.

Conclusions

In conclusion, the current study clearly demonstrated the protective effect of cortisol on LPS-induced inflammation injury in the RAW264.7 macrophage cell line.

Cortisol inhibited LPS-induced iNOS and COX-2 expression, as well as PGE_2 production, in the macrophages. Equally, it also inhibited the expression of pro-inflammatory cytokines, including IL-1β, IL-6 and TNFα. The anti-inflammatory effect of cortisol on macrophages is mediated through inhibition of the NF-κB and MAPK signaling pathways.

Abbreviations

COR: Cortisol; ELISA: Enzyme-linked immunosorbent assay; LPS: Lipopolysaccharide; MAPK: Mitogen-activated protein kinase; NF-κB: Nuclear factor kappa-light-chain-enhancer of activated B cells; qRT-PCR: Real-time quantitative reverse transcription PCR

Acknowledgements

None applicable.

Funding

The investigation was financially supported by the National Science Foundation of China (No.31672614, 31072176, 31302151), Natural Science Foundation of Jiangsu Province (No.BK20160062, BK2012265), Priority Academic Program Development of Jiangsu Higher Education Institutions (PAPD), Top-notch Academic Programs Project of Jiangsu Higher Education Institutions (TAPP), Natural Science Foundation of the Higher Education Institutions of Jiangsu Province, China (17KJB230007), The National Natural Science Foundation of China (No.31502127) and Students' Science and Technology Innovation Fund of Yangzhou University. The funders had no role in study design, data collection, analysis and interpretation, decision to publish, or preparation of the manuscript.

Authors' contributions

JD and JL conceived, designed and performed the majority of the experiments and drafted the manuscript. LC helped with experiments, provided valuable discussion and modified the final manuscript. YW, JL and YQ participated in experimental procedures and data analysis. HW conceived the study, participated in its coordination and helped to revise the manuscript. All authors read and approved the final manuscript.

Consent for publication

Not applicable.

Competing interests

The authors declare that they have no competing interests.

References

1. Sheldon IM, Noakes DE, Rycroft AN, Pfeiffer DU, Dobson H. Influence of uterine bacterial contamination after parturition on ovarian dominant follicle selection and follicle growth and function in cattle. Reproduction. 2002;123(6):837–45.
2. Sheldon IM, Cronin J, Goetze L, Donofrio G, Schuberth HJ. Defining postpartum uterine disease and the mechanisms of infection and immunity in the female reproductive tract in cattle. Biol Reprod. 2009;81(6):1025–32.
3. Kim KS, Lee DS, Bae GS, Park SJ, Kang DG, Lee HS, Oh H, Kim YC. The inhibition of JNK MAPK and NF-κB signaling by tenuifoliside a isolated from Polygala Tenuifolia in lipopolysaccharide-induced macrophages is associated with its anti-inflammatory effect. Eur J Pharmacol. 2013;721(1–3): 267–76.
4. Zhong Y, Chiou YS, Pan MH, Shahidi F. Anti-inflammatory activity of lipophilic epigallocatechin gallate (EGCG) derivatives in LPS-stimulated murine macrophages. Food Chem. 2012;134(2):742–8.
5. Okkyoung K, Meeyoung L, Jieun Y, Seiryang O, Youngwon C, Hyeongkyu L, Kyungseop A. Anti-inflammatory effects of methanol extracts of the root of Lilium Lancifolium on LPS-stimulated Raw264.7 cells. J Ethnopharmacol. 2010;130(1):28–34.
6. Fairweather D, Rose NR. Inflammatory heart disease: a role for cytokines. Lupus. 2005;14(14):646–51.
7. Chiu WT. Suppression of lipopolysaccharide-induced of inducible nitric oxide synthase and cyclooxygenase-2 by Sanguis Draconis, a dragon's blood resin, in RAW 264.7 cells. J Ethnopharmacol. 2008;115(3):455–62.
8. Kim KS, Cui X, Lee DS, Sohn JH, Yim JH, Kim YC, Oh H. Anti-inflammatory effect of neoechinulin a from the marine fungus Eurotium sp. SF-5989 through the suppression of NF-κB and p38 MAPK pathways in lipopolysaccharide-stimulated RAW264.7 macrophages. Molecules. 2013; 18(11):13245–59.
9. Lu Y, Yeh W, Ohashi P. LPS/TLR4 signal transduction pathway. Cytokine. 2008;42(2):145–51.
10. Shao J, Li Y, Wang Z, Xiao M, Yin P, Lu Y, Qian X, Xu Y, Liu J. 7b, a novel naphthalimide derivative, exhibited anti-inflammatory effects via targeted-inhibiting TAK1 following down-regulation of ERK1/2- and p38 MAPK-mediated activation of NF-κB in LPS-stimulated RAW264.7 macrophages. Int Immunopharmacol. 2013;17(2):216–28.
11. Hoffmann A, Baltimore D. Circuitry of nuclear factor κB signaling. Immunol Rev. 2006;210(1):171–86.
12. Yamamoto M, Yamazaki S, Uematsu S, Sato S, Hemmi H, Hoshino K, Kaisho T, Kuwata H, Takeuchi O, Takeshige K. Regulation of toll/IL-1-receptor-mediated gene expression by the inducible nuclear protein IkappaBzeta. Nature. 2004;430(6996):218–22.
13. Kamata H, Tsuchiya Y, Asano T. IκBβ is a positive and negative regulator of NF-κB activity during inflammation. Cell Res. 2010;20(11):1178–80.
14. Li F, Fu Y, Bo L, Liu Z, Li D, Liang D, Wen Z, Cao Y, Zhang N, Zhang X. Stevioside suppressed inflammatory cytokine secretion by Downregulation of NF-κB and MAPK signaling pathways in LPS-stimulated RAW264.7 cells. Inflammation. 2012;35(5):1669–75.
15. Carter AB, Knudtson KL, Monick MM, Hunninghake GW. The p38 mitogen-activated protein kinase is required for NF-kappaB-dependent gene expression. The role of TATA-binding protein (TBP). J Biol Chem. 1999; 274(43):30858–63.
16. Sapolsky RM, Romero LM, Munck AU. How do glucocorticoids influence stress responses? Integrating permissive, suppressive, stimulatory, and preparative actions. Endocr Rev. 2000;21(1):55–89.
17. Pontes GCS, Monteiro PLJ, Prata AB, Guardieiro MM, Pinto DAM, Fernandes GO, Wiltbank MC, Santos JEP, Sartori R. Effect of injectable vitamin E on incidence of retained fetal membranes and reproductive performance of dairy cows. J Dairy Sci. 2015;98(4):2437–49.
18. Pathak R, Prasad S, Kumaresan A, Kaur M, Manimaran A, Dang AK: Alterations in cortisol concentrations and expression of certain genes associated with neutrophil functions in cows developing retention of fetal membranes. Veterinary Immunology & Immunopathology 2015, 168(3–4):164-168.
19. Coutinho AE, Chapman KE. The anti-inflammatory and immunosuppressive effects of glucocorticoids, recent developments and mechanistic insights. Molecular & Cellular Endocrinology. 2011;335(1):2–13.
20. Hwang JH, Kim KJ, Ryu SJ, Lee BY. Caffeine prevents LPS-induced inflammatory responses in RAW264.7 cells and zebrafish. Chem Biol Interact. 2016;248:1.
21. Bi CL, Wang H, Wang YJ, Sun J, Dong JS, Meng X, Li JJ. Selenium inhibits Staphylococcus Aureus -induced inflammation by suppressing the activation of the NF-κB and MAPK signalling pathways in RAW264.7 macrophages. Eur J Pharmacol. 2016;780:159–65.
22. Shao CY, Wang H, Meng X, Zhu JQ, Wu YQ, Li JJ. Characterization of the innate immune response in goats after intrauterine infusion of E. Coli using histopathological, cytologic and molecular analyses. Theriogenology. 2012; 78(3):593–604.
23. Kaminska B. MAPK signalling pathways as molecular targets for anti-inflammatory therapy–from molecular mechanisms to therapeutic benefits. Biochimica et Biophysica Acta (BBA) - Proteins and Proteomics. 2005; 1754(1–2):253–62.
24. Mastorakos G, Ilias I. Maternal and fetal hypothalamic-pituitary-adrenal axes during pregnancy and postpartum. Ann N Y Acad Sci. 2003;997(997):136–49.
25. Patel OV, Takahashi T, Takenouchi N, Hirako M, Sasaki N, Domeki I. Peripheral cortisol levels throughout gestation in the cow: effect of stage of gestation and foetal number. Br Vet J. 1996;152(4):425–32.
26. Porcherie A, Cunha P, Trotereau A, Roussel P, Gilbert FB, Rainard P, Germon P. Repertoire of Escherichia Coli agonists sensed by innate immunity receptors of the bovine udder and mammary epithelial cells. Vet Res. 2012;43(1):1–8.
27. Shuster DE, Jr KM, Rainard P, Paape M. Complement fragment C5a and inflammatory cytokines in neutrophil recruitment during intramammary infection with Escherichia Coli. Infection & Immunity. 1997;65(8):3286–92.

28. Scheibel M, Klein B, Merkle H, Schulz M, Fritsch R, Greten FR, Arkan MC, Schneider G, Schmid RM. IκBβ is an essential co-activator for LPS-induced IL-1β transcription in vivo. J Exp Med. 2010;207(12):2621.
29. Karin M, Ben-Neriah Y. Phosphorylation meets Ubiquitination: the control of NF-κB activity. Immunology. 2000;18(18):621–63.
30. Ghosh S, Hayden MS. New regulators of NF-B in inflammation. Nat Rev Immunol. 2008;8(11):837–48.
31. Kearns JD. Distinct functions of negative regulators of NF-kappaB. Dissertations & Theses - Gradworks. 2009;
32. Hoffmann A, Levchenko A, Scott ML, Baltimore D. The IκB-NF-κB signaling module: temporal control and selective gene activation. Science. 2002; 298(5596):1241–5.
33. Liu M, Song S, Li H, Jiang X, Yin P, Wan C, Liu X, Liu F, Xu J. The protective effect of caffeic acid against inflammation injury of primary bovine mammary epithelial cells induced by lipopolysaccharide. J Dairy Sci. 2014; 97(5):2856–65.
34. Pratheeshkumar P, Kuttan G. Modulation of immune response by L. inhibits the proinflammatory cytokine profile, iNOS, and COX-2 expression in LPS-stimulated macrophages. Immunopharmacology & Immunotoxicology. 2011;33(1):73–83.
35. Berghe WV, Plaisance S, Boone E, Bosscher KD, Schmitz ML, Fiers W, Haegeman G. p38 and extracellular signal-regulated kinase mitogen-activated protein kinase pathways are required for nuclear factor-kappaB p65 transactivation mediated by tumor necrosis factor. J Biol Chem. 1998; 273(6):3285–90.
36. Caivano M. Role of MAP kinase cascades in inducing arginine transporters and nitric oxide synthetase in RAW264 macrophages. FEBS Lett. 1998;429(3):249–53.

6

Identification of amino acid variation in the prion protein associated with classical scrapie in Canadian dairy goats

Vythegi Srithayakumar[1,3*], Gordon B. Mitchell[2] and Bradley N. White[1]

Abstract

Background: A clear association of amino acid variation in the prion protein gene (*PRNP*) with susceptibility and resistance to classical scrapie exists in sheep, but not in goats. In this study we examined DNA sequence variation in the *PRNP* of 149 animals from two scrapie-infected herds of Saanen dairy goats, and identified 6 non-synonymous variants in the coding region.

Results: In the larger herd, all of the 54 scrapie-affected goats tested had at least one allele with the arginine (R) codon at position 211, with 52 being homozygous for that variant. No animal homozygous for the glutamine (Q) codon at 211 were affected and only two heterozygotes (R/Q) were affected. A weak association was found at position 146 and no significant associations were found with amino acid variation at the remaining four variant positions (142, 143, 222 and 240), however, the allelic variation was low. Similar patterns were observed in the second scrapie-affected herd.

Conclusion: We also evaluated previous studies on goat herds affected with scrapie and this relationship of R susceptibility and Q resistance at 211 was present independent of the genotypes at the other positions including 222. The fact that glutamine at 211 provides a significant protective property to scrapie irrespective of the other positions could be important for breeding strategies aimed at improving herd resistance to scrapie, while maintaining important productivity traits.

Keywords: Scrapie, Goat, Prion, Resistance, Susceptibility, Breeding, Variation

Background

Scrapie is a fatal neurodegenerative disease that affects sheep and goats. It is classified as a transmissible spongiform encephalopathy (TSE) belonging to a group of prion diseases, which includes Creutzfeldt-Jakob disease in humans, bovine spongiform encephalopathy (BSE) in cattle, and chronic wasting disease in cervids. Prion diseases not only have a serious impact on health and welfare, but control of these diseases impacts animal movement and trade. Prion diseases can occur sporadically as well as through heredity or infectious transmission routes [1] but pathogenesis is contingent on conversion of the normal host prion protein (PrP^{C}) to a disease-associated form (PrP^{Sc}). In sheep and goats, scrapie is characterized by the deposition of this abnormal, protease resistant prion protein (PrP^{Sc}) in the central nervous system and peripheral tissues [2, 3]. The host prion gene (*PRNP*) encodes the prion protein (PrP^{C}) and mutations within this gene have been associated with differential resistance and susceptibility to scrapie (as reviewed [4]).

Genetic resistance to scrapie is well established in sheep and three-codon *PRNP* genotypes have typically been utilized for risk assessment (as reviewed in [4]). The association of these three amino acid codons at positions 136, 154 and 171 produces 5 haplotypes (ARQ, VRQ, AHQ, ARR, ARH) and 15 genotypes [4, 5]. The ARR haplotype is associated with high resistance, whereas VRQ or ARQ are associated with higher susceptibility to scrapie [6, 7]. Sheep breeding programs selecting for the ARR genotype have been widely used for scrapie

* Correspondence: vythegisrithayakumar@trentu.ca
[1]Natural Resources DNA Profiling and Forensics Centre, DNA Building, Trent University, 2140 East Bank Drive, Peterborough, ON, Canada
[3]Livestock Gentec, Department of Agricultural, Food and Nutritional Science, University of Alberta, Edmonton, AB T6G2P5, Canada
Full list of author information is available at the end of the article

control and eradication programs in Europe and North America [8]. Patterns of associations to scrapie resistance and susceptibility based on the three codons used for sheep do not hold true in goats.

Although the natural occurrence of scrapie in goats appears generally lower than that in sheep, active surveillance reports suggest that the prevalence rates in goats may be underestimated in some countries [9]. There have been a number of studies that examined resistance and susceptibility to scrapie in goats from European countries (e.g., France, Italy, United Kingdom, Greece), where the goat populations are large and the incidence of scrapie has been high. These studies have characterized *PRNP* sequence variation in goats, and have reported a number of polymorphisms in goat populations worldwide [10–25]. In particular, thirty nine amino acid substitutions have been described, of which only six G127S [18], I142M [26], N146S/D [12], H154R [10, 12, 14, 19], R211Q [19], and Q222K [11, 14, 19] have been associated with resistance or susceptibility to scrapie in goats. At least 16 synonymous silent mutations have also been documented. To date, 39 out of the 256 codons have been shown to be polymorphic across goat breeds and generally 6 to 12 are described as polymorphic within a given goat breed. While some of the polymorphic sites are similar in sheep and goats, species-specific variation, like that observed for $A_{136}R_{154}R_{171}$, are also present (as reviewed in [4]).

Despite the high amount of variation observed in the *PRNP* in various goat breeds, clear associations with scrapie susceptibility or resistance are limited. In the studies that have suggested associations with *PRNP* polymorphisms, the power of the analyses is often weak because of the relatively low frequencies of the putative resistant allele and the emphasis on identifying resistance alleles rather than susceptibility alleles. The unaffected animals represent a mixture of resistant and susceptible animals of varying exposure status, while the affected animals are more homogeneous in susceptibility traits. The absence of clearly identified variation, which is associated with scrapie in goats, has limited the implementation of breeding strategies to eliminate scrapie sensitive alleles in goat herds.

The objectives of this study were: (i) to identify the variation in *PRNP* genotypes in 2 herds of Saanen dairy goats with scrapie affected animals; (ii) to examine associations between PrP^C amino acid variation and scrapie disease in the largest herd and (iii) to validate the associations using the second scrapie affected herd together with published associations.

Methods

Scrapie affected herds

The first scrapie affected dairy goat herd was composed of 331 Saanen goats greater than 12 months of age. The herd was depopulated as part of routine disease control measures following the detection and confirmation of a scrapie-positive goat. Jugular blood was sampled in EDTA-vacutainer tubes for DNA extraction and the obex and retropharyngeal lymph nodes were removed for scrapie diagnostic testing. Initial testing of obex and lymph node tissues was conducted using a commercially available ELISA (Bio-Rad TeSeE® ELISA, Bio-Rad Laboratories, Hercules, CA). Confirmatory testing was subsequently performed using immunohistochemistry with antibodies F89 and F99, and western immunoblot (Bio-Rad TeSeE® western immunoblotting kit, Bio-Rad Laboratories) as previously described [27]. An animal was considered positive for scrapie if it was positive in at least one tissue by the confirmatory tests. A total of 66 animals were found positive for classical scrapie. *PRNP* was sequenced for 54 of the 66 scrapie-affected and 56 of the unaffected animals.

A second, smaller scrapie-affected herd that comprised 130 Saanen goats greater than 12 months of age was also identified and subjected to depopulation. Following the testing regime described above, 13 animals were identified as classical scrapie positive. *PRNP* was sequenced for 10 scrapie-affected and 29 unaffected animals.

PRNP sequencing and genotyping

Genomic DNA was extracted from EDTA-preserved whole blood using the MagNa Pure DNA Isolation Kit for blood (Roche Applied Sciences) as per manufacturer's instructions, and was quantified using PicoGreen. The Goat *PRNP* reference sequence was obtained from Genbank (accession # X74758) and primers were designed to amplify the ORF of *PRNP* (PRNP2-F: AGCTGATGCCAGTGCTATGC and PRNP4R: GTGGCCTCCTTCCAGACTTG).

PCR buffer components were in the following concentrations: 1× PCR buffer (Invitrogen), 0.2 mM of each dNTP, 1.5 mM $MgCl_2$, 0.3 μM of forward primer, 0.3 μM of reverse primer, 0.05 U/μL of Taq DNA polymerase (Invitrogen), 10 ng of DNA and ddH2O for a total volume of 20 μl. PCR conditions used for amplification were as follows: initial denaturation at 94 °C for 5 min; [denaturation at 94 °C for 30 s, annealing at 60 °C for 1 min, extension at 72 °C for 1 min 30 s] × 35 cycles; final extension at 72 °C for 15 min; and 4 °C hold. Two μl of the amplified product were subjected to agarose gel electrophoresis to confirm amplification. PCR products were purified using ExoSAP (New England BioLabs) following the manufacturer's instructions. BigDye® Terminator v3.1 Cycle Sequencing Kit (Applied Biosystems) was used to sequence the fragments, using both the forward and reverse primers. Sequences were edited and aligned to Ovine and Caprine *PRNP* using MEGA™ 6.0 [28].

Analysis of genotypes in affected and unaffected samples

Comparison of the genotype frequencies of affected and unaffected groups were carried out using logistic regression models fitted to scrapie status of goats in the affected groups. When there were no positive scrapie cases, Fisher's exact test was used.

Relative susceptibility of genotypes to scrapie

Herd 1 consisted of 331 animals more than 12 months of age and 66 tested scrapie positive; the remaining 265 animals, consisted of animals that tested scrapie negative on ELISA. The scrapie negative animals may have been exposed to the disease and not succumbed or simply may not have been exposed to the disease. In order to estimate the relative susceptibility of different genotypes to scrapie, the frequencies of alleles and genotypes in the pre-exposed herd one were estimated. We assumed the genotypes of the 56 unaffected samples that were genotyped were representative of the 209 unaffected animals that were not genotyped and the genotypes of the 54 affected animals were representative of the12 affected animals that were not genotyped. The percentages of each genotype in the pre-exposed herd that became scrapie positive were compared.

Results

PRNP variation and allele frequencies in 2 scrapie affected Ontario Saanen herds

The open reading frames of the *PRNP* [5] were sequenced in 54/66 of the affected animals and 56/265 of the unaffected animals. A total of eight variable sites were found, with six non-synonymous substitutions (codon positions: 142, 143, 146, 211, 222 and 240) and two synonymous substitutions (Fig. 1). Amino acids are designated by the single letter codes that give six letter alleles or haplotypes. Initially allele haplotypes were assigned based on homozygous animals; heterozygous animals were genotyped based on the confirmed alleles in the homozygotes. This left haplotype uncertainty in three of the ten alleles that were in low frequency, (Table 1) and the genotypes with those alleles (Table 2). The haplotype frequencies for each of the two herds show a high prevalence of four alleles (Table 1).

Association of variants with scrapie susceptibility

We compared the genotypes of affected and unaffected animals in order to assess the association with scrapie susceptibility using logistic regression (Table 2). The genotypes IHNRQP/IHNRQP and IHNRQS/IHNRQP, were seen in significantly higher frequency in scrapie positive animals than scrapie negative animals. Conversely, IHNQQS/IHNRQP and IHNQQS/IHNRQS were present predominantly in scrapie negative animals. Next, we examined the relative susceptibility of the different genotypes to scrapie.

We focused on the affected animals, as the unaffected sample represents a mixture of potentially resistant animals as well as unexposed or less exposed animals (Table 2). The homozygous genotype IHNRQP/IHNRQP was estimated to be present in 53 individuals in the pre-exposed herd and 16 were found in the scrapie positive sample, while the heterozygote IHNQQS/IHNRQP was estimated to be present in 67 individuals prior to exposure and only one was found in the scrapie positive samples. In order to identify the amino acids conferring scrapie susceptibility we analysed the six variable positions individually.

The genotype frequencies of the affected animals were compared to the expected frequencies based on the pre-exposed herd and the null hypothesis that the affected animals represent a random sample of genotypes in the herd (Table 3). A significant difference was found at position 211, and a weaker significance at 146. Arginine (R) at 211 was clearly associated with susceptibility while glutamine (Q) was associated with resistance in the RQ_{211} heterozygote. There were fewer affected animals in herd 2 but all were RR_{211}, consistent with R being a susceptibility allele (Table 3).

When we examined the genotype estimates of the pre-exposed herd at position 211, the relative risk of developing the disease for the genotype RR_{211} and RQ_{211} was clear, with 64 of 182 RR_{211} being affected while only two of 125 RQ_{211} were affected. These data highlight the protective effect of Q in the heterozygote making it approximately 22 × more resistant than RR (Table 4).

The results in Tables 3 and 4 show a clear association between R_{211} and scrapie susceptibility. In the literature there are reported patterns of resistance for M_{142}, S_{146}, R_{211} and K_{222} [11, 14, 19, 21, 29, 30], but the power of the analyses for these codons are often weak because of low allele frequencies of the proposed resistant alleles (Table 5). We examined the 544 scrapie affected samples described in the literature, and this showed most are homozygous for the proposed susceptibility alleles at positions 142, 146, 211 and 222. This is because the resistant alleles are in very low frequency in most herds in Europe and the US [11, 13, 14, 19, 21, 29, 30] and emphasises the importance of establishing the allele frequencies in the herds when associating resistance and susceptibility with genotypes. As with the data presented here, the heterozygote RQ_{211} is found significantly less frequently than homozygous RR_{211}, and the QQ_{211} genotype was not seen in any affected animal.

Discussion

Scrapie poses a threat to the Canadian goat industry as herds on farms that test positive are entirely eradicated,

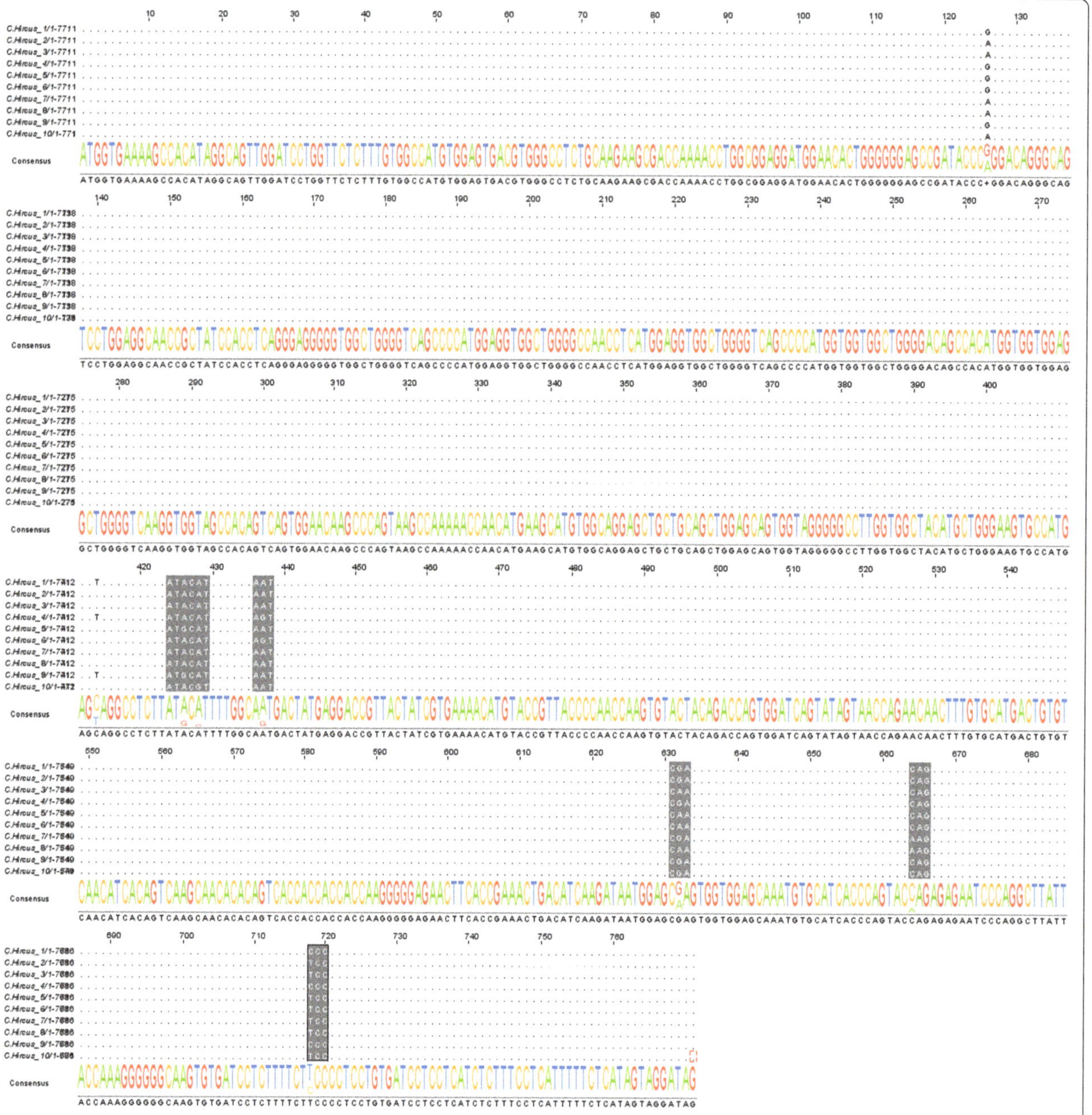

Fig. 1 Nucleotide sequences of the eight variable sites for the ten haplotypes observed in the Ontario Saanen herds. C. hircus *PRNP* (genbank accession # X74758.1) was used as the reference. Synonymous substitutions are denoted in lower case

and its presence limits the international export of goat genetics. Control of scrapie in sheep and goats requires knowledge of the *PRNP* genotypes which influence disease susceptibility. In sheep there is a clear association of amino acid variation at three positions in *PRNP* to susceptibility and resistance to scrapie. There are weaker associations with four positions in goats and this seems to be caused by low frequencies of variants, which in turn impact the power of the analyses. Analyses have also focused on identifying resistant alleles in unaffected animals that represent a mixture of resistant and unexposed animals. In this study, we had samples from a large herd of Saanen dairy goats with 66 affected individuals and a smaller herd with 13 affected animals.

The previous studies on goat scrapie have identified a number of variants in the different breeds and herds, which are associated with a lower incidence of scrapie [10–12, 14, 19, 26]. These include M_{142}, S_{146}, Q_{211} and K_{222}, which were present in the two herds we examined; but apart from Q_{211}, the frequencies of the other variants were low. However, these variants were only found in the unaffected animals in both herds.

Table 1 Allele frequencies for *PRNP* in herds 1 and 2

Amino acid position							Haplotype frequency	
	142	143	146	211	222	240	Herd 1	Herd 2
1	I	H	N	R	Q	P	0.459	0.526
2	–	–	–	–	–	S	0.305	0.231
3	–	–	–	Q	–	S	0.159	0.090
4	–	–	S	–	–	–	0.032	0.090
5¥	M	–	–	Q	–	S	0.005	0.000
6¥	–	–	S	Q	–	S	0.005	0.000
7¥	–	–	–	Q	K	S	0.005	0.000
8	–	–	–	–	K	S	0.014	0.000
9	M	–	–	–	–	–	0.014	0.064
10	–	R	–	–	–	S	0.005	0.000

¥ unconfirmed haplotypes

PRNP variation in Ontario Saanens

Thirty-seven amino acid substitutions have been previously described in the open reading frame of *PRNP* in goats around the world. The six polymorphisms found in the Saanens from Ontario have been previously described [11, 13, 14, 19, 26]. It is usual to observe only a subset of the global goat *PRNP* polymorphisms in a given breed or herd. Previous studies have found polymorphic positions ranging from 5 to 14 [11, 13, 18–20]. Saanens often have six polymorphic codons, with the specific position differing among different herds in different countries. Three of the six (142, 211, and 240) were found in Italian Saanens; four (142, 211, 222 and 240) were found in the French and Spanish Saanens. The remaining two polymorphic codons 143 and 146 that were seen in low allele frequencies (0.34 and 5 %, respectively), have been reported in Saanens crossbred with Hellenic and Damascus breeds, respectively [10, 12, 24]. As the Saanen breed originated from Switzerland, these similarities and differences are the result of founder effects combined with cross breeding with local goat breeds.

Of the ten haplotypes found in this study, three (IHNRQP, IHNRQS and IHNQQS) were present in 90 % of the sampled animals. The two major *PRNP* haplotypes (IHNRQP and IHNRQS) differed from one another by the presence of proline (P) or serine (S) at position 240. In both of the herds investigated, P_{240} was slightly higher in frequency (55 %) compared to S_{240} (45 %). This pattern has been noted in French Saanens [19, 21], Italian [11, 14], Spanish [20], British [18]; North American [13] and Greek goats [10, 23, 29, 30]. When we estimated the haplotype and genotype frequencies for all the animals in both herds, comparable trends were observed for the two main haplotypes (IHNRQP and IHNRQS); however, for the other haplotypes present in low frequencies, significant differences were present between the two herds. Low haplotypic frequencies are usually encountered when examining *PRNP* variation in goat herds [10, 11, 14, 19–21, 23].

Table 2 Comparison of genotypes among the complete herd, affected and unaffected animals in herd 1. Logistic regression was used to compare affected and unaffected animals

PRNP genotype	Pre-exposure[a]	Scrapie positive sample	Scrapie negative sample	OR (95 % CI)	*P*-value
IHNRQP/IHNRQP	53	16	7	2.947(1.089–7.973)	0.033
IHNRQS/IHNRQP	63	28	6	8.974(3.261–24.699)	<0.005
IHNRQS/IHNRQS	13	7	1	8.191(0.949–70.704)	0.055
MHNRQP/IHNRQS	14	0	3	ND[‡]	–
MHNQQS/IHNRQP	5	0	1	ND	–
IHNRQP/IRNRQS	1	1	0	ND	–
IHSRQP/IHNRQP	9	0	2	ND	–
IHSRQP/IHNRQS	14	0	3	ND	–
IHSRQP/IHSRQP	5	0	1	ND	–
ihsqqs/IHNRQP	5	0	1	ND	–
IHNQQS/IHNRQP	67	1	14	0.057(0.007–0.459)	0.007
IHNQQS/IHNRQS	44	1	9	0.098(0.012–0.827)	0.033
IHNQQS/IHNQQS	24	0	5	ND	–
IHNRKS/IHNRQP	5	0	1	ND	–
IHNRKS/IHNRQS	5	0	1	ND	–
IHNRKS/IHNQQS	5	0	1	ND	–
Total	331	54	56	–	–

[a]number of genotypes pre-exposure was estimated as described in the methods
[‡] OR could not be determined (ND) for genotypes without any scrapie positive cases

Table 3 Comparison of observed and expected numbers of scrapie positive samples for each amino acid variant. Allele frequencies of the pre-exposed herd are indicated in parentheses. *P*-values of <0.05 are indicated in bold. Chi-square analysis was used for herd 1, and Fisher's exact test was used for herd 2 as the sample size was low

		Numbers in Scrapie positive samples			
		Herd #1		Herd #2	
Codon	Genotype	Observed	Expected[ǂ]	Observed	Expected[ǂ]
142	II (0.915)	54	51	10	8
	IM (0.085)	0	3	0	2
	MM (0.000)	0	0	0	0
143	HH (0.997)	53	54	10	10
	HR (0.000)	0	0	0	0
	RR (0.003)	1	0	0	0
146	NN (0.876)	54	49	10	8
	NS (0.105)	0	5	0	2
	SS (0.019)	0	1	0	0
211	RR (0.615)	52	30	10	8
	RQ (0.333)	2	20	0	2
	QQ (0.051)	0	4	0	0
222	QQ (0.969)	54	52	10	10
	QK (0.021)	0	2	0	0
	KK (0.010)	0	1	0	0
240	PP (0.295)	16	12	3	5
	PS (0.465)	30	28	7	3
	SS (0.239)	8	15	0	2

[ǂ]Allele frequencies for the expected numbers were calculated using the frequencies of the pre-exposed herd

PRNP variants associated with scrapie susceptibility

When we compared the genotypes of affected and unaffected animals, we found two genotypes to be associated with scrapie susceptibility and two to be associated with resistance. Specifically, we compared the estimated allele and genotype frequencies of the herd prior to exposure with those of affected animals. Of the six variable positions, two (146 and 211) had genotype frequencies significantly different in the affected samples than those expected from the pre-exposed frequencies. There was weak significance for position 146 and similar results have been found in other studies, where the homozygous NN_{146} genotype has been associated with scrapie susceptibility in Cypriot goats [16, 24]. The homozygous RR_{211} genotype was strongly associated with scrapie susceptibility, with the heterozygous RQ_{211} genotype conferring a degree of protection to the disease.

Table 4 Comparison of the genotypes at position 211 to assess the relative scrapie risk

Genotype	Pre-exposed herd[ǂ]	Scrapie positive[ǂ]	Scrapie negative[ǂ]	Relative risk
RR	182	64	118	22
RQ	125	2	123	1
QQ	24	0	24	0

[ǂ]Total number of animals in each group, were calculated using the frequencies of the pre-exposed herd

Effect of position 211 on scrapie susceptibility and resistance

All the affected animals from herd 1 had at least one allele with arginine (R) at position 211. Specifically, of the 54 affected samples tested, 52 were RR_{211}, whereas, the remaining 2 were RQ_{211}. In the unaffected group, RR_{211} was seen in 25 of the 56 samples and RQ_{211} and QQ_{211} were seen in 26 and 5 animals, respectively. A similar result was seen in the second herd where all the affected samples were RR_{211}. These findings indicate that glutamine (Q) at position 211 confers protection to scrapie. The frequency of Q is usually low and therefore the homozygous QQ_{211} is seen in very low frequency in most herds, however it has not been seen any of the affected animals reported in the literature (Table 5).

Significant differences between the frequencies of the RQ_{211} in the unaffected samples compared to the affected samples shows heterozygous goats have a significantly lower susceptibility to scrapie than RR_{211} individuals. A comparison of the estimated genotypes of the herd prior to scrapie exposure to the affected animals shows 35 % of the homozygous RR_{211} individuals developed the disease, whereas only 1.6 % of the heterozygous RQ_{211} individuals succumbed. Similar patterns have been reported in the literature. In French goat populations, the risk of succumbing to scrapie for RR_{211} and RQ_{211} to be 30–39 % and 4.4–14 %, respectively [19, 21]. These results highlight the protection provided by the glutamine (Q) at position 211 and low prevalence of the disease in the heterozygous animals.

Although additional studies are required, it is interesting to compare the 211 variant position in goats to position 171, of the $A_{136}R_{154}Q_{171}$ allele in sheep. In sheep, homozygous ARR is considered resistant and homozygous ARQ is susceptible; whereas ARR/ARQ heterozygous sheep are relatively resistant to scrapie [7, 31]. Similar results were found in this study where homozygous QQ_{211} and RR_{211} were found resistant and highly susceptible respectively, while there was a very low prevalence of scrapie in RQ_{211} heterozygous animals. Although, scrapie has not been seen in Canadian ARR/ARQ sheep to date, it has been detected in Europe sheep [31–33]. Statistical models have estimated that genotypes with ARR haplotypes were 1000× more resistant to the disease compared to homozygous ARQ [33, 34]. In

Table 5 Comparison of genotypes affected with scrapie from the literature. A total of 544 scrapie cases from various breeds were assessed

Number of affected	142			146			211			222			References
	II	IM	MM	NN	NS	SS	RR	RQ	QQ	QQ	QK	KK	
30	83	1	0	-	-	-	83	7	0	87	3	0	Barillet et al. 2009
259	238	21	0	-	-	-	246	13	0	256	3	0	Corbiere et al. 2013
39	-	-	-	-	-	-	-	-	-	39	0	0	Vaccari et al. 2006
27	-	-	-	-	-	-	27	0	0	27	0	0	Bouzalas et al. 2010
25	-	-	-	-	-	-	-	-	-	25	0	0	Acutis et al. 2006
104	103	1	0	104	0	0	104	0	0	99	5	0	Fragkiadaki et al. 2011
Total:544	*424*	*23*	*0*	*104*	*0*	*0*	*460*	*20*	*0*	*533*	*11*	*0*	-

this study we estimated RQ_{211} were 22 × more resistant than homozygous RR_{211}, suggesting that RQ_{171} in sheep has a stronger influence on resistance than RQ_{211} in goats. Similar to homozygous ARR sheep, the QQ_{211} genotype in goats is seen in relatively low frequency [31–33].

Although the mechanism for the conversion of PrP^{C} to the pathogenic form (PrP^{Sc}) and its pathological effects are poorly understood, specific mutations in sheep *PRNP* associated with resistance are shown to destabilize the PrP^{C} protein [35]. Direct correlation of scrapie susceptibility and mutation induced changes are unlikely as the structural consequence of scrapie associated variants are local. As such, variations in scrapie susceptibility are likely caused by altering protein stability. Destabilizing the structure of PrP^{C} may increase protease sensitivity and slow amyloidogenesis, thereby preventing PrP^{C} from being converted to the pathogenic form. This is thought to account for the change from scrapie susceptibility to resistance following a single amino acid substitution (ARQ_{171} to ARR_{171}) in sheep [36], and perhaps also the inhibition of cell-free conversion following the substitution of R to Q at codon 211 in recombinant goat PrP [37].

Position 211 and developing scrapie resistant dairy goat herds in Ontario

An association of other variants (M_{142}, S_{146} and K_{222}) with protection towards scrapie was not tested in this study due to low allele frequencies of the potentially resistant allele. QQ_{211} provided protection regardless of the amino acids at the other positions such as 222. There is a suggestion from challenge studies that K_{222} may confer more resistance than Q_{211} [22]. In preliminary studies on Ontario dairy goat herds, the frequency of K_{222} has been found to be very low or absent, making elimination of Q_{222} very difficult. The extent to which elimination of R_{211} could result in scrapie resistant herds is of great interest and additional studies on surveying genotype frequencies in Ontario dairy herds are underway.

Conclusions

The data presented, highlights the association of arginine (R) at position 211 with susceptibility to scrapie and the protective effect of Q, irrespective of the amino acids at the other positions. To date there are no programs in place in Canada to select for resistance towards scrapie in goats. A major barrier may be the low frequency of scrapie resistant animals in Canadian goat herds. As the male to female ratio in breeding programs is often 1:20, the genotype of the bucks has a strong impact on the determination of the overall genetic profile of the herd. The most efficient strategy to select against high risk animals, at the same time minimizing the loss of production traits, would be to only utilize more resistant bucks or semen. This would be a particularly useful approach for goat producers associated with a farm previously infected with scrapie, and goat producers providing large numbers of breeding ewes to other producers.

In sheep, breeding for scrapie resistant with ARR/ARR rams has been a successful strategy for the prevention of scrapie outbreaks [38]. Furthermore, increased frequencies of the resistance associated ARR allele, as a result of selective breeding, is assumed to have a population effect by reducing the scrapie infection risk even for animals of susceptible genotypes [8, 39]. As the protection provided by R_{171} in heterozygous sheep seems to be several folds higher than the protection provided by Q_{211} in heterozygous goats, selection for homozygous QQ_{211} goats should be favoured. Breeding for resistance in goats by recruiting bucks with scrapie-resistant genetics, should increase the frequency of resistant alleles in the herd, with a resultant reduction in classical scrapie susceptibility.

Competing interests
The author(s) declare that they have no competing interests.

Authors' contribution
VS carried out the laboratory work, conducted the analyses and drafted the manuscript. GBM provided the samples for this study and helped draft the manuscript. BNW played a critical role in project development, manuscript preparation and obtained the funding to carry out this research. All the authors edited and approved the manuscript prior to submission.

Acknowledgments
The authors acknowledge the support staff in the Transmissible Spongiform Encephalopathy Unit at the CFIA Ottawa Laboratory Fallowfield for outstanding technical assistance, and the CFIA Operations staff involved in sample collection. We thank members of the Natural Resources DNA Profiling and Forensic Center, in particular Nguyen Thi Xuan Nguyen, for assistance with the scrapie genotyping.

Author details
[1]Natural Resources DNA Profiling and Forensics Centre, DNA Building, Trent University, 2140 East Bank Drive, Peterborough, ON, Canada. [2]National and OIE Reference Laboratory for Scrapie and CWD, Ottawa Laboratory Fallowfield, Canadian Food Inspection Agency, 3851 Fallowfield Road, Ottawa, ON, Canada. [3]Livestock Gentec, Department of Agricultural, Food and Nutritional Science, University of Alberta, Edmonton, AB T6G2P5, Canada.

References

1. Prusiner S. Prions. Proc Natl Acad Sci U S A. 1998;95:13363–83.
2. Andreoletti O, Berthon P, Marc D, Sarradin P, Grosclaude J, van Keulen L, Schelcher F, Elsen JM, Lantier F. Early accumulation of PrPSc in gut-associated lymphoid and nervous tissues of susceptible sheep from a Romanov flock with natural scrapie. J Gen Virol. 2000;81:3115–26.
3. Gonzalez L, Martin S, Siso S, Konold T, Ortiz-Pelaez A, Phelan L, Goldmann W. Stewart P. Saunders G. et al. High prevalence of scrapie in a dairy goat herd: tissue distribution of disease-associated PrP and effect of PRNP genotype and age. Vet Res. 2009;40:65.
4. Goldmann W. PrP genetics in ruminant transmissible spongiform encephalopathies. Vet Res. 2008;39:30.
5. Goldmann W, Hunter N, Foster JD, Salbaum JM, Beyreuther K, Hope J. Two alleles of a neural protein gene linked to scrapie in sheep. Proc Natl Acad Sci U S A. 1990;87(7):2476–80.
6. Baylis M, Goldmann W. The genetics of scrapie in sheep and goats. Curr Mol Med. 2004;4:385–96.
7. Harrington NP, O'Rourke KI, Feng Y, Rendulich J, DiFruscio C, Balachandran A. Prion genotypes of scrapie-infected canadian sheep 1998–2008. Canadian Journal of Veterinary Research-Revue Canadienne De Recherche Veterinaire. 2010;74:228–32.
8. European Food Safety Authority. Scientific Opinion on the scrapie situation in the EU after 10 years of monitoring and control in sheep and goats. EFSA Panel on Biological Hazards. EFSA Journal. 2014;12(7):3781.
9. Vaccari G, Panagiotidis CH, Acin C, Peletto S, Barillet F, Acutis P, Bossers A, Langeveld J, van Keulen L. et al. State-of-the-art review of goat TSE in the european union, with special emphasis on PRNP genetics and epidemiology. Vet Res. 2009;40:48.
10. Billinis C, Panagiotidis C, Psychas V, Argyroudis S, Nicolaou A, Leontides S, Papadopoulos O, Sklaviadis T. Prion protein gene polymorphisms in natural goat scrapie. J Gen Virol. 2002;83:713–21.
11. Acutis P, Bossers A, Priem J, Riina M, Peletto S, Mazza M, Casalone C, Forloni G, Ru G, Caramelli M, et al. Identification of prion protein gene polymorphisms in goats from italian scrapie outbreaks. J Gen Virol. 2006;87: 1029–33.
12. Papasavva-Stylianou P, Kleanthous M, Toumazos P, Mavrikiou P, Loucaides P. Novel polymorphisms at codons 146 and 151 in the prion protein gene of cyprus goats, and their association with natural scrapie. Vet J. 2007;173:459–62.
13. White S, Herrmann-Hoesing L, O'Rourke K, Waldron D, Rowe J, Alverson J. Prion gene (PRNP) haplotype variation in united states goat breeds (open access publication). Genet Sel Evol. 2008;40:553–61.
14. Vaccari G, Di Bari M, Morelli L, Nonno R, Chiappini B, Antonucci G, Marcon S, Esposito E, Fazzi P. et al. Identification of an allelic variant of the goat PrP gene associated with resistance to scrapie. J Gen Virol. 2006;87:1395–402.
15. Serrano C, Hammouchi M, Benomar A, Lyahyai J, Ranera B, Acin C, el Hamidi M, Monzon M, Badiola J. J. et al. PRNP haplotype distribution in moroccan goats. Anim Genet. 2009;40:565–8.
16. Papasavva-Stylianou P, Windl O, Saunders G, Mavrikiou P, Toumazos P, Kakoyiannis C. PrP gene polymorphisms in cyprus goats and their association with resistance or susceptibility to natural scrapie. Vet J. 2011; 187:245–50.
17. Hussain A, Babar ME, Imran M, Haq IU, Javed MM. Detection of four novel polymorphisms in PrP gene of pakistani sheep (damani and hashtnagri) and goats (kamori and local hairy) breeds. Virol J. 2011;8:246.
18. Goldmann W, Ryan K, Stewart P, Parnham D, Xicohtencatl R, Fernandez N, Saunders G, Windl O, Gonzalez L, Bossers A, Foster J, et al. Caprine prion gene polymorphisms are associated with decreased incidence of classical scrapie in goat herds in the united kingdom. Vet Res. 2011;42:110.
19. Barillet F, Mariat D, Amigues Y, Faugeras R, Caillat H, Moazami-Goudarzi K, Rupp R, Babilliot J. M, Lacroux C, et al. Identification of seven haplotypes of the caprine PrP gene at codons 127, 142, 154, 211, 222 and 240 in french alpine and saanen breeds and their association with classical scrapie. J Gen Virol. 2009;90:769–76.
20. Acin C, Martin-Burriel I, Monleon E, Lyahyai J, Luis Pitarch J, Serrano C, Monzon M, Zaragoza P, Jose Badiola J, et al. Prion protein gene variability in Spanish goats. Inference through susceptibility to classical scrapie strains and pathogenic distribution of peripheral PrPsc. Plos One. 2013;8, e61118.
21. Corbiere F, Perrin-Chauvineau C, Lacroux C, Costes P, Thomas M, Bremaud I, Martin S, Lugan S, Chartier C. et al. PrP-associated resistance to scrapie in five highly infected goat herds. J Gen Virol. 2013;94:241–5.
22. Lacroux C, Perrin-Chauvineau C, Corbiere F, Aron N, Aguilar-Calvo P, Maria Torres J, Costes P, Bremaud I, Lugan S. et al. Genetic resistance to scrapie infection in experimentally challenged goats. J Virol. 2014;88:2406–13.
23. Kanata E, Humphreys-Panagiotidis C, Giadinis ND, Papaioannou N, Arsenakis M, Sklaviadis T. Perspectives of a scrapie resistance breeding scheme targeting Q211, S146 and K222 caprine PRNP alleles in greek goats. Vet Res. 2014;45:43.
24. Ortiz-Pelaez A, Georgiadou S, Simmons MM, Windl O, Dawson M, Arnold ME, M. E, Neocleous P, Papasavva-Stylianou P, et al. Allelic variants at codon 146 in the PRNP gene show significant differences in the risk for natural scrapie in cypriot goats. Prion. 2014;8:65–6.
25. Aguilar-Calvo P, Andreoletti O, Fast C, Espinosa J, Lacroux C, Groschup MH, H, Maria Torres J, et al. Influence of prnp variants on the susceptibility of goats to scrapie and BSE. Prion. 2014;8:15–6.
26. Goldmann W, Martin T, Foster J, Hughes S, Smith G, Hughes K, Dawson M, Hunter N, et al. Novel polymorphisms in the caprine PrP gene: a codon 142 mutation associated with scrapie incubation period. J Gen Virol. 1996;77:2885–91.
27. Mitchell GB, O'Rourke KI, Harrington NP, Soutyrine A, Simmons MM, Dudas S, Zhuang D, Laude H, Balachandran A, et al. Identification of atypical scrapie in Canadian sheep. J Vet Diagn Invest. 2010;22:408–11.
28. Tamura K, Dudley J, Nei M, Kumar S. MEGA4: molecular evolutionary genetics analysis (MEGA) software version 4.0. Mol Biol Evol. 2007;24:1596–9.
29. Bouzalas IG, Dovas CI, Banos G, Papanastasopoulou M, Kritas S, Oevermann A, Papakostaki D, Evangelia C, Papadopoulos O, Seuberlich T, Koptopoulos G, et al. Caprine PRNP polymorphisms at codons 171, 211, 222 and 240 in a Greek herd and their association with classical scrapie. J Gen Virol. 2010;91:1629–34.
30. Fragkiadaki EG, Vaccari G, Ekateriniadou LV, Agrimi U, Giadinis ND, Chiappini B, Esposito E, Conte M, Nonno R, et al. PRNP genetic variability and molecular typing of natural goat scrapie isolates in a high number of infected flocks. Vet Res. 2011;42:104.
31. Baylis M, Chihota C, Stevenson E, Goldmann W, Smith A, Sivam K, Tongue S, Gravenor M, et al. Risk of scrapie in British sheep of different prion protein genotype. J Gen Virol. 2004;85:2735–40.
32. Tongue SC, Pfeiffer DU, Warner R, Elliott H, Vilas VD. Estimation of the relative risk of developing clinical scrapie: the role of prion protein (PrP) genotype and selection bias. Vet Rec. 2006;158:43.
33. Arnold M, Ortiz-Pelaez A. The evolution of the prevalence of classical scrapie in sheep in great Britain using surveillance data between 2005 and 2012. Prev Vet Med. 2014;117:242–50.
34. Dubois MA, Sabatier P, Durand B, Calavas D, Ducrot C, Chalvet-Monfray K. Multiplicative genetic effects in scrapie disease susceptibility. C R Biol. 2002;325:565–70.

35. Rezaei H, Choiset Y, Eghiaian F, Treguer E, Mentre P, Debey P, Grosclaude J, Haertle T, et al. Amyloidogenic unfolding intermediates differentiate sheep prion protein variants. J Mol Biol. 2002;322:799–814.
36. Eghiaian F, Grosclaude J, Lesceu S, Debey P, Doublet B, Treguer E, Rezaei H, Knossow M, et al. Insight into the PrPC-> PrPSc conversion from the structures of antibody-bound ovine prion scrapie-susceptibility variants. Proc Natl Acad Sci U S A. 2004;101:10254–9.
37. Eiden M, Soto EO, Mettenleiter TC, Groschup MH. Effects of polymorphisms in ovine and caprine prion protein alleles on cell-free conversion. Vet Res. 2011;42:30.
38. Nodelijk G, van Roermund HJW, van Keulen LJM, Engel B, Vellema P, Hagenaars TJ. Breeding with resistant rams leads to rapid control of classical scrapie in affected sheep flocks. Vet Res. 2011;42:5.
39. Hagenaars TJ, Melchior MB, Bossers A, Davidse A, Engel B, van Zijderveld FG. Scrapie prevalence in sheep of susceptible genotype is declining in a population subject to breeding for resistance. BMC Vet Res. 2010;6:25.

7

Histamine activates inflammatory response and depresses casein synthesis in mammary gland of dairy cows during SARA

Guangjun Chang[†], Lailai Wang[†], Nana Ma, Wenwen Zhang, Huanmin Zhang, Hongyu Dai and Xiangzhen Shen[*]

Abstract

Background: Mounting evidences observed that subacute ruminal acidosis (SARA) induced by high concentration (HC) diet increases the translocation of histamine from digestive tract into circulation causing a diverse of diseases in dairy cows. However, it is largely unknown how it does affect the function of mammary gland and milk quality. Hence, this study aims to observe the effects of histamine derived from the digestive tract on the inflammatory response and casein synthesis in the mammary glands during SARA. Twelve cows fitted rumen fistula were randomly divided into either control group administrated low concentration (LC) diet (60% forage, $n = 6$) or treatment group administrated HC diet (40% forage, $n = 6$) for 18 weeks.

Results: Our data showed that HC diet resulted in significant declines in rumen pH value, milk yield and milk quality, as well as longer duration of averaged pH value below 5.6 per day (more than 180 min) compared to LC diet, these findings confirmed SARA occurence. Our study also observed that SARA increased the content of histamine in rumen fluid, plasma, liver and mammary gland, and enhanced the mRNA expression of histamine specific receptor in the mammary gland. Additionally, we found that the mRNA expression of inflammatory response genes in mammary glands was increased, which was consistent with the protein expression results, showing that the protein kinase C(PKC) / nuclear factor kappa B (NF-κB) or protein kinase A (PKA) / NF-κB signalling pathways of the inflammatory response were activated. The mRNA expression of mTOR, P70S6K and αS1 in mammary glands were significantly decreased with the protein expression of mTOR, P70S6K and αS1-casein, and the phosphorylation levels of the mTOR and P70S6K proteins were also decreased.

Conclusions: Our study showed that the milk protein of lactating cows is depressed after long-term feeding of HC at the individual level, which was paralleled at the gene and protein levels. The inflammatory response in mammary gland caused by histamine derived from the digestive tract is related to the decline of casein synthesis. Our findings point to a new link between the inflammatory response and casein synthesis, but the understanding of the molecular mechanisms involved in this process will require further research.

Keywords: Mammary gland, Histamine receptor, NF-κB and mTOR signalling pathway, Casein, Inflammatory response

Background

Subacute ruminal acidosis (SARA) that is a common and frequent metabolic disorder disease can bring tremendous economic losses to dairy industry. The most important reason for SARA occurrence is that to maintain a high yield of milk, the nutrient density of dairy cow diets has increased [1]. The adjustment of the nutrient density is mainly through the feeding of HC diets. This feeding mode can increase the production of organic acids and cause a depression of the pH in the digestive tract over time [2, 3]. When the rumen pH value of cows falls below 5.6 for more than 3 h per day, SARA occurs. This decrease in pH may result in alterations in the type of fermentation and the composition of the microbiota in the rumen [4]. When the rumen has a low pH, the different types of bacteria make histidine for decarboxylization to produce histamine [5]. Histamine that is absorbed from the gastrointestinal tract can be

* Correspondence: xzshen@njau.edu.cn
Guangjun Chang and Lailai Wang contributed equally to this work.
College of Veterinary Medicine, Nanjing Agricultural University, Nanjing, Jiangsu, China

rapidly detoxified by amine oxidases in healthy subjects, and it also can be produced by epithelial cell to regulate many physiological processes, such as cell growth and differentiation [6]. The acid-induced impairment of epithelial barrier function facilitates the translocation of histamine from the digestive tract into the bloodstream [7]. The increase of histamine in blood can cause systemic histaminosis such as laminitis, which is a popular dieses in dairy farm [1], and it is also closely associated with inflammation and lactation of mammary gland [6, 8], those observations suggested that histamine originated from gastrointestinal tract entering mammary gland may elicits local inflammation, and then affects lactation of mammary gland and milk quality.

Histamine (2-[4-imodazole]-ethylamine, HIS) is synthesized by the decarboxylation of histidine by L-histidine decarboxylase (HDC), which is dependent on the cofactor pyridoxal-5′-phosphate [9]. After histamine interacts with its receptors (H_sR), of which there are four subtypes (H_1R, H_2R, H_3R and H_4R) that all belong to the family of G protein-coupled receptors, it will exert many (patho-) physiological effects [10]. H_1R-coupled $G\alpha_{q/11}$ can generate inositol-1, 4, 5-triphosphate and diacylglycerol and increase the intracellular calcium content through stimulating the inositol phospholipid signalling pathways, which has been demonstrated to activate NF-κB [11, 12]. H_2R-coupled $G\alpha_s$ can, through separate guanosine triphosphate-dependent mechanisms, activate adenylate cyclase and the phosphoinositide second messenger-cAMP to produce gastric acid and mucus [13, 14]. H_3R is mainly present in nervous system signal transduction, and it can lead to the inhibition of cAMP, the accumulation of Ca^{2+} and the activation of the mitogen-activated protein kinase pathway to release neurotransmitters when it is coupled with $G\alpha_{i/o}$ [15]. H_4R has been reported to couple with $G\alpha_{i/o}$ and play an important role in binding sites for interferon regulatory factor and NF-κB in the modulation of the immune system [16, 17].

Milk is a mixture of several ingredients, including water, fat, protein, lactose and inorganic salt. Meanwhile, casein is a complex particle containing calcium and phosphateand accounts for 80%~ 82% of the total amount of protein in milk. Casein has the following four groups: αS1-casein, αS2-casein, β-casein and κ-casein [18]. In dairy cows, the most abundantly expressed milk protein is αS1- casein. Janus kinase (JAK), the signal transducer and activator of transcription (STAT) signalling pathway and the mTOR signalling pathway are the primary pathways in milk protein expression. However, it has been reported that the role of the JAK-STAT signalling pathway is weak in regulating bovine milk protein expression [19], because the synthesis of αS1-casein is translational but not transcriptionally regulated. Previous studies have demonstrated that histamine can induce inflammation; however, it is unknown whether histamine derived from the digestive tract during SARA is capable of influencing inflammatory responses and casein synthesis in the mammary gland of dairy cows. Therefore, we hypothesized that the present study could investigate the effects of histamine derived from the digestive tract during SARA on the inflammatory response and casein synthesis in the mammary gland of lactating dairy cows fed high concentration (HC) diets.

Methods

All experimental procedures were approved by the Animal Experiment Committee of Nanjing Agricultural University, in accordance with the Regulations for the Administration of Affairs Concerning Experimental Animals (The State Science and Technology Commission of China, 1988). All the experimental protocols were performed in accordance with the approved guidelines and regulations.

Animals, diets, and experimental design

Twelve multiparous lactating Holstein cows were housed in individual tie- stalls in dairy farm of Nanjing Agricultural University and fitted with a rumen fistula (3–5 weeks post-partum, aged 3–5 years and averaged body weight 460 ± 25 kg) displayed an averaged milk yield 29 ± 3.17 kg/day at the beginning and 27 ± 2.76 kg/day at the end of the present experiment. All cows were fed low concentration (LC) diet for 1 month to reach a similar metabolic state, and then randomly divided into two groups. One group ($n = 6$) was administrated HC diet comprising 60% concentrate and 40% forage as treatment group, while another group ($n = 6$) was fed LC diet comprising 40% concentrate and 60% forage as control group. The experiment was lasted for 18 weeks. The details of the dietary ingredients and nutritional composition are presented in Table 1. Cows had fresh water for free drinking and were milked and fed at 04:00, 12:00, and 20:00 during the experiment.

Rumen pH and milk measurements

The pH value was measured by a pH-meterat1 h intervals for 24 h on the last 3 days of 18th week. Milk yield was measured at each of the 3 milkings on experimental days of the 1^{st}week and 18th week. Milk sample (50 mL) was also taken from the mixture of 3 milkings of each cow on these days, and then it was added potassium dichromate and stored at 4 °Cfor the measurement of milk protein and fat with the Milko Scan™ FT1device (Foss, Denmark). At 4 h after feeding on the last day of 18th week, Rumen content and fluid were mixed, and then the mixture was taken via rumen fistula, and

Table 1 Ingredients of low concentrate and high concentrate diet

Item	Ingredient	LC	HC
Forage	Corn silage	30.00	20.00
	Alfalfa hay	30.00	20.00
Concentrate	Corn	22.78	33.60
	Bran	5.15	15.00
	Soybean meal	9.80	9.00
	Calcium phosphate dibasic	0.92	0.53
	Limestone	0.00	0.52
	Salt	0.35	0.35
	Premix[a]	1.00	1.00
	Total	100.00	100.00
Nutrient composition	NE(MJ/Kg)	6.32	6.74
	CP (%)	16.00	16.20
	EE (%)	3.96	4.15
	NDF (%)	37.71	31.92
	ADF (%)	22.75	17.55
	NFC (%)	33.43	40.31
	Starch (%)	25.33	32.28
	Ca (%)	0.90	0.80
	P (%)	0.45	0.45

[a]Premix contained VA 1900 ku/kg; VD 250 ku/kg; VE 3000 mg/kg; Niacin 4000 mg/kg; Cu 1200 mg/kg; Fe 525 mg/kg; Zn 13,000 mg/kg; Mn 5500 mg/kg; I 170 mg/kg; Co 50 mg/kg; Se 27 mg/kg
LC low concentration diet, *HC* high concentration diet

squeezed to obtain rumen fluid (50 mL), which subsequently was filtered by four layer gauze. The filtrate was centrifuged at 10000×g for 30 min, and then the supernatant was stored at − 20 °C for further histamine determination.

Tissue and blood samples collection

On the last 3 days of the 18thweek, Blood were collected 4 h after feeding via the lacteal artery into 5-ml heparinised vacuum tubes, and plasma was isolated by centrifugation at 1900×g for 15 min and stored at − 20 °C for histamine determination. Approximate 10 g of Mammary gland tissue of each cow was harvested by biopsy from a deeper location of the right hind quarter. One small portion (about 0.5 g) supplemented with phosphate buffer saline (PBS, 0.5 mL) was homogenized using a Dounce homogenizer and centrifuged at 3000×g for 10 min, and then the supernatant was stored at − 20 °C for histamine determination. Another portion of mammary gland tissue was stored at − 80 °C. After sampling the mammary gland, the cows were slaughtered, and the liver tissue was collected aseptically within 5 to 10 min after killing, then liver tissue samples were cut into several smaller tissue pieces (about 0.5 g), which was prepared following the processing protocol of mammary gland for histamine determination and stored at. -20 °C.

Enzyme-linked immunosorbent assay

Histamine concentrations in rumen fluid, plasma, liver and mammary gland tissue were determined with a double-antibody sandwich enzyme-linked immunosorbent assay (cat.H171, JIANCHENG, China). According to the instruction of manufacturer, diluted histamine standard (1600 ng/mL to 100 ng/mL) and prepared samples was added to the monoclonal antibody enzyme well, which was pre-coated with bovine HIS monoclonal antibody, followed by incubation. Then, histamine antibody labelled with biotin was added and combined with Streptavidin-HRP to form an immune complex. The plate was incubated and washed 5 times to remove the uncombined enzyme. After the addition of Chromogen Solution A and B, the plate was read by the Microplate Reader (Bio-TekELX80, USA) to obtain the optical density value (OD value). The standard curve was made according to the OD value and corresponding concentration. Histamine concentration in the samples was calculated by the standard curve.

RNA extraction, cDNA synthesis and real-time polymerase chain reaction (PCR)

Total RNA was extracted using TRIzol following the instructions of the kit (cat.9108, Takara, Dalian, China). The concentration and the quality of the RNA were detected using a NanoDrop ND-1000 Spectrophotometer (ThermoFisher Scientific Inc., Waltham, USA). 400 ng of the total RNA was used to synthesize cDNA through the PrimeScript RT Master Mix Perfect Real Time kit (cat.RR036A, Takara, Dalian, China). The real-time PCR was conducted using the SYBR Premix Ex TaqTM kit (cat. DRR420A, Takara, Dalian, China) on an ABI 7300 Real-Time PCR System (Applied Biosystems, Foster City, CA, USA). The PCR protocol was denaturation at 95 °C for 15 s, then 40 cycles at 95 °C for 5 s, and 60 °C for 31 s. Glyceraldehyde phosphate dehydrogenase (GAPDH) was used as the internal reference gene. The primers were designed using Primer Premier Software 5.0 (Premier Biosoft International, USA) and synthesized by Generay Company (Shanghai, China), and the primer details are shown in Table 2. The $2^{-\Delta\Delta}$Ct method was used to analyse the results.

Western blotting

Total protein was extracted from frozen mammary gland tissues. One hundred mg of frozen minced mammary tissues and 1 mL ice-cold RIPA lysis buffer (cat. SN338, Sunshine Biotechnology Co., Ltd.,

Table 2 The primer list for reverse transcription and amplification of RT-qPCR

Gene	NCBI Accession	PCR product	Forward Primer (5'-3')	Reverse Primer (5' -3')
H_1R	NM_174083.4	245 *bp*	AGCCAGAACCAGCTTGAGAT	TTCATGTGCAAGCCAGACAC
H_2R	XM_015464941.1	198 *bp*	TTGGCAAGGTCTTCTGCAAC	GGAAGGACAGGGTGATGGAA
H_3R	XM_015466045.1	172 *bp*	CAGAAGATGGTGCTGGTGTG	AAGGGGTGAAGAACTCGAGG
H_4R	XM_001251983.5	221 *bp*	ATCCTTGCCCTCACGTTAGT	CGAAGGGATGCTGCTGATTC
IL-1β	NM_174093.1	167 *bp*	GGCCAAAGTCCCTGACCTCT	CTGCCACCATCACCACATTC
TNF-α	NM_173966.3	261 *bp*	CACATACCCTGCCACAAGGC	CTGGGGACTGCTCTTCCCTCT
NF-κB	NM_001045868.1	129 *bp*	ATACGTCGGCCGTGTCTAT	GGAACTGTGATCCGTGTAG
mTOR	XM_001788228.1	199 *bp*	ATGCTGTCCCTGGTCCTTATG	GGGTCAGAGAGTGGCCTTCAA
P70S6K	NM_205816.1	162 *bp*	GGACATGGCAGGGGTGTTT	GGTATTTGCTCCTGTTACTTTTCG
4EBP1	BC120290.1	177 *bp*	GGCAGGCAGTGAAGAGTC	CCTGGGCTGCGGGAT
αS1-casein	NM_181029.2	170 *bp*	ATTTTCAGACAATTCTACCAGCT	AATTCACTTGACTCCTCACCAC
GAPDH	NM_001034034.2	135 *bp*	TGTTGTGGATCTGACCTGCC	AAGTCGCAGGAGACAACCTG

Shanghai, China) were mixed and homogenized using a Dounce homogenizer. The supernatant was collected after centrifugation at 12,000 rpm for 15 min at 4 °C. The protein concentration was measured using a BCA Protein Assay kit (cat.23225, ThermoFisher, USA) and then diluted to the same concentration. Fifty μg of protein from each sample was loaded onto 8~12% sodium dodecyl sulphate polyacrylamide gel electrophoresis (SDS-PAGE) gels. After separation, the proteins were transferred to a nitrocellulose (NC) membrane (Pall Gelman Laboratory, USA). The membrane was blocked with 6~7% nonfat dry milk (dissolved in Tris-Buffered Saline with Tween, TBST) or 6~7% BSA (for phosphorylated protein) for 2 h at room temperature. Afterwards, the NC membrane was washed and incubated with TBST-diluted primary antibody at 4 °C overnight. The primary antibodies were used in previous study [20], and diluted at the following ratios: NF-κB p65 (1:500;cat.AN365, Beyotime, China), phosphorylated NF-κB p65 (1:500; cat.AN371, Beyotime, China), IL-6 (1:200; cat.sc-1265, Santa Cruz, USA), GAPDH (1:5000; cat.AP0066, Bioworld, USA), αS1-casein (1:400; cat.sc-376,961, Santa Cruz, USA), mTOR (1:500; cat.sc-8319, Santa Cruz, USA), phosphorylated mTOR (1:500; cat.sc-101,738, Santa Cruz, USA), P70S6K (1:500; cat.sc-9027, Santa Cruz, USA) and phosphorylated P70S6K (1:500; cat. sc-7984R, Santa Cruz, USA). The membranes were then washed and incubated with the corresponding secondary antibodies for 2 h at room temperature. The secondary antibodies were as follows: goat anti-rabbit (1:10000; cat. sc-2004, Santa Cruz, USA) for NF-κBp65, NF-κBpp65, IL-1β, P70S6K, mTOR, pP70S6K, and pmTOR; goat anti-mouse (1:10000; cat.sc-2005, Santa Cruz, USA) for αS1-casein; rabbit anti-goat (1:10000; cat.sc-2768, Santa Cruz, USA) for GAPDH. Finally, the membranes were treated with an enhanced chemiluminescence (ECL) detecting kit (E411–04, Vazyme Biotech Co.,Ltd., Nanjing, China). The signals were recorded through an imaging system (LAS4000, USA), and the results were analysed using Quantity One software (Bio-Rad, USA).

Statistical analysis

Data from milk and rumen pH was analyzed using repeated model of MIXED procedure of SAS (SAS version 9.4, SAS Institute Inc.). The effects of diet and days were considered as fixed factors, and goat was treated as random factor. Days with diet and goat was considered as repeated measure, and compound symmetry (CS) was used as the type of covariance. Data about histamine concentration, mRNA and protein expression was analyzed using ANOVA procedure of SAS. The significant difference was considered to occur when $P < 0.05$.

Results

Ruminal pH, milk yield and milk composition

Compared with LC diet, HC diet was significantly decreased the mean rumen pH value and increased the duration of rumen pH < 5.6/day, which was longer than 180 min per day after 18th week feeding.(Table 3). Additionally, the feeding of HC diet caused an increased tendency in milk yield and significant increase in milk fat in the 1st week, whereas resulted in significant decline in milk yield, milk protein and milk fat in the 18^{th}week compared to the feeding of LC diet.

Histamine content in rumen fluid, blood, liver tissues and mammary gland tissues in dairy cows

The mean histamine values in rumen fluid, blood, liver tissue and mammary gland tissue were measured and the results presented in Fig. 1 show that the values in the four tissues were significantly different between the LC and HC groups. The histamine content in the rumen

Table 3 Rumen pH, milk yield and composition of dairy cows fed LC or HC diets

Item	LC	HC	SEM	*p*-Value		
				Diet	Day	Diet × Day
Mean pH value	6.02	5.90	0.03	< 0.01	0.23	0.18
Minimum pH value	5.65	5.50	0.05	0.02	0.08	0.38
Time < pH 5.6, min/d	99.00	223.00	30.25	0.01	0.06	0. 57
Milk yield (kg/d) week1	28.7	30.9	0.47	0.08	0.23	0.76
Milk yield (kg/d) week18	28.05	26.92	0.30	< 0.01	0.05	0.41
Milk protein (%) week1	3.34	3.41	0.03	0.11	0.53	0.79
Milk protein (%) week18	3.02	2.88	0.04	0.01	0.51	0.31
Milk fat (%) week1	3.48	3.34	0.02	0.21	0.61	0.58
Milk fat (%) week18	3.54	2.94	0.04	0.02	0.12	0.81

fluid and blood of the HC group was significantly increased ($P < 0.01$; Fig. 1a) compared to the LC group. The histamine content in the liver tissues was higher compared with the liver tissues, and the differences in the liver tissues and mammary tissues between the LC and HC group were both significant ($P < 0.01$; Fig. 1b).

mRNA expression of histamine receptors in mammary gland tissues

The results showed that the different types of histamine receptors in the mammary gland tissues had heterogeneous expression, of which H_4R had the highest expression, while the expression of H_3R was the lowest (Fig. 2). H_1R, H_2R and H_4R had significantly different expressions in the LC group and HC group ($P < 0.01$). There was not a significant difference in H_3R expression between the LC and HC groups.

mRNA expression of inflammatory response genes in mammary gland tissues

The results showed that NF-κB mRNA expression in mammary tissues in the HC group was significantly high ($P < 0.01$) compared to the LC group (Fig. 3a). mRNA expression of tumour necrosis factor (TNF)-α and interleukin(IL)-1β were significantly up-regulated in the HC group compared with the LC group ($P < 0.05$).

mRNA expression of casein synthesis genes in mammary gland tissues

As shown in Fig. 3b, the mRNA expression of casein synthesis genes was changed by HC feeding. The mRNA expression of mTOR genes was significantly decreased in the HC group in mammary gland tissues compared with the LC group ($P < 0.01$). The mRNA expression of the S6K genes was significantly decreased in the HC group, but the mRNA expression of the 4EBP1 genes was markedly increased ($P < 0.01$) in the HC group. Importantly, the αS1-casein genes in the HC group had significantly decreased mRNA expression compared with the LC group in mammary gland tissues.

Protein expression of inflammatory factors in mammary gland tissues

Western blotting results (Fig. 4a) showed that the NF-κBp65 and NF-κB *p*p65protein in mammary glands were

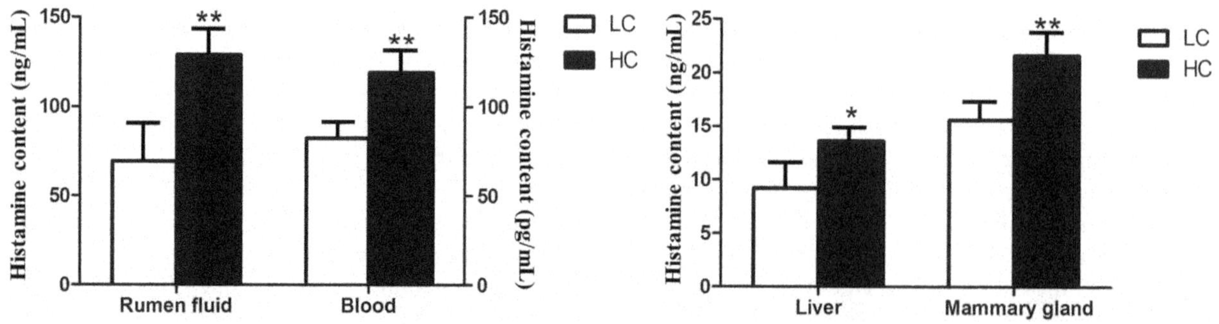

Fig. 1 Histamine concentrations in rumen, blood, liver and mammary tissues in the LC and HC group. The ordinate axis indicates the absolute histamine content in tissue ($n = 6$ in each group, mean ± SEM). The white packed column represents the LC group, while the black packed column represents the HC group. The significant differences are indicated (*$P < 0.05$, **$P < 0.01$)

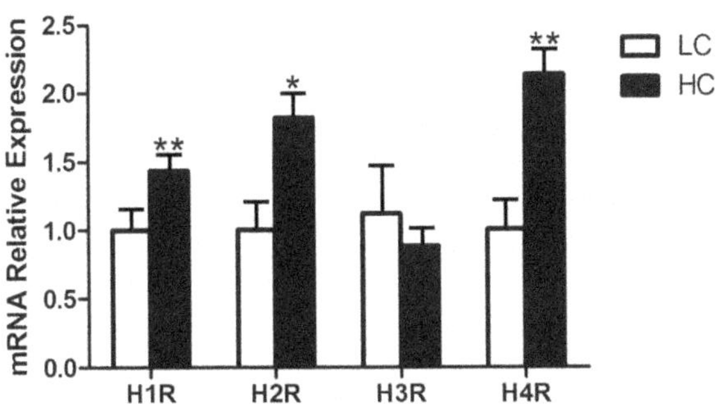

Fig. 2 mRNA expressions of the histamine receptor genes H_1R, H_2R, H_3R and H_4R in the mammary tissues of dairy cows in the LC and HC groups. The ordinate axis indicates the relative expression of histamine receptors in mammary gland tissues ($n = 6$ in each group, mean ± SEM). The white packed column represents the LC group, while the black packed column represents the HC group. The significant differences are indicated (*$P < 0.05$, **$P < 0.01$)

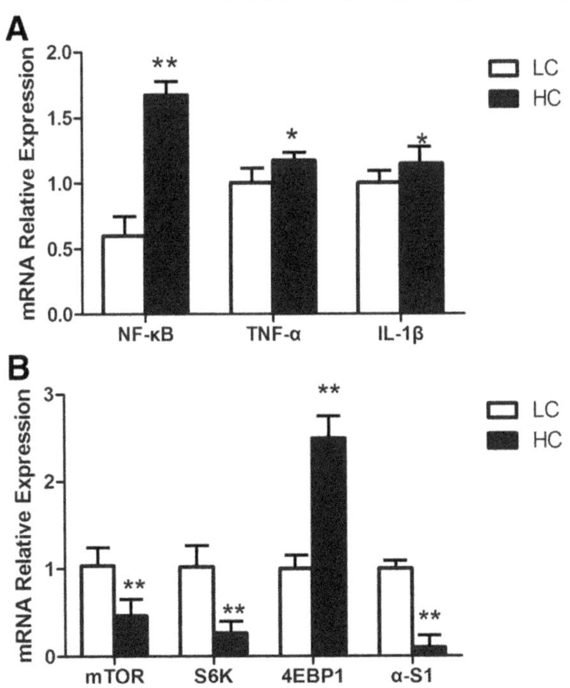

Fig. 3 mRNA expressions of inflammatory genes and casein synthesis in the mammary glands of dairy cows in the LC and HC groups. **a** shows the mRNA expression of inflammatory response genes in mammary gland tissue, while (**b**) shows the mRNA expression of casein synthesis genes. The ordinate axis indicates the relative expression of the genes in mammary gland tissues ($n = 6$ in each group, mean ± SEM). The white packed column represents the LC group, while the black packed column represents the HC group. The significant differences are indicated (*$P < 0.05$, **$P < 0.01$)

both significantly increased ($P < 0.01$) in the HC group. The significant increase of IL-1β ($P < 0.01$) further demonstrated the inflammation of the mammary gland tissues in the HC group.

Protein expression of casein synthesis in mammary gland tissues

Western blotting results (Fig. 4b) showed that the expression of the αS1-casein, P70S6K, and mTOR proteins in the mammary glands were all decreased in the HC group compared with the LC group ($P < 0.01$). Meanwhile, the phosphorylated proteins *p*P70S6K and *p*mTOR were both down-regulated in the HC group compared with the LC group ($P < 0.01$). The changes of protein levels were consistent with the change in gene expressions. The original western blotting bands for each target protein were shown in Additional file 1.

Discussion

SARA effects are caused by a combination of ruminal pH and diet type [21]. The pH results of the present study demonstrated that the dairy cows had experimentally SARA at the end of experiment through the feeding of an HC diet for 18 weeks, as the defined by the definition of previous study [22]. When SARA occurs, the histamine produced by rumen bacteria can easily pass through the damaged rumen epithelium into the bloodstream and cause inflammation [23, 24].

In fact, the concentrations of histamine in the rumen and plasma both increase during SARA. However, the absorption of histamine in the rumen is the direct cause of the increase in the concentration of histamine in the blood. Under normal conditions, dairy cows contain trace amounts of histamine [23], and when histamine is synthesized, it usually combines with polysaccharide sulphate, heparin and chondroitin sulphate to form inactive complexes stored in mast cell granules. The trace amount of histamine generally undergoes methylation or oxidation in the liver, where it is rapidly converted to a non-active substance in the urine. The results of the present study found that the histamine content is higher in the tissue of the mammary gland than in the tissue of the liver, which may be related to liver clearance. Meanwhile, the results illustrated that the liver has a higher tolerance to histamine than the mammary gland, as the mammary gland is more susceptible to the level of histamine [25].

Histamine can cause inflammation through the NF-κB signalling pathway, and this function is realized through combination with its receptors [26]. H_1R-coupled$G\alpha q_{/11}$, H_3R-coupled $G\alpha_{i/o}$ and H_4R-coupled $G\alpha_{i/o}$ can activate PKC and H_2R-coupled $G\alpha_s$ can activate PKA, both of which can further activate NF-κB through the classical pathway or the

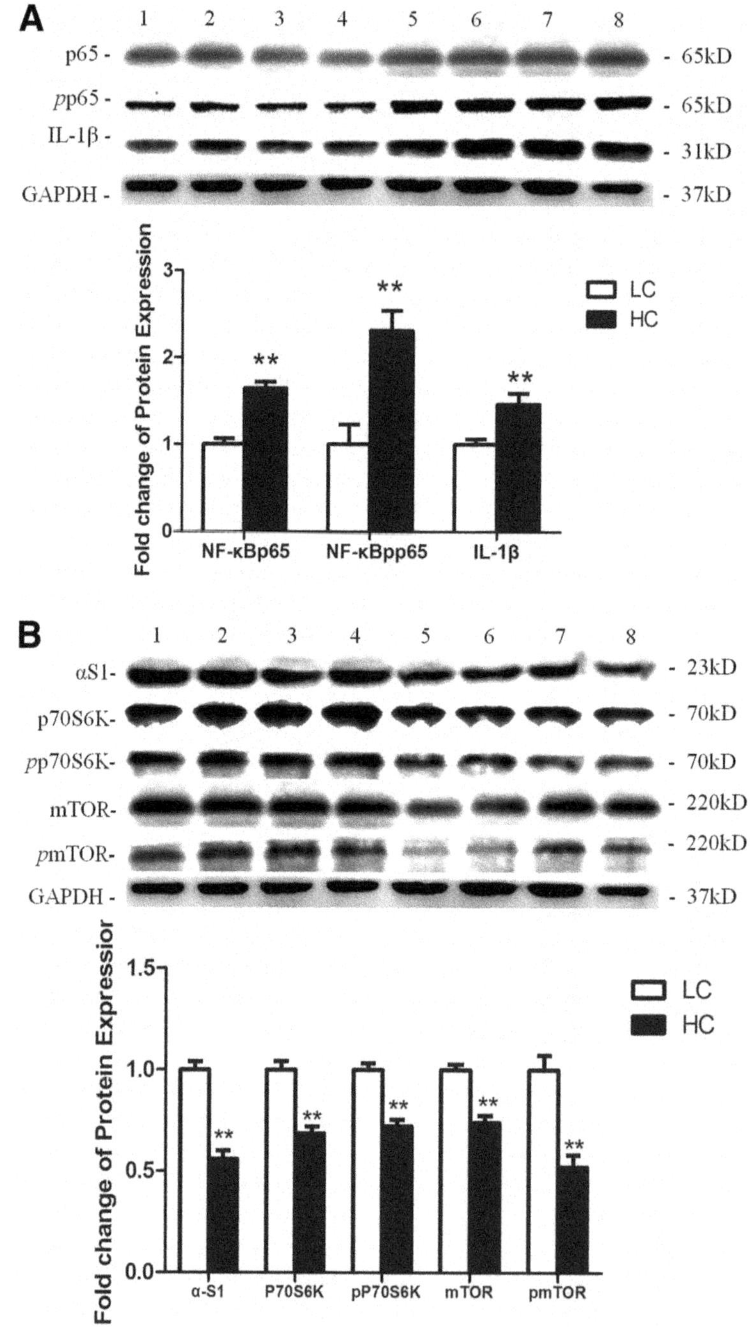

Fig. 4 Protein expressions of several inflammatory factors and casein synthesis in the mammary glands of dairy cows in the LC and HC groups. **a** shows the protein expression of several inflammatory factors in mammary gland tissue, while (**b**) shows expression of casein synthesis proteins in the mammary glands of dairy cows in the LC and HC groups. The samples used for western blot derived from the same experiment and the blots for target proteins were conducted in parallel. GAPDH was used as an internal reference protein, and the band density was used to quantify protein expression. The cropped bands 1–4 bands represent the LC group, while 5–8 bands represent the HC group. The significant differences are indicated (**$P < 0.01$)

alternative pathway [27]. Most inflammatory genes contain NF-κB binding sites within their promoters, and therefore once NF-κB has been activated, it will lead to the secretion of proinflammatory cytokines such as TNF-α and IL-1β [28, 29]. Our research found that three histamine receptors, H_1R, H_2R and H_4R, were significantly increased during SARA, which may be closely related to their function. Histamine activation of H_1R can change vascular permeability by acting on vascular endothelial cells and causing contraction of vascular smooth muscle cells; these features may exacerbate the inflammatory response [30]. Meanwhile, the expression of H_1R genes can be up-regulated by histamine [31]. Compared with H_1R, the absence of histamine can down-regulate the expression of H_2R genes in a tissue-specific manner [32]. However, unlike H_1R, histamine activation of H_2R has effects on the relaxation of smooth muscle cells in the blood vessels and inhibition of cytokine production [33]. Recent work has highlighted the major role of H_4R in eosinophils, mast cells, neutrophils, T-cells and dendritic cells as well as the production of cytokines influenced by these cells, which confirms the importance of this receptor in inflammatory responses [34–37]. We also found that H_3R expression was not different between the LC and HC groups. According to previous studies, H_3R is an autoreceptor regulating histamine release, and a knock-out of H_3R in mice resulted in an increased severity of neuro-inflammatory diseases with enhanced cytokine release [38, 39]. The mRNA expression of NF-κB, TNF-α and IL-1β genes was significantly increased in the HC group, which was consistent with the results of the protein expression levels of NF-κBp65, NF-κBpp65 and IL-1β in the HC group. This showed that the PKC- NF-κB or PKA- NF-κB signalling pathway in the HC group was activated.

It has been reported that the phosphatidyl inositol 3 kinase(PI3K)-serine and threonine protein kinase(Akt)-mTOR signalling pathway plays a major role in casein synthesis. PI3K is activated to convert phosphoinositide 4, 5-bisphosphate (PIP2) into phosphoinositide 3, 4, 5-triphosphate (PIP3) [40]. Then, the growth factor activated kinase Akt is phosphorylated by the products of PI3K in the activation loop (Thr308) and recruited to the plasma membrane [41]. Accordingly, activated Akt can phosphorylate the negative regulator tuberous sclerosis protein-2 (TSC2) and dissociate the TSC complex into TSC1 and TSC2 through the lysosome [42]. Ras homologue enriched in brain (Rheb) is a small guanosine triphosphatase (GTPase) that can activate mTORC1 in its GTP-loaded state, while the GTPase-activating protein (GAP) activity of TSC2 hydrolyses Rheb-GTP into Rheb-GDP to inactivate mTORC1 [43]. The activated mTORC1 can phosphorylate protein translational regulators P70S6K and eukaryotic cells and translation initiation 4 e binding protein 1(4EBP1) [44]. Phosphorylated P70S6K can stimulate cell growth, while the function of 4EBP1 to stimulate protein translation is inhibited when it is phosphorylated. We found that the mRNA expression of mTOR, P70S6K and α-S1 was significantly decreased in the HC group, in agreement with the protein expression of mTOR, P70S6K and α-S1. Meanwhile, the phosphorylation level of mTOR and P70S6K protein in the HC group was also decreased. Collectively, the present study indicated that histamine derived from rumen caused the inflammatory response and reduction of casein synthesis in the mammary gland via activating the NF-κB and mTOR signalling pathway during SARA. It is well known SARA can cause diverse and complex consequences [1], and the present result is only one of consequences and may provide a new way to control inflammatory response of mammary gland elicited by SARA.

Conclusion

Our research showed that the milk protein of lactating cows is depressed after long-term feeding of HC at the individual level, which was paralleled at the gene and protein levels. The inflammatory response caused by histamine derived from the rumen is related to reduce casein synthesis. Our findings suggest new relationships between the inflammatory response and casein synthesis, but the understanding of the molecular mechanisms involved in this process will require further research.

Abbreviations

HC: High concentration; IL-1β: Interleukin-1β; JAK: Janus kinase; LC: Low concentration; NF-κB: Nuclear factor kappa B; PKA: Protein kinase A; PKC: Protein kinase C; SARA: Subacute ruminal acidosis; STAT: The signal transducer and activator of transcription; TNF-α: Tumour necrosis factor-α

Funding

This study was supported by grants from the National Natural Science Foundation of China (31672618 and 31702301), and the Priority Academic Program Development of Jiangsu Higher Education Institutions (PAPD).

Authors' contributions

GC and LW carried out this experiment. GC drafted the manuscript, NM and WZ prepared the blood and the tissue samples. LW and HZ contributed to the statistical analyses and drafted the images, XS conceived the idea, and designed the experiment. XS finalized the manuscript. All authors contributed to interpret and discuss the manuscript. All authors read and approved the final manuscript.

Competing interests

The authors declare that they have no competing interests.

References

1. Plaizier JC, Krause DO, Gozho GN, McBride BW. Subacute ruminal acidosis in dairy cows: the physiological causes, incidence and consequences. Vet J. 2008;176(1):21–31.
2. Dong HB, Wang SQ, Jia YY, Ni YD, Zhang YS, Zhuang S, Shen XZ, Zhao RQ. Long-term effects of subacute ruminal acidosis (SARA) on milk quality and hepatic gene expression in lactating goats fed a high-concentrate diet. PLoS One. 2013;8(12):e82850.
3. Li S, Khafipour E, Krause DO, Kroeker A, Rodriguez-Lecompte JC, Gozho GN, Plaizier JC. Effects of subacute ruminal acidosis challenges on fermentation and endotoxins in the rumen and hindgut of dairy cows. J Dairy Sci. 2012;95(1):294–303.
4. Chang GJ, Zhang K, Xu TL, Jin D, Seyfert HM, Shen XZ, Zhuang S. Feeding a high-grain diet reduces the percentage of LPS clearance and enhances immune gene expression in goat liver. BMC Vet Res. 2015;11:11.
5. Koers WC, Britton R, Klopfenstein TJ, Woods WR. Ruminal histamine, lactate and animal performance. J Anim Sci. 1976;43(3):684–91.

6. Wagner W, Ichikawa A, Tanaka S, Panula P, Fogel WA. Mouse mammary epithelial histamine system. J Physiol Pharmacol. 2003;54(2):211–23.
7. Aschenbach JR, Gabel G. Effect and absorption of histamine in sheep rumen: significance of acidotic epithelial damage. J Anim Sci. 2000;78(2):464–70.
8. Gupta GC, Rajora VS, Pachauri SP. Histamine concentration in milk and blood during mastitis in buffaloes. Indian Vet Med J. 1999;23(1):65–6.
9. Endo Y. Simultaneous induction of histidine and ornithine decarboxylases and changes in their product amines following the injection of Escherichia coli lipopolysaccharide into mice. Biochem Pharmacol. 1982;31(8):1643–7.
10. Akdis CA, Blaser K. Histamine in the immune regulation of allergic inflammation. J Allergy Clin Immunol. 2003;112(1):15–22.
11. Aoki Y, Qiu DM, Zhao GH, Kao PN. Leukotriene B-4 mediates histamine induction of NF-kappa B and IL-8 in human bronchial epithelial cells. Am J Physiol-Lung Cell Mol Physiol. 1998;274(6):L1030–9.
12. Hill SJ, Ganellin CR, Timmerman H, Schwartz JC, Shankley NP, Young JM, Schunack W, Levi R, Haas HL. International union of pharmacology .13. Classification of histamine receptors. Pharmacol Rev. 1997;49(3):253–78.
13. Buckland KF, Williams TJ, Conroy DM. Histamine induces cytoskeletal changes in human eosinophils via the H-4 receptor. Br J Pharmacol. 2003;140(6):1117–27.
14. DelValle J, Gantz I. Novel insights into histamine H-2 receptor biology. Am J Physiol-Gastroint Liver Physiol. 1997;273(5):G987–96.
15. Lim HD, van Rijn RM, Ling P, Bakker RA, Thurmond RL, Leurs R. Evaluation of histamine H-1-, H-2-, and H-3-receptor ligands at the human histamine H-4 receptor: identification of 4-methylhistamine as the first potent and selective H-4 receptor agonist. J Pharmacol Experimental Ther. 2005;314(3):1310–21.
16. Zhu Y, Michalovich D, Wu HL, Tan KB, Dytko GM, Mannan IJ, Boyce R, Alston J, Tierney LA, Li XT, et al. Cloning, expression, and pharmacological characterization of a novel human histamine receptor. Mol Pharmacol. 2001;59(3):434–41.
17. Coge F, Guenin SP, Rique H, Boutin JA, Galizzi JP. Structure and expression of the human histamine H-4-receptor gene. Biochem Biophys Res Commun. 2001;284(2):301–9.
18. Groenen MAM, Vanderpoel JJ. REGULATION OF EXPRESSION OF MILK PROTEIN GENES - A REVIEW. Livest Prod Sci. 1994;38(2):61–78.
19. Wheeler TT, Broadhurst MK, Sadowski HB, Farr VC, Prosser CG. Stat5 phosphorylation status and DNA-binding activity in the bovine and murine mammary glands. Mol Cell Endocrin. 2001;176(1–2):39–48.
20. Zhang K, Chang GJ, Xu TL, Xu L, Guo JF, Jin D, Shen XZ. Lipopolysaccharide derived from the digestive tract activates inflammatory gene expression and inhibits casein synthesis in the mammary glands of lactating dairy cows. Oncotarget. 2016;7(9):9652–65.
21. Calsamiglia S, Cardozo PW, Ferret A, Bach A. Changes in rumen microbial fermentation are due to a combined effect of type of diet and pH. J Anim Sci. 2008;86(3):702–11.
22. Gozho GN, Plaizier JC, Krause DO, Kennedy AD, Wittenberg KM. Subacute ruminal acidosis induces ruminal lipopolysaccharide endotoxin release and triggers an inflammatory response. J Dairy Sci. 2005;88(4):1399–403.
23. Silberberg M, Chaucheyras-Durand F, Commun L, Mialon MM, Monteils V, Mosoni P, Morgavi DP, Martin C. Repeated acidosis challenges and live yeast supplementation shape rumen microbiota and fermentations and modulate inflammatory status in sheep. Animal. 2013;7(12):1910–20.
24. Hua CF, Tian J, Tian P, Cong RH, Luo YW, Geng YL, Tao SY, Ni YD, Zhao RQ. Feeding a high concentration diet induces unhealthy alterations in the composition and metabolism of ruminal microbiota and host response in a goat model. Front Microbiol. 2017;8:12.
25. Vesuna F, Raman V. Histamine - a potential therapeutic agent for breast cancer treatment? Cancer Biol Ther. 2006;5(11):1472–3.
26. Mitchell S, Vargas J, Hoffmann A. Signaling via the NF-κB system. Wires Syst Biol Med. 2016;8(3):227–41.
27. Bonizzi G, Karin M. The two NF-kappa B activation pathways and their role in innate and adaptive immunity. Trends Immunol. 2004;25(6):280–8.
28. Mohapatra DK, Reddy DS, Janaki Ramaiah M, Ghosh S, Pothula V, Lunavath S, Thomas S, Pushpa Valli SNCVL, Bhadra MP, Yadav JS. Rugulactone derivatives act as inhibitors of NF-κB activation and modulates the transcription of NF-κB dependent genes in MDA-MB-231cells. Bioorg Med Chem Lett. 2014;24(5):1389–96.
29. Kawai T, Akira S. Signaling to NF-kappa B by toll-like receptors. Trends Mol Med. 2007;13(11):460–9.
30. Akdis M, Verhagen J, Taylor A, Karamloo F, Karagiannidis C, Crameri R, Thunberg S, Deniz G, Valenta R, Fiebig H, et al. Immune responses in healthy and allergic individuals are characterized by a fine balance between allergen-specific T regulatory 1 and T helper 2 cells. J Exp Med. 2004;199(11):1567–75.
31. Kobayashi T, Inoue I, Jenkins NA, Gilbert DJ, Copeland NG, Watanabe T. Cloning, RNA expression and chromosomal location of a mouse histamine H2 receptor gene. Genomics. 1996;37(3):390–4.
32. Muller UR, Jutel M, Reimers A, Zumkehr J, Huber C, Kriegel C, Steiner U, Haeberli G, Akdis M, Helbling A, et al. Clinical and immunologic effects of H1 antihistamine preventive medication during honeybee venom immunotherapy. J Allergy Clin Immun. 2008;122(5):1001–7.
33. Teuscher C, Poynter ME, Offner H, Zamora A, Watanabe T, Fillmore PD, Zachary JF, Blankenhorn EP. Attenuation of Th1 effector cell responses and susceptibility to experimental allergic encephalomyelitis in histamine H-2 receptor knockout mice is due to dysregulation of cytokine production by antigen-presenting cells. Am J Pathol. 2004;164(3):883–92.
34. Varga C, Horvath K, Berko A, Thurmond RL, Dunford PJ, Whittle BJR. Inhibitory effects of histamine H-4 receptor antagonists on experimental colitis in the rat. Eur J Pharmacol. 2005;522(1–3):130–8.
35. Gutzmer R, Diestel C, Mommert S, Kother B, Stark H, Wittmann M, Werfel T. Histamine H4 receptor stimulation suppresses IL-12p70 production and mediates chemotaxis in human monocyte-derived dendritic cells. J Immunol. 2005;174(9):5224–32.
36. de Esch IJP, Thurmond RL, Jongejan A, Leurs R. The histamine H-4 receptor as a new therapeutic target for inflammation. Trends Pharmacol Sci. 2005;26(9):462–9.
37. Daugherty BL. Histamine H-4 antagonism: a therapy for chronic allergy? Brit J Pharmacol. 2004;142(1):5–7.
38. Ferstl R, Akdis CA, O'Mahony L. Histamine regulation of innate and adaptive immunity. Front Biosci. 2012;17:40–53.
39. Liu CL, Wilson SJ, Kuei C, Lovenberg TW. Comparison of human, mouse, rat, and Guinea pig histamine H-4 receptors reveals substantial pharmacological species variation. J Pharmacol Exp Ther. 2001;299(1):121–30.
40. Fruman DA, Meyers RE, Cantley LC. Phosphoinositide kinases. Annu Rev Biochem. 1998;67:481–507.
41. Sarbassov DD, Guertin DA, Ali SM, Sabatini DM. Phosphorylation and regulation of Akt/PKB by the rictor-mTOR complex. Science. 2005;307(5712):1098–101.
42. Menon S, Dibble CC, Talbott G, Hoxhaj G, Valvezan AJ, Takahashi H, Cantley LC, Manning BD. Spatial control of the TSC complex integrates insulin and nutrient regulation of mTORC1 at the lysosome. Cell. 2014;156(4):771–85.
43. Inoki K, Li Y, Xu T, Guan KL. Rheb GTPase is a direct target of TSC2 GAP activity and regulates mTOR signaling. Genes Dev. 2003;17(15):1829–34.
44. Jeong YJ, Cho HJ, Magae J, Lee IK, Park KG, Chang YC. Ascofuranone suppresses EGF-induced HIF-1 alpha protein synthesis by inhibition of the Akt/mTOR/p70S6K pathway in MDA-MB-231 breast cancer cells. Toxicol Appl Pharm. 2013;273(3):542–50.

8

Identifying individual animal factors associated with *Mycobacterium avium* subsp. *paratuberculosis* (MAP) milk ELISA positivity in dairy cattle in the Midwest region of the United States

Gustavo Machado[1,2,4*], Kaushi Kanankege[1], Val Schumann[3], Scott Wells[1], Andres Perez[1] and Julio Alvarez[1]

Abstract

Background: *Mycobacterium avium* subsp. *paratuberculosis* (MAP) is a widespread chronic disease of ruminants that causes severe economic losses to the dairy cattle industry worldwide. The objective of this study was to evaluate the association between individual cow MAP-ELISA and relevant milk production predictors in dairy cattle using data routinely collected as part of quality and disease control programs in the Midwest region of the U.S. Milk ELISA results of 45,652 animals from 691 herds from November 2014 to August 2016 were analyzed.

Results: The association between epidemiological and production factors and ELISA results for MAP in milk was quantified using four individual-level mixed multivariable logistic regression models that accounted for clustering of animals at the farm level. The four fitted models were one global model for all the animals assessed here, irrespective of age, and one for each of the categories of < 4 year-old, 4–8 year-old, and > 8 year-old cattle, respectively.

A small proportion (4.9%; $n = 2222$) of the 45,652 tested samples were MAP-seropositive. Increasing age of the animals and higher somatic cell count (SCC) were both associated with increased odds for MAP positive test result in the model that included all animals, while milk production, milk protein and days in milk were negatively associated with MAP milk ELISA. Somatic cell count was positively associated with an increased risk in the models fitted for < 4 year-old and 4–8 year-old cattle. Variables describing higher milk production, milk protein content and days in milk were associated with significantly lower risk in the models for 4–8 year-old cattle and for all cattle.

Conclusions: Our results suggest that testing cows with high SCC (> 26 × 1000/ml), low milk production and within the first 60 days of lactation may maximize the odds of detecting seropositive animals. These results could be useful in helping to design better surveillance strategies based in testing of milk.

Keywords: Dairy cattle, Mixed-model, *Mycobacterium avium* subsp. *paratuberculosis*, Somatic cell count, ELISA

* Correspondence: gmachad@ncsu.edu
[1]Department of Veterinary Population Medicine, University of Minnesota, St. Paul, MN, USA
[2]Department of Population Health and Pathobiology, College of Veterinary Medicine, North Carolina State University, 1060 William Moore Drive, Raleigh, NC, USA
Full list of author information is available at the end of the article

Background

Mycobacterium avium subsp. *paratuberculosis* (MAP), also known as paratuberculosis, is a chronic, gastrointestinal disease affecting domestic ruminants. MAP is caused by infection with MAP, an agent capable of infecting a wide range of domestic and wild animals, specially ruminants [1]. MAP is an OIE notifiable terrestrial disease (OIE-listed diseases, infections and infestations in force in 2016-http://www.oie.int/animal-health-in-the-world/oie-listed-diseases-2016/), and in the U.S., based on the most recent herd study, herd-level prevalence are as high as 90% [2], indeed MAP is a major problem for livestock health and productivity not only in U.S. but also globally.

The annual economic impact of MAP on the U.S. livestock industry has been estimated to of the order of hundreds of millions of dollars, and losses to the dairy industry alone are estimated to reach $200–250 million [3]. Absence of effective control measures to reduce MAP within-farm prevalence reduces farm income due to increased premature animal removal [4] and costs associated with MAP diagnostic testing [5]. In addition, MAP infection often leads to reduced milk production and quality, such as low fat and protein content [6], premature mortality, weight loss, early culling, and reduced slaughter value [7–9]. However, contradict results regarding the impact of MAP on milk yield and days in milk have been reported. This is especially true when serological tests are used to define MAP-infection status [10–12], and may be related to the long incubation period of the disease. The long incubation period also impairs the ability to use standard definitions for infected animals across studies difficult. Evaluating the impact of MAP infection is further complicated by differences in samples and diagnostic tests used in different studies.

At the individual level, presence of MAP antibodies has been linked to lactation number, lactation persistence (defined as the rate of decline in production after the peak milk production has been reached) [13] and milk production [11, 12, 14]. Age is another individual factor typically associated with increased odds of positivity [15], likely due to the course of MAP infection that is expected to be age- and infection-stage specific [16]. In addition, some found evidence of a seasonal pattern in birth of ELISA-positive cows in US dairy herds [17]. Nevertheless, due to the lack of periodic testing and standardized definitions of what an infected animal is, inconsistent epidemiological findings for MAP are quite common [18].

In the state of Minnesota there are 3470 licensed dairy herds and 460,000 dairy cattle (Minnesota Department of Agriculture- https://www.mda.state.mn.us/food/business/~/media/Files/food/business/economics/dairyind-profile.pdf), and the state ranks 7th in number of cows and 8th in milk production nationally, based on the latest data from the United States Department of Agriculture-National Agriculture Statistics Service (USDA-NASS). In MN, individual MAP prevalence has been estimated via diverse sample sources, ranging from 10% to 14% [19], whereas a within-herd mean of 13% positive animals, alternating from 0% to 67% was estimated based on ELISA [20].

The Dairy Herd Improvement Association (DHIA) stores information concerning milk production and quality, and offers testing for MAP to participant farms. In 2016 the participant farms included 1625 dairy herds in MN. Here, we used an individual-level mixed model that accounted for clustering of tested animals within each farm for the identification of characteristics associated with seropositivity in testing DHIA herds. The objectives of this study were a) to quantify the impact of MAP infection (defined by positive milk-ELISA results) on milk production indicators, and b) to identify cow-level possible risk factors associated with positive milk ELISA results in dairy cattle. These results can help to understand MAP dynamics in dairy cattle herds and design age-specific risk-based MAP management plans.

Methods

Study population and data definitions

Information on production and individual characteristics from dairy cattle (Minnesota (93%), Wisconsin (7%), South Dakota (1%), and Iowa (1%)) subjected to MAP testing between November 2014 and August 2016 was provided by the DHIA. Data included 55,298 tests performed on 46,114 cattle from 691 dairy herds. Because a subset of animals (18.21%) was tested more than once during the study period, if all results were negative, only the result from the first sample was retained for analysis. If one or more tests were positive, the first positive result was kept for further analysis. After that correction, we analyzed 46,114 individual observations.

Animals were classified as positive, inconclusive or negative based on a milk ELISA [IDEXX MAP Enzyme Linked Immunosorbent Assay (ELISA) kit (IDEXX Laboratories Inc., Maine, USA)] according to the manufacturer's instructions.

Explanatory variables

Variables determined in the same milk sample used for the MAP test that were available to us included sampling date, animal individual identification, date of birth, lactation number, milk production of the cow per day, fat and protein percentage in milk, reduced lactation persistency, milk urea nitrogen values, SCC, days in milk (DIM), and the 305-day mature equivalent milk production (305ME). No individual breed information was available, although overall ~ 90% of the animals in the studied population were Holstein. Age at sampling was categorized into three classes < 4 years, 4-to-8 years, and > 8 years. In addition, season at birth (Winter [December–January-February]; Spring [March–April-May]; Summer [June–July-August] and Fall [September–October-November]) was used in the analysis. Continuous variables were also categorized into quartiles and explored in the analysis to ensure satisfactory statistical power [21].

Data analysis

Due to the hierarchical structure of the data (i.e. multiple cows from each herd), we performed a logistic regression analysis using a generalized linear mixed model in which the output was the MAP ELISA result and herd was included as a random effect. The outcome variable was test result (positive/negative), and inconclusive results were removed from the database prior to fitting the models. First, all variables were screened in univariable models using a *p*-value ≤0.15 as a liberal threshold for consideration in the multivariable model [22]. Collinearity between pre-selected explanatory variables was then explored using Spearman's rank correlation coefficients for not normally distributed continuous variables. For variables with a correlation coefficient ≥ |0.50|, only the variable with the highest biological plausibility and/or with the strongest association to the output was maintained in the multivariable analysis.

Two-way biologically plausible interactions between pre-selected explanatory variables were investigated, and maintained in the multivariable selection when the *p*-value of the cross product was ≤0.20. Multivariable models were fit using a manual backward stepwise selection, so that non-significant variables ($p > 0.05$) that were not acting as confounders (a change > 40% in the coefficient of another covariate was observed when removed) or effect modifiers were removed from the multivariable model, until only explanatory variables were retained ($p \leq 0.05$).

The age and DIM were considered a priori confounders based on previous evidences [11, 23, 24] and forced into all models. In addition, in order to get a better understanding of the effect of the covariates in the different production stages of the animal, three independent multivariable models were also fitted for each of the categories of < 4 year-old (n = 26,357); 4–8 year old (n = 18,626) and > 8 year old (n = 1131), as described above.

Estimated odds ratios (OR) and 95% confidence intervals were obtained as measures of predictor effect. Model fit was assessed using the deviance, and significant (Mann-Whitney test) difference in the predicted probability of positive result between observed positive and negative observations was considered evidence of adequate goodness-of-fit. The predictive capacity of the final model by estimating the area under the receiver operating characteristic (ROC) curve analysis with 95% probability intervals computed using 2000 stratified bootstrap replicates.

Sensitivity of the results to the large proportion of negative samples was tested by repeating the analyses using a matching design, in which, from the 45,652 cattle, a *m:n* ratio of 1:4 was use, where for each positive test result of a given farm, four negative animals within the farm were selected as controls. The alternative matching design was tested for the four models (one global model and three age specific independent multivariable models).

The selection procedure of the final model in the matching study design was similar to that described above, with the exception that the random effect at the farm level was not incorporated because the matching adjusted for absence of independence.

All analyses were conducted using the R programming language version 3.3.2 [25]. Mixed models were fitted using the package lme4 [26] and the ROC analysis was performed using pROC [27].

Results

Descriptive results

Relatively few positive (n = 2865, 5.1%) and inconclusive (n = 470, 0.8%) test results were found among all 55,298 analyses performed during the study period. After exclusion of repeated tests and inclusive results, 45,652 assays were assessed, of which 2222 were positive (4.9%; $CI_{95\%}$: 4.6–6.0%). Samples originated from 691 herds, with a median of 30 animals samples per herd during the study period [interquartile range (IQR): 24–59]). Number of samples per month ranged from 1281 (October) to 5782 (November). Proportion of positive samples per month ranged from 3.0% ($CI_{95\%}$: 2.3–4.0%) to 6.1% ($CI_{95\%}$: 5.4–6.8%).

Multivariable analysis

Initially, eleven variables were preselected based on a *p*-value of association ≤0.15 in the univariable step. In step two, the lactation number and milk fat content were excluded due to correlation with age (r = 0.93) and milk protein content (r = 0.53), respectively (Table 1). Similarly, 305ME was correlated with lactation persistence (r = 0.57) and excluded from further analysis.

Variables included in the final global mixed effect logistic model were age, SCC, milk production, milk protein, and DIM (Table 1). Continuous variables were used as categorized in the final multivariable models since they provided better fit to the data. Increasing age was associated with higher odds of seropositivity, especially in higher age classes (> 4 to ≤8 year-old; OR = 1.99; > 8 years OR = 1.98). Higher values of SCC were also associated with seropositive results (> 26 to ≤39; OR = 1.36 and values greater than 39; OR = 1.66). Milk production,

Table 1 Estimated ratios along and coefficients with *p*-values of the fixed effects on factors associated with animal MAP-ELISA results from individual milk sample in 691 herds and 45,652 cows (tested ones) from four states in the north of USA

Variables	N	Positive (%)	Univariable analysis		Multivariable analysis			
			p-value*	odds ratio ($CI_{95\%}$)	β	S.E.	odds ratio ($CI_{95\%}$)	*p*-value*
Age (years)			< 0.001					< 0.001
1.25 to ≤4	26,152	3.5%	–	–	–	–	–	–
> 4 to ≤8	18,384	6.7%	< 0.001	2.20 (1.99–2.43)	0.724	0.053	1.99 (1.81–2.19)	< 0.001
> 8 to 15.29	1115	7.1%	< 0.001	2.43 (1.86–3.17)	0.728	0.137	1.98 (1.55–2.54)	< 0.001
Somatic cell count (×1000/ml)			< 0.001					< 0.001
1.0 to ≤ .4	11,711	3.2%	–	–	–	–	–	–
> 14 to ≤26	11,630	3.9%	0.05	1.15 (0.99–1.34)	0.058	0.078	1.05 (0.91–1.21)	0.45
> 26 to ≤39	10,622	5.3%	< 0.001	1.65 (1.43–1.91)	0.272	0.078	1.36 (1.18–1.56)	< 0.001
> 39 to 97	11,447	7.2%	< 0.001	2.33 (2.03–2.68)	0.467	0.077	1.66 (1.44–1.90)	< 0.001
Number of lactation			< 0.001					
0 to ≤1	14,072	3.3%	–	–				
> 1 to ≤2	15,521	4.4%	< 0.001	1.68 (1.47–1.92)				
> 2 to ≤3	8681	6.0%	< 0.001	2.36 (2.05–2.73)				
> 3 to 11	7378	7.4%	< 0.001	2.99 (2.58–3.46)				
Milk production (pounds)			< 0.001					< 0.001
0 to ≤52	12,220	7.7%	–	–	–	–	–	–
> 52 to ≤68	11,384	4.4%	< 0.001	0.53 (0.47–0.61)	– 0.507	0.066	0.66 (0.59–0.75)	< 0.001
> 68 to ≤86	11,465	3.7%	< 0.001	0.44 (0.38–0.51)	– 0.722	0.076	0.56 (0.49–0.65)	< 0.001
> 86 to 192	10,524	3.3%	< 0.001	0.35 (0.30–0.42)	–1.110	0.093	0.43 (0.36–0.51)	< 0.001
Milk fat (%)			0.03					
0 to ≤3.5	12,549	4.4%	–	–				
> 3.5 to ≤3.9	9903	4.5%	0.89	1.00 (0.88–1.15)				
> 3.9 to ≤4.5	12,657	4.7%	0.98	1.00 (0.87–1.13)				
> 4.5 to 9.9	10,543	5.9%	0.01	1.17 (1.03–1.34)				
Milk protein (%)			0.001					< 0.001
0 to ≤3.0	12,418	4.9%	–	–	–	–	–	–
> 3.0 to ≤3.3	13,952	4.1%	0.008	0.84 (0.73–0.95)	–0.262	0.072	0.76 (0.67–0.87)	< 0.001
> 3.3 to ≤3.5	8224	5.0%	0.61	0.96 (0.83–1.11)	– 0.261	0.083	0.74 (0.64–0.85)	0.001
> 3.5 to 7.4	11,058	5.7%	0.28	1.07 (0.93–1.23)	–0.380	0.083	0.67 (0.58–0.77)	< 0.001
Milk urea nitrogen (mg/dl)			0.17					
0 to ≤5	12,076	4.7%	–	–				
> 5 to ≤11	14,833	5.0%	0.77	1.02 (0.87–1.19)				
> 11 to ≤13	8087	5.0%	0.52	1.05 (0.88–1.26)				
> 13 to 81	10,656	4.7%	0.26	0.90 (0.76–1.07)				
Previous-305-d mature equivalent (305ME)(Kg)			< 0.001					
0 to ≤21,120	11,687	6.6%	–	–				
> 21,120 to ≤25,690	11,540	5.1%	< 0.001	0.69 (0.60–0.79)				
> 25,690 to ≤29,550	11,253	4.3%	< 0.001	0.55 (0.47–0.64)				
> 29,550 to 46,720	11,171	3.3%	< 0.001	0.42 (0.35–0.50)				
305-d mature equivalent (305ME)(Kg)			< 0.001					
0 to ≤20,220	11,718	6.8%	–	–				

Table 1 Estimated ratios along and coefficients with *p*-values of the fixed effects on factors associated with animal MAP-ELISA results from individual milk sample in 691 herds and 45,652 cows (tested ones) from four states in the north of USA *(Continued)*

Variables	N	Positive (%)	Univariable analysis		Multivariable analysis			
			p-value*	odds ratio ($CI_{95\%}$)	β	S.E.	odds ratio ($CI_{95\%}$)	*p*-value*
> 20,220 to ≤25,370	11,485	5.1%	< 0.001	0.65 (0.57–0.74)				
> 25,370 to ≤29,400	11,256	4.2%	< 0.001	0.53 (0.46–0.62)				
> 29,400 to 46,720	11,192	3.2%	< 0.001	0.39 (0.33–0.46)				
Lactation persistence (%)			< 0.001					
1 to ≤100	14,022	6.1%	–	–				
> 100 to ≤108	9976	5.0%	< 0.001	0.79 (0.70–0.90)				
> 108 to ≤116	11,286	4.0%	< 0.001	0.60 (0.52–0.69)				
> 116 to ≤225	10,368	4.0%	< 0.001	0.58 (0.51–0.67)				
Days in milk			< 0.001					< 0.001
< 30	3658	6.4%	–	–	–	–	–	–
–31-60	3280	5.3%	0.39	0.90 (0.72–1.13)	0.007	0.120	1.00 (0.79–1.27)	0.94
61–90	1950	3.1%	< 0.001	0.49 (0.35–0.68)	–0.580	0.169	0.55 (0.40–0.78)	< 0.001
91–120	1604	4.5%	0.01	0.67 (0.48–0.92)	–0.283	0.167	0.75 (0.54–1.04)	0.08
121–150	1351	4.3%	0.01	0.63 (0.45–0.89)	–0.403	0.180	0.66 (0.46–0.95)	0.02
151–180	1224	5.6%	0.07	0.74 (0.53–1.03)	–0.243	0.173	0.78 (0.55–1.10)	0.16
181–210	1862	5.0%	0.22	0.83 (0.62–1.11)	–0.152	0.154	0.85 (0.63–1.16)	0.32
211–240	3336	3.6%	< 0.001	0.61 (0.46–0.80)	–0.513	0.143	0.59 (0.45–0.79)	< 0.001
241–270	6743	3.6%	< 0.001	0.57 (0.45–0.73)	–0.615	0.124	0.54 (0.42–0.69)	< 0.001
271–300	6498	4.6%	0.001	0.68 (0.54–0.85)	–0.558	0.124	0.57 (0.44–0.72)	< 0.001
> 301	14,146	5.7%	0.12	0.85 (0.69–1.04)	–0.557	0.116	0.57 (0.45–0.71)	< 0.001
Season of cow's birth			0.07					
Winter (Dec-Feb)	12,339	5.3%	–	–				
Spring (Mar-May)	11,100	5.0%	0.16	0.91 (0.80–1.03)				
Summer (Jun-Aug)	12,134	4.5%	0.01	0.84 (0.74–0.96)				
Fall (Sep-Nov)	10,079	4.7%	0.05	0.88 (0.77–1.00)				

*p-values from the Likelihood-ratio test

percent of protein in milk and DIM were inversely associated with the risk of positive test results, so that animals with higher values in those variables had lower risk of being positive (Table 1).

The area under the ROC-curve for the final global model was 83.3% ($CI_{95\%}$: 82.5%–84.1%), indicative of a good predictive performance of the model. Fitted values from positive and negative observations were significantly different (Mann-Whitney test, *p*-value < 0.05).

Age-specific models were fitted separately, the model for both < 4 year-old and 4–8 year-old cattle included SCC, milk production, and DIM, while milk protein content was only selected in the model fitted on 4–8 year-old cattle (Additional file 1). Predictive power of both models was similar to the one determined in the general model, however the model fitted for young animals showed a better performance (AUC = 85.8 and 82.9 for young – < 4 years, and adult cattle 4 ≤ 8 years old, respectively). In contrast, no variables were significantly associated with the test result for the older cattle (> 8 years).

Results from the matched design *m:n* were similar to those reported here, suggesting that the large proportion of negative animals did not affect the model outputs.

Discussion

Although multiple studies have demonstrated the impact of MAP infection on production, there is much variability, in quantitative terms, in the results, which makes extrapolation of results to different settings difficult [6, 7, 10, 28]. The absence of official surveillance programs in the U.S. and standardized definitions for positive status to MAP further complicates the interpretation of epidemiological studies. The study here, which makes use of passive surveillance activities routinely and voluntarily performed by dairy herds, represents an attempt to identify the characteristics associated with seropositive animals in a high herd-prevalence setting. The small proportion of positive test results (4.9%) and large proportion of positive farms (55.4%; $CI_{95\%}$: 51.7–59.0%) is similar to figures obtained in previous studies, typically only a small portion of the animals in a herd test positive in serology-based results and hence individual seroprevalences are usually in the range of 3 to 10% [29–32], indeed other large serological studies perform outside U.S. reported occurrences of 2.3% and in a larger study MAP ranged from 5.6% in 2008, and 3.8% in 2015, peaking at 7% in 2009% both Denmark studies [33, 34], finally in China a 4.8% prevalence was reported using IDEXX ELISA [35].

Although SCC was here clearly associated with test result, with higher probability of positive results for those animals with increased counts (Table 1; Additional file 1), inconsistent results have been described in the literature [18, 36]. However, a strong association between somatic cells and antibodies against MAP has been found in Danish cows [37]. Somatic cell count was also shown to be strongly associated with higher risk for testing positive to MAP in 58,096 UK Holstein-Friesian cows, with higher SCC values in animals at higher and medium MAP risk compared with low-risk groups [12]. This effect was still evident once the possible influence of milk yield and age on probability of detection of MAP antibodies in milk was accounted for [12]. The SCC count could be useful when managing MAP, and should be interpreted altogether with other milk production parameters, such as milk yield and days in milk, that have been previously suggested to influence MAP milk ELISA results [11, 24].

The association between age of the tested animal and ELISA results (Table 1; Additional file 1) is similar to that described elsewhere [16] suggesting a limited ability of milk-ELISA for the discrimination of infected cows in young (< 3 year-old) animals. Age distribution in the tested population may therefore have a significant influence in the proportion of positive animals found. For example, some have found less than 0.33% < 2 year-old positive cattle, a proportion that increased to 0.94% for > 5 years-old animals [15]. For that reason, specific models were fitted for each age category. Although results were similar for < 4 and 4–8 year-old animals (Additional file 1), no associations were observed for > 8 year-old cattle, which may be explained, at least in part, by the reduced dataset ($n = 1115$), which may have affected the power to detect associations. The only difference between < 4 and 4–8 year-old cattle was the association of MAP milk ELISA result with milk protein content, which was not statistically significant for the younger cattle category, which may be due to limited impact of disease on production of animals in an early stage of MAP infection [6, 14].

Poorer performance has been observed in MAP-seropositive, compared to MAP-seronegative, animals [10]. Here, animals yielding > 86 pounds of milk per day were at a significantly lower risk of being positive (0.35, $CI_{95\%}$: 0.30–0.42) compared to the baseline category (Table 1). This could be due in part to a variation in the concentration of MAP-specific antibodies IgG in milk related with the milk production, so that animals with higher production experienced a dilution effect [11] Milk yield is also associated with age, but we observed the same association both in the multivariable model including age as a covariate and in the age-specific models, reinforcing the finding and suggesting an age-independent effect (Additional file 1). Although association was particularly strong for adult cattle, suggesting that older positive cows may have a more evident production decrease likely due to a more advanced stage of infection [6], young MAP-seropositive animals also show relatively poor performance, suggesting the effect of MAP infection was already significant at early stages of infection (Additional file 1). However, cows with higher milk yield are often under more stress, one should also consider the negative energy balance when analyzing animal performance and disease occurrences.

Days in milk associated negatively with MAP milk ELISA positive results especially after the first three months of lactation [11]. Our results from the global and the age-specific models are in agreement with these findings: both the global and the model fitted for the adult cows suggested that after 61 days in lactation the chances of finding a positive result decreased by almost half (Table 1, Additional file 1). Interestingly, in young cows this association was only identified after 211 days in milk, perhaps associated with a slower progress of infection at younger ages. This early negative association, for the global and adult cows model or later in the case of young animal, with days in milk is probably due to dilution of antibodies below the detection limit as suggested before [11]. The negative association between the number of liters produced and the sensitivity of MAP milk ELISA results should be further investigated. It would be interesting if data was collected in a higher resolution (i.e. weekly sampling instead of monthly), as this would allow one to capture smaller effects in the test result.

Higher protein content in milk was also associated with decreased odds of seropositivity as reported elsewhere [7]. Previously, this association was more evident for first lactation MAP-seropositive cows relative to cows that tested positive in later lactations [7], although other authors did not observe such association before 2 years of age [38]. Although an age-protein content interaction was not found in the model regardless of age, protein content was only identified as a significant covariate in the model for adult (4–8 year-old) age-specific model, suggesting a possible modulating effect of age on the association between MAP test result and protein content, at least for certain age classes.

No significant association between season of birth and odds of positivity for MAP was found here, in contrast with what has been previously described in 24 Jersey and 4 Holstein herds this may be related to different management strategies in the assessed populations [17]. Season temperatures in MN tend to fluctuate, according to MN department of natural resources Spring vary from 2 to 6 °C (north and south), Summer from 17 to 21 °C, Fall from 5 to 8 °C and Winter from – 13 to – 7 °C our model did not find an association as others [17] found a highest probability of infection during Summer months, such association should further be investigated especially in other countries with more favorable conditions of humidity and temperature.

A major limitation of the study here was the imperfect test accuracy of MAP milk ELISA, with the manufacturer indicating a test sensitivity at 0.52 and specificity at 0.98, which could lead to an underestimation of the prevalence and the existing associations in animals in earlier stages of infection. In addition, we only had access to dairy herds enrolled in the DHIA that were conducting MAP tests on some of their animals. These herds may have a higher level of awareness about the importance of the disease, which in turn could have result in different management strategies. However, the farms assessed here represent approximately 20% of all dairy herds in Minnesota (and 43% of all Minnesota DHIA herds) (Minnesota Department of Agriculture-https://www.mda.state.mn.us/food/business/~/media/Files/food/business/economics/dairyindprofile.pdf), but also representing a relatively good coverage for the Midwest region of the US. Finally, despite the limitations associated with the moderate correlation between results of milk and serum ELISA, milk samples are much more convenient, and more suitable to be use especially when considering large herds and countries with limited resources, where sampling is a big issue and to collect blood samples of such large population of cows for screening against MAP or any other disease is not possible.

Conclusion

Although the proportion of positive serological results in our study population was low, MAP infection still impacting dairy herds health in the study region. Our results suggest that cows were more likely to test positive when they had SCC above 26 (× 1000/ml) – according to the global model the model for young (< 4 years) cows – and above 39 (× 1000/ml) based on the model fitted for older cows (4–8 years). Cows with higher milk production had a lower likelihood of testing positive in all models, as well as cows that were more than 61 days into their lactation (and over 211 days for cows < 4 years). Our results provide guidelines to help design MAP sampling strategies when production data is available, upon evidences of the models sampling animals with SCC between 26 and 39 (× 1000/ml), low milk production, that are in the initial stages of lactation from 61 to 211 days' conditional on age would increase chances of identifying positive cows. However, results should be interpreted with care given the synergistic effect of several variables (age, milk production, days in milk), and the possible impact that other physiological and clinical condition (e.g. mastitis) may have, especially on the SCC counts.

Abbreviations

305ME: 305-day mature equivalent milk production; DHIA: Dairy Herd Improvement Association; DIM: days in milk; ELISA: Enzyme Linked Immunosorbent Assay; IQR: interquartile range; MAP: Mycobacterium avium subsp. paratuberculosis; OR: odds ratios; ROC: receiver operating characteristic; SCC: somatic cell count

Acknowledgements

The authors thank Dairy Herd Improvement Association (DHIA) at the state of Minnesota and a special thank-you is given to the Minnesota DHIA dairy producers participants for their valuable collaboration.

Funding

This work was supported by MN Drive-Global Food Venture initiative.

Authors' contributions

GM, KSTK and JA designed the study. VS and KSTK performed the data cleaning/collection. GM and JA performed the data analysis. All authors (GM, KSTK, VA, SJW, AP and JA) contributed to drafting and revising of the manuscript. All authors (GM, KSTK, VA, SJW, AP and JA) read and approved the final manuscript.

Consent for publication

Not applicable.

Competing interests

The authors declare that they have no competing interests.

Author details

[1]Department of Veterinary Population Medicine, University of Minnesota, St. Paul, MN, USA. [2]Department of Population Health and Pathobiology, College of Veterinary Medicine, North Carolina State University, 1060 William Moore Drive, Raleigh, NC, USA. [3]Minnesota DHIA, Minnesota Dairy Herd Improvement Association, Buffalo, MN, USA. [4]Raleigh, USA.

References

1. Daniels MJ, Hutchings MR, Beard PM, Henderson D, Greig A, Stevenson K, Sharp JM. Do non-ruminant wildlife pose a risk of paratuberculosis to domestic livestock and vice versa in Scotland? J of Wild Dise. 2003;39(1):10–5.
2. Lombard JE, Gardner IA, Jafarzadeh SR, Fossler CP, Harris B, Capsel RT, Wagner BA, Johnson WO. Herd-level prevalence of *Mycobacterium avium* subsp. *paratuberculosis* infection in United States dairy herds in 2007. Prev Vet Med. 2013;108(2–3):234–8.
3. Ott SL, Wells SJ, Wagner BA. Herd-level economic losses associated with Johne's disease on US dairy operations. Prev Vet Med. 1999;40(3–4):179–92.
4. Tiwari A, VanLeeuwen JA, Dohoo IR, Keefe GP, Weersink A. Estimate of the direct production losses in Canadian dairy herds with subclinical *Mycobacterium avium* subspecies *paratuberculosis* infection. Can Vet J. 2008; 49(6):569–76.
5. Raizman EA, Fetrow JP, Wells SJ. Loss of income from cows shedding *Mycobacterium avium* subspecies *paratuberculosis* prior to calving compared with cows not shedding the organism on two Minnesota dairy farms. J Dairy Sci. 2009;92(10):4929–36.
6. Johnson YJ, Kaneene JB, Gardiner JC, Lloyd JW, Sprecher DJ, Coe PH. The effect of subclinical *Mycobacterium paratuberculosis* infection on milk

7. Gonda MG, Chang YM, Shook GE, Collins MT, Kirkpatrick BW. Effect of *Mycobacterium paratuberculosis* infection on production, reproduction, and health traits in US Holsteins. Prev Vet Med. 2007;80(2–3):103–19.
8. Raizman EA, Wells SJ, Godden SM, Fetrow J, Oakes JM. The associations between culling due to clinical Johne's disease or the detection of *Mycobacterium avium* subsp. *paratuberculosis* fecal shedding and the diagnosis of clinical or subclinical diseases in two dairy herds in Minnesota, USA. Prev Vet Med. 2007;80(2–3):166–78.
9. Hendrick SH, Kelton DF, Leslie KE, Lissemore KD, Archambault M, Duffield TF. Effect of paratuberculosis on culling, milk production, and milk quality in dairy herds. Javma. 2005;227(8):1302–8.
10. McAloon CG, Whyte P, More SJ, Green MJ, O'Grady L, Garcia A, Doherty ML. The effect of paratuberculosis on milk yield-a systematic review and meta-analysis. J Dairy Sci. 2016;99(2):1449–60.
11. Eisenberg SWF, Veldman E, Rutten VPMG, Koets AP. A longitudinal study of factors influencing the result of a *Mycobacterium avium* ssp *paratuberculosis* antibody ELISA in milk of dairy cows. J Dairy Sci. 2015;98(4):2345–55.
12. Pritchard TC, Coffey MP, Bond KS, Hutchings MR, Wall E. Phenotypic effects of subclinical paratuberculosis (Johne's disease) in dairy cattle. J Dairy Sci. 2017;100(1):679–90.
13. Cole JB, Null DJ. Genetic evaluation of lactation persistency for five breeds of dairy cattle. J Dairy Sci. 2009;92(5):2248–58.
14. Kudahl A, Nielsen SS, Sorensen JT. Relationship between antibodies against *Mycobacterium avium* subsp *paratuberculosis* in milk and shape of lactation curves. Prev Vet Med. 2004;62(2):119–34.
15. Nielsen SS, Toft N, Okura H. Dynamics of specific anti-*Mycobacterium avium* subsp. *paratuberculosis* antibody response through age. PLoS One. 2013;8(4):e63009.
16. Kostoulas P, Browne WJ, Nielsen SS, Leontides L. Bayesian mixture models for partially verified data: age- and stage-specific discriminatory power of an antibody ELISA for paratuberculosis. Prev Vet Med. 2013;111(3–4):200–5.
17. Zare Y, Shook GE, Collins MT, Kirkpatrick BW. Evidence of birth seasonality and clustering of *Mycobacterium avium* subspecies *paratuberculosis* infection in US dairy herds. Prev Vet Med. 2013;112(3–4):276–84.
18. Donat K, Soschinka A, Erhardt G, Brandt HR. Paratuberculosis: decrease in milk production of German Holstein dairy cows shedding *Mycobacterium avium* Ssp. *paratuberculosis* depends on within-herd prevalence. Animal. 2014;8(5):852–8.
19. Raizman EA, Wells SJ, Munoz-Zanzi CA, Tavornpanich S. Estimated within-herd prevalence (WHP) of *Mycobacterium avium* subsp *paratuberculosis* in a sample of Minnesota dairy herds using bacterial culture of pooled fecal samples. Can J of Vet Research. 2011;75(2):112–6.
20. Raizman EA, Wells SJ, Godden SM, Fetrow J, Friendshuh K, Oakes JM. Characterization of Minnesota dairy herds participating in a Johne's disease control program and evaluation of the program risk assessment tool. Prev Vet Med. 2006;75(1–2):22–33.
21. Turner EL, Dobson JE, Pocock SJ. Categorisation of continuous risk factors in epidemiological publications: a survey of current practice. Epidemiol Perspect Innov. 2010;7:9.
22. Hosmer DW, Lemeshow S. Applied logistic regression. New York: Wiley; 1989.
23. Tiwari A, Vanleeuwen JA, Dohoo IR, Keefe GP, Haddad JP, Scott HM, Whiting T. Risk factors associated with *Mycobacterium avium* subspecies *paratuberculosis* seropositivity in Canadian dairy cows and herds. Prev Vet Med. 2009;88(1):32–41.
24. Lombard JE, Byrem TM, Wagner BA, McCluskey BJ. Comparison of milk and serum enzyme-linked immunosorbent assays for diagnosis of *Mycobacterium avium* subspecies *paratuberculosis* infection in dairy cattle. J Vet Diagn Investig. 2006;18(5):448–58.
25. Team RDC. R : a language and eviroment for statistical computing. Vienna: R Foundation for Statistical Computing; 2010.
26. Douglas B, Martin M, Ben B, Steven W. lme4: Linear mixed-effects models using Eigen and S4. R package version 10-6. 2014. http://CRANR-projectorg/package=lme4. Accessed Nov 2017.
27. Robin X, Turck N, Hainard A, Tiberti N, Lisacek F, Sanchez JC, Muller M. pROC: an open-source package for R and S plus to analyze and compare ROC curves. Bmc Bioinformatics. 2011;12:77.
28. Lombard JE, Garry FB, McCluskey BJ, Wagner BA. Risk of removal and effects on milk production associated with paratuberculosis status in dairy cows. Javma. 2005;227(12):1975–81.
29. Collins MT, Eggleston V, Manning EJB. Successful control of Johne's disease in nine dairy herds: results of a six-year field trial. J Dairy Sci. 2010;93(4):1638–43.
30. Schukken YH, Whitlock RH, Wolfgang D, Grohn Y, Beaver A, VanKessel J, Zurakowski M, Mitchell R. Longitudinal data collection of *Mycobacterium avium* subspecies *paratuberculosis* infections in dairy herds: the value of precise field data. Vet Res. 2015;46(1):65.
31. Lombard JE. Epidemiology and economics of Paratuberculosis. Vet Clin N Am-Food A. 2011;27(3):525–35.
32. Garry F. Control of Paratuberculosis in dairy herds. Vet Clin N Am-Food A. 2011;27(3):599–607.
33. Okura H, Nielsen SS, Tolf N. Prevalence of *Mycobacterium avium* subsp. *paratuberculosis* infection in adult Danish non-dairy cattle sampled at slaughter. Prev Vet Med. 2010;94(3–4):185–90.
34. Nielsen SS, Hansen KF, Kvist L, Kostoulas P. Dam's infection progress and within-herd prevalence as predictors of *Mycobacterium avium* subsp. *paratuberculosis* ELISA response in Danish Holstein cattle. Prev Vet Med. 2016;125:54–8.
35. Xianxia L, Jianjun L, Xueyun Y, Dengfeng W, Jianmei W, Jianyong W. The seroprevalence of *Mycobacterium avium* subspecies *paratuberculosis* in dairy cattle in Xinjiang Northwest China. Ir Vet J. 2017;70:1.
36. Hoogendam K, Richardson E, Mee JF. Paratuberculosis sero-status and milk production, SCC and calving interval in Irish dairy herds. Irish Vet J. 2009;62(4):265–71.
37. Baptista FM, Nielsen SS, Toft N. Association between the presence of antibodies to *Mycobacterium avium* subspecies *paratuberculosis* and somatic cell count. J Dairy Sci. 2008;91(1):109–18.
38. Pillars RB, Bolton MW, Grooms DL. Case-control study: productivity and longevity of dairy cows that tested positive for infection with *Mycobacterium avium* ssp. *paratuberculosis* as heifers compared to age-matched controls. J Dairy Sci. 2011;94(6):2825–31.

9

Evaluation of anaesthesia and analgesia quality during disbudding of goat kids by certified Swiss farmers

N. Wagmann[1], C. Spadavecchia[2], U. Morath-Huss[2], G. Schüpbach-Regula[3] and P. Zanolari[1*]

Abstract

Background: Certified Swiss farmers are allowed to disbud their goat kids using a standard intramuscularly administered anaesthetic mixture. This mixture, containing xylazine and ketamine, is officially distributed with the goal to provide a painless disbudding. This study aimed to evaluate the quality of analgesia and anaesthesia achieved during disbudding, when performed by Swiss farmers. To assess this, 174 goat kids at 31 different farms were observed and filmed during cautery disbudding.

Results: The standard anaesthetic mixture (0.05 mg/kg xylazine and 20 mg/kg ketamine) was used only in 71 goat kids. Fifty-eight goat kids were anaesthetised with different dosages of xylazine (median 0.18 mg/kg) and ketamine (median 10 mg/kg), 22 with xylazine only (median 0.61 mg/kg), 20 with xylazine (median 1.84 mg/kg) and perineural lidocaine (median 1.23 mg/kg), three with acepromazine (dosage unknown) and ketamine (10 mg/kg). Based on vocalisation, limb movement and head lifting during disbudding, a general reaction score was attributed to 168 goat kids (six were excluded due to firm restraint): 56.5% were scored zero (no limb movement, no vocalisation), 7.7% one, 17.3% two and 18.5% three (strong movements, vocalisation). Significant risk factors for higher reaction scores were the type of anaesthetic protocol and manipulation by the farmer during induction. Significant risk factors for longer recoveries were use of xylazine alone or xylazine in combination with perineural lidocaine, breed, younger age and recovery underneath heat lamp.

Conclusions: The present study indicates that anaesthesia and analgesia of goat kids disbudded by Swiss farmers is inadequate, as 35.8% of the animals showed moderate to strong behavioural reactions during the procedure. Unexpectedly, only 40.8% of the goat kids were anaesthetised with the standard anaesthetic mixture and several other protocols were used. A refinement of the recommended protocol is urgently needed to guarantee animal welfare.

Keywords: Cautery disbudding, Goat kid, Analgesia, Anaesthesia, Switzerland

Background

In dairy goat farms, disbudding of goat kids is a common husbandry practice. While horn absence in adult animals is desirable to reduce fights and potential trauma to animals and humans, the removal of the horn buds is painful and stressful for the young goats [1, 2]. In Australia, New Zealand and the United States, disbudding of goat kids is routinely practised [3]. In Europe, procedures which cause a significant amount of pain or distress are in general forbidden, but exceptions for disbudding are made if allowed under existing national legislations. The European recommendation states that due to the anatomy of the kids' skull, disbudding is a difficult procedure even if performed under anaesthesia; therefore, it should only be carried out by a veterinarian using an anaesthetic [4]. In Austria, disbudding is forbidden since 2005 [5], while in Germany and in the United Kingdom, disbudding can only be performed by a licensed veterinarian [6, 7].

In Switzerland, painful interventions can only be carried out under general or local anaesthesia by experienced personnel [8]. Swiss farmers who have acquired a certificate of competence by the Federal Food Safety and Veterinary Office (FSVO), are allowed to disbud their own goat kids under general anaesthesia until the age of

* Correspondence: patrik.zanolari@vetsuisse.unibe.ch
[1]Clinic for Ruminants, Vetsuisse Faculty, University of Bern, Bremgartenstrasse 109a, 3012 Bern, Switzerland
Full list of author information is available at the end of the article

3 weeks. In order to obtain the certificate, the farmers are required to participate in a theoretical course, followed by practical experience under the supervision of a veterinarian at their own farm. The acquired skills are then evaluated by the cantonal veterinary inspection office and a certificate is issued [9].

Private veterinarians are expected to deliver a standard anaesthetic mixture to the certified farmers to anaesthetise their own goat kids for disbudding [10]. The standard mixture, foreseen by the FSVO for intramuscular administration, contains xylazine (0.05 mg/kg) and ketamine (20 mg/kg).

Since the introduction of the certificate course in 2008, no data have been collected to evaluate whether the procedure is carried out under adequate conditions for the animals. The main aim of the present study was to assess the quality of analgesia and anaesthesia of goat kids disbudded by certified Swiss farmers. Based on field observation, it was hypothesised, that anaesthesia quality might not be adequate for disbudding of young goat kids. To this end, several behavioural responses during disbudding were quantified through direct observation and later analysis of video recordings.

Methods

Study design and farm selection

The current study was designed as a prospective, observational field study. The Swiss Goat Breeding Association provided a list of 68 certified farmers that perform disbudding of goat kids themselves. These farmers were contacted by phone and were informed about the study. Thirty-one farmers were selected based on their agreement to participate in the study. Farmers performing disbudding with veterinary assistance were excluded from the study. Farm data and individual goat kid data were separately recorded. Farm data was collected prior to disbudding as well as goat kid data during disbudding.

Farm data

Collected farm data included farm size, number of goat kids born per year, number of animals disbudded per year, average age at disbudding and past complications attributed to this intervention. Additionally noted was bedding type, heat lamp presence and infrastructure at the workplace.

Data collection at the farms took place from January to May 2017; time and date of disbudding were determined by the farmers. All farmers gave their informed written consent to have the disbudding procedure observed and recorded. Prior to disbudding, the farmers were interviewed using a standard questionnaire. Observation started at anaesthetic injection and ended when goat kids returned to a steady standing position after recovery from anaesthesia. Data were recorded with a standardized protocol (Additional files 1, 2). Farmers were asked to disbud their goat kids as per their routine farm procedure. There were two trained observers (NW and UM) and they did not interfere with the procedure at any stage. The video recordings[1] were analysed at a later time point by NW and UM.

Goat kid data

A total of 174 goat kids from 31 different farms were observed from the injection of the anaesthetic mixture to recovery. Individual goat kid data included ear tag number, breed, sex, age, weight, health condition and fasting duration. All goat-kids were observed from afar without being handled. Furthermore, composition, administration route and onset of action of the anaesthetic mixture, restraint method, skin preparation technique (clipping of the fur around the horn bud), disbudding technique (burning time, removing of horn bud, applying of dehorner), administration of analgesics and environmental temperature were recorded. To determine the dosage of drugs administered to goat kids of unknown body weight, a mean weight of 6.5 kg was assumed. During anaesthesia induction, time from injection to staggering, recumbency and loss of posture were measured. Recovery time was measured from the time from injection to first movement, to first attempt to stand and till steady standing (Additional file 3). Goat kids whose recovery could not be observed completely were excluded from further statistical analysis of recovery risk factors (eight for first movement, 22 for steady standing). The level of farmer care and monitoring of goat kids before and after disbudding were documented.

Reactions to disbudding

Video recordings were analysed starting from the application of the dehorner and ended when cautery disbudding was finished. Specific behavioural reactions during disbudding were documented according to predefined events (Additional file 4). Struggles (movements of legs or attempts to escape), vocalisation and tail movements during cautery disbudding of goat kids are already described as signs of pain and stress [11, 12].

Frequency of occurrence of specific behaviours during disbudding was graded from 1 to 4 as follows: 1) behaviour did not occur; 2) behaviour occurred 1–2 times; 3) behaviour occurred 3–6 times; 4) behaviour occurred > 7 times.

Based on the occurrence of vocalisation, limb-movement (paddling/kicking/pull-up limb) and head lifting during disbudding, a general reaction score was attributed to each goat kid (Additional file 5) [13]. Firmly restrained animals were not scored, as behaviours could not be clearly observed.

Direct scoring of pain intensity during disbudding was performed using a Visual Analogue Scale (VAS) on a 10 cm line (0 meaning no pain, 10 meaning the worst possible pain) by the main investigator (NW) and, if wished, by the farmer.

Risk factors

Most observed factors at farm and goat kid level (Additional file 6) were analysed to be potential risk factors for longer recovery and for higher general reaction score.

Statistical analysis

A sample size of 166 animals was calculated to estimate the prevalence of reactions during disbudding with a precision of 7%, assuming a prevalence of 50%, a population size of 1082 disbuddings per year, and a 95% confidence.[2] Data were recorded in Microsoft Excel spreadsheets.[3] Descriptive analysis and screening of risk factors was performed with the statistics program NCSS10[4] regression analysis with the software SAS 9.4.[5] Variables that were recorded on a continuous scale were checked for normality with the Shapiro Wilk W test. Non-normal risk factor variables were grouped into biologically meaningful categories (Additional file 6). The effect of different anaesthesia mixtures (excluded the acepromazine group due to small number of animals) was analysed for three different outcomes (time to first movement, time to steady standing and general reaction score).-Time to first movement and time to steady standing was recorded in minutes, and analysed as continuous, normally distributed variables. General reaction score was initially recorded on a 4-point scale and afterwards dichotomized into 0 (no reaction) and 1 (grade 1 to 3; Additional file 5).

For the screening of risk factors, ANOVA with Bonferroni correction for multiple comparisons was used for the outcomes time to first movement, time to steady standing and general reaction score. Associations with behavioural reaction during disbudding were screened with logistic regression. Potential risk factors which had a P-value < 0.1 in the screening were offered to a multiple linear or logistic regression model, respectively. None of the potential risk factors were correlated with each other (phi coefficients < 0.5). Generalized estimation equation was used to control for the effect of herd on the outcome [14]. SAS PROC GENMOD with herd as a REPEATED statement was used for this analysis. Variable selection was performed by stepwise backward selection until only significant ($P < 0.05$) risk factors remained in the model. No confounders (variables changing the effect of another risk factor by more than 20%) had to be included in the models. Two-way interaction terms had to be tested for significance but removed again from the final models to allow a more straightforward interpretation of the main effect of interest (anaesthesia protocol). Model fit was assessed using QIC and QICu. Model assumptions were checked by visual assessment of residuals versus predicted values. For the linear models, the assumption of normality of residuals was also formally checked with Shapiro Wilk W. Data were presented as median (range) or mean ± standard deviation (SD) unless otherwise specified.

Results

Farm data

Farm size

Herd size of included farms was 25 (4–260) goats that bred 35 (6–400) and disbudded 11 (2–80) goat kids per year. When asked about previously encountered adverse events during and after disbudding, 73.3% of the farmers reported to have never experienced any. The remaining reported cases from different farms of sudden death (6), excessive bleeding (2), prolonged recoveries (2), tetanus (1), split horn base (1) and intravenous injection of anaesthetics (1).

Infrastructure

For anaesthesia induction and recovery of goat kids 61.3% of the farmers utilized the same stalls; 38.7% had prepared separate places. A heat lamp was installed by 29.0% of the farmers at the induction area and by 41.9% at the recovery area. In 87.1% of the farms lighting was subjectively judged to be sufficient. Before disbudding, hair was clipped in 38.7% of the farms; 87.1% of the farmers used a disinfectant after disbudding, of which 74.1% contained an antibiotic substance (tetracycline or chloramphenicol). Environmental temperature during anaesthesia induction was 12.8 (4.0–25.0) °C and during recovery 14.7 (4.0–30.0) °C.

Goat kid data

Animals

Goat breeds included Chamois-Coloured Goat (116), Saanen (39), Toggenburg (14) and Grisons Striped (5). Most animals were female (158), 12.0 (3.0–31.0) days of age with a body weight of 6.4 (3.0–11.0) kg. The kids (24.1%) were not weighed before disbudding; 95.4% of the goat kids had a healthy appearance, 3.4% showed signs of diarrhoea and 1.1% were lame. Fasting time was 4.0 (1.5–12.0) h; 19.5% of the goat kids were not fasted.

Administration of anaesthetic mixture

The recommended xylazine-ketamine mixture was used in 40.8% of the goat kids; different dosages of xylazine and ketamine were administered to 33.3%; 12.6% of the kids received only xylazine, 11.5% of the kids xylazine and local lidocaine; 1.7% of the kids were administered acepromazine combined with ketamine as anaesthetic protocol (Table 1). Only 6.3% of the goat kids from two different farms were administered an additional analgesic (meloxicam/ tolefenamic acid) after disbudding.

Table 1 Median (range) dosage of injected anaesthesia-mixtures by certified Swiss farmers to goat kids for horn disbudding

Number of goat kids	Xylazine dose (mg/kg)	Ketamine dose (mg/kg)	Lidocaine dose (mg/kg)	Acepromazine dose (mg/kg)	Mean general reaction score (0–3)
71	0.05[a]	20[a]	0		1.24 (SD 1.27)[b]
58	0.18 (0.05–0.58)	10 (4–25)	0		0.69 (SD 1.06)
22	0.61 (0.36–1.23)	0	0		1.55 (SD 1.34)
20	1.84 (0.61–1.84)	0	1.2 (1.2–3.1)		0.05 (SD 0.23)[c]
3	0	10	0	Dosage unknown	2.33 (SD 1.15)

[a]recommended xylazine-ketamine mixture
[b]mean of 66 goat kids (5 excluded due to strong restraint)
[c]mean of 19 goat kids (1 excluded due to strong restraint)

Injections were performed in the hindlimbs (M. semitendinosus/ M. semimembranosus or M. biceps femoris) in 59.2% of the goat kids and in the forelimbs (M. supra-/infraspinatus, *M. deltoideus*, M. brachiocehalicus or M. omotransversarius) in 40.8%. Aspiration before injection was performed in 29.3% of the animals. No reaction at injection was observed in 69.5% of the kids. Only few animals showed vocalisation (12.1%) and limb movement (5.7%) or both at the same time (12.6%). Waiting time from injection to disbudding was 12 (3–46) min. Two of the 31 farmers used lidocaine for local anaesthesia. They inserted the needle close to the caudal ridge of the frontal process of the zygomatic bone to block the cornual branch of the zygomaticotemporal (lacrimal) nerve.

Anaesthesia induction phase

Farmers (90.3%) observed the goat kids until loss of posture, while 6.5% observed the animals at 5 min intervals. One farmer left the animals alone after the anaesthetic injection until disbudding was started. During anaesthetic induction, 40.8% of the goat kids were manipulated by the farmers before loss of posture. Induction times were 1.5 (0–5) min for staggering; 2.0 (0–13) min for sternal recumbency, 3.0 (1–13) min for lateral recumbency and 3.0 (1–13) min for loss of posture.

Recovery phase

Only 25.8% of the farmers observed their goat kids during the entire recovery phase; 35.8% sporadically looked after them (at minimum every 15 min), and 38.7% of the farmers did not observe the goat kids during recovery at all.

During the recovery phase, 38.5% of the goat kids were touched by the farmers, while 61.5% were left completely undisturbed. Overall, time from injection to first movement was 49.4 ± 27.6 min and to steady standing was 108.7 ± 45.2 min. Time to first movement for goat kids anaesthetised with the official ketamine-xylazine dosages was 35.3 ± 18.9 min and to steady standing 82.3 ± 26.0 min.

Reactions to disbudding

Video analysis

Cautery disbudding techniques varied: in 71.3% of the goat kids, a hot dehorner was rotated to burn larger circles around the horn bud region and in 28.7% of the animals it was held steadily, perpendicular to the skull. In 18.4% of the goat kids a pen knife or shears were used to remove the tip of the horn bud before cautery, depending on the size of the horn buds. After burning, the isolated central circle containing the horn bud was removed in 37.9% of the goat kids, while the remaining farmers left the bud in place. Burning time ranged from 7 to 120 s per horn (median 25 s), with the hot iron being applied between 1 to 35 times per horn (median 3 times). 25.8% of the farmers left the dehorner longer than 60 s per horn on the kid's heads. The use of restraints for disbudding, rendering the animal immobile was applied in 20.1% of the animals, other farmers just held the head of the goat kids. Due to strong reactions, three farmers decided to administer an additional dosage of anaesthetic to five goat kids during disbudding, three of them were anaesthetised with the recommended mixture and two with xylazine alone (additional dosage: once 0.3 mg/kg xylazine, once 0.17 mg/kg xylazine and three times a mixture of 0.01 mg/kg xylazine plus 3.5 mg/kg, 3.7 mg/kg, 4.3 mg/kg ketamine).

Frequency of occurrence of specific behaviours is presented in Table 2. General reaction scores during disbudding could be evaluated for 168 goat kids: score 0 was attributed to 56.5% of the animals, score 1 to 7.7%, score 2 to 17.3% and score 3 to 18.5%.

VAS-scoring

Mean VAS attributed by the observer was 3.2 ± 3.1. Farmer VAS was 2.1 ± 2.4. Twenty-seven farmers did not score a VAS. For both scores, values ranged between 0 and 10.

Risk factors

Longer recovery

Goat kids anaesthetised with non-standard protocols required longer time from injection to first movement and

Table 2 Number (n) of goat kids showing the following reactions during the disbudding procedure (Na = not adjustable)

Event	Grade 1 (0)	Grade 2 (1–2)	Grade 3 (3–6)	Grade 4 (> 6)	Na
Head lifting (n)	153 (88.4%)	5 (2.9%)	4 (2.3%)	6 (3.5%)	5 (2.9%)
Paddling (one limb) (n)	159 (91.9%)	8 (4.6%)	1 (0.6%)	0	5 (2.9%)
Paddling (several limbs) (n)	152 (87.9%)	5 (2.9%)	8 (4.6%)	3 (1.7%)	5 (2.9%)
Kicking (one limb) (n)	161 (93.1%)	5 (2.9%)	1 (0.6%)	1 (0.6%)	5 (2.9%)
Kicking (several limbs) (n)	162 (93.6%)	4 (2.3%)	2(1.2%)	0	5 (2.9%)
Pull up (one limb) (n)	136 (78.6%)	23 (13.3%)	9 (5.2%)	0	5 (2.9%)
Pull up (several limbs) (n)	154 (89.0%)	13 (7.5%)	1 (0.6%)	0	5 (2.9%)
Focused eye movement (n)	114 (65.9%)	11 (6.4%)	6 (3.5%)	1 (0.6%)	41 (23.7%)
Spontaneous blinking (n)	49 (28.3%)	32 (18.5%)	24 (13.9%)	27 (15.6%)	41 (23.7%)
Ear movement (n)	131 (75.7%)	8 (4.6%)	1 (0.6%)	0	33 (19.1%)
Tail movement (n)	87 (50.3%)	34 (19.7%)	14 (8.1%)	10 (5.8%)	28 (16.2%)
Nose Movement (n)	132 (76.3%)	3 (1.7%)	0	0	38 (22.0%)
Mouth Movement (n)	124 (71.7%)	8 (4.6%)	3 (1.7%)	1 (0.6%)	37 (21.4%)
Groaning (stimulus-associated) (n)	96 (55.5%)	28 (16.2%)	24 (13.9%)	25 (14.5%)	0
Vocalisation (n)	108 (62.4%)	17 (9.8%)	21 (12.1%)	27 (15.6%)	0

longer to stand. Furthermore, breed and age had a significant effect on recovery as younger goat kids (≤14 days) needed longer to stand than older ones. Goat kids recovering under a heat lamp needed longer to stand steady than goat kids recovering without heat lamp (Tables 3, 4). Goat kids (5.7%) showed head rubbing on objects or companions during the recovery phase.

Higher reaction grade

The anaesthetic mixture and manipulation of all animals during anaesthesia induction were identified as significant risk factors for behavioural reactions during disbudding (limb movement, head lifting, vocalisation). Goat kids anaesthetised with xylazine followed by lidocaine infiltration had a 10 times lower risk to show reactions than goat kids anesthetised with the recommended xylazine-ketamine mixture. Goat kids being manipulated during anaesthesia induction (before loss of posture) showed 3.5 times more reactions than goat kids left undisturbed (Table 5).

Discussion

Aim of this study was to assess the quality of analgesia and anaesthesia of goat kids disbudded by certified Swiss farmers. The findings show clearly that there is a need to improve the anaesthetic protocol used by certified Swiss farmers to anaesthetise their goat kids for disbudding. Nearly half of the goat kids (43.5%) had a general reaction score of 1–3 and therefore showed signs of consciousness or pain during the disbudding procedure; 18.5% of the goat kids had a general reaction score of 3, indicating strong movements and vocalisation.

Table 3 Analysis of risk factors for first movement after disbudding in goat kids anaesthetised with different protocols by certified Swiss farmers. Final generalized estimation equation model correcting for the effect of the herd, $n = 30$ herds and 166 goat kids (CI, confidence interval; Ref, reference). Eight goat kids were excluded, because first movement could not be exactly observed

Risk factor	Beta (in minutes)	95%CI	*p*-Value
Anaesthesia protocol			< 0.001
Recommended xylazine-ketamine mixture	Ref	Ref	Ref
Xylazine-ketamine (other dosage)	7.0	−4.2-18.1	0.223
Xylazine	17.4	1.4–33.4	0.033
Xylazine-Lidocaine	50.2	40.4–59.9	< 0.001
Breed			< 0.001
Chamois-Coloured	Ref	Ref	Ref
Saanen	20.7	8.3–33.1	0.001
Toggenburg	18.6	7.9–29.3	< 0.001
Grisons Striped	−7.1	−25.7-11.5	0.454

Table 4 Analysis of risk factors for being able to stand steady after disbudding in goat kids anaesthetised with different protocols by certified Swiss farmers. Final generalized estimation equation model correcting for the effect of the herd, $n = 29$ herds and 152 goat kids (CI, confidence interval; Ref, reference). Twenty-two goat kids were excluded, because recovery could not be visited to the end

Risk factor	Beta (in minutes)	95%CI	*p*-Value
Anaesthesia protocol			< 0.001
Recommended xylazine-ketamine mixture	Ref	Ref	Ref
Xylazine-ketamine (other dosage)	14.7	−6.0-35.3	0.165
Xylazine	71.7	49.5–93.8	< 0.001
Xylazine-Lidocaine	62.1	41.4–82.8	< 0.001
Breed			0.008
Chamois-Coloured	Ref	Ref	Ref
Saanen	−20.2	−41.0-0.7	0.058
Toggenburg	13.3	−8.0-34.6	0.222
Grisons Striped	24.9	1.4–48.4	0.038
Age of goat kids			< 0.001
> 14 days	Ref	Ref	Ref
≤ 14 days	17.7	8.9–26.4	< 0.001
Heat lamp in recovery box			0.008
No	Ref	Ref	Ref
Yes	28.4	7.6–49.3	0.008

The anaesthetic protocols used for disbudding varied significantly and were often not consistent with the official recommendations. In 2008, when the farmers certificate of competence for disbudding of own goat kids was introduced, the FSVO recommended to use a mixture of xylazine (0.05 mg/kg) and ketamine (20 mg/kg) based on an unpublished field study on 40 goat kids. At that time, this combination seemed to represent the best compromise for a good quality anaesthesia and reasonable recovery times, using drugs approved for use in goats. In the literature, only little information is available about anaesthesia of goats in general and in particular of goat kids. Most of the recommendations are inferred from experience in sheep. A mixture of xylazine (0.1–0.2 mg/kg) and ketamine (10 mg/kg) to be administered intramuscularly is reported for the use in goats [15]. While the ketamine doses used by the farmers, in the present study, were in the range of the above reported doses, xylazine was overdosed in several occasions. According to the farmers, four veterinarians seemed to refuse to provide ketamine to the farmers. This probably led to the use of increased xylazine doses (up to 1.84 mg/kg). This dose represents almost 20 times the recommended dose for goats (0.1 mg/kg IM or 0.05 mg/kg IV), that are considered more sensitive to xylazine than sheep [16]. Accidental intravenous injection could not be ruled out if aspiration was not performed (70.3%).

Table 5 Analysis of risk factors for limb movement, head lifting and/or vocalisation in goat kids anaesthetised with different anaesthesia protocols for horn disbudding. Final generalized estimation equation model correcting for the effect of the herd, n = 30 herds and 165 goat kids (CI, confidence interval; Ref, reference). The three goat kids of the acepromazine-ketamine group were excluded (to small group for statistical analysis). Six goat kids of the other four groups were excluded due to firm restraint

Risk factor	Odds Ratio	95%CI OR	*p*-Value
Anaesthesia protocol			0.003
Recommended xylazine-ketamine mixture	Ref	Ref	Ref
Xylazine-ketamine (other dosage)	1.1	0.3–3.4	0.915
Xylazine	1.8	0.3–9.4	0.503
Xylazine-Lidocaine	0.1	0.0–0.3	< 0.001
Intervention of goat owner during induction			0.002
No	Ref	Ref	Ref
Yes	3.5	1.6–7.9	< 0.001

A study on adult goats reported undesired cardiopulmonary effects such as hypoxemia, hypotension, hypoventilation and bradycardia following administration of intravenous xylazine (0.05 mg/kg) [17]. Clearly, side effects such as deep anaesthesia and long duration of action have to be expected after administration of the high doses used in the present study. Indeed, in those animals, complete absence of reaction to disbudding indicating deep anaesthesia and prolonged recoveries were observed. Surprisingly, goat kids recovering underneath a heat lamp took a longer time to steady standing. The improved cardiovascular and metabolic state should lead to a faster recovery from anaesthesia, but as the animals were left completely undisturbed, they might have felt comfortable underneath the heat lamp and they remained lying.

Goat kids that were administered xylazine followed by a lidocaine nerve block showed less behavioural reactions than goat kids anaesthetised with the other four anaesthetic protocols. Dosages of 1.2 mg/kg (0.2 ml lidocaine 2% per horn) and 3.1 mg/kg lidocaine (0.5 ml lidocaine 2% per horn) were used, respectively. Both farmers using lidocaine for local anaesthesia blocked only the cornual branch of the zygomaticotemporal (lacrimal) nerve. In goats, the sensory innervation of the horn is also provided by the cornual branch of infratrochlear nerve that can be blocked at the dorsomedial margin of the orbit [15]. Alvarez et al., 2015 showed that the cornual nerve block with lidocaine did not prevent stress (measured with cortisol-levels) and painful reactions such as vocalisation in goat kids during and after disbudding, even in higher dosages of 1 ml 2% lidocaine per horn [18]. Therefore, the most probable reason for the reduced reactions of goat kids disbudded under systemic xylazine and local lidocaine in this study was the high xylazine dose and not the local anaesthesia.

Further discussion point is the insufficient analgesia after the disbudding procedure. Non-steroidal anti-inflammatory drugs (meloxicam or tolefenamic acid) were administered to only 11 goat kids. Administration of meloxicam significantly reduced signs of pain, as measured with a VAS, in goat kids on the first day after disbudding [19]. One reason for the scarce administration of analgesics are the higher costs.

Not only the anaesthetic protocol, but also the disbudding technique varied considerably. Burning time per horn differed strongly (7–120 s). Short time application of the dehorner (< 20–25 s in total) and a proper disbudding technique are necessary, because the skull of goat kids is much thinner and the horn bud larger than in calves [20, 21]. The thin skullcap and the small sinus frontalis facilitate complications during thermal disbudding, such as necrosis of the frontal bone and the underlying frontal cortex of the brain [22]. A standard and well-defined disbudding technique will be necessary for future studies on anaesthesia and analgesia refinement. Further important point is a solid education of farmers to ensure adequate disbudding technique and proper administration of anaesthetics, what has to be realised by the FSVO.

Limitations of the present study were the small number of farmers actually disbudding without assistance of a veterinarian and therefore the limited number of observable animals. The influence of the various anaesthetic protocols was large due to the small sample size, so that many other possible risk factors for higher general reaction score (as for example waiting time from injection to disbudding, burning time or experience of disbudder by number of disbudded goat kids per year) did not become significant. Further limitations were the voluntary participation in the study and the observation at non-standardized field condition.

Conclusions

Anaesthesia and analgesia quality during disbudding performed by certified Swiss farmers was not sufficient, even in cases in which the officially recommended ketamine-xylazine mixture was administered. From the point of view of animal welfare, further studies to develop appropriate and safe anaesthesia protocols for young goat kids are definitely required, as well as improvement of farmer education to ensure proper administration of anaesthetic mixtures.

Endnotes

[1]GoPro Hero4 silver edition camera (https://gopro.com/update/hero4)

[2]WinEpiscope 2.0, Software for Quantitative Veterinary Epidemiology, Facultad de Veterinaria, Zaragoza, Spain, Agricultural University Wageningen, the Netherlands, University of Edinburgh, UK (http://www.winepi.net)

[3]Microsoft® Office Excel 2016, Microsoft Corporation (https://www.microsoft.com)

[4]NCSS 10 Data Analysis, Statistical Software, Kaysville, Utah, USA (https://www.ncss.com)

[5]SAS® 9.4 Software (http://support.sas.com/software/94/index.html)

Additional files

Additional file 1: Data collection recorded with a standardized protocol (goat kid protocol). Includes the goat-kid protocol.

Additional file 2: Data collection recorded with a standardized protocol (farm protocol). Includes the farm protocol.

Additional file 3: Description of the behavioural states and events observed during anaesthesia induction and recovery of goat kids before and after disbudding. Describes the following behaviours: Staggering, recumbency, loss of posture, first movement, attempt to stand, steady standing.

Additional file 4: Description of behavioural events observed during disbudding of anaesthetised goat kids. Describes the following events: Head lifting, limb movement, focused eye movement, spontaneous blinking, ear movement, tail movement, nose movement, mouth movement, vocalisation.

Additional file 5: General reaction score of goat kids during disbudding based on vocalisation, limb movement and head lifting. A score between 0 and 4, zero means no movement and no vocalisation, four means strong movements and vocalisation.

Additional file 6: Risk factors for the three outcomes grade, first movement and steady standing, divided in farm-level risk factors and goat kid-level risk factors. Description and categorization of the different risk factors on farm and goat kid level.

Abbreviations
ANOVA: Analysis of variance; FSVO: Federal Food Safety and Veterinary Office; NW, UM: N. Wagmann, U. Morath-Huss; QIC/ QICu: Quasi-likelihood under the independence model criterion; SD: Standard deviation; VAS: Visual Analogue Scale

Acknowledgements
We thank the Swiss goat breeding association for their collaboration and for the list of goat breeders. Special thanks equally go to the goat owners and breeders who agreed to participate in this study. The authors confirm that the funders had no influence over the study design, content of the article or selection of this journal and we thank Dr. Shannon Axiak Flammer for the careful English proofreading.

Funding
The study has been supported by the Federal Food Safety and Veterinary Office (FSVO). We thank the FSVO for funding the project.

Authors' contributions
PZ, CS designed, planned and supervised the study. NW, UM collected and analyzed data. NW, GS did statistical analysis. PZ, CS helped with editing the manuscript. All authors read and approved the final manuscript.

Consent for publication
Not applicable.

Competing interests
The authors declared that they have no competing interests.

Author details
[1]Clinic for Ruminants, Vetsuisse Faculty, University of Bern, Bremgartenstrasse 109a, 3012 Bern, Switzerland. [2]Department of Veterinary Clinical Science, Anaesthesiology and Pain Therapy Division, Vetsuisse Faculty, University of Bern, Länggassstrasse 124, 3012 Bern, Switzerland. [3]Veterinary Public Health Institute, Vetsuisse Faculty, University of Bern, Schwarzenburgstrasse 155, 3097 Liebefeld, Bern, Switzerland.

References
1. Fubini SL, Ducharme NG. Farm animal surgery. 2nd ed: Elsevier; 2017.
2. Hempstead MN, Waas JR, Stewart M, Cave VM, Sutherland MA. Behavioural response of dairy goat kids to cautery disbudding. Appl Anim Behav Sci. 2017;194:42–7.
3. Boyd JH. Disbudding goat kids. Vet Rec. 1988;122:494.
4. Council of Europe. Recommendation concerning goats. 1992. Art. 28: https://www.coe.int/t/e/legal_affairs/legal_co-operation/biological_safety_and_use_of_animals/farming/Rec%20goats%20E.asp. Accessed 8 Mar 2018.
5. Bundesgesetzblatt für die Republik Österreich. Nr. 118. 2004. https://www.ris.bka.gv.at/eli/bgbl/I/2004/118. Accessed 8 Mar 2018.
6. Tierschutzgesetz TschG § 6. https://www.gesetze-im-internet.de/tierschg/__6.html. Accessed 8 Mar 2018.
7. Veterinary Surgeons Act 1966. https://www.legislation.gov.uk/ukpga/1966/36. Accessed 8 Mar 2018.
8. Tierschutzgesetz (TschG) Art. 16. https://www.admin.ch/opc/de/classified-compilation/20022103/index.html (2005). Accessed 8 Mar 2018.
9. Tierschutzverordnung (TschV) Art. 32. https://www.admin.ch/opc/de/classified-compilation/20080796/index.html (2008). Accessed 8 Mar 2018.
10. Tierarzneimittelverordnung (TAMV) Art. 8/10. https://www.admin.ch/opc/de/classified-compilation/20030705/index.html (2004). Accessed 8 Mar 2018.
11. Alvarez L, Nava RA, Ramirez A, Ramirez E, Gutierrez J. Physiological and behavioural alterations in disbudded goat kids with and without local anaesthesia. Appl Anim Behav Sci. 2009;117(3–4):190–6.
12. Alvarez L, Gutierrez J. A first description of the physiological and behavioural responses to disbudding in goat kids. Anim Welf. 2010;19(1):55–9.
13. Wenger S, Jäggin N, Doherr M, Schatzmann U. Die Halothanästhesie zur Kastration des Saugferkels: Machbarkeitsstudie und Kosten-Nutzen-Analyse. Tierärztl Prax. 2002;30:164–70.
14. Liang KY, Zeger SL. Longitudinal data analysis using generalized linear models. Biometrika. 1986;73:13–22.
15. Clarke KW, Hall LW, Trim CM. Veterinary anaesthesia. 11th ed: Saunders; 2013.
16. Riebold TW. Ruminants. In: Grimm KU, Lamont LA, Tranquilli WJ, Greene SA, Robertson SA, editors. Veterinary anesthesia and analgesia. 5th ed: Wiley & Sons; 2015. p. 912–27.
17. Stegmann GF. Observations on some cardiopulmonary effects of midazolam, xylazine and a midazolam/ketamine combination in the goat. J S Afr Vet Assoc. 1999;70:122–6.
18. Alvarez L, De Luna JB, Gamboa D, Reyes M, Sanchez A, Terrazas A, Rojas S, Galindo F. Cortisol and pain-related behavior in disbudded goat kids with and without cornual nerve block. Physiol Behav. 2015;138:58–61.
19. Ingvast-Larsson C, Högberg M, Mengistu U, Olsen L, Bondesson U, Olsson K. Pharmacokinetics of meloxicam in adult goats and its analgesic effect in disbudded kids. J Vet Pharmacol Ther. 2011;34:64–9.
20. Matthews JG. Diseases of the goat. 3rd ed. Oxford: Blackwell Science; 2009.
21. Smith MC, Sherman DM. Goat medicine. 2nd ed. Ames, Iowa: Wiley-Blackwel; 2009.
22. Thompson KG, Bateman RS, Morris PJ. Cerebral infarction and meningoencephalitis following hot-iron disbudding of goat kids. N Z Vet J. 2005;53:368–70.

10

RNA-Sequencing for profiling goat milk transcriptome in colostrum and mature milk

Alessandra Crisà[1*], Fabrizio Ferrè[2], Giovanni Chillemi[3] and Bianca Moioli[1]

Abstract

Background: In this work we aimed at sequencing and assembling the goat milk transcriptome corresponding at colostrum and 120 days of lactation. To reconstruct transcripts we used both the genome as reference, and a *de novo* assembly approach. Additionally, we aimed at identifying the differentially expressed genes (DEGs) between the two lactation stages and at analyzing the expression of genes involved in oligosaccharides metabolism.

Results: A total of 44,635 different transcripts, organized in 33,757 tentative genes, were obtained using the goat genome as reference. A significant sequence similarity match was found for 40,353 transcripts (90%) against the NCBI NT and for 35,701 (80%) against the NR databases. 68% and 69% of the *de novo* assembled transcripts, in colostrum and 120 days of lactation samples respectively, have a significant match with the merged transcriptome obtained using Cufflinks/Cuffmerge. *CSN2, PAEP, CSN1S2, CSN3, LALBA, TPT1, FTH1, M-SAA3, SPP1, GLYCAM1, EEF1A1, CTSD, FASN, RPS29, CSN1S1, KRT19* and *CHEK1* were found between the top fifteen highly expressed genes. 418 loci were differentially expressed between lactation stages, among which 207 and 122 were significantly up- and down-regulated in colostrum, respectively. Functional annotation and pathway enrichment analysis showed that in goat colostrum somatic cells predominate biological processes involved in glycolysis, carbohydrate metabolism, defense response, cytokine activity, regulation of cell proliferation and cell death, vasculature development, while in mature milk, biological process associated with positive regulation of lymphocyte activation and anatomical structure morphogenesis are enriched. The analysis of 144 different oligosaccharide metabolism-related genes showed that most of these (64%) were more expressed in colostrum than in mature milk, with eight expressed at very high levels (*SLCA3, GMSD, NME2, SLC2A1, B4GALT1, B3GNT2, NANS, HEXB*).

Conclusions: To our knowledge, this is the first study comparing goat transcriptome of two lactation stages: colostrum and 120 days. Our findings suggest putative differences of expression between stages and can be envisioned as a base for further research in the topic. Moreover because a higher expression of genes involved in immune defense response, carbohydrate metabolism and related to oligosaccharide metabolism was identified in colostrum we here corroborate the potential of goat milk as a natural source of lactose-derived oligosaccharides and for the development of functional foods.

Keywords: Goat milk, Transcriptomic profile, RNA-seq, Lactation, Colostrum, Oligosaccharides, Milk somatic cells

* Correspondence: alessandra.crisa@crea.gov.it
[1]Consiglio per la ricerca in agricoltura e l'analisi dell'economia agraria (CREA) - Animal production research centre, Via Salaria 31, 00015 Monterotondo, Rome, Italy
Full list of author information is available at the end of the article

Background

One of the most important livestock and the oldest domesticated species raised worldwide is the goat (*Capra hircus*). Goats have long been used for their milk, meat, hair (including cashmere), and skins throughout much of the world and this species is also one of the best model organisms for mammary gland bioreactor studies [1]. Today, there are > 1000 goat breeds, and > 900 million goats are kept around the world according to a report by the UN Food and Agriculture Organization [2, 3]. Despite the agricultural and biological importance of goats, breeding and genetics studies have been delayed due to the lack of a reference genome. In [4] it was reported for the first time the sequencing and automated whole-genome optical mapping of the goat genome. Besides its nature of complete food able to provide nutritional subsistence of infants, healthy growth of children and nourishment of adult humans, milk contains a plethora of components that encode functional health benefits to consumers [5]. Among dairy livestock, goats produce milk containing free milk oligosaccharides (OS) at much higher levels than cattle or sheep [6, 7] and different studies have shown that their concentration is higher in colostrum than in the subsequent lactation stages [8–10]. A vast number of functional in vitro and in vivo studies gave evidence of the biological functions of OS in human health and nutrition [11–15]. Moreover, goat milk contains a significant variability of OS, similar to the one observed in human milk, so to hold high potential as a natural source of lactose-derived OS as supplement for infant formulae and for the development of functional foods, having been successfully isolated by membrane technology [7, 16]. Goat milk contains both neutral and acidic OS. Galactose (Gal), glucose (Glc), fucose (Fuc), Nacetylglucosamine (GlcNAc) and N-acetylgalactosamine (GalNAc) grafted on a lactose core make up neutral OS; these same monomers, plus N-acetylneuraminic acid (NeuAc) and N-glycolylneuraminic acid (NeuGc), comprise acidic OS [17, 18]. Oligosaccharides chains covalently linked to glycoprotein and glycolipids often confer specific biological functions to the molecules carrying them such as signaling, adherence and cellular migration through the body. Glycosylation of proteins and lipids plays a crucial role in numerous biological processes including the regulation of immune and inflammatory responses [19]. To understand the function of glycoconjugated it is important to investigate the regulation of expression of the genes encoding the "glycosylation-related" genes, such as the large families of glycosyltransferase and glycosidase [20–22].

Milk component synthesis and secretion by the mammary gland involve expression of a large number of genes. Formerly, many studies of gene expression were confined to the analysis of a limited number of selected genes of interest, e.g. the genes encoding the key lipogenic enzymes in the mammary gland [23, 24]. The recent development of Next-Generation Sequencing technologies allows a much broader and detailed knowledge on the biology of the mammary gland [25]. RNA-seq transcriptomic analysis has been widely used for model and non-model organisms and in particular for organisms for which limited genomic resources are available [26–29].

Principal aim of the present study aim was to profile the transcriptome of goat milk somatic cells (GMSCs) by using an RNA-seq approach. Because, currently, there are no official gene annotations for the goat genome, we decided to employ two strategies to reconstruct transcripts as accurate as possible: 1. using the genome as reference to build transcripts from the sequencing reads, and 2. using a *de novo*, genome-independent assembly approach. Additionally, we aimed to identify the differentially expressed genes (DEGs) between two lactation stages, colostrum (D1) and 120 days (D2); in fact, colostrum composition highly differs from milk composition once the lactation is established, being the colostrum richer in nutrients and characterized by high levels of bioactive components [30]. Finally, we also aimed to compare, in the two different lactation stages, the expression of genes involved in OS metabolism, which, to date, was never analyzed in goats, but calls upon utmost interest, as shown by the studies on the expression of glycosylation-related genes, in mouse [31], humans [20, 32] and cattle [33].

Methods

Sample collection and RNA extraction

RNA used in the present trial was obtained from goat milk somatic cells (GMSCs), in agreement with [34], who showed that GMSCs can be used to accurately reveal the cellular dynamics of mammary gland gene expression, and with other authors [35–37] who demonstrated an extensive similarity between the mammary gland and MSC/mammary epithelial cells transcriptome.

Milk was hand collected (both udder halves) from a Maltese goat, first parity, raised in a farm privately owned and traditionally managed. Two lactation stages were sampled: colostrum (D1) and 120 days milk (D2). Fifty mL of milk were transferred to falcon tubes and immediately centrifuged at 2000 g for 10 min at 20 °C in the presence of 50 µl EDTA 0,5M pH 8. The fat layer on the top of the supernatant was removed with a sterile pipette tip and the skim milk was discarded. The milk somatic cell-pellets were resuspended in 1 ml PBS 1× with 300 µl EDTA 0,5M pH 8. After centrifugation the pellets were lysed with 1 mL TRI REAGENT (Sigma-Aldrich, Milan, Italy) and total RNA was extracted following the TRI REAGENT protocol. Samples were digested with RNase-Free DNase Set (Qiagen, Milan, Italy) to remove DNA contamination and finally cleaned with RNeasy

Minelute kit (Qiagen, Milan, Italy). RNA quality and quantity was checked with spectrophotometric (NanoPhotometer™ Pearl, Implen GmbH, München, Germany) and microfluidic inspections (RNA 6000 Nano chip, 2100 Bioanalyzer, Agilent Technologies, Milan, Italy).

Illumina sequencing and transcriptome assembly

The RNA was sequenced by using the ARK-Genomic center service at The Roslin Institute, University of Edinburgh [38]. Samples taken on for processing had an RNA integrity number (RIN) value greater than 7. Briefly, the polyadenylated RNA was isolated onto beads to remove it from all of the ribosomal RNAs, was then fragmented to an average size of 180 bases and reverse-transcribed to cDNA using random oligonucleotide primers. Following second-strand cDNA synthesis, end repair, and addition of a single A base, Illumina RNA-seq primers were ligated to the A-tailed double stranded cDNA. The ligated adapters were extended by PCR to add the sample barcode sequences and the resulting PCR products were assessed for quality using the Agilent Bioanalyzer DNA 1000 chip (average insert length of 175 bases) and for quantity using the Kapa Biosciences library quantification kit (Illumina). Following single stranding, dilution, loading onto an Illumina V2 flow cell for the Genome Analyzer IIx instrument, using a cBot following the manufacturer's recommendations, the samples were finally sequenced for 79 bases paired end in a single lane of an Illumina Genome Analyzer IIx. The data was extracted to FastQ format using the Casava software from Illumina.

The raw read data from this study were deposited to the NCBI Sequence Read Archive (SRA) under accession number SRA261702. Read quality was examined using FastQC (v0.10.1) [39]. After adapter sequence removal, read ends were dynamically trimmed if base quality scores were lower than Q20 using Trimmomatic [40]. At the end of this process, reads were discarded if the total length of the read was shorter than 10 bases or if the fraction of undetermined bases was >2%.

The clean reads were subsequently analyzed with two approaches: a) by using the Tuxedo Suite, comprising of Bowtie (v2.1.0) [41], TopHat (v2.0.9) [42] and Cufflinks (v2.1.1) [43]. In particular, we aligned the sequence reads to the reference goat genome assembly AJPT0000000.1 [4] with a tolerance of two mismatches (Bowtie), we identified splicing junctions (TopHat), and we reconstructed transcripts and measured their expression levels (Cufflinks) reported as Fragments Per Kilobase of transcript per Million mapped reads (FPKM). b) by using the Inchworm RNA-Seq Assembler in the Trinity software [44], we performed a *de novo* transcriptome assembly, and we mapped the resulting contigs (corresponding to transcripts) to the goat genome using Gmap [45]. Additionally, Cufflinks was run in two different ways: a) using the available goat transcript sequences from the RefSeq database; b) without transcript information. For the latter, the Cufflinks generated transcriptomes obtained independently for the D1 and the D2 sample were merged using Cuffmerge, part of the Cufflinks distribution, to obtain a reference transcriptome. For the former, we started from the 27,946 RefSeq goat transcripts (as of July 2015), of which 27,649 reported as predicted, then we determined their genomic coordinates by using BLAT [46], converted them into General Feature Format (GFF), and feed this GFF file to Cufflinks to compute expression values, in FPKM units, using the Bowtie/TopHat mapped reads for the two samples.

Transcripts generated by Trinity and Cufflinks were compared to each other using BLASTN, requiring a significant match (E-value < = 10E-5) covering at least 50% of the longest transcript sequence.

Venn diagrams were obtained by using eulerAPE software [47].

Functional annotation and pathway enrichment analysis

The Cufflinks/Cuffmerge and the Trinity transcripts were BLASTed against the NCBI GenBank nucleotide (NT) or non-redundant protein (NR) databases to annotate them. A match was accepted if the BLAST top hit E-value was smaller than 10E-5. Transcripts for which no sufficiently good matches were found after this procedure were classified as 'unannotated'. Gene symbols were associated to each match using bioDBNet [48]. Gene ontology (GO) and KEGG pathways annotations were obtained using Blast2GO [49] against the NR database (E-value cutoff 10E-5). This method has been applied to the full set of transcripts obtained for the two samples.

Lists of highly expressed genes and lists of differentially expressed genes (DEGs) between stages were characterized using the free Database for Annotation, Visualization and Integrated Discovery (DAVID) v6.7 web-service [50, 51]. Enrichment in the GO three ontologies and in pathways was evaluated by DAVID by means of a modified Fisher Exact Test (EASE), using an EASE score threshold of 0.1. The DAVID Functional Annotation Clustering tool was used to group annotations, setting stringency to Medium and using default values for the other parameters. The Enrichment Score was used to rank functional clusters, and a representative annotation was chosen to represent each group as the one having lowest enrichment p-value. Clusters having no term with p-value < 0.05 were discarded.

Differential gene expression profiling

NOISeq was used to evaluate differential gene transcription between stages [52]. Read counts for each gene were compiled using HTSeq [53], normalized using the tmm (trimmed mean of M values) method, and analyzed by NOISeq using parameters nss = 5 (number of simulated replicates), pnr = 0.2 (size of simulated samples), v = 0.02 (simulated samples variability). Differentially expressed genes (DEGs) where selected as those with probability q > = 0.9. The WEB-based GEne SeT AnaLysis Toolkit (WebGestalt) software was used for functional enrichment analysis of the differential expressed genes [54, 55]. This tool integrates functional categories derived from centrally and publicly curated databases as well as computational analyses; it adds functional categories defined by hierarchical protein interaction network modules. The default software options were used with Benjamini-Hochberg multiple test adjustment. The WebGestalt software resulted in the visualization feature DAGs (directly acyclic graph) that not only revealed the hierarchical relationship of enriched GO terms but also extended to define relationship among genes in a network module. For the pathway collection, in addition to the KEGG database, WebGestalt has added data from Pathway Commons, which collects and integrates nine centrally curated biological pathway database, and WikiPathways, that is a primary source for open community-based curation. The Search Tool for the Retrieval of Interacting Genes/Proteins (STRING v9.1) database and web tool [56] was used to provide a critical assessment and integration of protein-protein interactions, including direct (physical) as well indirect (functional) associations. Interactions are derived from multiple sources. The probabilistic confidence score that is assigned to associations, is derived by separately benchmarking groups of associations against the manually curated functional classification scheme of the KEGG database. The various major sources of interaction/association data in STRING are benchmarked independently; a combined score is computed which indicates higher confidence when more than one type of information supports a given association. The "minimum required interaction score" puts a threshold on the confidence score, such that only interactions above this score are included in the predicted network. Lower thresholds lead to more interactions, but also to more false positives. The confidence score is the approximate probability that a predicted link exists between two enzymes in the same metabolic map in the KEGG database. For the present study, we considered *Bos Taurus* as the reference organism and a minimum required interaction score of 0.7, which corresponds to a high-confidence predicted network.

Enrichment of sugar metabolism genes

We edited a list of genes implicated in milk OS metabolism and of genes relative to the lactose synthesis pathway following [33] and [57] respectively, with integrations. A list of 144 genes belonging to different functional OS metabolism categories was examined in the milk goat transcriptome. Goat orthologs were identified by using BLASTX against the NCBI NR (E-value < 10E-5), and expression values of the orthologs were computed by Cufflinks, as described above.

Results

RNA-seq mapping statistics and *de novo* assembly

RNA sequencing produced a total of ~ 100 million reads per sample; the number of reads obtained before and after the quality filtering are reported in Table 1. The raw read data were deposited to the NCBI Sequence Read Archive (SRA) under accession number SRP057582.

Two strategies were employed to profile the goat milk transcriptome to overcome the limits of the currently available goat genome sequence and transcriptome annotations. We employed the putative transcripts provided by RefSeq through the Bowtie/TopHat/Cufflinks pipeline, but we also let the software reconstruct genes and transcripts directly from the read mapping. These pipelines were applied independently to both lactation samples, and the resulting transcripts were then merged and annotated. We also reconstructed the transcriptome using a *de novo* approach (i.e. without using a reference genome) by means of the software Trinity, with the aim of detecting transcripts encoded in missing or mis-assembled loci of the current genome assembly. Results of both approaches were compared and finally integrated.

The Cufflinks procedure when the RefSeq annotations were not provided generated 37,969 transcripts for the D1 sample, having an average length of 1,947 bp, organized in 31,311 putative genes, while the D2 sample generated 38,533 transcripts from 31,817 putative genes, 1,896 bp long on average. A total of 44,635 different transcripts, having average length of 2,296 bp (the longest transcript is 33,250 bp long) was obtained by merging the two datasets. The merging procedure, using Cuffmerge, also automatically discarded potential artifacts. These transcripts are organized in 33,757 clusters (i.e. tentative genes), which encode for 1.3 splicing variants on average (Additional file 1). All these transcripts were BLASTed against the NCBI nucleotide (NT) and non-redundant protein (NR) databases; a significant match was found for 40,353 of them (90%) against the NT and for 35,701 (80%) against the NR databases (Table 2).

The *de novo* transcriptome assembly for the D1 sample produced 136,420 assembled fragments from 75,147 clusters (i.e. putative genes), with 1,311 bp of fragment average length, of which 112,001 (82%) map to the goat

Table 1 Summary of read counts and alignments

Samples	Total reads	Clean reads	Mapped reads	Mapped/clean
D1	115,900,892	115,402,457	106,631,870	92.4%
D2	109,955,930	109,511,963	100,312,958	91.6%

Table 2 Statistics of the Cufflinks transcripts

Samples	Transcripts (Cufflinks)	Putative genes (Cufflinks)	Averaged fragments length (Cufflinks)	Total merged transcripts (Cuff merge)
D1	37,969	31,311	1,947	44,635 of which 40,353 (90% match against NT) 35,701 (80% match against NR)
D2	38,533	31,817	1,896	

genome assembly with at least 90% identity and 90% sequence coverage. 58,495 fragments (43%) found a significant match with the NCBI NR and only 5404 assembled fragments (0.03%) do not show any significant match with the assembly. A total of 142,344 fragments from 73,823 putative genes (average fragment length 1,363 bp) were assembled from the D2 sample reads; 115,686 fragments (82%) map well on the goat genome (only 0.05% do not have any significant genomic match), while 66,919 (47%) have a significant hit in the NCBI NR (Table 3).

The *de novo* assembled transcripts in the two lactation stages show a relevant correspondence with the merged transcriptome obtained using Cufflinks/Cuffmerge (44,635), as detected by BLASTing the former against the latter: 68% of the D1 sample *de novo* transcripts, and 69% of the D2 sample transcripts, have a significant match (E-value < 10E-5), and for 55% of both the D1 and the D2 sample transcripts, the match also covers at least 50% of the transcript sequence. While the assembled transcripts align remarkably well with the goat genome assembly, they must be taken cautiously since a large fraction of them is likely incomplete (i.e. not full length transcripts). To further validate them, we compared the *de novo* transcripts with those reconstructed by the Cufflinks/Cuffmerge pipeline, discarded those that find a good correspondence, and selected for further analysis those that, while not having a match with the genome, found a significant match in the NCBI NR database. The rationale is that this subset of *de novo* transcripts are more likely encoded by regions not present or mis-assembled in the current goat genome assembly and therefore missed by Cufflinks/Cuffmerge. Figure 1a reports as a Venn diagram a classification of the D1 *de novo* transcripts, that shows how many have a significant match with those reconstructed using the goat genome as reference by the Cufflinks pipeline, how many have a significant match with the NR database, and how many align well on the genome assembly, and the overlap between these groups. For example, a relevant portion of all transcripts (39,738, 29% of the total) shows significant matches with the Cufflinks transcripts, the NR and the genome, and corresponds to the region of overlap of the three groups at the center of the Venn diagram; another large group (35,939, 26%) aligns well on the genome and with the Cufflinks transcripts, but not with the NR, and corresponds to the overlap region between these two groups at the bottom of the Venn diagram. Figure 1b reports a similar Venn diagram for the D2 sample *de novo* transcripts. In this case, transcripts that have significant alignment with the Cufflinks transcripts, the NR and the genome are 45,641 (32% of the total), while those having a good match with the Cufflinks and the genome but not with the NR are 34,351 (24%). These might represent goat-specific transcripts, even if a fraction might not find a significant match in the NR because their sequence was not fully reconstructed, or reconstructed only or mainly in a non-translated region, or by the lack of an homologous protein in the database. In general, there is a good agreement between transcripts reconstructed using the genome as reference and those reconstructed using a *de novo* approach, and many of them can be annotated using a match with the NR dataset; this holds true for both the D1 and the D2 sample, and the relative size of these groups of transcripts is similar in the two samples. Yet, a large fraction of the *de novo* transcripts cannot be annotated.

A small fraction of transcripts (2699 for the D1 sample and 4098 for the D2 sample) find a significant match in the NR while not having correspondence either with the goat genome or with the Cufflinks transcripts. We suggest that those transcripts are encoded by loci not present or not well assembled in the current goat genome assembly. On the other hand, many *de novo* transcripts do not show any significant match with the Cufflinks/Cuffmerge transcripts, even though they can be mapped reliably to the goat genome assembly. Specifically, these are 36,324 for the D1 sample (the 34,119 that match only with the genome assembly plus the 2,205 that match with both the genome and the NR) and 35,794 for the D2 sample (33,902 plus 1,892), as shown in Fig. 1. Nevertheless, in the majority of cases (30,556 for the D1 and 30,672 for the D2 sample), even if the sequence of these transcript

Table 3 Statistics of the *de novo* transcripts

Samples	Transcripts *de novo*	Averaged fragments length *de novo*	Transcripts mapped to goat genome (90% identity)	Transcripts matched with BLAST NR (E-value 10E-5)
D1	136,420	1,311	112,001 (82%)	58,495 (43%)
D2	142,344	1,363	115,686 (82%)	66,919 (47%)

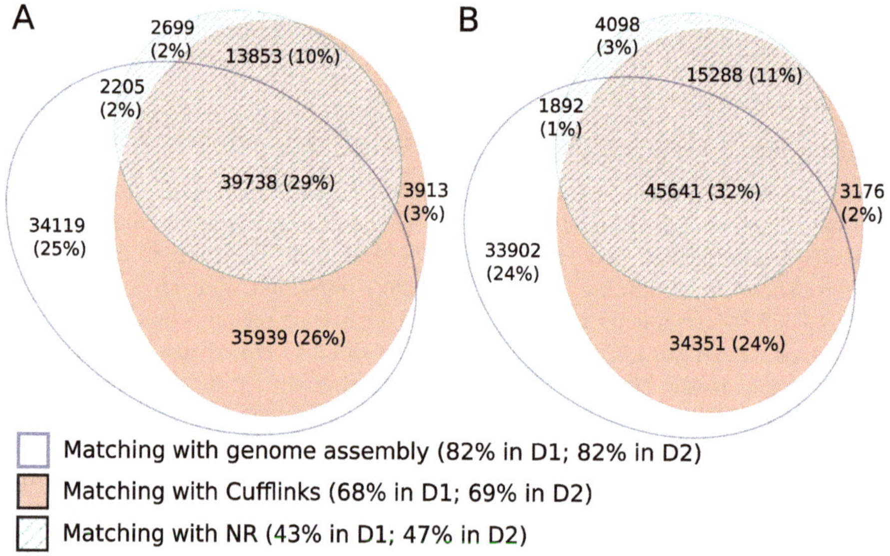

Fig. 1 Matches between the *de novo* transcripts with goat genome, Cufflinks/Cuffmerge transcripts, NR database. **a** *De novo* D1 transcripts match with goat genome, Cufflinks/Cuffmerge transcripts and NR database. **b** *De novo* D2 transcripts match with goat genome, Cufflinks/Cuffmerge transcripts and NR database

cannot be aligned significantly with any Cufflinks transcript, the mapped genomic coordinates of these *de novo* transcripts overlap with one of the loci reconstructed by Cufflinks with at least 200bp, suggesting that these transcripts are transcribed by the same loci. The lack of significant similarity as detected by BLASTing can be ascribed to errors in the transcript assembly, to the ability of Trinity and Cufflinks in reconstructing different portions of the transcript, or to incorrect reconstruction of alternative splicing events. The 3553 remaining *de novo* D1 and 3997 D2 sample transcripts need further investigation. A relevant fraction of the highly expressed transcripts (182 out of 798, 23% in the D1 sample; 180 out of 644, 28% in the D2 sample) did not show any significant match with the NCBI NR dataset. These un-annotated transcripts tend to be short, and nucleotide BLAST versus the NCBI NT collection showed that most of them find significant matches with ribosomal RNAs, suggesting that they might result from rRNA sample contamination.

Milk gene expression statistics

We first analyzed expression data, considering the transcripts obtained with the Cufflink procedure, setting four thresholds for the FPKM values (0.1, 1, 5, 15). Thereafter divided them into high (> = 500 FPKM), medium (> = 10 to 500 FPKM), and low expression (<10 FPKM) (Table 4). We observed a general agreement between the two samples (Table 4), both when using and when not using RefSeq predicted transcript annotations. In particular, there were 798 (3%) and 644 (2%) highly expressed genes in the D1 and in the D2 sample respectively when the Cufflinks procedure was employed without providing the RefSeq transcript coordinates, of which 496 genes are common between them. In general, gene expression resulted similar in the two considered samples, regardless of the analysis method.

In order to further categorize the genes with different level of expression, a multiphasic graph was obtained by plotting the log2 transformed FPKM value versus the

Table 4 Gene expressed in D1 and D2 or co-expressed (i.e. found expressed in both samples) as percentage of the total number of transcripts and following the categorization

	Without annotations			With RefSeq annotations		
Category	D1	D2	Co-expressed	D1	D2	Co-expressed
FPKM value > =0.1	31,040 (95%)	31,136 (95%)	29,392	16,646 (60%)	16,200 (58%)	15,242
FPKM value > =1	17,188 (52%)	17,751 (54%)	15,768	11,327 (41%)	10,634 (38%)	10,203
FPKM value > =5	8,800 (27%)	8,472 (26%)	7,801	6,028 (22%)	4,983 (18%)	4,826
FPKM value > =15	4,465 (14%)	3,876 (12%)	3,643	3,017 (11%)	2,153 (8%)	2,095
Highly expressed genes (> = 500 FPKM)	798 (3%)	644 (2%)	594	164 (0.06%)	104 (0.03%)	103
Medium expressed genes (> = 10 FPKM to 500 FPKM)	7,733 (24%)	7,771 (24%)	6,856	3,782 (14%)	2,888 (11%)	2,750
Lowly expressed genes (<10FPKM)	24,453 (74%)	24,512 (75%)	23,663	23,460 (86%)	24,414 (89%)	23,383

expressed genes (Fig. 2). Two peaks, at 0.2 and 1 FPKM, respectively, are visible.

Global milk transcriptome functional annotation and pathway analysis

To look at the different pattern of biological functions in the two studied lactation stages, transcripts reconstructed by Cufflinks in the D1 (31,311) and D2 sample (31,817) were annotated independently using the Blast2GO pipeline, starting from the BLASTX against the NR results. A total of 14,179 transcripts from the D1 and 14,620 from the D2 sample can be associated with at least one GO term. Distribution of GO terms and KEGG pathways were investigated, resulting quite similar. Both D1 and D2 sample GO terms lists, ranked by the Blast2GO score show at the top rather generic GO terms related to transcription regulation, signal transduction, transport, metabolic processes, cell division, phosphorylation, and others. It is to be expected that these two samples, that share the expression of the large majority of genes, also show similar GO terms distributions.

We also assessed the enrichment in the number of genes associated to a specific GO term or pathway comparing one sample to the other (Fisher exact test, p-value < = 0.05, FDR 5%). In the D1 sample, there was enrichment in 67 terms associated to cellular metabolic processes, cellular organization, biosynthesis, and response to stimuli (Additional file 2 and Fig. 3a). In the D2 sample, instead, the 20 enriched GO terms included several terms related to apoptosis (positive regulation of cell death, positive regulation of apoptotic process, induction of programmed cell death, developmental programmed cell death) (Additional file 2 and Fig. 3b). The examination of KEGG pathway categories and subcategories indicated that a large number of pathways (124) results expressed in both samples, and the most represented ones are related to purine and pyrimidine metabolism, aminoacyl-tRNA biosynthesis, inositol and phosphatidylinositol metabolism and signaling, glycerolipid, glycerophospholipid and sphingolipid metabolism, lysine degradation, oxidative phosphorylation, amino- and nucleotide-sugar metabolism, and T cell receptor signaling. Three pathways (amino sugar and nucleotide sugar metabolism, lysine degradation, steroid hormone biosynthesis) are enriched in the D1 with respect to the D2 sample in terms of number of expressed pathway members (Fisher exact test, p-value < = 0.05, FDR 5%). Two pathways, the naphthalene degradation and the lipoic acid metabolism pathways, are specific for the D1 sample, even though represented by very few expressed genes, while no pathway appears specific for the D2 sample (Additional file 3).

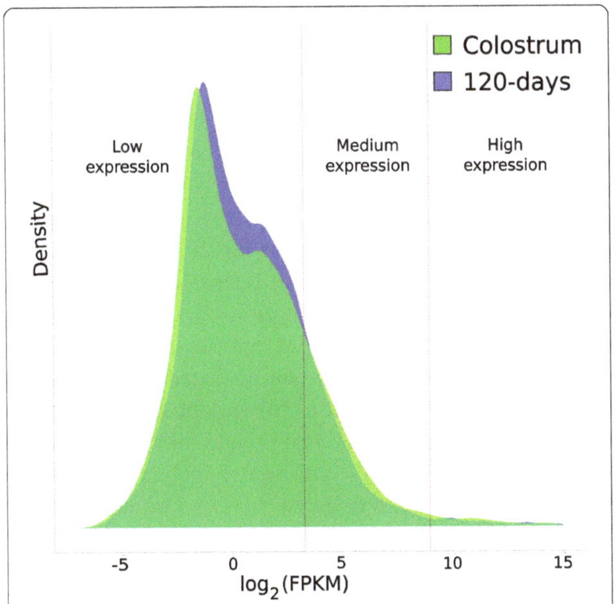

Fig. 2 Plot of the log2 (FPKM) value distribution in the D1 and D2 samples. Vertical *red* lines show the gene expression groups

Top genes expressed in colostrum and mid lactation

In general the dynamic range of gene expression intensities covered several orders of magnitude and gene expression values in the two samples were highly correlated (Pearson correlation coefficient 0.93). The fifteen genes with the highest FPKM values in D1 and/or D2 sample were β-casein (*CSN2*), progestagen-associated endometrial protein (*PAEP*), α-S2 casein (*CSN1S2*), κ-casein (*CSN3*), α-lactalbumin (*LALBA*), tumor protein translationally-controlled 1 (*TPT1*), ferritin heavy chain 1 (*FTH1*), serum amyloid A3 (*M-SAA3*, goat locus LOC100860781), secreted phosphoprotein 1 (*SPP1*), translationally-controlled tumor protein-like (*LOC102179679*), glycosylation dependent cell adhesion molecule-1 (*GLYCAM1*), eukaryotic translation elongation factor 1 alpha 1 (*EEF1A1*), cathepsin D (*CTSD*), fatty acid synthase (FASN), ribosomal protein S29 (*RPS29*), α-S1 casein (*CSN1S1*), keratin 19 (*KRT19*) and checkpoint kinase 1 (*CHEK1* or *CHK1*) (Table 5).

Ten of the fifteen highly expressed genes in D1 (*PAEP, TPT1, FTH1, M-SAA3, SPP1, LOC102179679, EEF1A1, CTSD, FASN, RPS29*) showed a decrease in expression in the D2 sample, while five (*CSN2, CSN1S2, CSN3, LALBA, GLYCAM1*) showed a gene expression increase from D1 to D2 lactation stages. In D2 sample three genes, *CSN1S1, KRT19, CHEK1*, showed a higher expression relative to D1 sample. Only three genes out of fifteen (*M-SAA3, CTSD, RPS29)* are highly expressed in one condition but not in the other.

Functional annotation analysis of highly expressed genes during lactation

To better characterize only the most abundantly transcribed genes, the groups of highly expressed genes described before (FPKM ≥ 500) in the two lactation stages

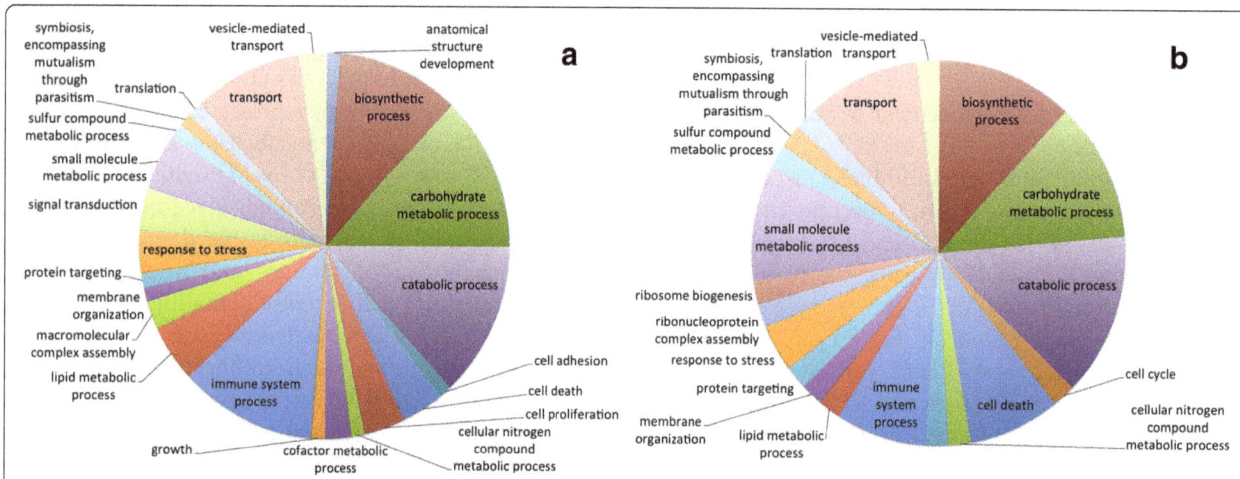

Fig. 3 Pie charts of the GO terms associated with the expressed genes in the two analyzed samples, D1 (**a**) and D2 (**b**). Each GO term detected by blast2go on the results of the BLAST against the NCBI NR database was mapped to its corresponding GOSlim (a selection of representative GO terms maintained by the Gene Ontology Consortium)

were subjected to gene ontology classification and functional analysis using DAVID (Additional file 4). Annotations results showed more significant terms in the three GO categories and in the annotation cluster for D1 sample than for D2. Overall, the data indicated that in both samples highly expressed genes are involved in mRNA translation, protein localization, transport and modification, ATP biosynthesis, regulation of apoptosis and cell proliferation. In the GO apoptosis relative terms, some gene were ubiquitously expressed both with positive and negative regulation action (*HMOX1, GRX1, UBB, UBA52, RPS27A, CIDEA, NOTCH1, CDKN1A, SQSTM1*), some were implicated in the positive regulation (*LALBA, TXNIP, PRDX1, RPS3, CD44, JUN, RPL11*) and some in the negative regulation (*CCL2, NFKBIA, TPT1, HSPA5, GSTP1*).

Table 5 List of the fifteen highly expressed genes (in bold) in the D1 and D2 samples, with their respective FPKM values

Gene symbol	D1 (FPKM)	D2 (FPKM)
CSN2	**121511**	**320950**
PAEP or BLG	**65569**	**45561**
CSN1S2	**48645**	**103105**
CSN3	**47454**	**111404**
LALBA	**39379**	**60454**
TPT1	**12008**	**7603**
FTH1	**11827**	**6563**
M-SAA3	**10640**	3141
SPP1	**10047**	**7286**
LOC102179679	**7751**	**3769**
GLYCAM1	**7034**	**18498**
EEF1A1	**6294**	**3463**
CTSD	**5532**	2974
FASN	**4995**	**3580**
RPS29	**4775**	1949
CSN1S1	2113	**5089**
KRT19	4103	**4806**
CHEK1	2325	**3923**

Both in colostrum and in late lactation we found statistically significant ($P < 0.001$) enrichment of genes associated with GO terms related to nervous system (long-term strengthening of neuromuscular junction, axonogenesis, regulation of transmission of nerve impulse).

In the D1 sample most genes were specifically involved in carbohydrates metabolism (e.g. glucose and hexose catabolism), oxidative phosphorylation, regulation of unfolded protein, defense response, cellular homeostasis. We identified the solute carrier family 2 (facilitated glucose transporter), member 3, gene (*SLC2A3*) highly expressed in both lactation stages.

Heat shock protein genes (*HSPA5, HSPA6, HSPA8, HSP90AA1, HSP 70.1*) were found between the highly expressed genes with *HSPA1A* and *HSPB1* identified between the differentially expressed genes up-regulated in colostrum. GO-BP classification for *HSP* genes showed association with the following terms: response to unfolded protein, negative regulation of programmed cell death, negative regulation of apoptosis, response to extracellular stimulus, chaperone mediated protein folding requiring cofactor, response to endoplasmic reticulum stress, '*de novo*' posttranslational protein folding, cellular response to stress, negative regulation of peptidase activity.

The highly expressed genes in D2 sample included genes involved in positive regulation of apoptosis, in hydrogen peroxide catabolic process, in regulation of protein transport, modification process and localization, in negative regulation of response to stimulus and of cytokine secretion.

Differentially expressed genes between lactation stages

Differentially expressed loci (DEGs) between the D1 and the D2 samples were identified using expression levels of the 33,757 transcripts reconstructed by the Cufflinks/ Cuffmerge procedure and the goat genome assembly as reference. A total of 418 transcripts, of which 401 were annotated to 329 distinct gene symbols, have NOISeq differential expression probability $q >= 0.9$. Of these, 275 reconstructed transcripts associated to 207 different gene symbols were up-regulated in D1, and 143 transcripts associated to 122 different genes were down-regulated (Additional file 5). The DAVID and WebGestalt tools were used for the functional enrichment analysis of these two gene lists. These web-services do not support all the organisms and gene identifiers, so we set the *Homo sapiens* species as reference. 169 out of 207 and 111 out of 122 were the up and down regulated genes in D1 sample included in the final analyses. Most GO categories in which DEGs were annotated correspond to those already described for the highly expressed genes.

Functional enrichment analysis of DEGs in D1 sample

The functional characterization of the annotated up-regulated genes in D1 showed enriched genes associated with GO-BP categories involved in glycolysis, carbohydrate metabolism and catabolism, defense response, cytokine activity, regulation of growth, cell death, vasculature development. The most important GO-MF terms were associated with protein binding and chemokine activity and the most important GO-CC terms were associated with cytoplasm and extracellular space (Additional files 6, 7).

Between the DEGs up-regulated in D1 sample we found the carbohydrate transporter encoding genes *SLC2A1* and *SLC2A10* and the nucleotide sugar transporter gene *SLC35E1*.

Higher expression of genes annotated in immune and defense terms *(IL1R2, CCL3, CCL2, OLR1, C3, CLU, CCL5, CCL4, FCRL4, CCR9, GPI, BPI, PPBP, FCAR, CCL20, SEMA3C, IL12B, LBP, CFI, NFIL3, PTX3, DMBT1)* was also here identified.

Many DEGs up-regulated in D1 were found in GO terms relative to positive (*CCL2, PRTN3, CSF1, CLU, IGF1, ADM, F3, JUN, TGIF1, IL12B, FGF1, IL13RA1, SCG2*) and negative (*INHBA, ACVR1B, CDA, ING1, ENO1*) regulation of cell proliferation and cell growth. Otherwise other genes were identified in terms relative to apoptosis and programmed cell death (*FASTKD3, DLC1, EGLN3, DDIT4, MAGED1, ECE1, JUN, SIAH2, PPP1R15A, FGF2, PHLDA1, SOX9, PPIF, INHBA, ACVR1B, F3, IL12B, IGFBP3*) with some as negative regulators (*NOL3, CCL2, BAG3, F3, CLU, SERPINB2, TNFRSF18, HSPB1, IGF1, HSPA1A, SCG2). CCL2, CSF1, JUN, DDIT4* and *PPP1R15a* were also found between the highly expressed genes during lactation.

In D1 sample many DEGs were found in the statistically significant ($P < 0.01$) enriched GO terms associated with blood vessel development and morphogenesis (*GPI, PLXDCI, JUN, APOLD1, SEMAC3C, FGF1, FGF2, SCG2*); *GPI* and *JUN* were present also in the highly expressed gene list.

The WebGestalt KEGG/DAVID pathway enrichment analyses showed the glycolysis/gluconeogenesis and the metabolism of carbohydrates as specific for the D1 up-regulated genes (*GPI, TPI1, LDHA, SLC2A1, PGAM1, HK2, HK1, PGK1, GAPDH, ENO1*) while the cytokine-cytokine receptor interaction as the common pathway between the up (*IL1R2, CCL3, CCL2, CSF1, CCL5, CCL4, CCR9, ACVR1B, INHBA, CCL20, PPBP, TNFRSF18, IL12B, IL13RA1*) and down (*IL20RB, LIFR, CNTFR, IL7R, CD27, TGFB2*) regulated gene analysis (Additional file 6). Even the WebGestalt Pathway Commons analysis showed some pathway in common between D1 (*FN1, CCL2, NDRG1, SLC2A1, KAT5, RCAN1, IL13RA1, BHLHE40, CCL5, HSPA1A, DDIT4, JUN, FOSB, IGF1, TNFRSF18, EGLN3, ENO1, NT5E, SAP30, IGFBP3, TGIF1, MT2A, GAPDH, IGFBP1, LDHA, MMP1, ADM, MDK, FGF1, PGK1, CSF1, PPP1R15A, HK2, HK1, LBP, CLU, HSPB1*) and D2 (*PLCG1, PAOX, LEF1, CTLA4, CD4, GJA1, RET, NDRG2, MAPK13, COL4A1, TGFB2, ESR1, ETV1, FABP4, MYB*) up-regulated genes: integrin family cell surface interactions, glypican1 network and nectin.

Given the several biological functions and pathways identified through the analysis of gene expression in colostrum, we looked at the list of up regulated genes to search for predicted protein-protein interactions by using the STRING web resource. In the network view nodes are proteins while edges represent the predicted functional associations; the disconnected nodes were hidden (Additional file 8). Out of the 207 genes in the list, for 180 of them STRING found 97 protein interaction with a predicted confidence score > 0.7. The interaction enrichment analysis results showed that the network have significantly more interactions than expected.

Functional enrichment analysis of DEGs in D2 sample

The down-regulated DEGs in colostrum and consequently up-regulated in mid lactation stage showed enrichment of the GO-BP terms associated with positive regulation of lymphocyte activation and anatomical structure morphogenesis. The most important GO-MF terms were associated with actin binding and molecular transducer activity; the most important GO-CC terms were associated with plasma membrane and cell surface (Additional files 6, 9).

Between DEGs many transporters were identified: solute carrier family 15 (H+/peptide transporter), member 2 (*SLC15A2*), solute carrier family 28 (concentrative nucleoside transporter), member 1 (*SLC28A1*), solute carrier family 5 (sodium/sugar co-transporter), member 9 (*SLC5A9*), solute carrier family 5 (sodium/glucose co-transporter), member 1 (*SLC5A1*).

The WebGestalt KEGG/DAVID pathway enrichment analyses showed the T cell receptor signaling pathway (*MAPK13, PLCG1, CTLA4, CD4, ICOS, CD28*), cell adhesion molecules (CAMs) (*ICOS, CTLA4, CD4, CD28*), Jak-STAT signaling pathway (*IL20RB, LIFR, CNTFR, IL7R*), VEGF signaling pathway (*PLCG1, MAPK13, PLA2G2C*) specific for the D2 up-regulated genes. WebGestalt KEGG identified two more terms in Fc epsilon RI signaling pathway and hematopoietic cell lineage. The WebGestalt Pathway Commons analysis showed statistically significant gene enrichment (*PLCG1, LEF1, CTLA4, CD4, GJA1, RET, NDRG2, MAPK13, TGFB2, ESR1, ETV1, FABP4, MYB*) of the class I PI3K signaling events mediated by Akt, signaling events mediated by Hepatocyte Growth factor Receptor (c-Met), EGFR-dependent Endothelial signaling events, IL5-mediated signaling events, IFN-gamma pathway (Additional file 6).

Enrichment of oligosaccharides and lactose metabolism related genes

To identify the OS metabolism genes in goat milk we looked at the expression of 144 genes belonging to different functional glycosyltransferases commonly found in mammals, namely fucosyltransferases, galatosaminyl transferases, galactosyltransferases, sialyltransferases, mannosyltransferases, N-acetylglucosaminyl transferases. We included the glycosidases, fucose synthesis genes, sialic acid (SIA) synthesis genes, sugar transporters, and the genes for the lactose pathway in consideration that lactose L (Galb1-4Glc) forms the reducing end of milk oligosaccharides.

One hundred-ten genes out of 144 were expressed in the two milk samples with a minimum of 0.02 FPKM (*ST6GALNAC3*) in the D1 samples and a maximum of 60,453 FPKM (*LALBA*) in the D2 sample (Additional file 10). All genes of lactose synthesis, SIA synthesis genes, and fucose synthesis, were expressed in goat milk, whereas 34 genes belonging to the other categories were lacking, indicating a selective expression of some of these. Half of the genes showed a low expression (FPKM < 10) and half had medium expression value (>10 and <100 FPKM) while nine genes were expressed at higher levels, namely *LALBA, SLCA3*, GDP-mannose-4, 6-dehydratase (*GMSD*), NME/NM23 nucleoside diphosphate kinase 2 (*NME2*), *SLC2A1, B4GALT1*, UDP-GlcNAc:betaGal beta-1,3-N-acetylglucosaminyltransferase 2 (*B3GNT2*), N-acetylneuraminic acid synthase (sialic acid synthase) (*NANS*), hexosaminidase subunit beta (*HEXB*).

In Fig. 4 the log2 FPKM values for the expressed genes in D1 and D2 samples were graphically plotted. Most of the genes (64%) showed higher expression value in colostrum than in the later lactation stage, which can be better seen by plotting the log2 of the expression fold change (FC) (Additional file 11). The highest FC was for the *ST6GALNAC3* gene with a value of 69 in D2 sample even with a small FPKM expression value (1.38).

In general, almost all the genes coding for enzymes included in the lactose pathway showed an up-regulation in D1 stage; in particular, and increased *LALBA* and *B4GALT1* gene expression was revealed between D1 and D2.

Considering the other galactosyltransferase genes, the *B3GALT* family was expressed at lower level respect to *B4GALT* as found in bovine [33].

Among sialyltransferases genes, both α2,3-sialyltransferases and α2,6- sialyltransferases showed higher levels of expression in D1 than D2 sample except for *ST3GAL2, ST3GAL6, ST6GALNAC2, ST6GALNAC3, ST8SIA1. ST3GALT1, ST3GALT4, ST3GALT6, ST6GALNAC2, ST6GALNAC4* genes showed the highest expression and *ST3GAL1* showed a 4-fold change FPKM value with respect to *ST6GAL1*.

Among the *ST8SIA* family genes, only *ST8SIA1* was expressed in D1 and D2 samples.

Considering fucose synthesis, *GMSD* and *TSTA3* gene were highly expressed relative to *FPGT* and *FUK* both in D1 and D2 samples. Among fucosyltransferases, *FUT1, 2, 4, 7, 8, 10, 11* and *POFUT1* were found expressed even if at low levels, while *POFUT2* at medium level.

Among the N-acetylglucosaminyl transferase category, the *B3GNT2* gene was one of the higher expressed in D1 sample.

In the galactosaminyl transferase category, the *B3GALNT1* gene was expressed at higher level in D1 relative to D2 with a 4-fold change value.

A number of genes encoding proteins involved in complex N-glycan degradation were identified in goat milk, lysosomal sialidase (*NEU1, NEU3*), β-galactosidase (*GLB1*), β-hexosaminidase subunit α/β (*HEXA/HEXB*), α-L-fucosidase (*FUCA1, FUCA2*) able to catalyze the removal of Neu/NeuNAc, β-D-Gal, GlcNAc and Fuc, respectively, from the non-core structure of N-glycans. The *HEXB* was one of the most expressed genes overall and had high FPKM value with a two fold change ratio in D1 relative to D2 sample.

Discussion

RNA-seq mapping and gene expression statistics

In our experiment we choose to sequence the colostrum and 120 days samples at a resolution never used before in similar studies during lactation: i.e. more than 100 million mapped reads. Actually, forty millions have been considered sufficient to provide reliable measurement of a single transcript per cell and for the detection of major splice isoforms for abundant and moderately abundant transcripts [58]. However, greater sequencing depth allows more accurate

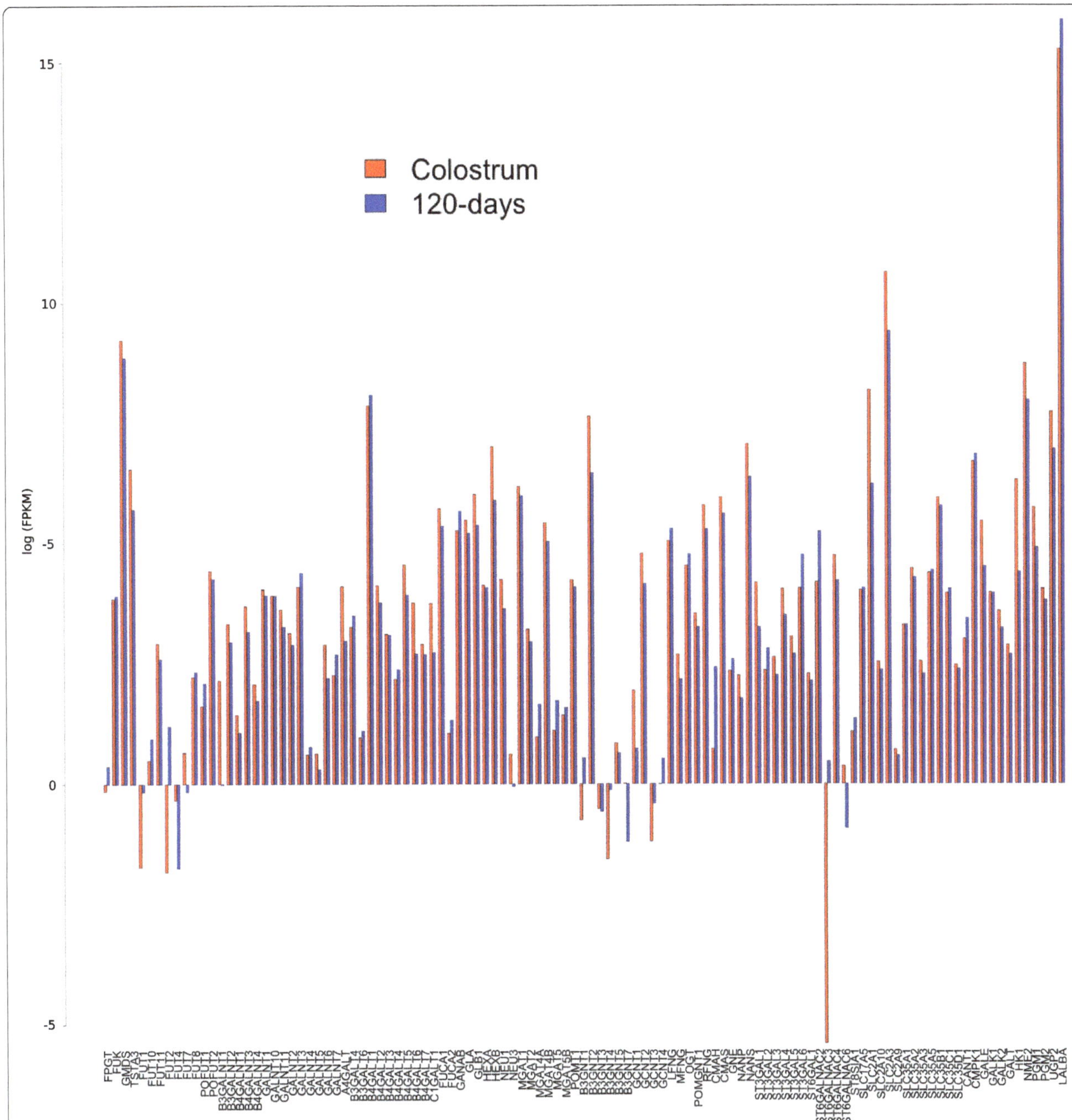

Fig. 4 FPKM values of genes involved in goat oligosaccharides metabolism belonging to eleven functional categories

assessment of the relative abundance of low-expression transcripts as well as the detection, by at least one read mapping, of a greater number of genes [59, 60]. In this work we wanted to explore as much as in detail the expressed genes in very different physiological stages, therefore we privileged having high sequencing depth and we chose to not pool samples and do not include biological replicates.

Considering the *de novo* assembly approach we compared our results with a similar study on goat mammary gland in which 98,864 transcripts and 51, 361 unigenes with a mean length of 1,219 bp were obtained [61]. Numbers are comparable, even if smaller, to those obtained by us. Moreover the authors used RNA pools of tissue at four life times and constructed a reference dataset for gene annotation without using a reference genome.

We compared our results with similar RNA-seq profiling in bovine milk somatic cells. In all these studies different experimental design, sampling, and sequencing depth were described but from a global point of view a media of 15,000-24,000 expressed genes have been identified

[27, 28, 36]. In a previous work a threshold value of 0.2 RPKM was found for detectable gene expression in milk somatic cells [58]. If we consider the category of a more conservative FPKM value > 1, we get a number of co-expressed genes comparable to that found in bovine MSCs, here defined as those genes expressed at the FPKM threshold in both samples. In an analogous work in sheep a minor number of expressed genes was found, i.e. about 10,000 [29].

Global milk transcriptome functional annotation and pathway analysis

Through the analysis of global milk transcriptome we noted that pathways involving nucleosides and ribonucleosides were enriched in colostrum. These compounds belong to the non-protein-nitrogen fraction, with biological role which is not limited to act as metabolites but to work as bioactive substances in the regulation of body functions, contributing to cell renewal or restoration [62]. Moreover nucleotide sugars act as donors for sugar residues in the glycosylation reactions that produce polysaccharides They are the glycosyl donors for glycosyltransferases and the precursors of glycoproteins, glycolipids and OS [63]. On the other hand, also the lysine degradation pathway was highly represented, in agreement with the findings of [64] suggesting that when Lys is in excess of requirements, the mammary gland appears to dispose of the extra supply via the oxidative mechanism. An increase in the catabolism of several amino acids was indicated in different studies on bovine mammary gland [65, 66], because amino acids could be an important pool of energy during lactation. The results of the BLAST2GO pipeline obtained in the present study confirm therefore the ability of the mammary gland to adapt different molecular functions according to the biological need of the animal.

Top genes expressed in colostrum and mid lactation

Among the top expressed genes in colostrum and mid lactation, we detected *CSN2, LALBA, CSN1S1, CSN3 FTH1, SPP1, FASN* and *EEF1A1*, also identified by [57] among the top 20 expressed genes in the milk fat layer of mature human milk. The high expression of casein (*CN*) genes, all increasing in expression from D1 to D2 lactation stages, was not surprising because caseins, together with the whey proteins, represent the 90% of milk proteins fraction [67]. In this work, we found that *CSN2* and the whey protein α-lactalbumin (α-LA) coding gene (*LALBA*) have the highest expression values, as already detected by [61] in goat, [33] in cow and [507] in human.

The *PAEP* gene, also known as BLG, encodes the beta-lactoglobulin, the major whey protein in bovine, buffalo, goat, and horse milk [68], while *EEF1A1* high expression is justified by the considerable request of milk protein synthesis during lactation, being *EEF1A1* one of the most abundant protein synthesis factors [69, 70].

As for the *CTSD* (cathepsin D) gene, studies on rat mammary gland demonstrated that *CTSD* dependent activity is required to effect the cleavage of prolactin, the fundamental hormone for the function of the mammary gland, influencing also the expression of a number of genes as well the secretion of lipids and milk proteins [71–73].

The *FTH1* gene encodes for ferritin, and has been identified as an anti-apoptotic factor [74]. Ferritin is present in blood and extracellular fluids, including milk. In cows it was found that ferritin concentrations were higher in colostrum than in serum [75], in agreement with [57] who reported a higher ferritin genes expression in human colostrum with respect to mature milk. The expression of *M-SAA3.2* gene in the bovine mammary gland was reported by [76] who confirmed the role of this gene in host defense, likely in the neonate gastrointestinal tract. Moreover it has been suggested that *M-SAA3* represent one of the most promising candidates affecting milk protein and fat percentage of dairy cattle [27].

The *SPP1* gene, encoding for ospeopontin protein, is essential for mammary gland development, milk production, local mammary gland immunity and it has a significant role in the modulation of milk protein gene expression [77, 78]. The *FASN* gene is directly involved in *de novo* synthesis of most of the short and medium-chain fatty acids in milk and is essential for the development, functional competence, and maintenance of the lactating mammary gland [61, 79, 80]. *KRT19* gene expression has been associated with an intact cytoskeleton required for autophagic cell death in goat mammary gland [81]. The *GLYCAM1* gene encodes for a milk fat globule glycoprotein member of the mucin family. *GLYCAM1* mRNA was reported to be highly expressed during lactation in goat mammary gland [82], so suggesting a protective role of this gene.

High expression of the *RPS29* gene in goat mammary gland was already reported by [83] and in human mammary gland epithelium by [84].

Functional annotation analysis of highly expressed genes during lactation

The finding of most statistically significant GO terms associated with translation, ribosomial biogenesis, rRNA processing is in agreement with the 25% of the highly expressed ribosomal protein genes. This is biologically meaningful because an increased translational requirement is needed to support caseins and other milk proteins synthesis. The gene encoding for the EEF2 translational elongation factor has been found expressed during goat lactation as previously reported in bovine [66]. However in ruminants mammary gland only a minor portion of the synthetic machinery appears to be used for production of milk proteins but for the biological needs of mammary tissue during lactation [85].

Many highly expressed genes during lactation were also associated to GO terms implicated in regulation of programmed cell death and cell proliferation relative terms. Mammary apoptosis was first demonstrated during tissue involution after lactation, but it has been detected also during lactation, in mammary tissue of lactating mice, goats and cattle [86, 87]. In bovine mammary tissue a high level of apoptosis was found in early lactation [88–90]. These authors speculated that their results were an effect of tissue edema and increased apoptotic leucocytes or that may be due to death of nonfunctional or senescent cells or to removal of excess capacity of newly synthesized and yet undifferentiated cells. In colostrum samples many different somatic cells type (leukocytes and macrophages) are present beyond milk epithelial cells, so we can suppose that both hypothesis could be corroborated by our results; on the other hand, we did not find, in the present trial, any of the most apoptotic markers (i.e. *Bax, Bcl-2, TGF-β1, CPP-32, Fas, CASP* genes) [90–92].

UBB, UBA 52 and *RPS27A* genes represent three of the four ubiquitin genes found in mammals, expressing the main non-lysosomal route for intracellular protein degradation in eukaryotes. Both [93] and [28] found genes belonging to the ubiquitin-proteasome pathway expressed during early involution of mouse mammary gland and at peak lactation in bovine MSC, respectively. Recent findings also substantiate a pivotal role of the ubiquitin/proteasome pathway in the regulation of apoptosis [94, 95]. Also the NFκBIA gene, here found as highly expressed both in D1 and D2 samples, could have a role in the regulation of cell proliferation and programmed cell death because involved in activation of stressful situations requiring rapid reprogramming of gene expression [96]. It was reported to be expressed in bovine mammary gland during lactation by [90].

Heat shock protein genes (*HSPA5, HSPA6, HSPA8, HSP90AA1, HSP 70.1*), found between the highly expressed genes in the lactation, have a protective function, allowing cells to survive lethal conditions, and also interact with various components of the programmed cell death machinery; on the other hand, membrane bound HSPs mediate immunological functions [97, 98]. Our finding confirms what reported in literature about HSP functions and demonstrates how MSCs, in goat colostrum, can support a high level of protein synthesis act to operate in both immunity and apoptosis.

Another highly expressed GO category of genes included the nervous system and the regulation of transmission of nerve impulse, already reported by [65] in cow lactation given its role in milk ejection and control of blood flow.

Most of the highly expressed identified genes were specifically involved in carbohydrates metabolism. A previous study on goat mammary gland showed how glucose is oxidized about twice as rapidly as acetate and is essential for milk secretion and that the glucose uptake by the lactating gland was much greater than that required for milk lactose synthesis, suggesting that substantial amounts are oxidized under normal conditions, as in the isolated gland [99]. On the other hand the glucose transporter *SLC2A3* was highly expressed in both lactation stages, in agreement with the findings of [66] on cow lactation; these authors suggested that the high expression of genes involved in energy metabolism was likely related to the high-energy demand for milk synthesis and secretion [65].

Differentially expressed genes between lactation stages

One of the major focuses of the present trial was to evaluate the differentially expressed genes between lactation stages, to assess and confirm the high nutritional value of goat colostrum and milk. Colostrum is a source of immunomodulatory molecules that positively influence the immune status of the neonate, including polymorphonuclear leukocytes and macrophages, T-lymphocytes, B-lymphocytes, plasma cells and epithelial cells [100, 101]. In the colostrum stage, the present study identified as up-regulated genes encoding carbohydrate transporter (*SLC2A1* and *SLC2A10*) and the nucleotide sugar transporter gene *SLC35E1.* Glucose transport across the plasma membranes of mammalian cells is carried out by two distinct processes employing the passive, facilitative, energy-independent glucose transporters GLUT, encoded by *SLC2A* genes, and the active sodium-dependent and energy-dependent glucose transporters SGLT, encoded by *SLC5A* genes. The member of the solute carrier family *SLC35* encode antiporters transporting nucleotide sugars pooled in the cytosol into the lumen of the Golgi and/or the ER in exchange for the corresponding nucleoside monophosphates. The transported nucleotide sugars are utilized as sugar donors by glycosyltransferases for the synthesis of sugar chains of glycoproteins, glycolipids and polysaccharides [102]. Currently, the SLC35 human family comprises 31 hydrophobic and homologous proteins, which are divided into 7 subfamilies, from SLC35A to SLC35G. Among the 31 SLC35 family members, only about one-third of them have been studied in detail. Little is known about the remaining members of the family, particularly the members of the SLC35E, SLC35F and SLC35G subfamilies and to date there are no known reports on the biological functions and the subcellular localizations of these proteins [103]. *SLC35E1* gene expression was found to be ubiquitous in various human adult and fetal tissues and cell lines with the highest value in mammary gland [104]. So our results confirm the expression of this gene in GMSC. Moreover we found the polysaccharides and glycosaminoglycan binding molecular function enriched in colostrum sample. This suggests a possible support to the oligosaccharides metabolism whose concentration has been reported to decrease from colostrum to late lactation stage in goat milk [105].

The finding of genes annotated in immune and defense terms in D1 sample is consistent with the general conception that milk after parturition provides passive protection for the neonate and has immunostimulatory capabilities. The chemokinesis and chemotaxis properties of goat milk toward different leukocyte populations were studied during lactation; mononuclear cells (MNCs) were the major cell found in colostrum and polymorphonuclear leukocytes (PMNs) the predominant cell type found during the late lactation stages (>weak 19) [106].

Many DEGs up-regulated in D1 were found in GO terms relative to either positive or negative regulation of cell proliferation and cell growth, as well as to apoptosis and programmed cell death. In the colostrum sample the protein encoded by chemokine (C-C Motif) Ligand 2 (*CCL2)* gene seems to have a positive effects on cell proliferation and a negative effects on apoptosis; this cytokine displays chemotactic activity for monocytes and basophils but not for neutrophils or eosinophils [107]. Among the genes involved in the positive regulation of cell proliferation, the fundamental role of the Colony Stimulating Factor 1 (*CSF1)* gene has to be underlined. Being the protein encoded by this gene a cytokine controlling production, differentiation, and function of macrophages, it promotes the release of proinflammatory chemokines, so playing an important role in innate immunity and in inflammatory processes [108]. Hence, we can argue that in colostrum cells the expression of both *CCL2* and *CSF1* genes could be involved in the recruitment of monocyte/macrophage to participate in immune responses, development, and tissue homeostasis and in monocyte recruitment in mammary gland tissue where collateral vessels are actively [109]. Furthermore the finding of DEGs in enriched GO terms associated with blood vessel development and morphogenesis agree with a study in bovine lactation [58] and confirm an important role for angiogenesis at the onset of goat lactation, while increasing blood flow across the mammary gland [110].

Of the common pathway of the up-regulated genes identified in the two lactation stages, integrins occupy an important role. Through integrins, the extra-cellular matrix influences growth, differentiation, and survival of mammary epithelial cells. Both in mouse [93] and bovine mammary gland [111] these proteins has been found expressed at different level during lactation and involution. In D1 sample enrichment of *IGF-1, IGFBP1* and *IGFBP3* genes was here revealed. The IGF system plays a pivotal role in mammary tissue homeostasis, regulating cell proliferation and differentiation during lactogenesis [112]; furthermore. *IGF-1* and *IGFBP3* were found expressed in bovine mammary gland at higher level during early lactation [89], while in goat mammary gland they have been found expressed during involution [113].

In the predicted protein interaction network the up-regulated genes in D1, obtained with STRING, the central role of the Jun Proto-Oncogene (*JUN)* is outlined by its multiple connections with other proteins. *JUN*, which was also one of the highly expressed genes, encodes a member of the activator protein-1 (*AP-1*) transcription factor family, a key component of many signal transduction pathways. AP-1 is composed of dimers of Jun (cJun, JunB, and JunD), Fos (cFos, FosB, Fra-1, and Fra-2), or other closely related factors such as ATF proteins. The Jun members homodimerize or heterodimerize with different Fos or ATF members, while Fos members only heterodimerize with different Jun members [114]. So the predicted *JUN* interaction with *FOSB* in the network has experimental biological relevance. Differential expression and activation of Jun and Fos members allow AP-1 complexes to control a wide variety of cellular functions. The AP-1 factor was indicated as a regulator of postnatal proliferation of mouse mammary epithelial cells [114] and of the early and late phases of mouse mammary gland involution [115]. The genes that are activated by FosB are unknown, but one excellent candidate is the oxytocin receptor gene which contains within its promoter an AP-1 site. A requirement for FosB in the nurturing response in mice was demonstrated [116], raising the hypothesis that FosB might be induced by oxytocin binding to receptors on preoptic area of the hypothalamus (POA) neurons, this leading to increased expression of the oxytocin receptor gene. Thus, by a feed-forward mechanism, POA neurons would express more oxytocin receptors and would therefore have enhanced sensitivity to oxytocin.

Some of *JUN* interactions could be relevant in cells physiology, as the interaction with the various *CCL* proteins and *IL12B*, which plays a role in cell proliferation/cell growth and immunity. The interaction with *FGF2/FGF1* could be explained by their inclusion in the same GO terms functional enrichment, the blood vessel development and morphogenesis. *JUN* is predict to link to *HSPB1*, which is in turn strongly associated to *HSPA1A/DNAJB1/BAG3.* All these genes are involved in cell maintenance and cellular assembly and organization. Small heat shock proteins (sHsp), of which *HSPB1* is a member, comprise the most widespread but also the most poorly conserved family of molecular chaperones and are involved in protecting cells from various unfavorable conditions, therefore playing a fundamental role in cell survival. At a molecular level, sHsp are involved in preventing aggregation of partially misfolded proteins and have been shown to intervene in modulating cell death pathways by interacting with various components of the cell death machinery upstream and downstream of the mitochondrial apoptotic events so preventing apoptosis in different lethal stress situations [117, 118]. At the bottom side of the network graph, the nine genes (*GAPDH, PGK1, TPI1, ENO1, LDHA, PGAM1, GPI, HK1, HK2*) linked together with many interaction edges belong to the glycolysis/gluconeogenesis KEGG pathway and carbohydrate metabolism biological process, which are two of the functional enrichment terms that significantly differentiate D1 from D2 lactation stage. The link between *JUN* and *GAPDH* derives from text-mining, because co-mentioned in a PubMed abstracts (*GAPDH* was used as gene reference gene).

Considering the up-regulated DEGs in D2 samples, some of these belong to the solute carrier family (*SLC*), coding for the SGLT family proteins. SGLT is a family of solute-linked carriers that contain sodium-coupled transporters for several nutrients. The Na +–electrochemical gradient provided by the Na +–K+ ATPase pump is utilized to transport substrates into cells against its concentration gradient. The co-transported substrates are sugars, inositol, proline, pantothenate, iodide, urea and other undetermined solutes [119]. In cattle, the glucose transporter SGLT1, encoded by *SLC5A1* gene, is responsible for the high-affinity, conservative uptake of most monosaccharides (glucose and galactose), except fructose; in human, SGLT4, encoded by *SLC5A9* gene, is an essential transporter for mannose, 1,5-anhydro-D-glucitol, and fructose [120]. In mammary gland epithelia, the protein encoded by the *SLC15A2* gene (PEPT2) is reported to be involved in the reuptake of small peptides accumulated from the hydrolysis of milk proteins [121]. The *SLC28A1* gene encodes for sodium-dependent, concentrative nucleoside transporter, pyrimidine- nucleoside preferring, epithelia localized, CNT1. We might conclude therefore that, both in D1 and D2 stages, sugar and nucleotide sugar transporters were differentially expressed to take account of the different physiological need of the mammary gland during lactation.

Within the more specific pathways of the up-regulated genes in D2 stage, the T cell receptor signaling, cell adhesion molecules, Jak-STAT signaling and VEGF signaling pathways must be mentioned.

In particular, the JAK-STAT signaling, which regulates many cellular processes, including innate and adaptive immune function, development, cell proliferation, differentiation, and apoptosis, has been found to be crucial in the mammary gland with STAT5 factor been important for controlling expression of milk protein genes in rodent [122]. However, in cattle and sheep mammary gland, the role of JaK2-STAT5 signaling pathway seemed to support a minor role for milk protein synthesis [29, 66]. In the present work we did not find *STAT5* between the highly expressed genes either in D1 or D2 samples but we found the transcription factor E74-like factor 5 (*ELF5*) expressed. This factor has been postulated to work as regulator of milk protein expression, encompassing STAT5 as a key player and its expression has been found increased during cow and sheep lactation [29, 66]. Therefore our findings allow to hypothesize a similar physiological role of this gene during lactation in goats.

In conclusion, the results obtained for the genes up-regulated in mature milk depict a scenario in which cells are ready to respond at external and immunological stimuli, to transduce signals inside cells and to orchestrate appropriate actions (e.g. cytokine gene transcription, JAK activation responsible for apoptosis execution) [123–126].

Enrichment of oligosaccharides and lactose metabolism related genes

The synthesis of free OS and OS chain on glycoconjugates is mediated by the concerted action of many glycosyltransferases (GTs), glycosidases, sulfotransferase and accessory molecules; GTs typically catalyze the transfer of a monosaccharide from a nucleotide-sugar to a growing acceptor-OS attached to a carrier (in some cases directly to the carrier molecule). The OS product of one reaction is the substrate in the next reaction, and the array of glycosylation-related genes expressed in a particular cell is an important factor in the control of OS synthesis [20].

In this work we identified many expressed genes encoding for protein involved at different steps in the OS metabolism.

α-lactalbumin (*LALBA*) has important biological and functional properties and, in the mammary gland, interacts with β1,4-Galactosyltransferase (*B4GALT1*) to form the lactose synthase complex that produce lactose, the core unit from which all the other OS are synthesized [127, 128]. The increased *LALBA* expression that we found between D1 and D2 is consistent both with an increase of milk protein output and with lactose synthesis. The *B4GALT1* gene showed a high expression during goat lactation suggesting that the protein essential for the lactose synthase complex formation is ensured. Almost all the other genes coding for enzymes included in the lactose pathway showed an up-regulation in D1 stage similarly to what reported for human milk [57].

The *B4GALT1* encoded protein is one of the best-studied galactosyltransferase enzymes and has been identified on the plasma membrane, so that roles in cell-cell or cell-substrate recognition were hypothesized [129].

The genes implicated genes in sialylated OS metabolism were found expressed at medium values in our experiment. Considering the SIA synthesis genes [130], the *NANS* and the cytidine monophosphate N-acetylneuraminic acid synthetase (*CMAS*) were the most expressed, similarly to what found in bovine [33]. Goat milk has been reported to contain a high concentration of sialylated OS, specifically 3′sialyllactose (3′-SL), 6′sialyllactose (6′-SL); their biosynthesis starts from Gal in lactose with the addition of N-acetylneuraminic acid (NeuAc) or N-glycolylneuraminic acid (NeuGc), in α2-3 aor α2-6 linkages respectively. A detailed qualitative and quantitative characterization of goat colostrum oligosaccharides has been carried out by [131]. The authors identified 78 oligosaccharides, of which 40 (51.3%) were neutral non-fucosylated, 3 (3.8%) neutral fucosylated and 35 (44.9%) corresponded to sialylated (Neu5Ac/Neu5Gc) oligosaccharides. The predominant OS were sialyl-lactoses followed by Hex-HexNAc-Neu5Ac and Hex-HexNAc-Neu5Gc residues. The most abundant OS were galactosyl-lactoses.

The sialyltransferase enzymes *ST6GAL1* and *ST3GAL4* have been shown to be responsible for the production of 6′-SL and 3′SL in mouse milk respectively [132]. So we can suggest that both in goat *ST6GAL1* could act in the biosynthesis of 6′SL and *ST3GAL4* plus *ST3GAL1* and *ST3GAL6* in the biosynthesis of 3′SL. Our results agree with the literature both because higher OS concentrations have been reported in colostrum than in late lactation in different animal species and given that some oligosaccharide profile analyses indicated how 3′-sialylactose is more abundant than the 6′-sialylactose in colostrum and late lactation goat milk [105, 133].

ST3GAL5 encodes the GM3 synthase responsible for the biosynthesis of ganglio-series gangliosides that are sialic acid-bearing glycosphingolipids, the major glycans in nerve cells. It was recently demonstrated that ST3Gal5 catalyzes GM4 synthesis in vivo [134–136] suggesting that the sialic acid moieties of gangliosides and glycoproteins in the brain frontal cortex probably participate in a variety of cellular events including the formation of memory and learning. It was shown that it may be feasible to use sialyllactose, separated from cheese whey or colostrum, as a functional food for brain activation [137].

The protein encoded by one of the higher expressed gene in D1 sample (*B3GNT2*) synthesizes a unique structure known as poly-N-acetyllactosamine (polyLacNAc) a linear carbohydrate polymer composed of alternating N-acetylglucosamine and galactose residues. Recently, results obtained with knockout B3gnt2−/− mice strains showed that hyperactivation of lymphocytes occurred due to a lack of polylactosamine on receptor molecules, indicating that polylactosamine has an important role in immunological biofunctions [138, 139]. The analysis of gene expression profiles in human monocytes, dendritic cells and macrophages found expressed *B3GNT2* suggesting that monocytes synthesize N-glycans carrying di, tri, and tetra-antennary structures, possibly elongated by poly-N-acetyllactosamine chains. This agrees with our results of genes annotated in immune and defense terms in D1 somatic cell and with the finding that mononuclear cells (MNCs) are the major cells found in colostrum [106].

The expressed *B3GALNT1* gene encodes for the globoside synthase an enzyme that catalyze the addition of N-acetylgalactosamine to globotriaosylceramide (Gb3) to form globoside (Gb4), a globo-series glycosphingolipid (GSL) member [140]. Because of the relationship to many human pathogens and diseases the interest in glycolipid-based blood group antigens remains high. Moreover milk glycosphingolipids have been reported to participate in the newborn's defense against pathogens. Ovine milk has been found to have a high content of complex neutral GSL, such as Gb3 and Gb4, which are very abundant in human milk but not in bovine milk and which are able to bind to some *E. coli* strains to prevent infection [141].

In our samples many genes both for fucose synthesis and fucosyltransferase were found expressed. In contrast to human milk, the content of fucosylated oligosaccharides is rather low in the milk of domestic animal species [18]. In goat colostrum 2′FL was found by [142], while in [131] 2′-FL plus fucosyl-lactosamine and lacto-N- fuco-pentaose were detected even if at low abundance. During fucosylation, activated fucose is transferred to various lactosamine acceptors by specific fucosyltransferases (Fuc-T) encoded by the fucosyltransferase gene (*FUT*). The biosynthesis of GDP-L-fucose consists of two cytosolic pathways. The constitutively active *de novo* pathway involves two enzymes (GMD, encoded by *GMSD* gene and FX, encoded by *TSTA3* gene). In the alternative salvage metabolism, that utilizes L-Fuc obtained from extracellular sources or from intracellular degradation of glycoproteins and glycolipids, L-fucokinase (*FUK* gene) and GDP-L-fucose pyrophosphorylase (*FPGT* gene) are involved. After synthesis in the cytosol, GDP-L-Fuc is translocated into the Golgi apparatus via a specific GDP-fucose transporter encoded by gene *FUCT1/SLC35C1* [143, 144]. Because in D1 and D2 samples *GMSD* and *TSTA3* gene were highly expressed relative to *FPGT* and *FUK* we suggesting that in GMSC the *de novo* pathway is preferentially activated and that a high amount of fucose is synthesized in cytoplasm ready to be imported in Golgi apparatus for fucosylation. Among the fucosyltransferases, *POFUT2* gene was the highest expressed in D1 and D2 samples. The encoded soluble enzymes is localized in the endoplasmic reticulum (ER) and its ability to distinguish between folded and unfolded structures has led to the proposal that POFUT2 function as chaperones or in the folding and secretion of their target proteins [145].

Among glycosidase genes, the *HEXB* gene has been found highly expressed in both samples with a two fold change ratio in D1 relative to D2 sample. Two major lysosomal isoenzymes exist in human tissues, which are the products of the assembly of two subunits, α and β, encoded by the homologous genes *HEXA* and *HEXB*. The two isoenzymes Hex A (αβ) and Hex B (ββ) differ in their substrate specificity. The active site of the β-subunit hydrolyzes uncharged substrates, whereas the α-subunit, in addition, cleaves negatively charged substrates [146]. The greater expression of *HEXB* could be associated with the presence of higher percentage of neutral OS in goat milk. In bovine MSCs higher expression values for *HEXA* gene were found [33]. Moreover because lysosomal hexosaminidases release terminal β-glycosidically linked N-acetylglucosamine and N-acetylgalactosamineresidues from a number of glycoconjugates [147] we can speculate that the higher oligosaccharides concentration in colostrum could be due both to glycan degradation activities and glycosyltransferase activities.

Considering neuraminidase genes, *NEU2* and *NEU4* were not found expressed, as in bovine [33].

All the investigated sugar transporter genes were found expressed in goat milk, except *SLC2A8* and *SLC35A4*. These last two genes, contrary to goat, were present in bovine mammary gland and somatic cells [33, 66]. We already discussed *SLC2A1*, *SLC2A10* and *SLC2A3* as DEG and highly expressed genes respectively. Among the nucleotide sugars transporters, *SLC35A2*, *SLC35A5*, *SLC35B1* and *SLC35C1* were expressed at higher levels. Both the CMP-Sialic Acid Transporter (*SLC35A1*) [148] and GDP-Fucose Transporter 1 (*SLC35C1*) [149] were found expressed at low and medium levels, respectively. *SLC17A5* (also known as AST), encode a protein (sialin) with a sialic acid transport function whose mutations are the primary cause of lysosomal sialic acid storage diseases [150].

Conclusion

Transcriptional profiling is a powerful approach for identification of genes globally and functionally expressed in various tissue including mammary gland and milk. In this work we characterized the goat milk somatic cells transcriptome at two lactation stages by using RNA-seq and by integrating two bioinformatic methodologies. We looked at the gene expression, using software applications that integrate functional enrichment analysis and statistical analysis of large set of genes. We identified distinct genes and pathway underlying the differences between colostrum and 120 days milk. Finally, for the first time, a study of glycosylation-related genes was carried out in goat milk to get a picture of genes possibly implicated in oligosaccharides biosynthesis.

Additional files

Additional file 1: Cufflink/Cuffmerge transcriptome in D1 and D2 lactation samples. Information on gene annotation, Gene Ontology terms and Kegg pathways are included.

Additional file 2: Enriched Gene Ontology term analysis in D1 and D2 lactation samples.

Additional file 3: Enriched Kegg pathway analysis in D1 and D2 lactation samples.

Additional file 4: DAVID Functional Annotations and Clustering for highly expressed genes in D1 and D2 lactation samples. (XLS 154 kb)

Additional file 5: Differentially Expressed Genes (DEGs) between D1 and D2 lactation samples.

Additional file 6: DAVID Functional Annotations and Clustering for up-regulated and down-regulated genes in D1 sample. The most significant enriched GO categories under Biological Process, Molecular Function, and Cellular Component are included in a summary sheet.

Additional file 7: Directed Acyclic Graphs (DAGs) for the Biological Process, Molecular Function and Cellular Component categories of up-regulated genes in D1 sample. Each node shows the number of genes in the category and the adjusted *p*-value indicating the significance of enrichment. GO categories in red are the enriched GO categories while the black ones are their non-enriched parents. GO categories in the top 10 that also have a p-value < 0.05 are colored in red.

Additional file 8: Network view of the up-regulated gene list in D1. In this view the network nodes are proteins and the edges represent the predicted functional associations. The edges are draw according to the view settings. In evidence mode, an edge may be drawn with up to 7 differently colored lines that represent the existence of the seven types of evidence used in predicting the associations. The fusion view shows the individual gene fusion events per species; the neighborhood view shows runs of genes that occur repeatedly in close neighborhood in (prokaryotic) genomes; the occurrence view shows the presence or absence of linked proteins across species; the experiments view shows a list of significant protein interaction datasets, gathered from other protein-protein interaction databases; the text mining view shows a list of significant protein interaction groups, extracted from the abstracts of scientific literature; the database view shows a list of significant protein interaction groups, gathered from curated databases; the coexpression view shows the genes that are co-expressed in the same or in other species (transferred by homology).

Additional file 9: Directed Acyclic Graphs (DAGs) for the Biological Process, Molecular Function and Cellular Component categories of down-regulated genes in D2 sample. Each node shows the number of genes in the category and the adjusted *p*-value indicating the significance of enrichment. GO categories in red are the enriched GO categories while the black ones are their non-enriched parents. GO categories in the top 10 that also have a p-value < 0.05 are colored in red. GO categories in the top 10 that have a p-value > 0.05 are colored brown, and the black ones are the parents of the top 10 categories.

Additional file 10: List of the 110 genes belonging to eleven functional oligosaccharides metabolism categories with FPKM expression in D1 and D2 samples.

Additional file 11: Graphical representation of the log2 (D1/D2 FPKM).

Abbreviations

DEG: Differentially expressed gene; GO-BP: Gene ontology biological processes; GO-CC: Gene ontology cellular component; GO-MF: Gene ontology molecular function

Acknowledgments

We thanks Carmela Tripaldi that kindly provided milk samples.

Funding

This study was supported by the research programme "Improvement of Italian livestock through the use of innovative biotechnologies: functional genomics, transcriptomics and proteomics (GENZOOT)" funded by the Italian Ministry of Agriculture.

Authors' contributions

AC conceived the study, carried out the experiments and drafted the manuscript. FF analyzed sequencing data and helped in writing the manuscript. GC helped with the preparation/submission of the additional manuscript materials/sequencing data. BM was a contributor in the manuscript final version. All authors read and approved the final manuscript.

Competing interests

The authors declare that they have no competing interests.

Consent for publication

Not applicable.

Ethics approval and consent to participate

Milk samples were collected by the owner of a private farm as part of routine milking; colostrum was also collected by the owner. Collection protocol and animal care were in accordance with Directive 2010/63/EU of the European Parliament and the Council of 22 September 2010, on the protection of animals used for scientific purposes [European Commission 2010 Directive 2010/63/EU of the European Parliament and of the Council of 22 September 2010 on the protection of animals used for scientific purposes. Official Journal of the European Union L 276: 33–79].

Author details

[1]Consiglio per la ricerca in agricoltura e l'analisi dell'economia agraria (CREA) - Animal production research centre, Via Salaria 31, 00015 Monterotondo, Rome, Italy. [2]Department of Pharmacy and Biotechnology (FaBiT), University of Bologna Alma Mater, Via Belmeloro 6, 40126 Bologna, Italy. [3]Applications and Innovation Department, CINECA, SCAI SuperComputing, Via dei Tizii 6, 00185 Rome, Italy.

References

1. MacHugh DE, Bradley DG. Livestock genetic origins: goat buck the trend. Proc Natl Acad Sci. 2001;98:5382–4.
2. Food and Agriculture Organization of the United nations. http://www.fao.org/corp/statistics/en. Accessed 11 Oct 2016.
3. Devendra C. Dairy goats in Asia: multifunctional relevance and contribution to food and nutrition security. In: E-Proceeding of the First Asia Dairy Goat Conference. Serdang: University Putra Malaysia and the Food and Agricultural Organization of the United Nation Publ; 2012. p. 1–6.
4. Dong Y, Xie M, Jiang Y, Xiao Y, Du N, Zhang X, Tosser-Klopp W, Wang G, Yang J, Liang S, Chen J, Chen W, Zeng J, Hou P, Bian Y, Pan C, Li S, Liu Y, Wang X, Servin W, Sayre B, Zhu B, Sweeney B, Moore D, Nie R, Shen W, Zhao Y, Zhang R, Li G, Faraut J, Womack T, Zhang J, Kijas Y, Cockett J, Xu N, Zhao X, Wang S, Wang J, Wang W. Sequencing and automated whole-genome optical mapping of the genome of a domestic goat (Capra hircus). Nat Biotechnol. 2013;31:135–41.
5. Mills S, Ross RP, Hill C, Fitzgerald GF, Stanton C. Milk intelligence: Mining milk for bioactive substances associated with human health. Int Dairy J. 2011;21:377–401.
6. Nakamura T, Urashima T. The milk oligosaccharides of domestic farm animals. Trends Glycosci Glyc. 2004;16:135–42.
7. Martinez-Ferez A, Rudloff S, Guadix A, Henkel CA, Pohlentz G, Boza JJ, Guadix EM, Kunz C. Goats' milk as a natural source of lactose-derived oligosaccharidess: Isolation by membrane technology. Int Dairy J. 2006;16:173–81.
8. Urashima T, Fukuda K, Kitaoka M, Ohnishi M, Terabayashi T and Kobata. Milk Oligosaccharides. Nutrition and Diet Research Progress. New York: Nova Biomedical Books; 2011.
9. Barile D, Marotta M, Chu C, Mehra R, Grimm R, Lebrilla CB, German JB. Neutral and acidic oligosaccharides in Holstein-Friesian D1 during the first 3 days of lactation measured by high performance liquid chromatography on a microfluidic chip and time-of-flight mass spectrometry. J Dairy Sci. 2010;93:3940–9.
10. Nakamura T, Kawase H, Kimura K, Watanabe Y, Ohtani M, Arai I, Urashima T. Concentrations of Sialyloligosaccharides in Bovine D1 and Milk during the Prepartum and Early Lactation. J Dairy Sci. 2003;86:1315–2.
11. Lars B. Recent advances on structure, metabolism, and function of human milk oligosaccharides. J Nut. 2006;136:2127–30.
12. Chichlowski M, German JB, Lebrilla CB, Mills DA. The influence of milk oligosaccharides on microbiota of infants: opportunities for formulas. Ann rev food sci technol. 2011;2:331–51.
13. Hickey RM. The role of oligosaccharides from human milk and other sources in prevention of pathogen adhesion. Internat Dairy J. 2012;22:141–6.
14. Urashima T, Taufik E, Fukuda K, Asakuma S. Recent advances in studies on milk oligosaccharides of cows and other domestic farm animals. Biosci Biotech Bioch. 2013;77:455–66.
15. Crisà A. Milk carbohydrates and oligosacchharides. In: Park YW, Haenlein GFW, editors. Milk dairy products in human nutrition. Chichester: John Wiley & Sons, Ltd; 2013. p. 129–47.
16. Martinez-Ferez A, Guadix A, Zapata-Montoya JE, Guadix EM. Influence of transmembrane pressure on the separation of caprine milk oligosaccharides from protein by cross-flow ultrafiltration. Int J Dairy Technol. 2008;61:333–9.
17. Kiskini A, Defilippo E. Oligosaccharides in goat milk: structure, health effects and isolation. Cell Mol Biol. 2013;1:25–30.
18. Albrecht S, Lane JA, Mariño K, Al Busadah KA, Carrington SD, Rita M, Hickey RM, Rudd PM. A comparative study of free oligosaccharides in the milk of domestic animals. Br J Nutr. 2014;111:1313–28.
19. Marth JD, Grewal PK. Mammalian glycosylation in immunity. Nat Rev Immunol. 2008;8:874–87.
20. García-Vallejo JJ, Gringhuis SI, Dijk W, Die I. Gene Expression Analysis of Glycosylation-Related Genes by Real-Time Polymerase Chain Reaction. Glycobiology protocols. New Jersey: Human Press Inc; 2007. p. 187–209.
21. Hashimoto K, Tokimatsu T, Kawano S, Yoshizawa AC, Okuda S, Goto S, Kanehisa M. Comprehensive analysis of glycosyltransferases in eukaryotic genomes for structural and functional characterization of glycans. Carbohydrate Res. 2009;344:881–7.
22. Harduin-Leper A. Comprehensive Analysis of Sialyltransferases in Vertebrate Genomes. Glycobiology Insights. 2010;2:29–61.
23. Bionaz M, Loor JJ. Gene networks driving bovine milk fat synthesis during the lactation cycle. BMC Genomics. 2008;9:366.
24. Harvatine KJ, Boisclair YR, Bauman DE. Recent advances in the regulation of milk fat synthesis. Animal. 2009;3:40–54.
25. Ferreira AM, Bislev SL, Bendixen E, Almeida AM. The Mammary Gland in Domestic Ruminants: A Systems Biology Perspective. J Proteom. 2013;94:110–23.
26. Roy NC, Altermann E, Park Z, McNabb WC. A comparison of analog and Next-Generation transcriptomic tools for mammalian studies. Brief funct genomics. 2011;10:135–50.
27. Cui X, Hou Y, Yang S, Xie Y, Zhang S, Zhang Y, Zhang Q, Lu X, Liu GE, Sun D. Transcriptional profiling of mammary gland in Holstein cows with extremely different milk protein and fat percentage using RNA sequencing. BMC Genomics. 2014;15:226.
28. Wickramasinghe S, Gonzalo Rincon G, Alma I-TA, Medrano JF. Transcriptional profiling of bovine milk using RNA sequencing. BMC Genomics. 2012;13:45.
29. Paten AM, Duncan EJ, Pain SJ, Peterson SW, Kenyon PR, Blair HT, Dearden P. Functional Development of the Adult Ovine Mammary Gland—insights from Gene Expression Profiling. BMC Genomics. 2015;16:748.
30. Elfstrand L, Lindmark-Månsson H, Paulsson M, Nyberg L, Àkesson B. Immunoglobulins, Growth Factors and Growth Hormone in Bovine D1 and the Effects of Processing. Int Dairy J. 2002;12:879–87.
31. Nairn AV, York WS, Harris K, Hall EM, Pierce JM, Moremen KW. Regulation of glycan structures in animal tissues: transcript profiling of glycan-related genes. J Biol Chem. 2008;283:17298–313.
32. Clark KJ-R, Griffiths J, Bailey KM, Harcum SW. Gene-expression profiles for five key glycosylation genes for galactose-fed CHO cells expressing recombinant IL-4/13 cytokine trap. Biotechnol Bioeng. 2005;90:568–77.
33. Wickramasinghe S, Hua S, Rincon G, Islas-Trejo A, German JB, Lebrilla CB, Medrano JF. Transcriptome Profiling of Bovine Milk Oligosaccharide Metabolism Genes Using RNA-Sequencing. PlosOne. 2011;11, e18895.
34. Boutinaud M, Jammes H. Potential uses of milk epithelial cells: a review. Reprod Nutr Dev. 2002;42:133–47.

35. Medrano JF, Rincon G. Comparative Analysis Of Bovine Milk And Mammary Gland Transcriptome Using RNA-Seq. Leipzig: 9th World Congress on Genetics applied to livestock production; 2009. p. 852.
36. Canovas A, Rincòn G, Bevilacqua C, Islas-Trejo A, Brenaut P, Hovey RC, Boutinaud M, Morgenthaler C, VanKlompenberg MK, Martin P, Medrano JF. Comparison of five different RNA sources to examine the lactating bovine mammary gland transcriptome using RNA-Sequencing. Scientific Reports. 2014;4:5297.
37. Boutinaud M, Herve L, Lollivier V. Mammary epithelial cells isolated from milk are a valuable, non-invasive source of mammary transcripts. Front Genet. 2015;6:323.
38. The ArkDB. http://www.thearkdb.org/arkdb/). Accessed 11 Oct 2016.
39. The Babraham Bioinformatics group. http://www.bioinformatics.babraham.ac.uk/projects/fastqc. Accessed 11 Oct 2016.
40. Bolger AM, Lohse M, Usadel B. Genome Analysis Trimmomatic : A Flexible Trimmer for Illumina Sequence Data. Bioinformatics. 2014;1–7.
41. Langmead B, Trapnell C, Pop M, Salzberg SL. Ultrafast and memory-efficient alignment of short DNA sequences to the human genome. Genome Biol. 2009;10:R25.
42. Trapnell C, Pachter L, Salzberg SL. TopHat: discovering splice junctions with RNA-Seq. Bioinformatics. 2009;25:1105–11.
43. Trapnell C, Williams BA, Pertea G, Mortazavi A, Kwan G, Van Baren MJ, Salzberg SL, Wold BJ, Pachter L. Transcript assembly and quantification by RNA-Seq reveals unannotated transcripts and isoform switching during cell differentiation. Nat Biotechnol. 2010;28:511–5.
44. Grabherr MG, Haas BJ, Yassour M, Levin JZ, Thompson DA, Amit I, Adiconis X, Fan L, Raychowdhury R, Zeng Q, Chen Z, Mauceli E, Hacohen N, Gnirke A, Rhind N, Di Palma F, Birren BW, Nusbaum C, Lindblad-Toh K, Friedman N, Regev A. Full-length transcriptome assembly from RNA-Seq data without a reference genome. Nature Biotechnol. 2011;29:644–52.
45. Wu TD, Watanabe CK. GMAP: a genomic mapping and alignment program for mRNA and EST sequences. Bioinformatics. 2005;21:1859–75.
46. Kent WJ. BLAT - The BLAST-like alignment tool. Genome Res. 2002;12:656–64.
47. Micallef L, Rodgers P. eulerAPE : Drawing Area-Proportional 3-Venn Diagrams Using Ellipses. PLoS ONE. 2014;9:e101717.
48. Mudunuri U, Che A, Yi M, Stephens RM. bioDBnet: the biological database network. Bioinformatics. 2009;25:555–6.
49. Götz S, García-Gómez JM, Terol J, Williams TD, Nagaraj SH, Nueda MJ, Robles M, Talón M, Dopazo J, Ana Conesa A. High-throughput functional annotation and data mining with the Blast2GO suite. Nucleic Acids Res. 2008;36:3420–35.
50. The DAVID Bioinformatic database. http://david.abcc.ncifcrf.gov. Accessed 11 Oct 2016.
51. Huang DW, Sherman BT, Lempicki RA. Bioinformatics enrichment tools: paths toward the comprehensive functional analysis of large gene lists. Nucleic Acids Res. 2009;37:1–13.
52. Tarazona S, García-Alcalde F, Dopazo J, Ferrer A, Conesa A. Differential expression in RNA-seq: a matter of depth. Genome Res. 2011;21:2213–23.
53. HTSeq: Analysing high-throughput sequencing data with Python. http://www-huber.embl.de/users/anders/HTSeq/doc/overview.html. Accessed 11 Oct 2016.
54. The WEB-based GEne SeT AnaLysis Toolkit. http://www.webgestalt.org. Accessed 11 Oct 2016.
55. Wang J, Duncan D, Shi Z, Zhang B. WEB-based GEne SeT AnaLysis Toolkit (WebGestalt): update 2013. Nucleic Acids Res. 2013;41:W77–83.
56. Szklarczyk D, Franceschini A, Wyder S, Forslund K, Heller D, Huerta-Cepas J, Simonovic M, Roth A, Santos A, Tsafou KP, Kuhn M, Bork P, Jensen LJ, Von Mering C. STRING v10: protein-protein interaction networks, integrated over the tree of life. Nucleic Acids Res. 2015;43:D447–52.
57. Lemay DG, Ballard O, Hughes M, Morrow AL, Horseman ND, Nommsen-Rivers L. RNA Sequencing of the Human Milk Fat Layer Transcriptome Reveals Distinct Gene Expression Profiles at Three Stages of Lactation. PLoS ONE. 2013;8, e67531.
58. Mortazavi A, Williams BA, Mccue K, Schaeffer L, Wold B. Mapping and Quantifying Mammalian Transcriptomes by RNA-Seq. Nat Methods. 2008;5:1–8.
59. Łabaj PP, Leparc GG, Linggi BE, Markillie LM, Wiley HS, Kreil DP. Characterization and improvement of RNA-Seq precision in quantitative transcript expression profiling. Bioinformatics. 2011;27:i383–91.
60. Toung JM, Morley M, Li M, Cheung VG. RNA-sequence analysis of human B-cells. Genome Res. 2011;21:991–8.
61. Shi H, Jiangjiang Zhu J, Jun Luo J, Wenting Cao W, Huaiping Shi H, Dawei Yao D, Jun Li J, Sun Y, Xu H, Yu K, Loor JJ. Genes Regulating Lipid and Protein Metabolism Are Highly Expressed in Mammary Gland of Lactating Dairy Goats. Funct Integr Genomics. 2014;15:309–21.
62. Schlimme E, Martin D, Meisel H. Nucleosides and nucleotides: natural bioactive substances in milk and D1. British J Nut. 2000;84:S59–68.
63. Park YW. Rheological characteristics of goat and sheep milk. Small Rumin Res. 2007;68:73–87.
64. Mabjeesh SJ, Kyle CE, Macrae JC, Bequette BJ. Lysine Metabolism by the Mammary Gland of Lactating Goats at Two Stages of Lactation. J Dairy Sci. 2000;83:996–1003.
65. Bionaz M, Periasamy K, Rodriguez-Zas SL, Everts RR, Lewin H, Hurley WL, Loor JJ. Old and new stories: revelations from functional analysis of the bovine mammary transcriptome during the lactation cycle. PLoS ONE. 2012;7:e33268.
66. Bionaz M, Loor JJ. Gene Networks Driving Bovine Mammary Protein Synthesis during the Lactation Cycle. Bioinformatics and Biology Insights. 2011;5:83–98.
67. Farrell HM, Jimenez-Flores R, Bleck GT, Brown EM, Butler JE, Creamer LK, Hicks CL, Hollar CM, Ng-Kwai-Hang KF, Swaisgood HE. Nomenclature of the Proteins of Cows' Milk—Sixth Revision. J Dairy Sci. 2004;87:1641–74.
68. Hinz K, O'Connor PM, Huppertz T, Ross RP, Kelly AL. Comparison of the principal proteins in bovine, caprine, buffalo, equine and camel milk. J Dairy Res. 2012;79:185–91.
69. Aggarwal J, Sharma A, Kishore A, Mishra BP, Yadav A, Mohanty A, Sodhi M, Kataria RS, Malakar D, Mukesh M. Identification of Suitable Housekeeping Genes for Normalization of Quantitative Real-Time PCR Data during Different Physiological Stages of Mammary Gland in Riverine Buffaloes (Bubalus Bubalis). J Anim Physiol An N. 2013;97:1132–41.
70. Pradeep J, Monika S, Ankita S, Umesh KS, Amit K, Ashok M, Mishra BP, Sandeep M, Kataria RS, Kaushik J, Mukesh M. Expression Analysis of Solute Carrier (SLC2A) Genes in Milk Derived Mammary Epithelial Cells during Different Stages of Lactation in Sahiwal (Bos indicus) Cows. Research Article. J Adv Dairy Res. 2014;2:117.
71. Lkhider M, Castino R, Bouguyon E, Isidoro C, Nd Ollivier-Bousquet M, Cathepsin D. Released by Lactating Rat Mammary Epithelial Cells Is Involved in Prolactin Cleavage under Physiological Conditions. J Cell Sci. 2004;117:5155–64.
72. Ishida M, Maehara M, Watanabe T, Yanagisawa Y, Takata Y, Nakajima R, Suzuki M, Harigaya T. Vasoinhibins, N-Terminal Mouse Prolactin Fragments, Participate in Mammary Gland Involution. J Mol Endocrinol. 2014;52:279–87.
73. CastinoR DS, Bouguyon E, Demoz M, Isidoro C, Ollivier-Bousquet M. Prolactin Promotes the Secretion of Active Cathepsin D at the Basal Side of Rat Mammary Acini. Endocrinology. 2008;149:4095–105.
74. Berberat PO, Katori M, Kaczmarek E, Anselmo D, Lassman C, Ke B, Shen X, Busuttil RW, Yamashita K, Csizmadia E, Tyagi S, Ottobein LE, Brouard S, Tobiash E, Bach FH, Kupiec-Weglinski JW, Soares MP. Heavy Chain Ferritin Acts as an Antiapoptotic Gene That Protects Livers from Ischemia Reperfusion Injury. FASEB J. 2003;17:1724–26.
75. Kanno Y, Ohtsuka H, Yoshikawa Y, Watanabe K, Orino K. Measurement of ferritin and anti-ferritin autoantibodies in serum and colostrum of Holstein and Japanese Black cows. Animal Sci J. 2013;84:556–61.
76. Molenaar AJ, Harris DP, Rajan GH, Pearson ML, Callaghan MR, Sommer L, Farr VC, Oden KE, Miles MC, Petrova RS, Good LL, Singh K, McLaren RD, Prosser CG, Kim KS, Wieliczko RJ, Dines MH, Johannessen KM, Grigor MR, Davis SR, Stelwagen K. The acute-phase protein serum amyloid A3 is expressed in the bovine mammary gland and plays a role in host defence. Biomarkers. 2009;14:26–37.
77. Dudemaine PL, Thibault C, Alain K, Bissonnette N. Genetic Variations in the SPP1 Promoter Affect Gene Expression and the Level of Osteopontin Secretion into Bovine Milk. Anim Genet. 2014;45:629–40.
78. Sheehy PA, Riley LG LG, HRaadsma HW, Williamson P, Wynn PC. A Functional Genomics Approach to Evaluate Candidate Genes Located in a QTL Interval for Milk Production Traits on BTA6. Anim Genet. 2009; 40:492–98.
79. Crisà A, Marchitelli C, Pariset L, Contarini G, Signorelli F, Napolitano F, Catillo G, Valentini A, Moioli B. Exploring Polymorphisms and Effects of Candidate Genes on Milk Fat Quality in Dairy Sheep. J Dairy Sci. 2010;93:834–45.
80. Suburu J, Shi L, Wu JJ, Wang S, Samuel M, Thomas MJ, Kock ND, Yang G, Kridel S, Chen YO. Fatty acid synthase is required for mammary gland

development and milk production during lactation. Am Physiol - Endoc M. 2014;306:E1132–43.
81. Ollier S, Robert-Granié C, Bernard L, Chilliard Y, Leroux C. Mammary transcriptome analysis of food-deprived lactating goats highlights genes involved in milk secretion and programmed cell death. J Nut. 2007;137:560–7.
82. Le Provost F, Cassy S, Hayes H, Martin P. Structure and Expression of Goat GLYCAM1 Gene: Lactogenic-Dependent Expression in Ruminant Mammary Gland and Interspecies Conservation of the Proximal Promoter. Gene. 2003;313:83–9.
83. Faucon F, Rebours E, Bevilacqua C, Helbling JC, Aubert J, Makhzami S, Dhorne-Pollet S, Robin S, Martin P. Terminal Differentiation of Goat Mammary Tissue during Pregnancy Requires the Expression of Genes Involved in Immune Functions. Physiol Genomics. 2009;40:61–82.
84. Maningat D, Sen P, Rijnkels M, Sunehag AL, Hadsell DL, Bray M, Haymond MW. Gene Expression in the Human Mammary Epithelium during Lactation: The Milk Fat Globule Transcriptome. Physiol Genomics. 2009;37(1):12–22.
85. Hanigan MD, France J, Mabjeesh SJ, McNabb WC, Bequette BJ. High Rates of Mammary Tissue Protein Turnover in Lactating Goats Are Energetically Costly. J Nut. 2009;139:1118–27.
86. Knight CH, Peaker M. Mammary Development and Regression during Lactation in Goats in Relation to Milk Secretion. Quarterly J Exp Physiol. 1984;69:331–38.
87. Stefanon B, Colitti M, Gabai G, Knight CH, Wilde CJ. Mammary Apoptosis and Lactation Persistency in Dairy Animals. J Dairy Res. 2002;69:37–52.
88. Capuco AV, Wood DL, Baldwin R, McLeod K, Papee MJ. Mammary Cell Number, Proliferation, and Apoptosis During a Bovine Lactation: Relation to Milk Production and Effect of bST. J Dairy Sci. 2001;84:2177–87.
89. Sorensen MT, JNørgaard JV, Theil PK, Vestergaard M, Sejrsen K. Cell Turnover and Activity in Mammary Tissue during Lactation and the Dry Period in Dairy Cows. J Dairy Sci. 2006;89:4632–39.
90. Nørgaard JV, Thaareil PK, Sørensen MT, Sejrsen K. Cellular Mechanisms in Regulating Mammary Cell Turnover during Lactation and Dry Period in Dairy Cows. J Dairy Sci. 2008;91:2319–27.
91. Wareski P, Motyl T, Ryniewicz Z, Orzechowski A, Gajkowska B, Wojewodzka U, Ploszaj T. Expression of Apoptosis-Related Proteins in Mammary Gland of Goat. Small Ruminant Res. 2001;40:279–89.
92. Zarzynska J, Motyl T. Apoptosis and autophagy in involuting bovine mammary gland. J Physiol Pharmacol. 2008;59:275–88.
93. Lemay DG, Neville MC, Rudolph MC, Pollard KS, German JB. Gene Regulatory Networks in Lactation: Identification of Global Principles Using Bioinformatics. BMC Syst Biol. 2007;1:56.
94. Jesenberger V, Jenstch S. Deadly encounter: ubiquitin meets apoptosis. Nature Reviews Mol Cell Biol. 2002;3:112–21.
95. Yang Y, Yu X. Regulation of Apoptosis: The Ubiquitous Way. FASEB J. 2003;17:790–99.
96. Karin M, Ben-Neriah Y. Phosphorylation Meets Ubiquitination: The Control of NF-κB Activity. Annual Rev Immunol. 2000;18:621–63.
97. Dubey A, Prajapati KS, Swamy M, Pachauri V. Heat Shock Proteins : A Therapeutic Target Worth to Consider. Veterinary World. 2015;8:46–51.
98. Schmitt E, Gehrmann M, Brunet M, Multhoff G, Garrido C. Intracellular and Extracellular Functions of Heat Shock Proteins: Repercussions in Cancer Therapy. J Leukocyte Biol. 2006;81:15–27.
99. Annison EF, Linzell JL. The oxidation and utilization of glucose and acetate by the mammary gland of the goat in relation to their over-all metabolism and to milk formation. J Physiol. 1964;175:372–85.
100. Hagiwara K, Kataoka S, Yamanaka H, Kirisawa R, Iwai H. Detection of Cytokines in Bovine Colostrum. Vet Immunol Immunopat. 2000;76:183–90.
101. Uruakpa FO, Ismond MAH, Akobundu ENT. Oligosaccharide and Its Benefits : A Review. Breast. 2002;22:755–67.
102. Ishida N, Kawakita M. Molecular physiology and pathology of the nucleotide sugar transporter family (SLC35). Eur J Physiol. 2004;447:768–75.
103. Song Z. Roles of the nucleotide sugar transporters (SLC35 family) in health and disease. Mol Aspects Med. 2013;34:590–600.
104. Nishimura M, Suzuki S, Satoh T, Naito S. Tissue-Specific mRNA Expression Profiles of Human Solute Carrier 35 Transporters. Drug Metab Pharmacokinet. 2009;24:91–9.
105. Claps S, Di Napoli MA, Sepe L, Caputo AR, Rufrano D, Di Trana A, Annicchiarico G, Fedele V. Sialyloligosaccharides content in colostrum and milk of twogoat breeds. Small Rumin Res. 2014;121:116–9.
106. Manlongat N, Yang TJ, Hinckley LS, Bendel RB, Krider HM. Physiologic-Chemoattractant-Induced Migration of Polymorphonuclear Leukocytes in Milk. Clini Diag L Immun. 1998;5:375–81.
107. Deshmane SL, Kremlev S, Amini S, Sawaya BE. Monocyte Chemoattractant Protein-1 (MCP-1): An Overview. J Interf Cyt Res. 2009;29:313–26.
108. Sapi E. The role of CSF-1 in normal physiology of mammary gland and breast cancer: an update. Exp Biol Med. 2004;229:1–11.
109. Schwertfeger KL, Rosen JM, Cohen DA. Mammary Gland Macrophages: Pleiotropic Functions in Mammary Development. J Mammary Gland Biol Neoplasia. 2006;11:229–38.
110. Mepham TB. Amino Acid Utilization by Lactating Mammary Gland. J Dairy Scie. 1982;65:287–98.
111. Singh K, Dobson J, Phyn CVC, Davis SR, Farr VC, Molenaar AJ, Stelwagen K. Milk Accumulation Decreases Expression of Genes Involved in Cell–extracellular Matrix Communication and Is Associated with Induction of Apoptosis in the Bovine Mammary Gland. Livest Prod Sci. 2005;98:67–78.
112. Fleming JM, Brandimarto JA, Cohick WS. The Mitogen-Activated Protein Kinase Pathway Tonically Inhibits Both Basal and IGF-I-Stimulated IGF-Binding Protein-5 Production in Mammary Epithelial Cells. J Endocrinol. 2007;194:349–59.
113. Lin J, Bao ZK, Zhang Q, Hu WW, Yu QH, Yang Q. Transcriptome Analysis of the Mammary Gland from GH Transgenic Goats during Involution. Gene. 2015;565:228–34.
114. Shen O, Zhang Y, Uray IP, Hill JH, Kim HT, Lu C, Young MR, Gunther EJ, Hilsenbeck SG, Chodosh LA, Colburn NH, Brown PH. The AP-1 transcription factor regulates postnatal mammary gland development. Dev Biol. 2006;295:589–603.
115. Marti A, Lazar H, Ritter P, Jaggi R. Transcription Factor Activities and Gene Expression During Mouse Mammary Gland Involution. J Mammary Gland Biol. 1999;4:154–2.
116. Brown JR, Ye H, Bronson RT, Dikkes P, Greenberg ME. A Defect in Nurturing in Mice Lacking the Immediate Early Gene fosB. Cell. 1996;86:297–309.
117. Acunzo J, Katsogiannou M, Rocchi P. Small heat shock proteins HSP27 (HspB1), αB-crystallin (HspB5) and HSP22 (HspB8) as regulators of cell death. Int J Biochem Cell B. 2012;44:1622–31.
118. Vos MJ, Hageman J, Carra S, Kampinga HH. Structural and Functional Diversities between Members of the Human HSPB, HSPH, HSPA, and DNAJ Chaperone Families. Biochemistry. 2008;47:7001–11.
119. Ostrowska M, Jarczak J, Zwierzchowski L. Glucose transporters in cattle - a review. Anim Sci P. 2015;33:191–212.
120. Tazawa S, Yamato T, Fujikura H, Hiratochi M, Itoh F, Tomae M, Takemura Y, Maruyama H, Sugiyama T, Wakamatsu A, Isogai T, Isaji M. SLC5A9/SGLT4, a new Na+–dependent glucose transporter, is an essential transporter for mannose, 1,5-anhydro-D-glucitol, and fructose. Life Sci. 2005;76:1039–50.
121. Groneberg D, Döring F, Theis S, Nickolaus M, Fischer A, Daniel H. Peptide Transport in the Mammary Gland: Expression and Distribution of PEPT2 mRNA and Protein. Am J Physiol Endocrinol Metab. 2002;282:E1172–79.
122. Mertens C, Darnell JE. SnapShot: JAK-STAT Signaling. Cell. 2007;131:612–612.e1.
123. Rawlings JS, Rosler KM, Harrison D. The JAK/STAT Signaling Pathway. J Cell Sci. 2004;117:1281–83.
124. Jung ID, Lee HS, Lee HY, Choi OH. FcεRI-Mediated Mast Cell Migration: Signaling Pathways and Dependence on Cytosolic Free Ca2+ Concentration. Cell Signal. 2009;21:1698–705.
125. Hemmings BA, Restuccia DF. PI3K-PKB/Akt Pathway. In: Cantley L, Hunter T, Sever R, Thorner J, editors. Cold Spring Harb Perspect Biol, vol. 4. 2012. p. a011.
126. Organ SL, Tsao MS. An Overview of the c-MET Signaling Pathway. Therapeutic Advances in Medical Oncology. 2011;3:S7–19.
127. Stinnakre MG, Vilotte JL, Soulier S, Mercier JC. Creation and phenotypic analysis of a-lactalbumin-deficient mice. Proc Natl Acad Sci U S A. 1994;91:6544–8.
128. Asano M, Furukawa K, Kido M, Matsumoto S, Umesaki Y, Kochibe N, Iwakura Y. Growth retardation and early death of β-1, 4-galactosyltransferase knockout mice with augmented proliferation and abnormal differentiation of epithelial cells. EMBO J. 1997;16:1850–7.
129. Qasba PK, Ramakrishnan B, Boeggeman E. Structure and function of –1,4-galactosyltransferase. Curr Drug Targets. 2008;9:292.
130. Tanner ME. The enzymes of sialic acid biosynthesis. Bioorg Chem. 2005;33:216–28.
131. Martín-Ortiz A, Salcedo J, Barile D, Bunyatratchata A, Moreno FJ, Martin-García I, Clemente A, Sanz ML, Ruiz-Matute AI. Characterization of goat colostrum oligosaccharides by nano-liquid chromatography on chip quadrupole time-of-flight mass spectrometry and hydrophilic interaction liquid chromatography-quadrupole mass spectrometry. J Chromatogr A. 2016;1428:143–53.
132. Fuhrer A, Sprenger N, Kurakevich E, Borsig L, Chassard C, Hennet T. Milk sialyllactose influences colitis in mice through selective intestinal bacterial colonization. J Exp Med. 2010;13:2843–54.

133. Monti L, Cattaneo TMP, Orlandic M, Curadic MC. Capillary electrophoresis of sialylated oligosaccharides in milk from different species. J Chromatogr A. 2015;1409:288–91.
134. Schnaar RL. Brain gangliosides in axon-myelin stability and axon regeneration. FEBS Lett. 2010;584:1741–7.
135. Inokuchi J, Uemura S. ST3 Beta-Galactoside Alpha-2,3-Sialyltransferase 5 (ST3GAL5). In: Taniguchi N, Honke K, Fukuda M, Narimatsu H, Yamaguchi Y, Angata T, editors. Handbook of Glycosyltransferases and Related Genes, Springer Japan. 2014. p. 675–86.
136. Wang B, Brand-Miller J. The role and potential of sialic acid in human nutrition. Eur J Clin Nutr. 2003;57:1351–69.
137. Sakai F, Ikeuchi Y, Urashima T, Fujihara M. Effects of Feeding Sialyllactose and Galactosylated N-Acetylneuraminic Acid on Swimming Learning Ability and Brain Lipid Composition in Adult Rats. Components. 2006;254:249.
138. Togayachi A, Kozono Y, Kuno A, Ohkura T, Sato T, Hirabayashi J, Ikehara Y, Narimatsu H. β3GnT2 (B3GNT2), a Major Polylactosamine Synthase: Analysis of B3gnt2-Deficient Mice. Method Enzymol. 2010;479:185–204.
139. Trottein F, Schaffer L, Ivanov S, Paget C, Vendeville C, Cazet A, Groux-Degroote S, Lee S, Krzewinski-Recchi MA, Faveeuw C, Head SR, Gosset P, Delannoy P. Glycosyltransferase and sulfotransferase gene expression profiles in human monocytes, dendritic cells and macrophages. Glycoconj J. 2009;26:1259–74.
140. Honke K. Beta1,3-N-Acetylgalactosaminyltransferase 1 (B3GALNT1). In: Taniguchi N, Honke K, Fukuda M, Narimatsu H, Yamaguchi Y, Angata T, editors. Handbook of Glycosyltransferases and Related Genes, Springer Japan. 2014. p. 447–53.
141. Zancada L, Sánchez-Juanes F, Alonso JM, Hueso P. Neutral glycosphingolipid content of ovine milk. J Dairy Sci. 2010;93:19–26.
142. Urashima T, Bubb WA, Messer M, Tsuji Y, Taneda Y. Studies of the neutral trisaccharides of goat (Capra hircus) colostrum and of the one- and two-dimensional 1H and 13C NMR spectra of 6'-N acetylglucosaminyllactose. Carbohydr Res. 1994;262:173–84.
143. Niittymäki J, Mattila P, Renkonen R. Differential gene expression of GDP-L-fucose synthesizing enzymes, GDP-fucose transporter and fucosyltransferase VII. APMIS. 2006;114:539–48.
144. Becker DJ, Lowe JB. Fucose: biosynthesis and biological function in mammals. Glycobiology. 2003;13:41R–53R.
145. Chen CI, Keusch JJ, Klein D, Hess D, Hofsteenge J, Gut H. Structure of human POFUT2: insights into thrombospondin type 1 repeat fold and O-fucosylation. EMBO J. 2012;31:3183–97.
146. Kytzia HJ, Sandhoff K. Evidence for two different active sites on human beta-hexosaminidase A. Interaction of GM2 activator protein with beta-hexosaminidase A. J Biol Chem. 1985;260:7568–72.
147. Gravel RA, Kaback MM, Proia RL, Sandhoff K, Suzuki K, Suzuki K. The GM2 gangliosidoses. In: Scriver CR, Beaudet AL, Valle D, Sly WS, editors. The Metabolic and Molecular Bases of Inherited Diseas. 8th ed. 76: McGraw-Hill; 2001. p. 3827–76.
148. Nishihara S. Solute Carrier Family 35 (CMP-Sialic Acid Transporter), Member A1 (SLC35A1). In: Taniguchi N, Honke K, Fukuda M, Narimatsu H, Yamaguchi Y, Angata T, editors. Handbook of Glycosyltransferases and Related Genes, Springer Japan. 2014. p. 1369–77.
149. Bakker A, Ashikov A, Routier FH, Gerardy-Schahn R. GDP-Fucose Transporter 1 (2014) (*SLC35C1*). In: Taniguchi N, Honke K, Fukuda M, Narimatsu H, Yamaguchi Y, Angata T, editors. Handbook of Glycosyltransferases and Related Genes, Springer Japan. 2014. p. 1403–11.
150. Verheijen FW, Verbeek E, Aula N, Beerens CEMT, Havelaar AC, Joosse M, Peltonen L, Aula P, Galjaard H, Van der Spek PJ, Mancini GMS. A new gene, encoding an anion transporter, is mutated in sialic acid storage diseases. Nat Genet. 1999;23:462–5.

Glucose concentration in capillary blood of dairy cows obtained by a minimally invasive lancet technique and determined with three different hand-held devices

B. Mair[1], M. Drillich[1], D. Klein-Jöbstl[1], P. Kanz[1], S. Borchardt[2,4], L. Meyer[2], I. Schwendenwein[3] and M. Iwersen[1*]

Abstract

Background: Dairy cows have a massive demand for glucose at the onset of lactation. A poor adaption to this period leads to an excessive negative energy balance with an increased risk for ketosis and impaired animal health and production.

Besides the measurement of ketones, analysing the glucose concentration in blood is reported as helpful instrument for diagnosis and differentiation of ketosis. Monitoring metabolic parameters requires multiple blood sampling. In other species, new blood sampling techniques have been introduced in which small amounts of blood are rapidly analysed using electronic hand-held devices.

The objective of this study was to evaluate the suitability of capillary blood for blood glucose measurement in dairy cows using the hand-held devices FreeStyle Precision (FSP, Abbott), GlucoMen LX Plus (GLX, A. Menarini) and the WellionVet GLUCO CALEA, (WGC, MED TRUST). In total, 240 capillary blood samples were obtained from dry and fresh lactating Holstein-Friesian cows. Blood was collected from the skin of the exterior vulva by using a lancet. For method comparison, additional blood samples were taken from a coccygeal vessel and analyzed in a laboratory. Glucose concentrations measured by a standard laboratory method were defined as the criterion standard.

Results: The Pearson correlation coefficients between the glucose concentrations analyzed in capillary blood with the devices and the reference were 73 % for the FSP, 81 % for the GLX and 41 % for the WGC. Bland-Altman plots showed biases of −18.8 mg/dL for the FSP, -11.2 mg/dL for the GLX and +20.82 mg/dL for the WGC. The optimized threshold determined by a Receiver Operating Characteristics analysis to detect hyperglycemia using the FSP was 43 mg/dL with a sensitivity (Se) and specificity (Sp) of 76 and 80 %. Using the GLX and WGC optimized thresholds were 49 mg/dL (Se = 92 %, Sp = 85 %) and 95 mg/dL (Se = 39 %, Sp = 92 %).

Conclusions: The results of this study demonstrate good performance characteristics for the GLX and moderate for the FSP to detect hyperglycemia in dairy cows using capillary blood. With the study settings, the WGC was not suitable for determination of glucose concentrations.

Keywords: Dairy cow, Glucose, Capillary blood, Hand-held meter

* Correspondence: michael.iwersen@vetmeduni.ac.at

[1]Department for Farm Animals and Veterinary Public Health, Clinical Unit for Herd Health Management in Ruminants, University Clinic for Ruminants, University of Veterinary Medicine Vienna, 1210 Vienna, Austria

Full list of author information is available at the end of the article

Background

Glucose is an essential nutrient for all higher organisms. Certain vitally important cells and tissues like the brain, erythrocytes and the mammary gland are dependent on the energy provided by glucose. Hence, it is important that the concentration of glucose in the blood is kept at a physiological level [1]. Ruminants and particularly dairy cows have a special position with respect to the glucose metabolism, because of the massive demand of their mammary gland for glucose at the onset of lactation and the permanent alteration between pregnancy and lactation. Furthermore, most of the glucose is generated by gluconeogenesis using volatile glucogenic precursors (e.g. propionate, lactate, alanine, valerate and isobutyrate) [2].

Oikonomou et al. [3] reported a significant influence of the glucose concentration on reproductive performance, as glucose is essential as major energy source for ovarian cyclicity [4]. Additionally, a study by Galvão et al. [5] has shown that cows suffering from uterine infections within the first 3 weeks after parturition had a decreased polymorphonuclear neutrophils (PMN) glycogen concentration and hence, a decrease in PMN functions. It was reported that the PMN glycogen concentration is positively associated with the blood glucose concentration. Furthermore, Garverick et al. [6] found greater blood glucose concentration in cows that conceived at first insemination in comparison with unsuccessfully bred cows, and hypothesized that blood glucose concentration within the first 3 weeks after parturition can be used as predictor for the likelihood of pregnancy after first insemination.

For most tissues, the supply and utilization of nutrients is regulated by hormones. Also in Ruminants insulin is a key player in the energy metabolism, influencing various metabolic processes in muscles, adipose tissue, liver and the mammary gland [7]. Ruminants are reported to suffer from 'insulin resistance' (IR) at the end of gestation and during early lactation [8–10]. The phenomenon of IR is characterized by a less than normal biological response to a physiological concentration of insulin, as a result of decreased sensitivity of the tissues, a decreased responsiveness to insulin or a combination of both [11].

Insulin resistance at the end of gestation and at early lactation are considered as physiologic mechanism to ensure the glucose and hence energy supply for the gravid uterus and lactating mammary gland. However, IR can transgress into a pathologic state, leading to ketosis for instance with chronically increased concentrations of ß-hydroxybutyric acid (BHBA) in cows [2]. From an aetiological point of view two different types of ketosis exist: First the hypoglycaemic-hypoinsulinaemic form, 'Type I'-ketosis, which occurs three to 6 weeks after calving in high-yielding cows where the energy demand in the udder exceeds the capacity of gluconeogenesis. Second, the hyperglycaemic-hyperinsulinaemic form, 'Type II'-ketosis exists, which occurs earlier in lactation as a consequence of overfeeding the animals in the dry period [12]. Besides the measurement of the BHBA concentration, analysing the glucose concentration in blood is reported as helpful instrument for diagnosis and especially differentiation between these two types of ketosis [13, 14], although Herdt [15] described the glucose concentration as an insensitive measure of the energy status of an animal.

Monitoring metabolic parameters requires multiple blood sampling, which is usually done by repeated venepuncture in cattle. This procedure requires handling or restraining of the animal, which may result in stress-induced hyperglycemia [16]. In other species, new techniques have been introduced in which small amounts of blood are rapidly analysed for glucose concentrations using electronic hand-held devices [17–19]. Applying these minimally invasive techniques to obtain capillary blood in dairy cattle might be beneficial in generating reliable and valid glucose concentrations.

The primary objective of this study was to test whether capillary blood can be used to determine the glucose concentrations in dairy cows using minimally invasive techniques. A second objective was to evaluate test characteristics of three commercially available hand-held devices for determination of blood glucose concentrations. Therefore a capillary blood drop, obtained from the skin of the exterior vulva, was tested with three different hand-held devices. Additionally, glucose concentrations in coccygeal blood were analysed in a laboratory to be used as 'criterion standard' [20] and analysed with the hand-held devices, too to distinguish between the effect of the device and the type of blood sample used for the on-site test.

Methods

The study was approved by the institutional ethics committee of the University of Veterinary Medicine, Vienna (04/12/97/2013; date of approval 17 December 2013) according to the Good Scientific Practice guidelines and the national authority according to § 26 of Law for Animal Experiments [Tierversuchsgesetz 2012 – (TVG 2012); BMWF GZ 68.205/0007-II/3b/2014] as well as by the Slovakian Regional Veterinary Food Administration (428/2014) and was conducted between March and April 2014. A written informed consent was obtained from the farm owner to collect samples and to publish the study results. The farm involved in the study adheres to a high standard of veterinary care based on standard operating procedures.

Animals

A commercial dairy farm in Slovakia with approximately 2700 Holstein-Friesian cows and additional youngstock was chosen as study site. The dairy herd was kept in free stall barns with rubber mats on concrete floors. Before calving the cows were moved in a free stall barn with straw beddings. A total mixed ration was fed twice per day and pushed up frequently. The average energy corrected milk yield (based on 4.0 % butterfat and 3.4 % protein) was 9165 kg in 2013.

A sample size calculation (type I error $\alpha = 0.05$, type II error $\beta = 0.2$) was performed to detect a maximum irrelevant difference in the glucose concentration of 3 mg/dL between the methods evaluated in the study, resulting in 182 animals needed. To compensate for potential data losses due to necessary exclusions because of pre-analytical or analytical problems 240 animals were enrolled in the study. Animals of all lactations within 2 weeks ante-partum up to 4 weeks post-partum were eligible for enrollment.

Sampling procedures

The sampling procedures are illustrated in Fig. 1. To obtain capillary blood, the skin of the exterior vulva was dry-cleaned and punctured with one of three different available lancets, in random order. The lancets used in the study [Microtainer Contact-Activated Lancet (MT, Becton-Dickinson), SafetyLancets special (SL, MED TRUST Handelsges.m.b.H.), MiniCollect Safety Lancets (MC, Greiner Bio-One International AG)] have a penetration depth of 2 mm with differing blade widths between 0.8 (SL) to 1.5 mm (MT and MC). If the volume of the obtained blood drop was insufficient for determination of the glucose concentration, another puncture was made. The quantity of punctures was recorded on a pre-assembled data capture form. Three different electronic hand-held meters [FreeStyle Precision, (FSP, Abbott GmbH & Co. KG), GlucoMen LX Plus, (GLX, A. Menarini GmbH), WellionVet GLUCO CALEA, (WGC, MED TRUST Handelsges.m.b.H.)] and the associated electrochemical test strips [FreeStyle

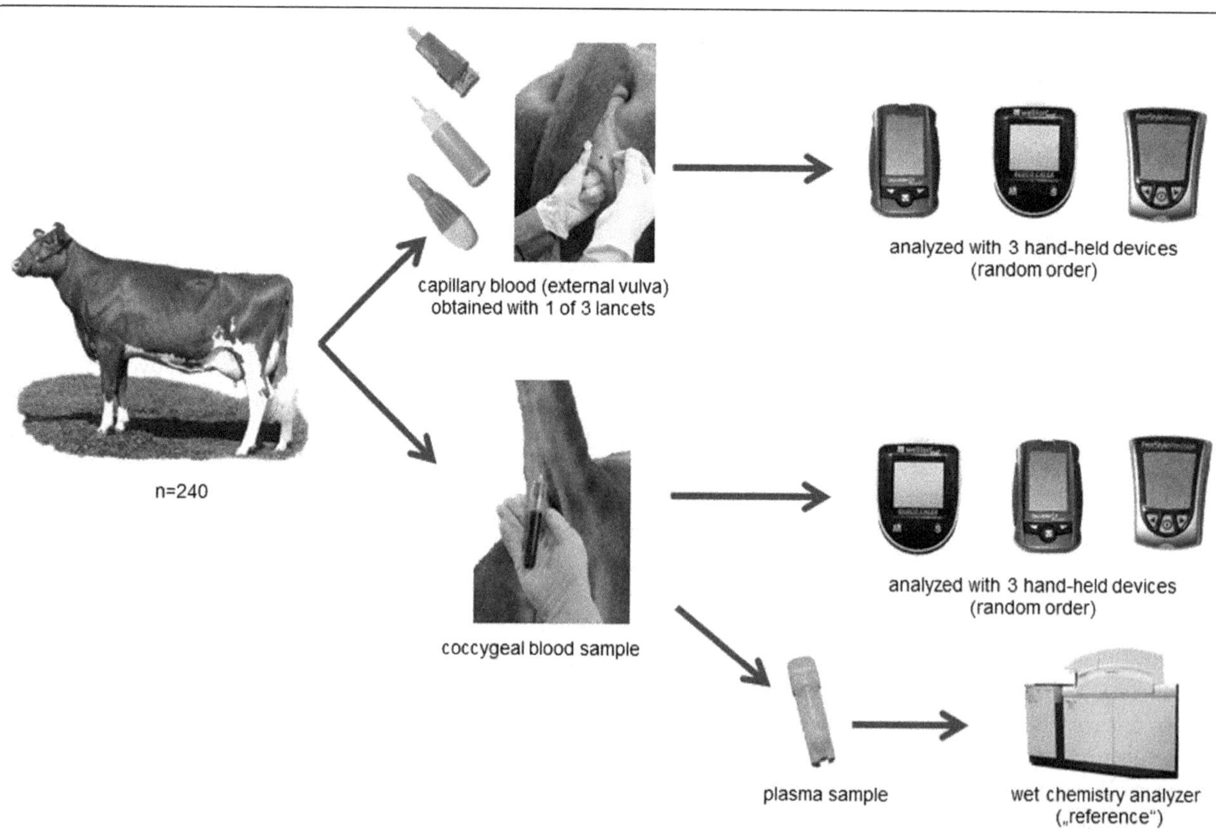

Fig. 1 Sampling procedures. In total, 240 dairy cows were enrolled. Capillary blood was obtained from the external vulva by using 1of 3 lancet types. The glucose concentrations in the capillary blood drop as well as in a coccygeal blood sample were analyzed each, by using 3 different hand-held devices. Plasma obtained from the coccygeal blood sample was analyzed for glucose in the laboratory of the University using a wet chemistry analyzer and was regarded as "reference" in the study

Precision blood glucose, (Abbott Diabetes Care), GlucoMen LX Sensor, (A. Menarini GmbH), WellionVet GLUCO CALEA test strips, (MED TRUST Handelsges.m.b.H.)] were used in the study to determine the blood glucose concentration. The WGC was already validated for use in dogs, cats and horses by using preprogrammed species-specific chips, offered with each batch of test strips. Because of best possible match in terms of red blood cell parameters, the chip initially designed for use in cats was chosen for measurements in this study.

Operating principle of the hand-held devices and laboratory analyses

The test principle of all 3 hand held devices is based on an amperometric biosensor technology. The meter systems consist of two components: The meter itself processes the data by applying integrated algorithms and presents the results on an integrated display. The associated biosensor test strips are designated to analyze a small amount of a blood sample (i.e. for the GLX: 0.3 μL; WGC: 0.5 μL; FSP: 0.6 μL) by initiating further enzymatic reactions.

After the application of the blood on the sensor, an electrochemical reaction starts using either glucose dehydrogenase (FSP) or glucose oxidase (GLX and WGC) as catalysator. The enzymes are oxidoreductases and oxidize glucose to gluconolactone. By this, electrons from the glucose are transferred to the oxidized form of a mediator, thereby converting it to the reduced form. By re-oxidation of the mediator electrons are transferred to the electrode surface. The resulting current is monitored by the meter and directly proportional to the glucose concentration in the blood sample. Using meter specific software algorithms, the current is converted into a measure of the glucose concentration which is displayed as digital value.

The sensors of the 3 hand-held devices were directly dipped onto the surface of the capillary blood drop. After approximately 5 s each, the blood glucose concentrations were presented on the display of the devices.

Besides the manufacturer's intended and certified use of testing capillary blood, the devices were used "off-label" for measuring the glucose concentration in venous and/or arterial blood. For this, an additional blood sample was drawn from a coccygeal vessel using a blood-collection tube system (Vacuette, Greiner Bio-One GmbH) consisting of a Sodium Fluorid vacuum tube (Vacuette, FX Sodium Fluoride, Greiner Bio-One GmbH) and a 20-gauge needle (Vacuette 0.9 × 38 mm, Greiner Bio-One GmbH). Blood obtained from the coccygeal vessel was tested as well with all three devices as previously described. Approximately 2 h after collection, the coccygeal blood samples were centrifuged at 2200 × *g*, at a temperature of 18 °C for 5 min. (Eppendorf Centrifuge 5804, Eppendorf AG). Supernatant plasma was split into two aliquots in microtubes of 2 ml each (Microtube, Sarstedt), as reference and back-up sample and were stored at −18 °C until further analyses. The glucose concentration of this plasma sample was analyzed at the Central Clinical Pathology Unit (CCPU) of the University of Veterinary Medicine, Vienna and was considered as the reference value (i.e. the criterion standard) in the present study. Plasma was analyzed with an automated wet chemistry analyzer Cobas 6000/c501 (Roche Diagnostics GmbH, Vienna, Austria) using a colorimetric hexokinase method. Hexokinase catalyzes the phosphorylation of glucose to glucose-6-phosphate and adenosine-diphosphate by ATP. This reaction is coupled with an NADP colorimetric indicator system.

To evaluate the intra-assay variability of the analyses performed at the CCPU, a subset of 20 aliquots of plasma, obtained from a blood sample of one cow, were randomly placed among the samples obtained from the study animals. Furthermore, intra- and inter-assay coefficients of variations (CV) were calculated for each specific hand-held device. For this, three blood samples with different glucose concentrations based on FSP measurements with low (30 mg/dL), medium (56 mg/dL) and high (70 mgl/dL) glucose concentrations were tested ten times with one device (intra-assay) and additionally with ten different devices of the same type (inter-assay).

Statistical analyses

The data were analyzed using SPSS statistics for Windows (Version 20.0; IBM Deutschland GmbH) and BiAS for Windows (Version 10.06; Epsilon-Verlag). Data were tested for normal distribution using the Kolmogorov-Smirnov-Test. For each tested hand-held device, descriptive statistics were estimated for the glucose concentrations analyzed in capillary and coccygeal blood. Additionally, the Pearson correlation coefficients were calculated between the reference and the glucose concentrations measured with each specific device in capillary and coccygeal blood. The level of significance for all statistical tests was set at $\alpha = 0.05$.

A Passing-Bablok regression [21] was performed to compare the glucose concentration in the reference sample with the concentrations measured with the three hand-held devices in capillary and coccygeal blood. For this, the slopes and the intercepts of the regression lines and their 95 % confidence intervals were determined. The intercept (a) reflects the constant differences and the slope (b) the proportional differences between the two methods. If the confidence interval for the intercept includes the value 0, no constant difference between the two methods exists. If the interval for the slope includes the value 1, no proportional difference occurs. If neither a constant nor a proportional difference could be observed, both methods can be used interchangeably [22]. In addition, the agreement between the reference and the hand-held meters were graphically depicted using the method as reported by Bland and Altman [23].

Based on the glucose concentrations analyzed at the CCPU, samples were classified as hypo- (<40 mg/dL), normo- (40–60 mg/dL) or hyperglycemic (>60 mg/dL). According to these classifications sensitivities (Se) and specificities (Sp) for each hand-held device to detect hypo- and hyperglycemia were calculated. To determine optimized thresholds to identify hypo- and hyperglycemia using the hand-held meters, Receiver Operating Characteristics (ROC) analyses were performed. The closer the resulting graph of the ROC analysis is to the left upper angle of the coordinate system, the greater is the accuracy of the test [24]. The resulting area under the ROC curve (AUC) is a measure of the discriminatory power of a test to identify animals as normoglycemic and hypo- or hyperglycemic, respectively. An AUC of 1 represents a perfect test; an AUC of 0.5 and below represents a worthless test. ROC analyses were based on maximizing the Youden-Index (J), calculated as (Se + Sp - 1). The J reflects the number of all correctly identified outcomes and can range from 0 to 1. A value close tol indicates a valuable diagnostic test, whereas a value close to 0 implies a worthless test [25].

Results

The mean glucose concentration of the 20 samples taken from one cow to evaluate the intra-assay variability of measurements at the CCPU was 44.45 ± 0.83 mmol/L, resulting in a CV of 1.86 %.

In total, 240 Holstein-Friesian cows were tested. Thirty four (14.2 %) of the animals were in first lactation, 97 (40.4 %) in second lactation, and 109 (45.4 %) in third or greater lactation. All animals were sampled between 21 days ante-partum and 29 days post-partum (median 7 days in milk; 25 % Percentile 3 and 75 % Percentile 13), resulting in 199 (83 %) samples from lactating cows and 41 (17 %) samples from dry cows.

The reference samples comprised two samples (0.8 %) classified as hypoglycemic (glucose < 40 mg/dL), 149 samples (62.2 %) as normoglycemic (glucose between 40 to 60 mg/dL) and 89 samples (37.1 %) as hyperglycemic (glucose > 60 mg/dL). All lancets used in the study were eligible for capillary blood specimen collection, and only small numerical differences could be observed in the total number of incisions needed. Capillary blood could be obtained with first incision in 94 % (n = 75) using the SL, 96 % (n = 77) using the MT and in 96 % (n = 77) using the MC lancet. An additional second (and third) incision was necessary for the MT in three (two) cases, using the SL in two (zero) and the MC in three (zero) cases. Measurements of four (1.7 %) capillary blood samples using the GLX device resulted in an "E4"-failure. According to the manufacturer's instructions this error message indicates a "damaged sensor, incorrect application or quantity of the blood sample". Additionally, an "E3"-failure was shown by the FSP in two (0.8 %) measurements of capillary blood, indicating "a test error or glucose concentrations below the detection limit". These six measurements were repeated.

Descriptive parameters for the analyzed glucose concentrations are presented in Table 1. The Pearson correlation coefficients between the glucose concentrations analyzed in capillary blood using the hand-held meters and the reference were 73.3 % for the FSP, 80.5 % for the GLX and 41.2 % for the WGC (P < 0.01, for all three devices). Using coccygeal blood, the corresponding correlation coefficients were 86.6 % for the FSP, 78.8 % for the GLX and 50.5 % for the WGC (P < 0.01, for all three devices).

The differences between the glucose concentrations analyzed in plasma at the laboratory and the concentrations measured with the hand-held devices either in whole blood obtained from a coccygeal vessel or in capillary blood are presented in Fig. 2. The glucose concentrations analyzed with the FSP and the GLX in the capillary blood as well as in the coccygeal blood resulted in smaller glucose concentrations compared with the laboratory results in plasma. Measuring the glucose concentrations using the WGC overestimated the laboratory results.

Bland-Altman plots (Fig. 3) between the reference test and coccygeal blood demonstrated a negative bias of −4.89 ± 6.31 mg/dL for the FSP and −9.98 ± 6.67 mg/dL for the GLX, whereas a positive bias for the WGC of +24.64 ± 15.19 mg/dL was determined. Using capillary blood for testing, biases of −18.80 ± 7.96 mg/dL for the FSP, −11.20 ± 6.42 mg/dL for the GLX and +20.82 ± 15.41 mg/dL for the WGC were detected (Table 2 and Fig. 3).

To compare the specific results of the hand-held devices with the reference, a Passing-Bablok regression was performed (Table 2). Using the GLX for measurements in coccygeal blood, the confidence interval for the slope includes the value 1, but the confidence interval for the intercept did not include 0. Hence, no proportional but a constant difference was detected. Using the FSP and the WGC in coccygeal blood, a proportional and a constant difference exist. Similar findings for measurements using the three hand-held devices in capillary blood were detected.

Receiver Operating Characteristics analyses were intended to calculate the optimal thresholds to differentiate between normo- and hyperglycemia using the electronic hand-held devices. Because only two reference samples were classified as hypoglycemic, no reliable

Table 1 Descriptive statistics of the glucose concentrations analyzed in coccygeal and capillary blood using three different hand-held devices as well as in plasma analyzed at the laboratory (reference)

Parameter	Plasma[a]	Capillary blood			Coccygeal blood		
	Laboratory	FSP[b]	GLX[c]	WGC[d]	FSP	GLX	WGC
Number of samples (n)	240	240	240	240	240	240	240
Mean (mg/dL)	58.17	39.56	46.96	78.99	53.28	48.18	82.86
SD (mg/dL)	10.48	11.65	10.75	16.69	12.33	10.48	17.59
Median (mg/dL)	58.00	38.00	46.00	79.00	53.00	49.00	83.00
Interquartile range (mg/dL)	12.50	18.00	16.00	26.00	16.00	13.00	26.00

[a]obtained from coccygeal blood (reference)
[b]FSP: FreeStyle Precision, (Abbott GmbH & Co. KG, Wiesbaden, Germany)
[c]GLX: GlucoMen LX Plus (A. Menarini GmbH, Vienna Austria)
[d]WGC: WellionVet GLUCO CALEA (MED TRUST Handelsges.m.b.H., Marz, Austria)

ROC analyses for those samples were possible. Hence, the results are not presented. Detailed results of the ROC analyses for the three different hand-held devices to detect hyperglycemia as well as further test characteristics are presented in Table 3 and Fig. 4.

The calculated AUCs for measurements in capillary and coccygeal blood indicate that the GLX (93 and 85 %) and the FSP (87 and 92 %) were more eligible in determining hyperglycemia than the WGC (71 and 70 %). With a calculated AUC of 93 % and a Youden-Index of 78 % the GLX was the most capable device to detect hyperglycemia in capillary blood. Sensitivities and specificities using the optimized thresholds for all three devices are shown in Table 3. Identifying hyperglycemia in capillary blood, Se was greatest for the GLX (93 %), however Sp was greatest for the WGC (92 %).

The calculated inter- and intra-assay CV for analyzing the glucose concentration using the different devices are presented in Table 4. The average inter- and intra-assay CVs were acceptable for the FSP with 4.2 and 5.2 %, but less satisfying for the GLX with 19.9 and 14.4 %, and for the WGC with 12.6 and 15.7 %.

Discussion

To our knowledge, this is the first study evaluating the suitability of the three hand-held meters to determine the glucose concentration in capillary blood in dairy cows. Using a minimally invasive technique to obtain capillary blood, as already used in small animal medicine

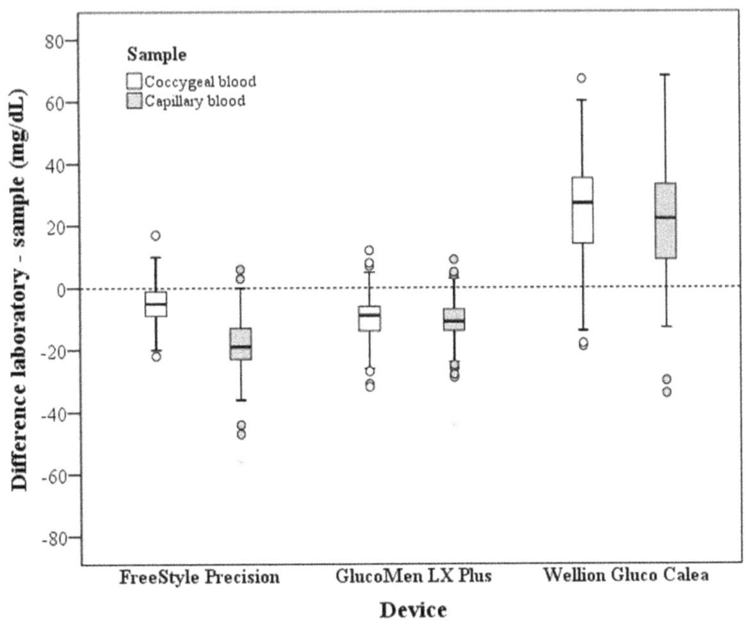

Fig. 2 Parallel boxplots showing the differences between the glucose concentrations in plasma analyzed at the laboratory (reference) and the concentrations measured with 3 different hand-held devices in either whole blood obtained from a tail vessel or in capillary blood. The heavy black line inside each box marks the median (50th percentile); lower and upper hinges mark the 25th and 75th percentiles. Whiskers end at the smallest and largest statistical values that are not outliers; outliers and extreme values are designated by o and *

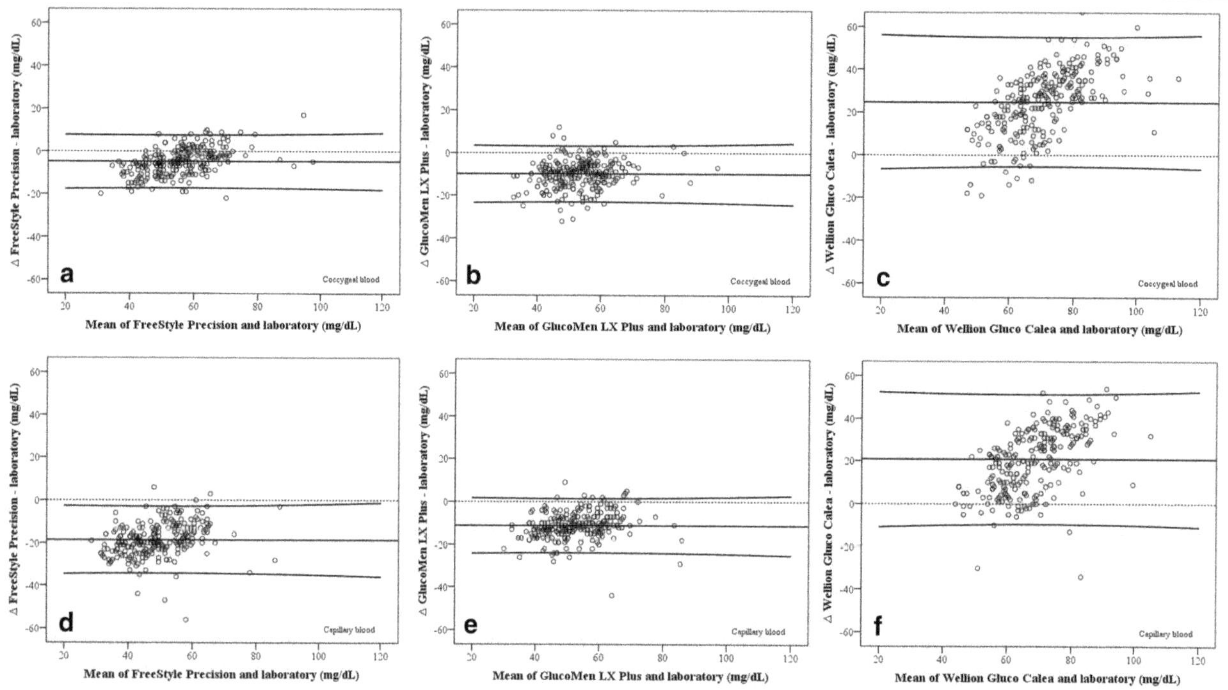

Fig. 3 Bland-Altman plots of differences in glucose concentrations measured with the FreeStyle Precision (panels **a** and **d**), GlucoMen LX Plus (panels **b** and **e**) and Wellion GLUCO CALEA (panels **c** and **f**) hand-held devices in coccygeal (*top*) or capillary (*bottom*) blood and the reference test against their mean. The solid line in the middle represents the mean; the solid upper and lower lines represent the mean ± 2 SD

[17–19], is considered as useful tool for monitoring the energy metabolism in dairy cows [26]. Additionally, measurements of the glucose concentrations in cows already suffering from hyperketonemia is beneficial to gain further insights in the type of existing ketosis [13] in order to optimize the animals' treatment. For these purposes, a reliable and accurate test system is essential.

Based on laboratory thresholds, the number of plasma samples ($n = 2$, 0.8 %) classified as hypoglycaemic was low. This was not expected because several cows were sampled in the early lactation where a negative energy balance is expected. It can be speculated if these results support the hypothesis of a physiological insulin-resistance in dairy cows [12]. Hence, further studies are planned to investigate the usefulness of population based clinical decision limits to detect a hypoglycaemic condition in cows during the transition period.

Whereas the FSP and the GLX were already validated for BHBA measurements to diagnose subclinical ketosis in dairy cows [27, 28], the WGC has never been used for any studies in bovines. It should be mentioned, that the chip used in this study was intended for use in cats and not certified by the manufacturer for use in bovines. The suitability of different hand-held devices to analyze glucose concentrations in blood obtained from the jugular vein or a coccygeal vessel in dairy cows were already evaluated by others [13, 29, 30]. For the FreeStyle Precision Xtra device (Abbott Diabetes Care Inc.), a previous version of the FSP, Wittrock et al. [30] determined a Pearson correlation coefficient between coccygeal whole

Table 2 Differences between the glucose concentrations analyzed by three different hand-held devices and the laboratory results using the Bland-Altman analysis method and Passing-Bablok regression analysis

Device[a]	Capillary blood						Coccygeal blood					
	Passing-Bablok				Bland-Altman		Passing-Bablok				Bland-Altman	
	Slope (*b*)	CI_{95}^{b} for *b*	Intercept (*a*)	CI_{95} for *a*	Bias (*d*)	SD of *d*	Slope (*b*)	CI_{95} for *b*	Intercept (*a*)	CI_{95} for *a*	Bias (*d*)	SD of *d*
FSP	0.73	0.67–0.80	29.30	26.70–31.67	−18.80	7.96	0.72	0.66–0.78	19.43	16.28–22.33	−4.89	6.31
GLX	0.82	0.75–0.88	19.57	16.67–23.00	−11.20	6.42	0.88	0.80–1.00	14.93	9.00–19.00	−9.98	6.67
WGC	0.41	0.34–0.50	24.53	17.50–30.64	20.82	15.41	0.40	0.34–0.47	23.00	17.16–28.09	24.64	15.19

[a]FSP: FreeStyle Precision, (Abbott GmbH & Co. KG, Wiesbaden, Germany), GLX: GlucoMen LX Plus (A. Menarini GmbH, Vienna, Austria), WGC: WellionVet GLUCO CALEA (MED TRUST Handelsges.m.b.H, Marz, Austria)

[b]CI_{95}: 95 % confidence interval

Table 3 Thresholds and corresponding test characteristics of 3 different hand-held meters to detect hyperglycemia in capillary and coccygeal blood based on Receiver Operating characteristics (ROC) analysis using a laboratory plasma glucose concentration of ≥ 60 mg/dL as reference

Device[a]	Capillary blood					Coccygeal blood				
	Optimized device threshold[b] [mg/dL]	AUC[c] [%(CI^{d}_{95})]	Se[e] [%(CI_{95})]	Sp[f] [%(CI_{95})]	Youden-Index[g]	Optimized device threshold[c] [mg/dL]	AUC [%(CI_{95})]	Se [%(CI_{95})]	Sp [%(CI_{95})]	Youden-Index
FSP	43	87.4 (82.5–91.3)	76.4 (66.2–84.8)	84.1 (77.3–89.5)	0.61	59	91.7 (87.5–94.9)	75.3 (65–83.8)	91.4 (85.7–95.3)	0.67
GLX	49	93.4 (89.5–96.2)	92.1 (84.5–96.8)	85.4 (78.8–90.6)	0.78	49	85.2 (80.1–89.5)	85.4 (76.3–92)	69.5 (61.5–76.8)	0.55
WGC	95	70.5 (64.3–76.3)	39.3 (29.1–50.3)	92.1 (86.5–95.8)	0.31	94	70.0 (63.8–75.7)	55.1 (44.1–65.6)	84.8 (78–90.1)	0.40

[a]FSP: FreeStyle Precision, (Abbott GmbH & Co. KG, Wiesbaden, Germany), GLX: GlucoMen LX Plus (A. Menarini GmbH, Vienna, Austria), WGC: WellionVet GLUCO CALEA (MED TRUST Handelsges.m.b.H, Marz, Austria)
[b]based on Receiver Operating Characteristics (ROC) analyses
[c]AUC: area under the receiver operating characteristics (ROC) curve
[d]CI_{95}: 95 % confidence interval
[e]Se: sensitivity
[f]Sp; specificity
[g]Youden-Index = (Se + Sp − 1)

blood samples and the corresponding plasma glucose concentration analysed at a laboratory of 95 % with a mean difference of -0.03 ± 1.96 mmol/L (equivalent to −0.54 ± 35 mg/dL). Voyvoda and Erdogan [13] determined a correlation coefficient between the glucose concentration in jugular whole blood analyzed with the Optium Xceed device (Abbott Diabetes Care) and its corresponding laboratory plasma concentration of 63 % with a mean difference of 0.27 mmol/L (4.86 mg/dL). Evaluating the One Touch II (Lifescan Inc) Roeder et al. [29] reported a correlation coefficient of 94 % with a mean difference of 12.95 mg/dL for samples obtained either from the jugular vein in calves, or by coccygeal

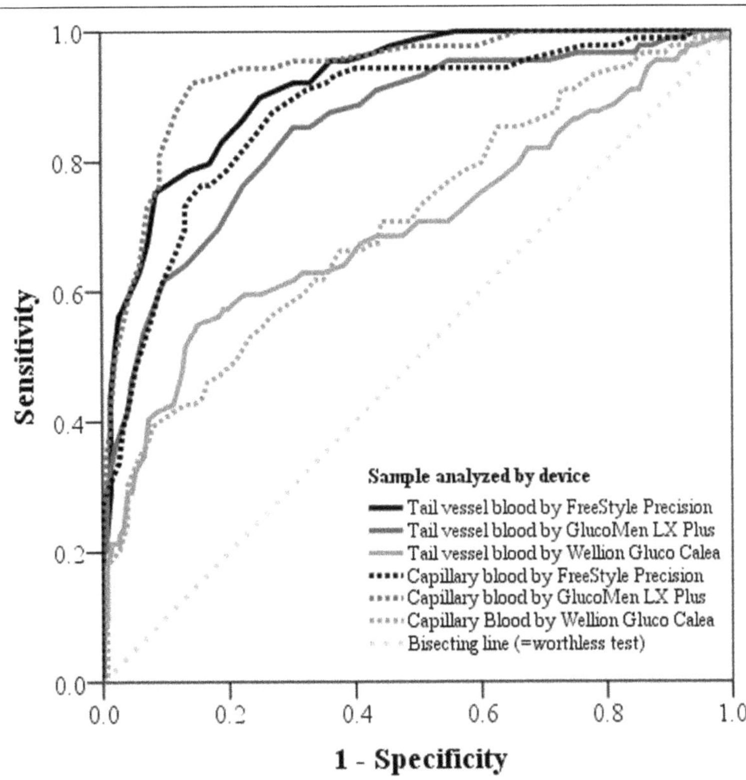

Fig. 4 Receiver operating characteristics curves for three different hand-held devices for diagnosis of hyperketonemia either in tail vessel or capillary blood, based on laboratory glucose concentration in plasma > 60 mg/dL as threshold for hyperketonemia

Table 4 Intra-assay and inter-assay coefficients of variation for hand-held meters FreeStyle Precision (FSP), GlucoMen LX Plus (GLX) and WellionVet GLUCO CALEA (WGC) within low, medium and high glucose concentrations

Parameter	Glucose concentration[a]											
	FSP[c]				GLX[d]				WGC[e]			
	Low	Medium	High	Average	Low	Medium	High	Average	Low	Medium	High	Average
Intra-assay												
Mean [mg/dL]	26.80	46.60	63.50	45.64	28.90	57.40	73.30	53.20	109.50	122.20	119.60	117.10
SD [mg/dL]	1.99	2.07	2.27	2.11	2.64	11.30	10.58	8.17	19.70	20.20	15.20	18.37
CV[b] [%]	7.43	4.44	3.57	5.15	9.13	19.69	14.43	14.42	17.99	16.53	12.71	15.74
Inter-assay												
Mean [mg/dL]	26.40	48.40	66.10	46.70	40.00	58.70	74.50	57.73	93.40	113.30	114.30	107.00
SD [mg/dL]	1.58	1.65	2.18	1.80	8.94	12.96	11.35	11.08	14.39	16.53	8.94	13.29
CV[b] [%]	5.98	3.41	3.30	4.23	22.35	22.08	15.23	19.89	15.41	14.59	7.82	12.61

[a]Based on measurements using FreeStyle Precision (low = 30 mg/dL; medium = 56 mg/dL; high = 70 mg/dL)
[b] Coefficient of variation = SD × 100/mean
[c]FSP = FreeStyle Precision (Abbott GmbH & Co. KG, Wiesbaden, Germany)
[d] GLX = GlucoMen LX Plus (A. Menarini GmbH, Vienna, Austria)
[e] WGC = WellionVet GLUCO CALEA (MED TRUST Handelsges.m.b.H, Marz, Austria)

venepuncture in cows and their plasma glucose concentrations, analysed at a laboratory. The results of our study, using the FSP and the GLX devices for analysing the glucose concentration in coccygeal blood showed similar deviations from the laboratory results as reported by Roeder et al. [29] and Voyvoda and Erdogan [13]. A lower deviation, however, was found by Wittrock et al. [30]. The reason for this finding remains speculative, but might be caused by the repeated measurements of 81 cows resulting in 709 blood samples analysed in the study by Wittrock et al. [30].

The highest deviation with a mean difference of approximately +21 mg/dL for measurements in capillary and +25 mg/dL in coccygeal blood compared with the reference was found for the WGC. It is very likely that the chip's calibration curve for cats led to this higher deviation. Hence, the development of a cow-specific chip is recommended for this device before further studies in cattle should be conducted.

The determined average inter- and intra-assay coefficients of variation for the FSP (4.2 and 5.2 %) were below 15 % as requested by the European Medicines Agency (EMEA, 2011) for on-site tests. The intra-assay CV of the GLX (14.4 %) and the inter-assay CV of the WGC (12.6 %) were within the upper limit of the requested range, whereas the inter-assay CV of the GLX (19.9 %) and the intra-assay CV of the WGC (15.7 %) exceeded the requested threshold in the bovine species.

The determined correlation coefficients for the glucose concentrations analyzed in capillary blood and the reference test were high, but not excellent for the FSP (73.3 %) and the GLX (80.5 %), whereas the correlation for the WGC was poor (41.2 %). The corresponding results for coccygeal blood were approximately 13 percentage points higher for the FSP, 9 percentage points higher for the WGC, but 3 percentage points lower for the GLX. A possible explanation for the poorer performance of these two devices for measurements in capillary blood might be that sometimes it was necessary to squeeze the skin of the vulva to obtain an adequate amount of blood. Squeezing the skin might cause the excretion of tissue fluid, which mixes up with the blood drop and thereby influences the glucose concentration. This, however, does not explain the smaller difference in the calculated correlation coefficients between capillary and coccygeal blood, respectively, compared with the reference as observed for the other devices. Another possible impact on the results might be the amount of blood applied on the test strips. According to the manufacturer of the WGC, the test strip must remain sufficiently immerged into the blood after the acoustic signal of the device occurs. It should be noticed that this signal indicates the start of the measurement with the WGC, whereas for other devices this signal indicates the end of the measurement procedure.

Bland-Altman plots revealed a negative bias for the FSP and the GLX and a positive bias for the WGC. Glucose concentrations analyzed in plasma and serum are reported to be greater than the corresponding concentrations in whole blood [14, 31]. Kuwa et al. [31] found 13.2 % greater glucose concentrations in plasma compared with those analyzed from venous whole blood. Additionally, significant greater glucose concentrations were found in capillary blood compared with the corresponding concentrations determined in venous blood [31]. This might be a major reason for the observed biases of the FSP and the GLX, but does not explain the differences for the WGC.

With thresholds optimized by ROC analysis, the GLX showed the best performance to detect hyperglycemia using capillary blood as test substrate. The results for the GLX were even better in the capillary blood than in the coccygeal blood. Remarkable lower sensitivities were found for WGC compared with the other devices. The Youden-Index presented in Table 3 confirms that the GLX is the most capable device to analyze glucose concentrations in capillary blood, as this value is the closest to 1, i.e. a 'perfect test' [25]. Glucose concentrations determined by the FSP, however, should be interpreted with care, whereas the WGC with the settings for cats is unsuitable.

This study is an investigation of measuring glucose in dairy cows using electronic hand-held devices in capillary and coccygeal blood, respectively. The analyzed glucose concentrations were influenced by the test system and to some extent to the site of sampling. A potential weakness of this study is the composition of the sample set. Samples were collected from animals kept on one commercial farm. Additional field studies are required also to confirm the hypothesized beneficial effects of monitoring the glucose concentration as predictor of the animal's metabolic status [13], immune competence [3, 5] or reproductive performance [6]. According to the manufacturers manuals the devices deliver reliable measurements for glucose concentrations between 20 and 500 mg/dL (600 mg/dL for the GLX). In our study, the maximum glucose concentration in plasma analyzed at the laboratory was 100 mg/dL. Hence, further studies are necessary to test if the devices lead to reliable results with greater glucose concentrations. Furthermore the validity of clinical decision limits to address hyper- or hypoglycemic statuses, especially for cows during the transition period, has to be assessed in further clinical studies.

Conclusions

The objective of this study was to test whether capillary blood can be used to determine the glucose concentration in dairy cows using three different electronic hand-held devices. Good test characteristics for determining hyperglycemia by an on-site test were found for the GLX and moderate for the FSP. Hence, these devices are eligible for use in veterinary practice. With the settings for cats, the WGC was not suitable for determination of glucose concentrations neither in capillary nor in coccygeal blood. For this device, the development of a cattle-specific chip is recommended. Capillary blood obtained by minimally invasive puncture of the skin of the exterior vulva is considered as suitable for analyzing glucose concentrations.

Abbreviations

AUC: area under the receiver operating characteristics curve; BHBA: ß-hydroxybutyric acid; CCPU: Central Clinical Pathology Unit; CV: coefficient of variation; FSP: FreeStyle Precision; GLX: GlucoMen LX Plus; IR: insulin resistance; *J*: Youden-Index; MC: MiniCollect safety lancet; MT: Microtainer contact-activated lancet; PMN: polymorphnuclear neutrophils; ROC: receiver operating characteristics; SE: sensitivity; SL: SafetyLancets special; SP: specificity; WGC: WellionVet GLUCO CALEA.

Competing interests

MED TRUST Handelsges.m.b.H. supplied the WellionVet GLUCO CALEA hand-held devices as well as the associated test strips and supported the study, in part, by a grant. MED TRUST Handelsges.m.b.H. played no role in the study design nor in the collection, analysis and interpretation of data, nor in the decision to submit the manuscript for publication. None of the authors has any financial or personal relationships that could inappropriately influence or bias the content of the manuscript.

Authors' contributions

BM assisted in study design, performed sampling and sample preperation, and wrote the manuscript. MD supported the study development and statistical analyses, and reviewed the manuscript. DKJ supported the study design, development and statistical analyses. PK performed sampling and sampling preparation. SB and LM supported the study development and sampling procedures. IS supported the study, supervised laboratory analyses and reviewed the manuscript. MI initiated, planned and supervised the study, supported sampling procedures, performed statistical analyses and provided valuable references and suggestions during the preparation of the manuscript.

Acknowledgements

This study was supported, in part, by a grant from MED TRUST Handelsges.m.b.H., Marz, Austria. We wish to thank the staff from FirstFarms Slovakia, Alexander Erlen, Armin Hofer, Markus Suda und David Süss for their technical assistance. Preliminary results were presented as a poster at the 10th Berlin-Brandenburgischer Rindertag, Berlin, Germany, 3rd–4th October 2014. Benedikt Mair submitted this study as diploma thesis at the University of Veterinary Medicine Vienna, Austria.

Author details

[1]Department for Farm Animals and Veterinary Public Health, Clinical Unit for Herd Health Management in Ruminants, University Clinic for Ruminants, University of Veterinary Medicine Vienna, 1210 Vienna, Austria. [2]FirstFarms Slovakia, 900 68 Plavecký Štvrtok, Slovakia. [3]Department for Pathobiology, Central Clinical Pathology Unit, University of Veterinary Medicine Vienna, 1210 Vienna, Austria. [4]Current address: Clinic for Animal Reproduction, Faculty of Veterinary Medicine, Freie Universität Berlin, Berlin, Germany.

References

1. Aschenbach JR, Kristensen NB, Donkin SS, Hammon HM, Penner GB. Gluconeogenesis in dairy cows: The secret of making sweet milk from sour dough. IUBMB Life. 2010;62(12):869–77.
2. De Koster JD, Opsomer G. Insulin resistance in dairy Cows. Vet Clin North Am Food Anim Pract. 2013;29(2):299–322.
3. Oikonomou G, Arsenos G, Valergakis GE, Tsiaras A, Zygoyiannis D, Banos G. Genetic relationship of body energy and blood metabolites with reproduction in Holstein cows. J Dairy Sci. 2008;91(11):4323–32.
4. Rabiee AR, Lean IJ, Gooden JM, Miller BG. Relationships among metabolites influencing ovarian function in the dairy cow. J Dairy Sci. 1999;82(1):39–44.
5. Galvão KN, Flaminio MJBF, Brittin SB, Sper R, Fraga M, Caixeta L, Ricci A, Guard CL, Butler WR, Gilbert RO. Association between uterine disease and indicators of neutrophil and systemic energy status in lactating Holstein cows. J Dairy Sci. 2010;93(7):2926–37.
6. Garverick HA, Harris MN, Vogel-Bluel R, Sampson JD, Bader J, Lamberson WR, Spain JN, Lucy MC, Youngquist RS. Concentrations of nonesterified fatty acids and glucose in blood of periparturient dairy cows are indicative of pregnancy success at first insemination. J Dairy Sci. 2013;96(1):181–8.

7. Sasaki S-I. Mechanism of insulin action on glucose metabolism in ruminants. Anim Sci J. 2002;73(6):423–33.
8. Petterson JA, Dunshea FR, Ehrhardt RA, Bell AW. Pregnancy and undernutrition alter glucose metabolic responses to insulin in sheep. J Nutr. 1993;123(7):1286–95.
9. Bell AW, Bauman DE. Adaptations of glucose metabolism during pregnancy and lactation. J Mammary Gland Biol Neoplasia. 1997;2(3):265–78.
10. Hayirli A. The role of exogenous insulin in the complex of hepatic lipidosis and ketosis associated with insulin resistance phenomenon in postpartum dairy cattle. Vet Res Commun. 2006;30(7):749–74.
11. Kahn CR. Insulin resistance, insulin insensitivity, and insulin unresponsiveness: A necessary distinction. Metab Clin Exp. 1978;27(12 SUPP. 2):1893–902.
12. Holtenius P, Holtenius K. New aspects of ketone bodies in energy metabolism of dairy cows: a review. Zbl Vetmed A. 1996;43(10):579–87.
13. Voyvoda H, Erdogan H. Use of a hand-held meter for detecting subclinical ketosis in dairy cows. Res Vet Sci. 2010;89(3):344–51.
14. Kraft W, Dürr UM. Klinische Labordiagnostik in der Tiermedizin. Stuttgart: Schattauer; 2005.
15. Herdt TH. Variability characteristics and test selection in herd-level nutritional and metabolic profile testing. Vet Clin North Am Food Anim Pract. 2000;16(2):387–403.
16. Wiedmeyer CE, Johnson PJ, Cohn LA, Meadows RL, Kerl ME, Tessman RK, Perlis J, Declue AE. Evaluation of a continuous glucose monitoring system for use in veterinary medicine. Diabetes Technol Ther. 2005;7(6):885–95.
17. Wess G, Reusch C. Capillary blood sampling from the ear of dogs and cats and use of portable meters to measure glucose concentration. J Small Anim Pract. 2000;41(2):60–6.
18. Zeugswetter FK, Karlovitz S. Buccal glucose measurements in dogs using safety lancets for blood sampling. Tieraerztliche Praxis Ausgabe Kleintiere Heimtiere. 2014;42(3):174–9.
19. Zeugswetter FK, Rebuzzi L, Karlovits S. Alternative sampling site for blood glucose testing in cats: Giving the ears a rest. J Feline Med Surg. 2010;12(9):710–3.
20. American medical association (AMA). American medical association manual of style: a guide for authors and editors. JAMA. 1998;279(21):1752–3.
21. Passing H. Bablok: A new biometrical procedure for testing the equality of measurements from two different analytical methods. Application of linear regression procedures for method comparison studies in clinical chemistry, Part I. J Clin Chem Clin Biochem. 1983;21(11):709–20.
22. Bilić-Zulle L. Comparison of methods: Passing and Bablok regression. Biochem Med. 2011;21(1):49–52.
23. Bland JM, Altman DG. Statistical methods for assessing agreement between two methods of clinical measurement. Lancet. 1986;1(8476):307–10.
24. Swets JA. Measuring the accuracy of diagnostic systems. Science. 1988; 240(4857):1285–93.
25. Youden WJ. Index for rating diagnostic tests. Cancer. 1950;3(1):32–5.
26. Kanz P, Drillich M, Klein-Jobstl D, Mair B, Borchardt S, Meyer L, Schwendenwein I, Iwersen M. Suitability of capillary blood obtained by a minimally invasive lancet technique to detect subclinical ketosis in dairy cows by using 3 different electronic hand-held devices. J Dairy Sci. 2015; 98(9):6108–18.
27. Iwersen M, Falkenberg U, Voigtsberger R, Forderung D, Heuwieser W. Evaluation of an electronic cowside test to detect subclinical ketosis in dairy cows. J Dairy Sci. 2009;92(6):2618–24.
28. Iwersen M, Klein-Jöbstl D, Pichler M, Roland L, Fidlschuster B, Schwendenwein I, Drillich M. Comparison of 2 electronic cowside tests to detect subclinical ketosis in dairy cows and the influence of the temperature and type of blood sample on the test results. J Dairy Sci. 2013;96(12):7719–30.
29. Roeder BL, Schaalje GB, Kelly EJ, Clark FD. A rapid method for determination of blood glucose concentration in cattle. J Am Vet Med Assoc. 1996;208(5):707–10.
30. Wittrock JAM, Duffield TF, LeBlanc SJ. Short communication: Validation of a point-of-care glucometer for use in dairy cows. J Dairy Sci. 2013; 96(7):4514–8.
31. Kuwa K, Nakayama T, Hoshino T, Tominaga M. Relationships of glucose concentrations in capillary whole blood, venous whole blood and venous plasma. Clin Chim Acta. 2001;307(1–2):187–92.

Mycoplasma bovis antibody dynamics in naturally exposed dairy calves according to two diagnostic tests

Mette Bisgaard Petersen[1*], Nadeeka K. Wawegama[2], Matthew Denwood[1], Philip F. Markham[2], Glenn F. Browning[2] and Liza Rosenbaum Nielsen[1]

Abstract

Background: Inexpensive and convenient diagnostic tests for use in clinical work and for the surveillance of infection with *Mycoplasma bovis* are in demand. The objective of this longitudinal field study was to gain knowledge about the dynamics of antibodies against *M. bovis* in sera from naturally exposed calves with and without different clinical signs, measured by two different ELISA tests.

Results: A total of 83 calves were subject to between one and five blood samples and clinical examinations using a standard protocol during five herd visits to each of four outbreak dairy herds. The blood samples were analysed for the presence of antibodies against *M. bovis* using the commercial IgG ELISA test BioX K302 (BioX) and an in-house indirect IgG ELISA test (MilA ELISA).

Linear mixed models were used to describe and compare the antibody dynamics as measured by the two tests in relation to the disease status and age of the animals.

The BioX ELISA response was below the recommended cut-off (37 ODC%) for the entire study period in many of the calves. The estimated mean ODC% increased slowly but did not reach the recommended individual animal cut-off in three of the four herds. The highest estimated ODC% was not reached until the calf was 110–130 days old. The MilA ELISA response rose above the recommended cut-off (135 antibody units (AU)) in almost all calves, and in two herds, the estimated mean was above the individual animal cut-off shortly after the birth of the calf. The highest estimated antibody concentration was reached when the calf was approximately 60 days old. Disease status of the calf was not significantly associated with the results of either test.

Conclusions: We conclude that the BioX ELISA cannot be recommended for use in calves below 3 months of age. The MilA ELISA was able to detect antibodies shortly after birth (i.e. from approximately 3 weeks of age and onwards) and is therefore a more sensitive test for *M. bovis* exposure in young calves. Neither ELISA seemed able to differentiate between calves with arthritis and/or otitis media, and respiratory disease.

Keywords: *Mycoplasma bovis*, ELISA, BioX bio K 302, MilA ELISA, Dairy calves, Antibody

Background

Mycoplasma bovis causes severe disease in cattle worldwide. The typical clinical manifestations in calves are pneumonia, otitis media and arthritis [1]. The primary diagnostic tool used in calves is bacterial culture of body fluid samples [2], but this is too expensive and time-consuming for use in group or herd diagnostics or for surveillance purposes. Although bacterial DNA-detection tests (such as PCR assays) are becoming more popular, the diagnostic material used for this technique is more difficult to obtain and process than a blood sample. An ELISA for antibody-detection is easy to perform on serum samples and is often less expensive, and these assays are already commonly used for the diagnosis of other diseases in cattle. Knowledge about the dynamics of antibody response in infected animals in relation to disease and age is essential when using an ELISA as a diagnostic

* Correspondence: mbp@sund.ku.dk
[1]Department of Veterinary and Animal Sciences, Faculty of Health and Medical Sciences, University of Copenhagen, Grønnegårdsvej 8, 1870 Frederiksberg, Denmark
Full list of author information is available at the end of the article

tool. This knowledge requires longitudinal studies of naturally exposed calves, involving repeated observations of clinical signs combined with samples being taken for laboratory testing. However, this is time-consuming, inconvenient and expensive, and therefore rarely implemented, meaning that our existing knowledge about antibody dynamics in calves comes mostly from experimental studies. Calves vaccinated with an experimental aerosol vaccine against *M. bovis* at three to 4 weeks and five to 6 months of age appeared to have a detectable immunoglobulin G (IgG) response against *M. bovis* within 14 days, and the antibody concentrations in serum remained at a high level for at least 42 and 30 days, respectively [3].

However, neither the *M. bovis* antibody response to systemic disease syndromes, such as arthritis, or the dynamics of the antibody response over time in naturally infected calves is clear. It is also crucial to know how to interpret ELISA results in young calves, since maternally derived antibodies against *M. bovis* might be present in uninfected calves. Furthermore, very young calves may not be able to generate an antibody response to bacterial infections [4]. Other authors have found the antibody titres in young dairy calves to be low, suggesting low levels of passive transfer of antibodies from the dam [5, 6]. No correlation has yet been found between clinical signs and antibody response in individual calves, but seroconversion to *M. bovis* has been shown to be predictive of disease at a group level in feedlot cattle [7, 8]. To date, there have been few evaluations of the use and interpretation of different *M. bovis* ELISA tests under field conditions. A recent study found that the antibody response in cows was very dynamic, of short duration and dependent on clinical signs in the cow [9], but similar studies in calves under different herd and disease conditions are lacking.

An in-house IgG-detection ELISA (MilA ELISA) developed by Wawegama et al. [10] has an estimated animal-level sensitivity and specificity of 94.3% and 94.4%, respectively using a cut-off of 105 antibody units (AU) [8]. This study compared the results from the MilA IgG ELISA with those obtained from the BioX K302 and K260 ELISA assays in two small groups of experimentally infected calves, and found that both BioX tests had low sensitivity.

To the best of our knowledge, there has been no comparison of the antibody responses measured using the MilA ELISA and the BioX K302 ELISA (BioX) in dairy calves, and knowledge about the generation of antibodies in serum in naturally exposed calves with and without different clinical signs is lacking. Therefore, the objective of this longitudinal field study was to describe and compare the dynamics of antibody responses to *M. bovis* in the serum of dairy calves with different disease manifestations using two different ELISA tests.

Methods

Data for this study were collected from four dairy herds, from which both cows and calves were sampled. A description and an analysis of the data from cows are presented in Petersen et al. [9], while analyses of the data from calves are presented here. The basic study design and herd selection were the same for the two studies. All farm owners were informed about the procedures in the study and gave written consent to use of their animals and farm data before study start. The study design was approved by the veterinary department of the agricultural advisory services, SEGES, before initiation.

Study design

A longitudinal observational study was carried out between 1st July 2015 and 5th April 2016 in four Danish dairy cattle herds with acute outbreaks of *M. bovis*-associated disease. Each herd was visited five times with an interval of approximately 3 weeks between each visit. The first visit was as close as possible to the onset of the disease outbreak. The clinical status of selected calves was assessed at each visit, and blood samples were collected from them. Where possible, the same animals were sampled again at each subsequent visit, allowing both between- and within-animal analysis over time.

Study population

The study herds were selected based on the detection of *M. bovis*-associated clinical signs by the herd advisory veterinarian and by diagnostic test results (positive in PCR assays on milk samples or in ELISA tests on sera from cows or calves). All herds had a recent history of sudden-onset of clinical signs indicative of *M. bovis* infection among the cows and/or calves, and several strongly positive ELISA and/or PCR test results for *M. bovis*.

All herds tested positive in both an ELISA and a PCR assay on at least one occasion during the study period, but the clinical signs present in the herds and the affected age groups differed. Information about the farms is presented in Table 1.

Laboratory analysis of calves that were euthanised and necropsied outside the planned project activities revealed additional information about the study herds. One four-month-old calf from Herd 2 had chronic degenerative arthrosis in several joints and bronchopneumonia with overlying pleuritis; *M. bovis* was cultured from joint fluid, and both joint fluid and lung tissue were PCR positive for *M. bovis*. Two one-month-old calves in Herd 4 had chronic omphalitis, bronchopneumonia, synovitis in several joints and bilateral otitis media; *Mycoplasma spp.* were cultured from these calves and identified as *M. bovis* by PCR.

Table 1 Summary description of the four Danish dairy herds and diagnostic test results prior to and during the study period

Herd no.	1	2	3	4
Prior to enrolment				
Herd size (No. of cows)	177	174	182	391
Estimated start of outbreak	Early Jun 2015	Early Jul 2015	Late Nov 2015	Mid Dec 2015
After enrolment				
Data collection	1 Jul 2015–16 Sep 2016	20 Jul 2015–6 Oct 2015	8 Dec 2015–23 Feb 2016	20 Jan 2016–5 Apr 2016
Age group primarily affected	Cows	Cows and calves	Cows	Cows and calves
Primary clinical signs				
Cows	Mastitis	Mastitis, arthritis	Arthritis	Mastitis, arthritis
Calves	Pneumonia, otitis media	Arthritis, otitis media, pneumonia	Few cases of arthritis and otitis media	Arthritis, otitis media, pneumonia
Diagnostic tests				
Number of calves	15	22	20	26
Number of samples	51	101	89	93
Positive samples[a]				
BioX[b]	16	26	14	13
MilA[c]	48	98	41	88
Positive cultures from necropsied calves (total necropsied calves)	0 (0)	1 (1)	0 (2)	2 (2)

[a]Number of seropositive samples out of all sera obtained during the study period
[b]ODC% values > 37 in BioX Bio K 302 ELISA (BioX Diagnostics, Belgium)
[c]AU > 135 in the MilA ELISA

Bacterial culture was negative for two seven-month-old calves from Herd 3 that were necropsied, even though arthritis and otitis media were observed in both animals. However, the presence of typical clinical signs (including arthritis/swelling of the limbs, and very high serum antibody titres against *M. bovis*) in multiple cows, and the failure to detect any other pathogens suggest that it is highly likely that the clinical signs were associated with *M. bovis*.

No animals from Herd 1 were necropsied, but positive PCR results from milk from cows with mastitis were obtained before and during the study period.

Sample collection

The project budget allowed for approximately 400 tests in total, corresponding to five tests on each of 80 calves. However, during an outbreak of *M. bovis* some calves may die or be euthanized because of clinical disease, so we initially enrolled more than 80 calves in the study. During the first visit, the investigating veterinarian and the farmer sampled the calves strategically to ensure that a sufficient number of calves suspected to have *M. bovis*-associated disease were included, as well as a sample of calves without evidence of clinical disease. The rationale behind this non-random sampling strategy was to maximize the likelihood of including sufficient numbers of animals with and without different clinical signs.

During the herd visits, each calf underwent a clinical examination with a focus on the respiratory and musculoskeletal system, using a standardised clinical protocol, which is available from the corresponding author. One of three veterinarians performed the clinical examinations, and at least one of the authors of the clinical examination protocol was present at each visit.

In addition, a venous blood sample was collected from each calf using a 10 ml plain Vacutainer tube (Kruuse, Denmark). Blood samples were stored in a cool environment and delivered to the Eurofins Steins laboratory (Vejen, Denmark) within 36 h of collection. The serum samples were frozen at – 18 °C.

Detection of antibodies

The serum samples were thawed at the National Veterinary Institute, in the Technical University of Denmark (DTU), Copenhagen, and analysed for antibodies against *M. bovis* using the commercial kit BioX Bio K 302 ELISA (BioX Diagnostics, Rochefort, Belgium) and the in-house indirect IgG ELISA test (MilA ELISA) developed at the University of Melbourne, Australia, by Wawegama et al. [10].

The BioX assay was performed according to the manufacturer's instructions. The test outcome was calculated as:

$$ODC\% = ((OD_{sample} - OD_{negative\ control}) / (OD_{positive\ control} - OD_{negative\ control})) \times 100.$$

Where OD is the optical density. An ODC% > 37 was considered positive, as recommended by the manufacturer. The manufacturer reported the sensitivity and specificity of the test at this cut-off to be 100% in a small sample of experimentally infected and negative control calves [11]. However, other authors have found that the assay has a low sensitivity in experimentally infected animals [12] and studies in cows under field conditions has supported this observation [9].

The MilA ELISA was performed as described by Wawegama et al. [10]. The mean antibody concentration in antibody units (AU) was calculated by plotting the OD values on a standard curve derived from a set of known positive-control sera included on each plate. In feedlot cattle with BRD the sensitivity and specificity of this assay have been estimated at 94.3% (95% confidence interval: 89.9–99.6%) and 94.4% (95% confidence interval: 90.3–99.6%), respectively, using 105 AU as cut-off [8]. However, the authors recommend using AU > 135 as cut-off for a test positive interpretation.

Statistical analysis

Individual calves were categorised into the following disease groups:

1) Likely *M. bovis*-associated disease ('*M. bovis*').
2) Respiratory disease only ('Respiratory').
3) No clinical signs of disease ('Healthy').

This classification was based on the recorded clinical signs and photographs of each calf taken during the herd visits, using the following specific inclusion criteria:

- *M. bovis:* calves with clinical signs indicating infection with *M. bovis.* All calves in this group had arthritis and/or otitis media at one or more herd visits, and some of these calves also had signs of respiratory disease.
- Respiratory: calves with only clinical signs of respiratory disease that did not fall into the category given above. All calves had one or more of the following clinical signs at one or more herd visits: dyspnoea, abnormal lung sounds on auscultation, discharge from the nares or eyes and coughing.
- Healthy: calves with no clinical signs of disease.

Modelling of antibody responses

Separate linear mixed models were used, with BioX and MilA ELISA results as the outcome variables. The ODC% and AU measurements were log transformed to improve the normality of the residuals, and a fixed constant of 1 was added to all results to enable log transformation of all values, including the small number of zero values.

For each of the two outcomes, model selection was used to find the most parsimonious model based on the potential explanatory variables and their two-way interactions. The fixed effects to be tested were: the disease group (categorical variable), the herd (categorical variable) and the age (in days) of the calf on the sampling date (pseudo-continuous variable). An additional quadratic effect of age was included in order to allow for a non-linear relationship between age and ELISA response. The final model was obtained using backward stepwise elimination based on Akaike's information criterion (AIC). A random effect of calf identification number was included in all models in order to account for repeated samples from the same calf. Confidence intervals for the predicted mean ODC% and AU values (for an "average" calf) were calculated for each herd using parametric bootstrapping.

Age intervals containing fewer than three observations from different calves in the same herd were removed. This was done because different age groups were sampled across the different herds, and to restrict the models to regions of parameter space with enough observations. As a result, herd-specific graphs of predicted ELISA responses span different age intervals for each herd. Dashed vertical lines in Figs. 1 and 2 illustrate the herd-specific age ranges included in the modelling.

The goodness of fit of the model was estimated using marginal and conditional pseudo R^2 for mixed models, which was estimated using the method described by Nakagawa and Schielzeth [13]. Using this method, the marginal R^2 describes the variation explained by the fixed effects alone, and the conditional R^2 describes the variation explained by both fixed and random effects. All data management and analyses were done in R version 3.2.2 (R Core Team, 2016), with linear mixed models implemented using the lme4 package [14].

Results

Descriptive statistics

In total, 83 calves were enrolled in the study and 334 blood samples were collected. Table 2 shows the number of calves stratified by herd and disease group.

Of the 83 calves, 52 were sampled on five occasions, 16 on four occasions, 1 on three occasions, 6 on two occasions and 8 on one occasion. The primary reason for calves dropping out of the study was euthanasia (N = 13), and eight calves were moved to another property during the study.

Due to a laboratory error, samples from the first visit to Herd 1 were not analysed.

The BioX ELISA response was below the recommended cut-off of 37 ODC% for the entire study period for 48 of the calves, above the cut-off for the entire study period for eight calves, and 27 changed status (Fig. 1). Very few calves had an ODC% above the cut-off before they were 40–60 days old (Fig. 1).

The MilA ELISA response was above the recommended cut-off of 135 AU throughout the entire study period for the majority of the calves (Fig. 2). Only one calf was below the cut-off at the end of the study period.

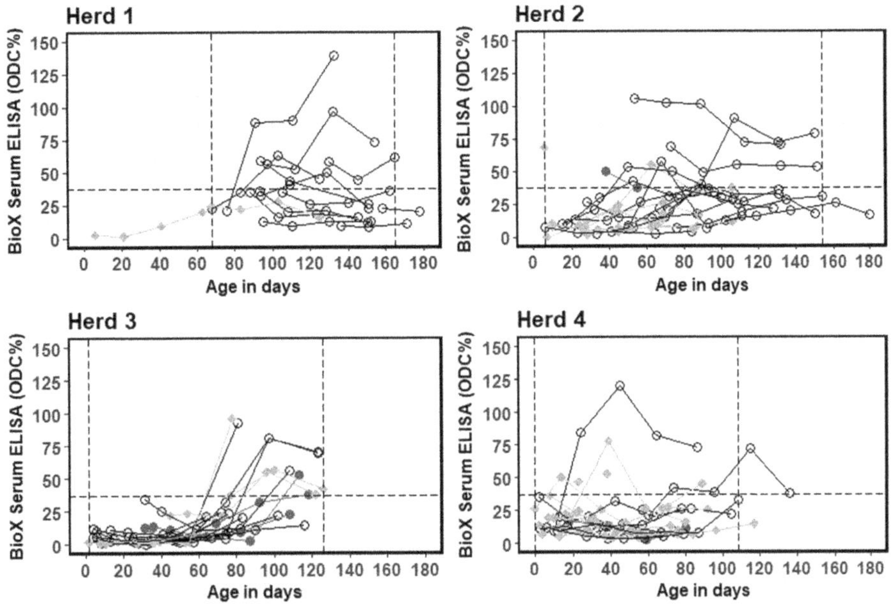

Fig. 1 Distribution of ELISA measurements in the BioX ELISA Bio K302 assay (ODC% = sample coefficient) of serum antibodies against *M. bovis* in four Danish dairy herds. Grey squares = "*M. bovis*"; black circles = "Respiratory"; grey dots = "Healthy". Horizontal dashed lines show the recommended ELISA cut-off (37 ODC%). Vertical dotted lines indicate the limits for including observations in the modelling of antibody response dynamics. Results from the same calf are linked by lines

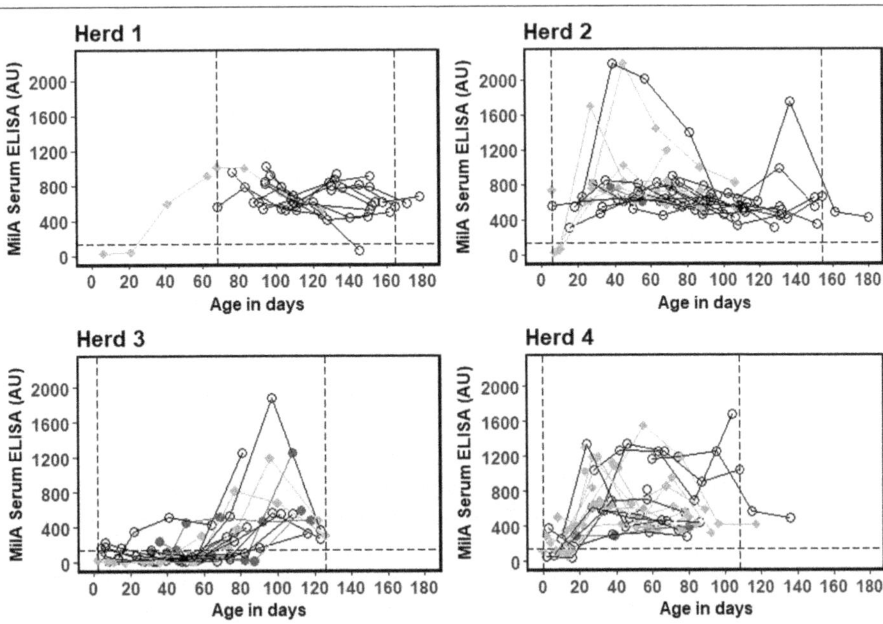

Fig. 2 Distribution of serum ELISA measurements in the MilA assay (AU = sample antibody units) of serum antibodies against *M. bovis* in four Danish dairy herds. Grey squares = "*M. bovis*"; black circles = "Respiratory"; grey dots = "Healthy". Horizontal dashed lines show the recommended ELISA cut-off (135 AU). Vertical dotted lines indicate the limits for including observations in the modelling of antibody response dynamics. Results from the same calf are linked by lines

Table 2 Distribution of calves by disease group and herd in a Danish longitudinal field study of four dairy herds with an outbreak of *M. bovis*-associated disease

Disease group	*M. bovis*	Respiratory	Healthy	Total
Herd 1				
Calves	2	13	0	15
Samples	8	43	0	51
Herd 2				
Calves	7	14	1	22
Samples	29	70	2	101
Herd 3				
Calves	6	11	3	20
Samples	23	51	15	89
Herd 4				
Calves	15	10	1	26
Samples	49	40	4	93
Total				
Calves	30	48	5	83
Samples	109	204	21	334

The MilA ELISA detected antibodies soon after birth (i.e. at approximately 20 days of age), but as was evident in Herd 3, the antibodies did not increase quickly in all herds. The responses varied, but remained above the cut-off once it had been reached.

Based on the raw data plots, there seems to be no association between disease group and antibody responses for either ELISA test (Figs. 1 and 2), so disease group was not included as an explanatory variable in the statistical models.

Modelling of antibody responses

The final model with log transformed BioX ODC% as the outcome included the linear and quadratic effects of age, the fixed effect of herd, and the two-way interactions between herd and both the linear & quadratic effects of age (Table 3). The variance associated with the random effect of animal was considerable, although less so than the residual variance. There was a positive linear effect of age in all herds, but the sign of the quadratic effect was dependent on herd. Based on Fig. 3, the mean estimate of ODC% in three of the four herds increased gradually with age and did not reach the recommended individual animal cut-off. In the remaining herd, the rate of increase in ODC% increased with the age of the calf.

Table 3 Final model describing explanatory variables and random effects of log transformed BioX ELISA optical density measurements (ODC%). The marginal R^2 was 39% and conditional R^2 was 61%

Variables	Variance	95% confidence interval		
Random effects				
Animal	0.28	0.13–0.38		
Residuals	0.47	0.37–0.54		
		Estimate	SE	*p*-value
Fixed effects				
Intercept		2.55	0.26	< 0.001
Age days (linear)		14.68	3.93	< 0.001
Age days (quadratic)		−8.16	2.27	< 0.001
Herd 1		0		–
Herd 2		0.30	0.30	0.311
Herd 3		−0.10	0.30	0.753
Herd 4		0.37	0.31	0.229
Age days (linear)*Herd 1		0		–
Age days (linear)*Herd 2		−5.93	4.27	0.165
Age days (linear)*Herd 3		7.11	4.59	0.123
Age days (linear)*Herd 4		−8.15	5.06	0.108
Age days (quadratic)*Herd 1		0		–
Age days (quadratic)*Herd 2		4.80	2.65	0.072
Age days (quadratic)*Herd 3		13.42	3.19	< 0.001
Age days (quadratic)*Herd 4		13.95	3.31	< 0.001

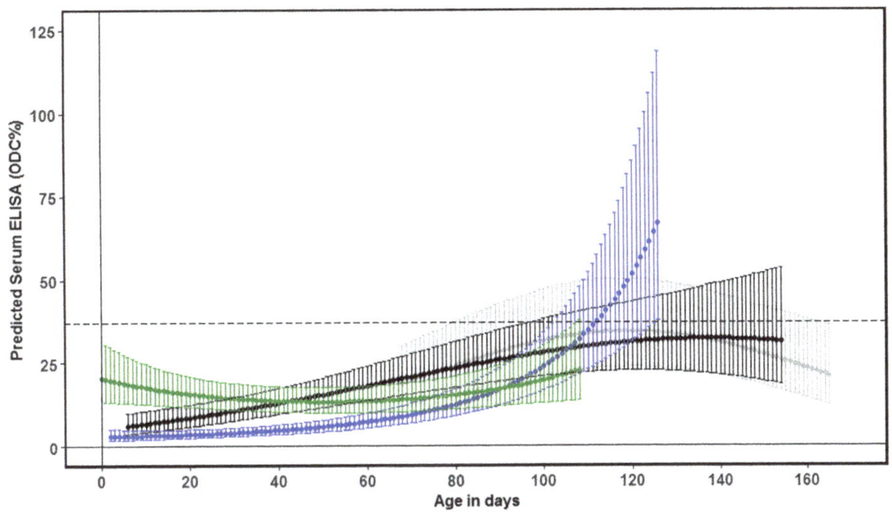

Fig. 3 Estimated mean antibody response in sera (solid line) and 95% confidence intervals (shaded area) as measured by the BioX ELISA Bio K302 assay for the herd-specific age ranges for which observations were available. Herd 1 is grey, Herd 2 is black, Herd 3 is blue and Herd 4 is green. The dashed line shows the recommended individual animal ELISA cut-off (37 ODC%)

The highest mean estimate of ODC% was not reached until the calf was approximately 110–130 days old (depending on the herd), with some suggestion of a plateau and eventual decline above this age in three of the herds (Fig. 3). However, a comparison of this relationship among herds is complicated by the difficulty in extrapolating the polynomial effect of age outside the observed parameter space, which was further compounded by the small differences in the ages of the calves among herds.

The final model with log transformed MilA AU as the outcome also included the linear and quadratic effects of age, the fixed effect of herd, and the two-way interactions between herd and both the linear & quadratic effects of age (Table 4). The variance associated with the random effect of animal was less than that estimated for the BioX ELISA. Again, there was a positive linear effect of age in all herds, but as for the BioX ELISA, the sign of the quadratic effect was dependent on herd. Based on Fig. 4, it appears that the MilA ELISA detected antibodies in younger calves, and for two herds, the mean estimate of MilA AU was above the recommended animal cut-off value for animals less than 20 days old. The overall shape of the relationship between age and estimated AU was similar in three of the four herds (Fig. 4). It can be characterised by an initial phase of increase followed by a plateau and an eventual decrease, although the peak was reached at the older age of 110–120 days in Herd 1, compared to approximately 60–80 days in Herds 2 and 4. A significantly different pattern, which was more similar to an exponential increase from an initially low MilA AU value, was estimated for Herd 3 (Fig. 4).

Discussion

This is the first observational study to illustrate and analyse the dynamics of serum antibody responses against *M. bovis* in naturally exposed and infected calves. In addition to a large variation in antibody responses among calves both within and between herds, the key findings were that the BioX ELISA rarely detected antibodies in calves under the age of 2 months, while the MilA ELISA was able to detect antibodies in the same calves soon after birth. Since the calves in these herds were all very likely to be truly exposed, we conclude that the MilA ELISA is a potentially useful test of *M. bovis* exposure. However, neither ELISA could differentiate between calves with arthritis and/or otitis, and respiratory disease, which indicates that the tests may be less useful for distinguishing animals with *M. bovis*-related diseases from those that have just been exposed to the pathogen.

Antibody dynamics measured by the BioX ELISA

In general, very few calves seroconverted to values above the recommended cut-off of 37 ODC%, and the majority remained below the cut-off for the BioX ELISA for the entire study period, despite the fact that these calves either showed signs of *M. bovis*-associated disease or were housed with diseased calves during acute outbreaks of disease caused by *M. bovis*. In Herd 2, several calves had severe arthritis, and *M. bovis* was cultured from one necropsied calf. Despite this, the BioX ELISA did not detect antibodies above the recommended cut-off in sera from these calves (Fig. 1; grey triangles). It is possible that because these calves were very young when they were infected, their immune system was not yet able to produce antibodies against the *M. bovis* antigen used in

Table 4 Final model describing explanatory variables and random effects of log transformed MilA ELISA antibody units (AU). The marginal R^2 was 59% and conditional R^2 was 65%

Variables	Variance	95% confidence interval		
Random effect				
Animal	0.09	0.007–0.150		
Residuals	0.57	0.463–0.655		
		Estimate	SE	*P*-value
Fixed effects				
Intercept		5.77	0.22	< 0.001
Age days (linear)		12.91	3.53	< 0.001
Age days (quadratic)		−8.25	2.18	< 0.001
Herd 1		0		–
Herd 2		0.61	0.25	0.015
Herd 3		−0.83	0.25	0.001
Herd 4		0.32	0.26	0.231
Age days (linear)*Herd 1		0		–
Age days (linear)*Herd 2		−11.90	3.89	0.003
Age days (linear)*Herd 3		10.79	4.35	0.014
Age days (linear)*Herd 4		−15.88	4.84	0.001
Age days (quadratic)*Herd 1		0		–
Age days (quadratic)*Herd 2		4.13	2.64	0.118
Age days (quadratic)*Herd 3		14.71	3.29	< 0.001
Age days (quadratic)*Herd 4		−2.19	3.40	0.520

the BioX ELISA. A similar phenomenon has been seen with *Salmonella* Dublin [4] and Virtala et al. [5] also found that young calves often fail to seroconvert to common respiratory pathogens including *M. bovis*, although this was assessed using a different assay. The BioX ELISA was also used in a vaccine challenge study by Dudek et al. [15]. In the positive control group, which consisted of five- to six-week-old heifers intratracheally challenged with *M. bovis*, the antibody response increased only slightly to a maximum of approximately 50

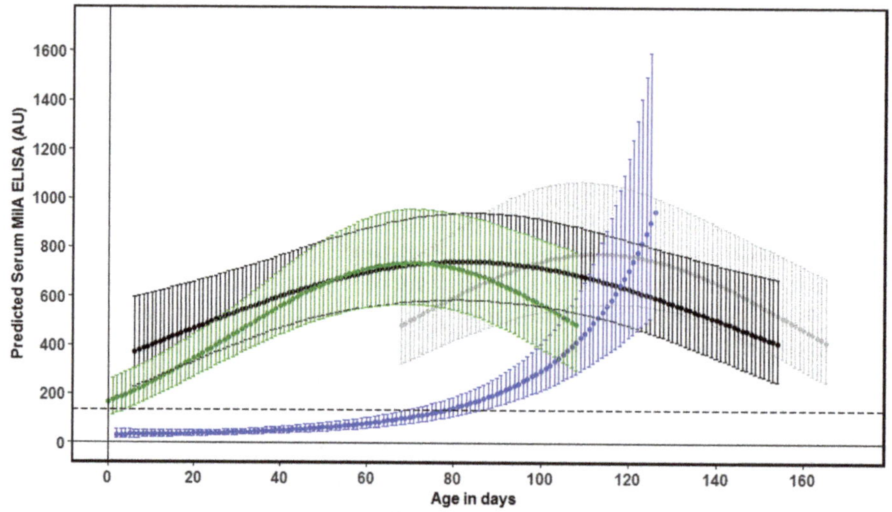

Fig. 4 Estimated mean antibody response in serum (solid line) and 95% confidence intervals (shaded area) as measured by the in-house MilA ELISA for the herd-specific age ranges for which observations were available. Herd 1 is grey, Herd 2 is black, Herd 3 is blue and Herd 4 is green. The dashed line shows the recommended individual animal ELISA cut-off (135 AU)

ODC% at 4 weeks after challenge, and thereafter declined slightly. By contrast, in the vaccinated group, which was inoculated subcutaneously with inactivated *M. bovis* mixed with two adjuvants, serum antibody concentrations rapidly increased within 2 weeks of vaccination and reached a maximum of around 200 ODC% in 4 weeks. This difference is likely to reflect the greater stimulation of the systemic immune response following inoculation with an adjuvanted whole cell vaccine compared to natural infection.

Our study is the first to evaluate the dynamics of antibody responses as measured by the BioX ELISA in calves under field conditions and we have shown that it is not a suitable test for reliable diagnosis of *M. bovis* in calves under 3 months of age. However, this does not exclude the possibility that this test may be useful for group diagnostics in younger calves with an adjusted cut-off. Additional studies are needed to explore this possibility in herds with differing disease status.

Antibody dynamics measured by MilA ELISA

The MilA ELISA detected antibodies above the cut-off in calves as young as approximately 20 days of age, with antibody concentrations rising markedly over a short timeframe, and at the end of the study period, all but one of the calves were above the cut-off. However, the MilA ELISA did not detect antibodies early in all four herds. In Herd 3, the antibody levels did not start to rise until the calves were 60–80 days of age, and then increased rapidly. No *M. bovis*-associated disease was found among the calves in this herd during the first two herd visits, indicating that they had managed to prevent transmission between cows and calves and had kept the infection pressure low around the calves. In addition, the first visit to Herd 3 was only 1 week after the appearance of clinical signs in the herd, and the farmer did not feed any waste milk to the calves. The later response detected by the MilA ELISA in this herd could be a result of initial infection among the cows and later transmission to the calves (i.e. around 40–60 days after the outbreak had started). The other herds were visited three to 4 weeks after the outbreak had started, and transmission to the calves had already occurred by this time. This could explain the high serum antibody concentrations in young calves in these herds. It can therefore be concluded that the MilA assay does not detect young calves as positive in non-infected groups of calves, which supports a reasonable specificity and therefore the usefulness of this test for confirmation and surveillance purposes. To substantiate this finding, control herds with no known *M. bovis*-associated diseases should be assessed for comparison. The high antibody concentrations detected in most calves in this study, including the healthy calves, suggests that the MilA test is very sensitive and probably detects exposure to *M. bovis* rather than *M. bovis*-associated disease. Both experimental and field studies of the MilA ELISA also suggest that this test has a high level of sensitivity [8].

The use of antibodies to detect infection with *M. bovis*

Our study has shown that the *M. bovis* BioX and MilA ELISAs give very different results for calves with the same exposure and disease status in calves younger than 2 months old indicating different immune reactions to the underlying infection (Herd 2 and 4 in Figs. 1 and 2). In herds with clinical signs in older calves, the modelled antibody responses were similar in shape, indicating similar immune reactions to the underlying infection (Herd 1 and 3 in Figs. 1 and 2). This is likely associated with the age-related development of immune competence in the calves, which is important for the interpretation of the test results for diagnosis.

In addition to the disease burdens in the calves studied here, severe clinical signs associated with infection with *M. bovis* where found among the cows in all four herds, which is highly likely to have resulted in transmission to the calves in some of the herds. Fewer calves had severe signs of *M. bovis*-associated disease in Herds 1 and 3 than in Herds 2 and 4, i.e. herds with little or no segregation of cows and calves and feeding of unpasteurised whole milk to the calves. As *M. bovis* spreads by direct contact between animals and through ingestion of milk contaminated with *M. bovis* [1], this is likely to have contributed to the severity of the disease in these two herds. In Herds 1 and 3 the calves were fed milk replacer and housed in a building that physically separated them from the cows.

Clinical disease did not seem to correlate with the antibody response detected by either ELISA assay. The effect of "Herd" was retained during model selection, indicating that the different ELISA responses were mostly influenced by differences among the herds, and not differences in the underlying disease status of individual animals. Although Martin et al. [7] performed their study using an indirect haemagglutination assay, they also concluded that serum antibodies against *M. bovis* were not indicative of disease at an individual level, only at group level.

Only five calves were classified as being healthy, which is a small number to include in the models as a separate group, yet disease group was not found to be significant. To assess the robustness of the results, models were generated both without the 'Healthy' group, and with the healthy calves included in the 'Respiratory' group, and none of these variations altered the conclusions.

The 'Respiratory' group consisted of many calves, and it is not possible to know whether the disease in this group was caused by *M. bovis* alone or in conjunction with other respiratory pathogens. At the time of sampling, the calves were housed in herds with an active or recent spread of *M. bovis*, so it is likely that the disease seen was at least partly attributable to infection with *M. bovis*. In Herds 2 and 4, *M. bovis* was isolated from necropsied calves, and all herds were free of the likely differential diagnosis *Salmonella* Dublin throughout the study period, making it likely that the arthritis was caused by *M. bovis.*

Conclusions

This is the first study to evaluate the dynamics of antibody responses using the BioX ELISA in calves under field conditions. Based on our data, we cannot recommend the use of this test in calves under 3 months of age. The MilA ELISA was able to detect antibodies shortly after birth (i.e. from approximately 3 weeks of age and onwards) and is likely to be a good assay for detecting exposure to *M. bovis.* Neither ELISA could differentiate between calves with arthritis and/or otitis media, and respiratory disease.

Abbreviations

AU: antibody units; BioX: BioX Bio K 302 ELISA test; ELISA: enzyme-linked immunosorbent assay; IgG: immunoglobulin G; *M. bovis* : *Mycoplasma bovis*; MilA: in-house indirect IgG ELISA test; OD: optic density; ODC%: corrected optic density measurement; PCR: polymerase chain reaction

Acknowledgements

The authors would like to thank the veterinarians Ulla Torpe and Martin Lund (Dyrlægerne Egtved), Lars Pedersen (SEGES) and the veterinarians/ veterinary students Annie Nielsen, Eva Elisabeth Toft-Petersen, Franziska Helene Pedersen and Signe Rejnhardt Olsen for their assistance with the herd visits and veterinarians Dinah Lerdahl Holm and Jeanette Petersen and the National Veterinary Institute at Technical University of Denmark for assistance and facilities for the laboratory testing of samples.

Funding

The study was funded by the Danish Milk Levy Fund, the Danish Cattle Levy Fund, the Danish Pesticide Fund and the University of Copenhagen. NKW's visit to Denmark to perform MilA ELISA was jointly funded by a Dyson Fellowship and a University of Melbourne Engagement and Partnerships grant from the Faculty of Veterinary and Agricultural Sciences, University of Melbourne.

Authors' contributions

MBP: Planned the study design and the clinical protocol, planned and performed the field sampling and clinical recordings, data management and statistical analysis, drafted the manuscript. NKW: Planned and developed the MilA ELISA, planned and performed the MilA ELISA testing in Denmark, lead all laboratory analyses and assisted with interpretation of laboratory results, commented on and approved the manuscript. MD: Supervised and assisted with the statistical analyses and interpretation of results, commented on and approved the manuscript. PFM: Planned and developed the MilA ELISA, commented on and approved the manuscript. GFB: Planned the development and assessment of the MilA ELISA, guided the interpretation of results, commented on and approved the manuscript. LRN: Planned the study design and the field study, supervised data management and statistical analysis, commented on and approved the manuscript. All authors approved the final version to be published and agreed to be accountable for all aspects of the work.

Consent for publication

Not applicable.

Competing interests

The authors declare that they have no competing interests.

Author details

[1]Department of Veterinary and Animal Sciences, Faculty of Health and Medical Sciences, University of Copenhagen, Grønnegårdsvej 8, 1870 Frederiksberg, Denmark. [2]Asia-Pacific Centre for Animal Health, Melbourne Veterinary School, Faculty of Veterinary and Agricultural Sciences, The University of Melbourne, Parkville, VIC 3010, Australia.

References

1. Maunsell FP, Donovan GA. *Mycoplasma bovis* infections in young calves. Vet Clin North Am Food Anim Pract. 2009;25:139–77.
2. Sachse K, Salam HS, Diller R, Schubert E, Hoffmann B, Hotzel H. Use of a novel real-time PCR technique to monitor and quantitate mycoplasma bovis infection in cattle herds with mastitis and respiratory disease. Vet J. 2010;186:299–303.
3. Zhang R, Han X, Chen Y, Mustafa R, Qi J, Chen X, et al. Attenuated mycoplasma bovis strains provide protection against virulent infection in calves. Vaccine. 2014;32:3107–14.
4. Ld R, Smith BP, Spier SJ, Dilling GW. Effect of calf age and Salmonella bacterin type on ability to produce immunoglobulins directed against Salmonella whole cells or lipopolysaccharide. Am J Vet Res. 1992;53:1895–9.
5. Virtala AMK, Grohn YT, Mechor GD, Erb HN, Dubovi EJ. Association of seroconversion with isolation of agents in transtracheal wash fluids collected from pneumonic calves less than three months of age. Bovine Practitioner. 2000;34:77–80.
6. Van Donkersgoed J, Ribble CS, Boyer LG, Townsend HG. Epidemiological study of enzootic pneumonia in dairy calves in Saskatchewan. Can J Vet Res. 1993;57:247–54.
7. Martin SW, Bateman KG, Shewen PE, Rosendal S, Bohac JG, Thorburn M. A group level analysis of the associations between antibodies to seven putative pathogens and respiratory disease and weight gain in Ontario feedlot calves. Can J Vet Res. 1990;54:337–42.
8. Wawegama NK, Markham PF, Kanci A, Schibrowski M, Oswin S, Barnes TS, et al. Evaluation of an IgG enzyme-linked immunosorbent assay as a serological assay for detection of *Mycoplasma bovis* infection in feedlot cattle. J Clin Microbiol. 2016;54:1269–75.
9. Petersen MB, Pedersen J, Holm DL, Denwood M, Nielsen LR. A longitudinal observational study of the dynamics of *Mycoplasma bovis* antibodies in naturally exposed dairy cows. J Dairy Sci. 2018 https://doi.org/10.3168/jds.2017-14340.
10. Wawegama NK, Browning GF, Kanci A, Marenda MS, Markham PF. Development of a recombinant protein-based enzyme-linked immunosorbent assay for diagnosis of mycoplasma bovis infection in cattle. Clin Vaccine Immunol. 2014;21:196–202.
11. Anonymous. https://www.biox.com/en/bio-k-302-monoscreen-abelisa-mycoplasma-bovis-indirect-monowell-p-250/. 2017. Assessed 17-11-2017.
12. Schibrowski ML, Barnes TS, Wawwgama NK, Vance ME, Markham PF, Mansell PD, et al. The performance of three immune assays to assess the serological status of cattle experimentally exposed to *Mycoplasma bovis*. Vet Sci. 2018; 5(27). https://doi.org/10.3390/vetsci5010027.
13. Nakagawa S, Schielzeth H. A general and simple method for obtaining R2 from generalized linear mixed-effects models. Methods Ecol Evol. 2013;4:133–42.
14. Bates D, Maechler M, Bolker B, Walker S. Fitting linear mixed-effects models using lme4. J Stat Softw. 2015;67:1–48.
15. Dudek K, Bednarek D, Ayling RD, Kycko A, Szacawa E, Karpinska TA. An experimental vaccine composed of two adjuvants gives protection against mycoplasma bovis in calves. Vaccine. 2016;34:3051–8.

13

Variation in fat content between liver lobes and comparison with histopathological scores in dairy cows with fatty liver

C. Gerspach[1*], S. Imhasly[2], R. Klingler[1], M. Hilbe[3], S. Hartnack[4] and M. Ruetten[3]

Abstract

Background: The assessment of a liver biopsy remains the gold standard for diagnosing and staging fatty liver in dairy cows, which is often necessary for diagnostic and research purposes. Accuracy of the diagnosis relays on the quality of the biopsy, the assumed representativeness of a small tissue sample for a disease process throughout the liver and accurate human evaluation of histologic specimens. The objective of the present study was to assess the distribution of triacylglycerol (TAG) infiltration throughout the parenchyma of livers with different degrees of fatty liver in dairy cows. In addition, histopathological scores from the corresponding specimens were compared to a quantitative measurement of TAG, as well as the agreement between two observers.

Methods: Thirty livers with different degrees of lipid infiltration were selected and 10 different locations throughout the liver were assessed. The TAG content was measured enzymatically, calculated in % or mg/g wet weight, and assigned to a scoring system. Corresponding tissue specimens were stained with hematoxylin-eosin (H&E) and Oil red O (ORO) for histopathological evaluation, using a scoring system.

Results: The difference in TAG content between any locations was less than 2%. Based on the scoring system the TAG concentration was even distributed in 79.3% of the livers. Based on kappa statistics the agreement between two pathologists and staining technique in scoring histological specimens was moderate to fair.

Conclusions: Overall the distribution of TAG throughout the liver and the accuracy of human evaluation of liver biopsies may lead to acceptable diagnoses for clinical purposes. Within the liver lobules a common pattern of lipid distribution depending on severity could be observed. For the staging of lipid infiltration for research projects, some degree of variation needs to be considered.

Background

Fatty liver is a metabolic disease occurring during the transition period in dairy cattle. Diagnosis and staging of fatty liver is often necessary for diagnostic and research purposes. Lipid infiltration can reach from a physiological state, in response to negative energy balance in early lactation, to a progressed state, with altered liver function and clinical disease. The clinical symptoms are unspecific [1]. Therefore, the main reasons to evaluate liver biopsies are establishing a morphological diagnosis and to assess the severity of disease. The lipid content of liver tissue can be evaluated by biochemically, histologically or by specific gravity [1]. For quantitative determination of triacylglycerol (TAG) or total lipid content, the TAG content of liver tissue can be quantitatively measured enzymatically using an automated analyzer [2]. The histologic evaluation of liver tissue is a technique, routinely used for clinical but also for research purposes.

Two issues influencing an accurate diagnosis have been reported in different species. First the tissue sample should be representative for the disease which depends on the method of collection, size and number of percutaneous biopsies [3–7]. In addition, the assessment of a biopsy is only successful in diffuse hepatopathies. A percutaneous liver biopsy is always taken from the right lobe and represents 0.002 to 0.1% of the whole liver parenchyma, depending on the size of the biopsy needle [8].

* Correspondence: cgerspach@vetclinics.uzh.ch
[1]Department of Farm Animals, Vetsuisse Faculty, University of Zurich, Winterthurerstrasse 260, Zurich, Switzerland
Full list of author information is available at the end of the article

Fatty liver in cattle is reported to be diffuse [8]. However, in beef bulls concentrations of cholesterol and phospholipids were higher in the intermediate lobe, compared to the right and left lobes [9]. Local foci of fatty degeneration in livers of dairy cows have been reported, which are due to tension and resulting in ischemia at the insertion of a serosal attachment (tension lipidosis) [10]. Focal fatty liver with unknown etiology has been diagnosed in a heifer [11]. In dogs and cats liver lesions of different etiology were partially uneven distributed in disorders usually regarded as diffuse [5].

Second, histological evaluation of liver tissue is prone to bias [12–15]. In a study investigating human evaluation of liver steatosis in people revealed a high level of variability between human observation and standardized computer measurement. At the same time a disagreement of up to 37% between observers was determined [13].

Despite several attempts to find and develop alternative and less invasive diagnostic methods in the past, the evaluation of a liver biopsy remains the current gold standard in diagnosing and investigating fatty liver in dairy cows.

The goal of our study was to investigate the distribution of TAG infiltration throughout livers with different stages of fatty liver, by comparing biopsies from 10 different localizations. In addition, we compared histopathological scores to a quantitative measurement of triacylglycerol in liver tissue and the agreement between two observers, using two different 2 embedding methods.

Methods

Collection of liver tissue

Wedge biopsy specimens of livers were taken at an abattoir. The selection was focused on livers with potential lipid infiltration. Liver samples were excised immediately after slaughter from 10 different localizations of each liver, using a scalpel (Fig. 1). The tissue samples from the corresponding localizations were placed into appropriate tubes for histopathologic examination and measurement of TAG.

Histopathology

For histopathology, tissue samples were collected with a size of 1 cm × 1 cm × 1 cm approximately 1 cm under the surface and fixed in 10% buffered formalin, and small samples were snap frozen in liquid nitrogen. The samples fixed in formalin were dehydrated by an ascending alcohol series ending in xylol and finally embedded in paraffin. The specimens were sectioned at a thickness of 2–3 μm and sections were stained with hematoxylin-eosin (H&E), periodic acid-Schiff stain (PAS) or Oil red O (ORO). The snap frozen samples were cut directly in 3–4 μm sections and stained with PAS and ORO without any prior fixation (alcohol) applied. The histological lesions were staged, according to a scoring system, into 4 categories: 0 = no abnormalities; 1 = moderate fatty liver (only cells from one zone are affected; periportal, midzonar or zentrolobular); 2 = severe fatty liver (the periportal and midzonal or midzonal to centrolobular

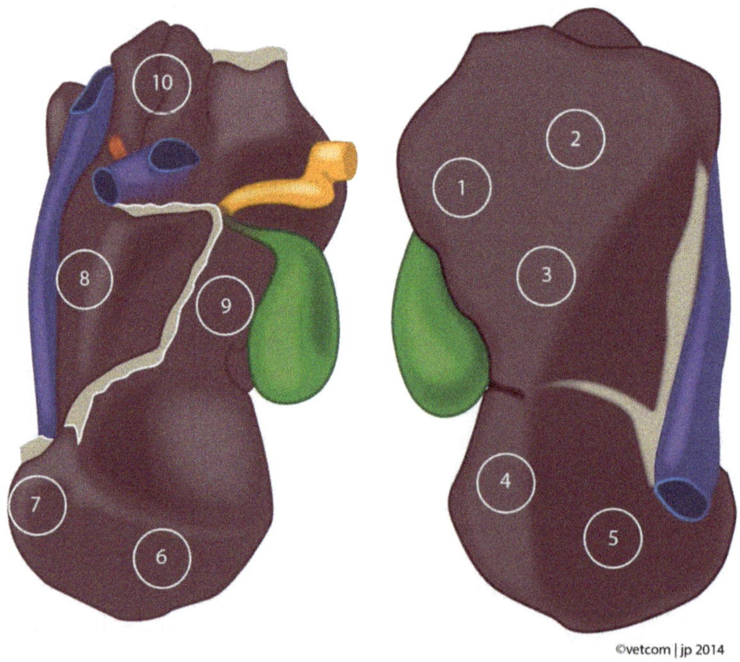

Fig. 1 Liver with 10 locations for tissue sampling (1–3 = right lobe; 4–5 = left lobe, diaphragmatic surface; 6–8 = left lobe, visceral surface; 9 = quadrate lobe; 10 = caudate lobe)

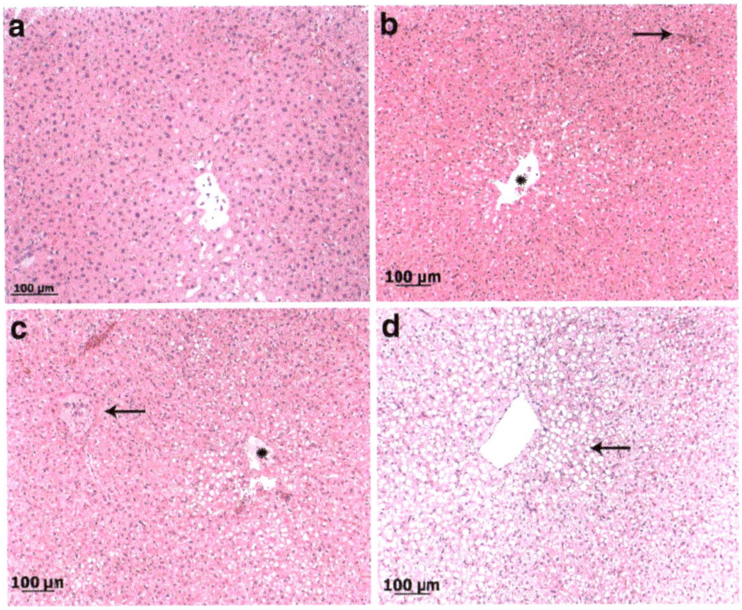

Fig. 2 Bright field images of different histological grades of fatty liver degeneration. **a** Grade 0: one lobule with a central vein of a normal liver, no lipid vacuoles in the cytoplasm. **b** Grade 1: small rim of hepatocytes with clear demarcated fatty vacuoles round a central vein (star). The portal field marking the edge of the liver lobule is marked with an arrow. **c** Grade 2: half of the hepatocytes show fatty vacuoles, central vein (*star*) and portal field (*arrows*) are marked. **d** Grade 3: all hepatocytes of the lobule show lipid vacuoles. The vacuoles in hepatocytes around the central vein are bigger (*arrow*)

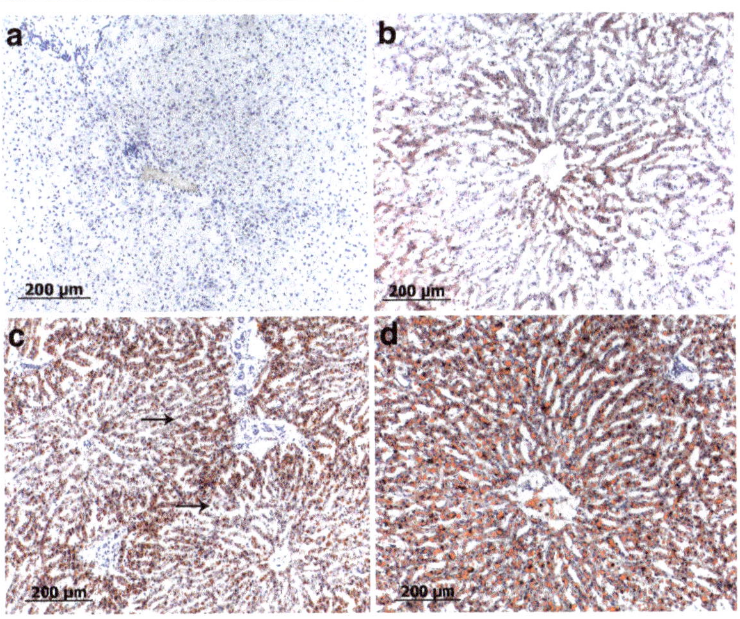

Fig. 3 Bright field images ORO staining of different histological grades of fatty liver degeneration. **a** Grad 0: normal liver, no fat deposition visible. **b** Grade 1: Positive red signal intracytoplasmatically in hepatocytes around central vein. **c** Grade 2: half of hepatocytes of a liver lobule show clear red stained fatty vacuoles. The edge to an area with only finely red stained lipid droplets is marked with arrows. **d** Grade 3: entire liver lobule shows red stained, clear demarcated fatty vacuoles in hepatocytes

areas are affected); 3 = very severe fatty liver (all three zones are affected, including the Kupffer cells) for both, H&E (Fig. 2) and ORO (Fig. 3). Respectively, the same grading scale was used for intracellular glycogen depositions. All specimens were evaluated by two board-certified pathologists (MR, MH) with a high level of experience, blinded to liver localizations and animal identification.

The intracellular vacuoles were classified according to morphology either as fatty vacuoles or glycogen deposition. Clearly demarcated vacuoles with or without displacement of the nucleus, was classified as fatty vacuole. Foamy cytoplasm and a centrally placed nucleus within slightly swollen hepatocytes was considered as glycogen deposition. These findings were verified by using either PAS (glycogen) or ORO (fat) staining.

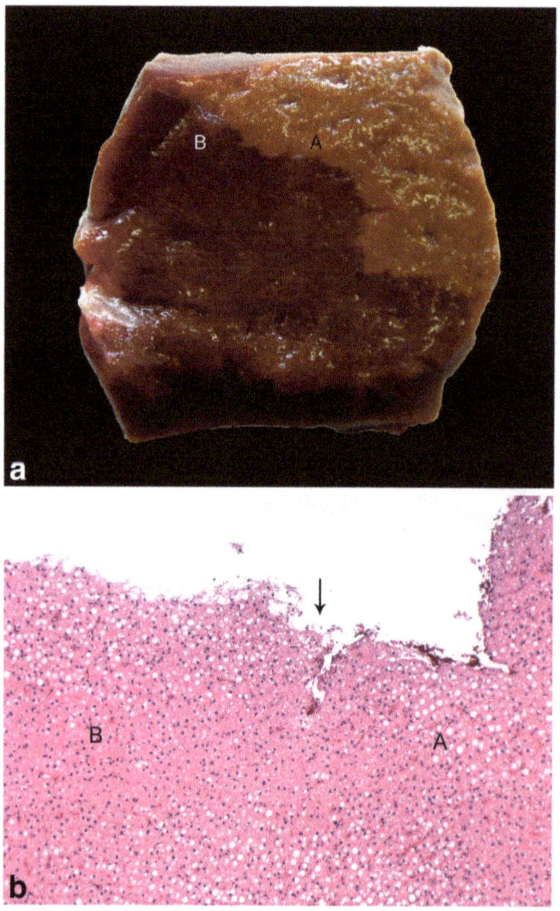

Fig. 4 (a) Liver tissue with heterogeneous distribution of lipid infiltration (A: normal liver, B: area with suspected fatty degeneration), (b) H&E stained tissue from corresponding specimen (A: normal liver, B: area with suspected fatty degeneration), arrow: border between macroscopically different appearance; no difference of fat deposition could be found within the hepatocytes

Liver tissue triglyceride measurement

Small samples of liver tissue were placed in 2 ml Eppendorf Safe-Lock tubes (Eppendorf, Germany) immediately snap frozen in liquid nitrogen, and stored at −85 °C for later determination of TAG content. After thawing, analysis was performed by tissue saponification in ethanolic KOH as described [16]. In short, approximately 100 mg of tissue samples were transferred to new pre-weighed 2 ml tubes, after removing non-parenchymatos tissue, and the exact weight recorded. By adding 350 µl of ethanolic KOH (2 parts EtOH: 1 part 30% KOH) incubation was performed overnight, at 55 °C in a thermos shaker until tissue was

Table 1 Overall scores for 30 livers. HE, H&E stain; ORO, Oil Red Stain; MR and MH, pathologists

	H&E		ORO	
liver	MR	MH	MR	MH
1	0	0	0	1
2	0	0	0	1
3	0	0	0	1
4	3	3	3	3
5	0	0	0	0
6	0	0	0	0
7	0	0	0	0
8	0	0	0	0
9	0	0	0	1
10	0	0	0	0
11	0	0	0	2
12	0	0	0	1
13	1	2	2	2
14	0	0	0	1
15	1	2	1	3
16	2	3	3	3
17	0	1	1	3
18	0	1	1	1
19	3	3	3	3
20	1	2	1	3
21	1	1	2	2
22	0	0	0	1
23	1	1	1	2
24	0	0	0	1
25	1	2	2	2
26	3	3	3	3
27	3	3	3	3
28	0	1	0	1
29	2	2	3	3
30	2	2	3	3

κ kappa reliability test, *CI* confidence interval

completely digested. 650 µl water:EtOH (1:1) were added and centrifuged at a speed of 16′000 G for 5 min. Supernatant was transferred in a new tube and 200 µl water:EtOH (1:1) were added and vortexed. 200 µl of the mixture were transferred in a new tube and 215 µl 1 M $MgCl_2$ were added and vortexed and left on ice for 10 min. Finally, the tubes were centrifuged at 16′000 G for 5 min and the supernatant was transferred in a new tube and stored at −20 °C until measurement.

The TAG content was measured enzymatically with the free glycerol reagent (Sigma-Aldrich). As triolein equivalent (CTE) standard curve were produced by glycerol standard (Sigma-Aldrich). In a 96 well plate, 6 µl of the sample were added to 200 µl of the free glycerol reagent, incubated for 15 min at room temperature and read at 540 nm. The blank were subtracted from the samples/standards and the concentration was calculated comparing to the standard curve. Each sample was measured in triplicates.

The results were also calculated in mg/g wet liver and assigned to a scoring system as used by Haudum et al. (2011): 0 = ≤ 50 mg/g (mild); 1 = ≥ 51–100 mg/g (moderate); 2 = ≥ 100–150 mg/g (severe); 3 = ≥ 150 mg/g (very severe).

Data analysis

Descriptive statistics was performed using GraphPad Prism (Prism 6.05, GraphPad Software, Inc., La Jolla, CA, USA). The agreement between two observers, assessing the same samples, was calculated using the κ test (unweighted kappa, www.vassarstats.net). Values of 0–0.2 represented a slight agreement, 0.21–0.4 a fair, 0.41–0.60 a moderate, 0.61–0.8 substantial and >0.8 almost perfect agreement [17].

Linear mixed effects models were utilized to assess if TAG concentration differed significantly between locations.

Generalized linear mixed effect models were used to assess if disagreement between the two observers could be explained by the TAG concentration. Animal was considered as random effect, TAG concentration and localisation as fixed effects. Model selection was based on AIC (Akaike's information criterion) with lower values indicating a better model fit. The analysis was performed with the software R (R Core Team 2015) and the package lme4 [18].

The intra-assay coefficient of variation (CV) expressed as a percent value was calculated as CV = within-assay standard deviation/mean × 100. The intra-assay CV was 4.8%.

Results

Animals

A total of 300 liver samples from 30 animals were examined for histologic findings and TAG content. The cows were between 2.5 and 9 years old and 4 to 500 days post-partum. They belonged to the breeds Red Holstein (*n* = 11), Brown Swiss (*n* = 6), Holstein Frisian (*n* = 6), and Montbeliard (*n* = 1). In 3 animals breed, age, and number of days in milk could not be determined.

Pathological assessment of liver tissue

Macroscopically 29 of 30 livers had a homogenous appearance with a slightly brighter brownish coloration than normal liver tissue. One liver showed few, randomly distributed clearly demarcated, light brown to yellowish foci that were not associated to any ligament insertion site (Fig. 4a). The histopathological examination did not reveal obvious differences in the degree of lipid infiltration in that liver (Fig. 4b).

Table 2 Agreement between observers for each localization and overall (whole livers)

Liver	H&E		ORO	
	κ	95% CI	κ	95% CI
1	0.70	0.49–0.90	0.29	0.08–0.50
2	0.60	0.38–0.81	0.29	0.08–0.50
3	0.54	0.31–0.76	0.37	0.15–0.59
4	0.56	0.33–0.78	0.30	0.09–0.51
5	0.62	0.40–0.85	0.29	0.07–0.50
6	0.64	0.41–0.86	0.29	0.08–0.51
7	0.45	0.22–0.68	0.21	0.01–0.41
8	0.54	0.32–0.76	0.22	0.01–0.42
9	0.49	0.27–0.71	0.23	0.03–0.44
10	0.74	0.54–0.93	0.29	0.08–0.49
overall	0.58	0.51–0.65	0.28	0.21–0.35

Table 3 Agreement between H&E and ORO for each localization and overall (whole liver)

Liver	MR		MH	
	κ	95% CI	κ	95% CI
1	0.64	0.40–0.87	0.38	0.14–0.62
2	0.59	0.34–0.83	0.43	0.20–0.66
3	0.53	0.28–0.79	0.48	0.25–0.71
4	0.47	0.20–0.74	0.20	0.00–0.42
5	0.57	0.31–0.82	0.34	0.12–0.58
6	0.57	0.32–0.83	0.35	0.11–0.58
7	0.57	0.32–0.83	0.34	0.11–0.58
8	0.58	0.34–0.83	0.36	0.13–0.59
9	0.54	0.29–0.79	0.52	0.29–0.74
10	0.53	0.28–0.79	0.29	0.06–0.51
overall	0.56	0.48–0.64	0.34	0.29–0.44

κ kappa reliability test, *CI* confidence interval

Based on the average score of 10 different localizations from each liver, the histological assessment revealed an overall diagnosis of fatty liver in up to 25/30 livers, depending on the pathologist and staining technique (Table 1). No concomitant liver lesions ore evidence of inflammation were detected. In specimens stained with HE, the scores were evenly distributed throughout all 10 localizations in agreement of both observers, in 22/30 livers (73.3%). In 4/30 (13.3%) livers the observers disagreed in their grading.

The scores were even distributed in 17/30 (56.7%) livers stained with ORO throughout all 10 localizations and were in agreement of both observers. In 1/30 (3.3%) livers, findings were not distributed diffuse throughout all localizations, observed by both pathologists. In 11/30 (36.7%) livers, there was a disagreement between observers. Overall the agreement between the two observers was fair to moderate in evaluating samples stained with H&E and fair in samples stained with ORO (Table 2). The agreement between H&E and ORO was fair to moderate (Table 3). Both observers gave slightly higher scores when evaluating ORO stained samples.

The evaluation of H&E stained specimens revealed 76/300 (25.3%) samples being diagnosed with different scores. Of those only 2/300 (0.7%) differed by 2 scores. In liver tissue stained with ORO, there was a disagreement in 163/300 (54%) specimens. In 35/300 (11.7%) specimens there was a difference of 2 scores.

The distribution of intracellular glycogen was negatively correlated with the distribution of fatty depositions (data not shown). There were no significant differences found between formalin fixed and snap frozen samples.

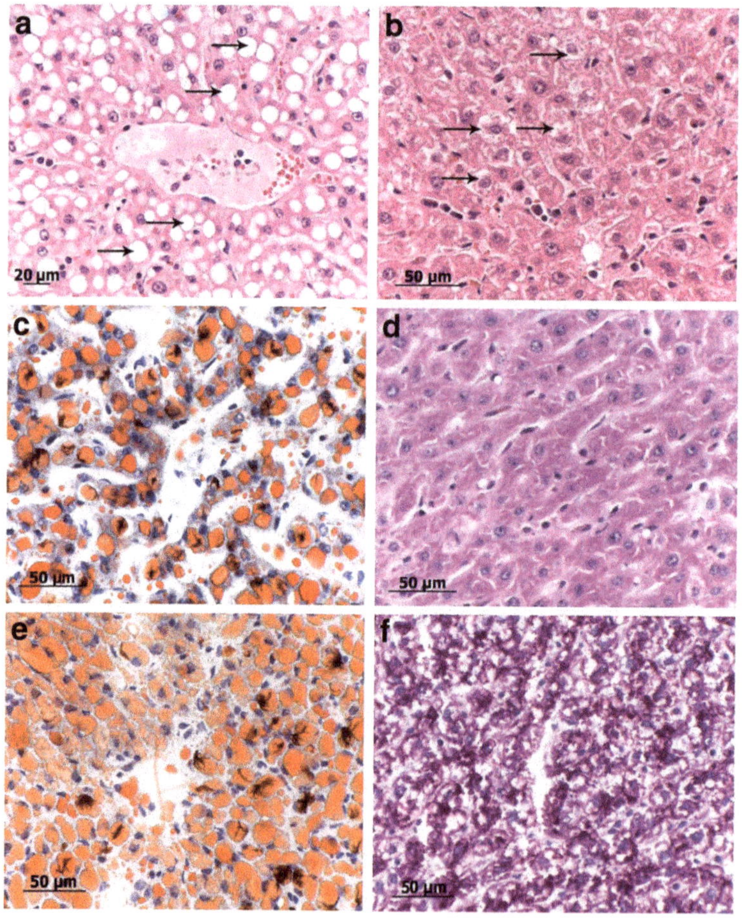

Fig. 5 Bright field images, comparison different morphology of fatty vacuoles and glycogen depositions of formalin fixed and snap frozen liver samples, H&E, PAS and ORO staining. **a** large, clear demarcated fatty vacuoles intracytoplasmatically in hepatocytes displacing the nucleus excentrically. **b** Intracytoplasmatical glycogen deposition (arrows) in swollen hepatocytes showing a foamy brightened cytoplasm. **c** ORO staining of formalin fixed tissue sample, showing clear red stained fatty vacuoles diffusely distributed in the liver lobule. **d** PAS staining of glycogen deposition of formalin fixed tissue samples. **e** ORO staining of snap frozen liver sample, diffuse red coloration of hepatocytes, unclear cell borders, sinusoids indistinct and difficult to localize. Distribution of fatty vacuoles similar to formalin fixed tissue. **f** PAS staining of snap frozen liver samples, diffuse pinkish coloration of hepatocytes, unclear cell borders, sinusoids indistinct and difficult to localize. Distribution of glycogen depositions similar to formalin fixed tissue

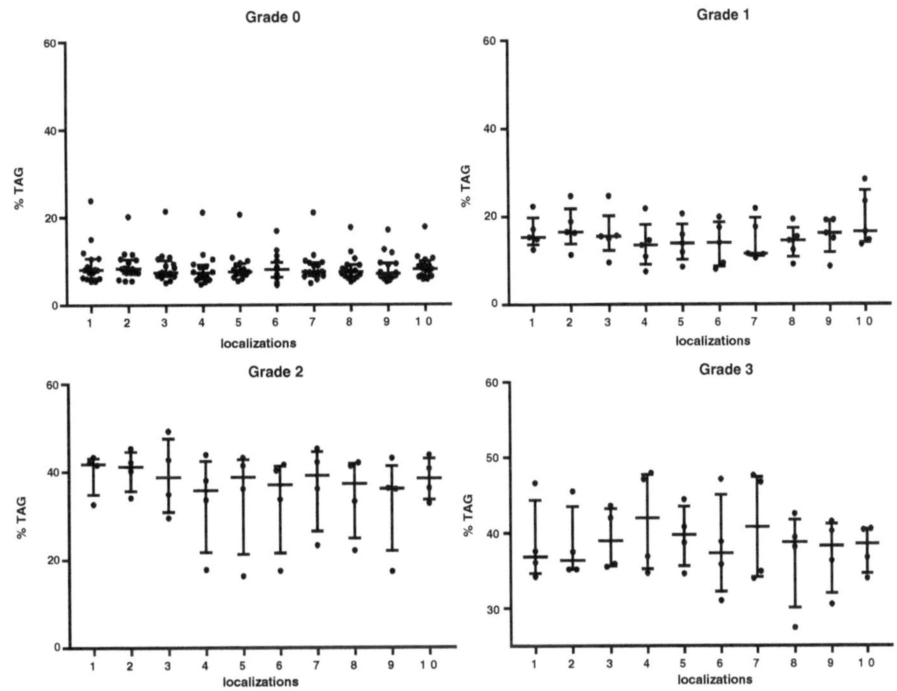

Fig. 6 TAG content (% wet weight) of 10 localizations from 29 livers with degree 0 to 3

However, the morphology of the cell borders, vacuoles and nucleus of hepatocytes in the slightly thicker, not fixed sections were more difficult to assess. The intensity and distribution of coloration of the special staining was comparable in either method (Fig. 5).

Biochemical analysis of hepatic TAG content

The percentage of TAG from wet liver of 10 locations from 29 livers was measured (Fig. 6). Tissue from liver no. 25 was not available for TAG measurement. Based on a linear mixed effect model approach, location was found to be significantly associated with TAG concentration (p = 0.003), albeit the difference between any two localizations was smaller than 2%. Table 4 demonstrates the comparison of localization 1, which is usually the side for percutaneous biopsy, with localizations 2 to 10. Localizations 4,5,6,8, and 9 had significantly different TAG concentrations, all of them being less than 2%. Scoring of 290 liver tissue samples revealed 124 specimens with moderate, 49 with severe, and 117 specimens with very severe fatty liver. Based on the scoring system the TAG concentration was even distributed in 79.3% of the livers. In 4 of 29 livers, the scores differed by 1 and in 2 livers by 2 scores. Mean, standard deviation and range of the measured values of each liver are shown in Table 5. The highest variation in TAG content within one liver was 25% (liver no. 13). Since the TAG content was >150 mg/g in all 10 locations, the scores were equally throughout the liver. The same liver was diagnosed with marked variation, based on histopathological evaluation. The variation between locations was significantly higher in livers diagnosed with scores 2 or 3 compared to livers with a score of 1.

Table 4 Comparison of the localizations 2–10 to localization 1 (baseline; usual side for percutaneous biopsy) using a linear mixed effect model approach

Localization	Value	Std. Error	p-value
1 (intercept)	18.8	2.5	0.0000
2	0.1	0.7	0.8892
3	−0.4	0.7	0.5607
4	−1.5	0.7	0.0279
5	−1.4	0.7	0.0389
6	−2.0	0.7	0.0048
7	−0.8	0.7	0.2532
8	−1.9	0.7	0.0056
9	−1.8	0.7	0.0090
10	−0.2	0.7	0.7654

In Table 6 the mean TAG content is listed for each grade of fatty liver, using H&E and ORO stains (also Fig. 7).

The TAG concentration did not influence the agreement between observers, which was tested using a generalized linear mixed effect model.

Discussion

The examination of a liver biopsy remains the gold standard for diagnosing and staging fatty liver. Important

Table 5 TAG content (% wet weight) in 29 livers. Mean, standard deviation, and range of measured values for 29 livers

Liver	Mean	Std Dev.	Range
1	6.6	0.67	5.5–7.7
2	7.3	0.93	5.8–8.3
3	7.6	0.72	6.8–9.3
4	33.4	2.88	27.4–36.9
5	5.4	0.37	4.9–6.2
6	5.5	0.62	4.5–6.4
7	6.5	0.41	5.9–7.3
8	6.2	0.55	5.5–7.4
9	11.3	1.0	9.1–12.4
10	10.0	0.83	8.9–11.2
11	11.3	1.53	10.0–15.0
12	7.7	0.54	6.9–8.5
13	26.9	9.92	16.4–41.5
14	9.1	0.75	8.2–10.7
15	21.8	2.05	19.1–24.7
16	34.3	1.37	32.6–36.1
17	19.72	2.32	16.8–23.8
18	7.93	0.91	6.6–9.7
19	36.4	1.83	34.7–40.8
20	16.0	1.13	14.5–17.6
21	11.0	3.48	7.5–19.2
22	8.9	0.57	8.1–10.3
23	14.5	5.91	8.7–28.4
24	7.2	0.44	6.5–8.0
25	-	-	-
26	44.7	2.64	40.5–47.9
27	40.1	3.46	37.5–47.1
28	14.5	1.36	11.6–16.5
29	42.9	2.51	40.3–49.3
30	41.9	2.95	36.2–45.4

Table 6 Mean and SD of TAG content for each score. HE, H&E stain; ORO, Oil Red Stain; MR and MH, pathologists

	Liver TAG (% wet weight)			
	MR		MH	
	HE	ORO	HE	ORO
Grade 0	9.0 (± 3.7)	8.3 (± 2.6)	7.9 (± 2.0)	6.7 (± 1.7)
Grade 1	17.5 (± 4.2)	15.5 (± 5.6)	15.0 (± 4.7)	9.1 (± 2.7)
Grade 2	39.6 (± 5.0)	34.5 (± 1.6)	31.2 (± 12.7)	20.1 (± 12.6)
Grade 3	38.2 (± 5.1)	39.7 (± 4.8)	38.2 (± 4.9)	33.2 (±9.6)

factors affect an appropriate diagnosis of liver biopsies, including the quality of the biopsy, the small tissue sample representing the lesion of the whole organ, and the interpretation of histological findings. Accurate assessment of the degree of fatty liver is a key requirement for staging, required for research purposes.

The results of this study demonstrate minor variation in the distribution of lipids in 10 different locations within the liver of dairy cows. The TAG content was up to 25% different within one liver. The difference in TAG content between any location was less than 2%. Therefore, the TAG content of a liver biopsy taken either at the right lobe (percutaneous) or at the quadrate lobe (during laparotomy) differs less than 2% from other locations throughout the liver.

In humans, lesions of nonalcoholic fatty liver disease are unevenly distributed [7]. However, caution must be used in extrapolating results in humans to cows. In the study of Ratziu (2005) the steatosis grade displayed substantial agreement between 2 biopsies, whereas inflammation and fibrosis did not. In cattle, hepatic lipidosis is a reversible process and inflammation and fibrosis are usually not seen [1, 19]. Inflammatory cells or fibroblasts could not be detected on our samples.

One common feature with human NAFLD was the distribution of lipid infiltration within the hepatic lobules (Fig. 8). In mild cases we could observe infiltrations close to the central vein. With increasing severity the lipid infiltration extended to all zones. In two cases with mild fatty liver, infiltration could be found more periportal (Table 7). Because of the cross-sectional nature of this study, we could not demonstrate the changes in lipid infiltration over time.

Based on histopathologic scores the pathologists agreed with an even distribution of lipid infiltration in 73.3% of the livers. The histopathological evaluation of the liver tissue revealed some disagreement between the pathologists and between staining technique. Overall ORO staining revealed slightly higher scores compared to H&E staining. Differences in histopathologic interpretation between observers is well known [3]. Human evaluation of biopsies is prone to bias [13], but to date there is also no consensus for an objective grading and staging system for the histological evaluation of fatty liver in cattle. Therefore, there might be differences in the prevalence of hepatic changes between studies [3]. In addition, without accurate and widely accepted methods it is difficult to compare findings related to TAG content in liver tissue between studies [20].

In the literature, different scores are used for lipid infiltration in fatty liver [21, 22]. Therefore, the comparison between scores based on histopathology and measured TAG is difficult.

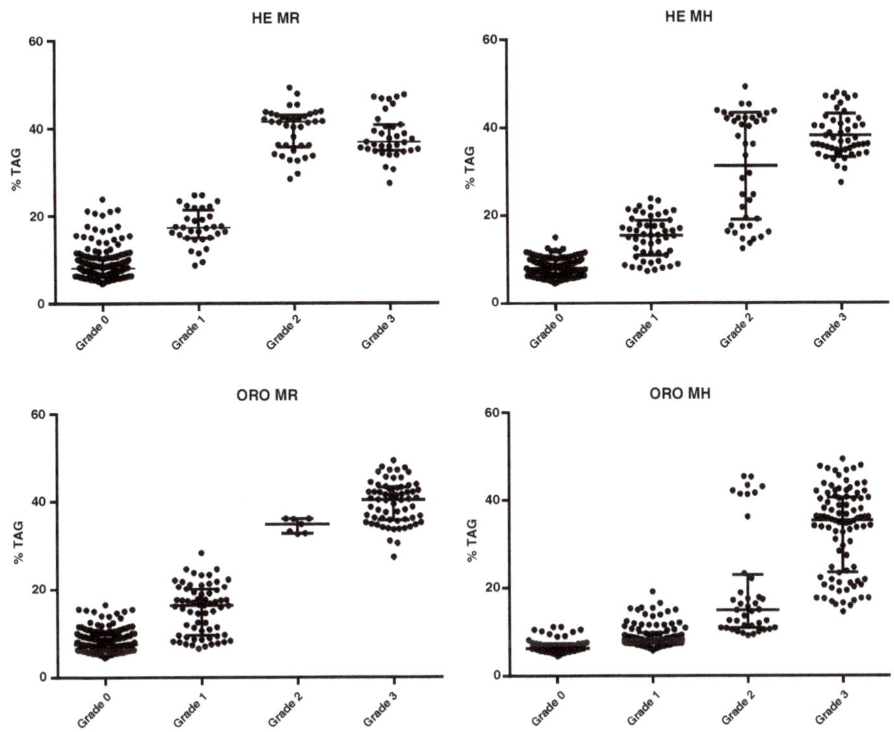

Fig. 7 TAG content (% wet weight) of 29 livers, according to staining technique (H&E, ORO) and pathologists (MR, MH)

The content of TAG was significantly higher in livers diagnosed with fatty liver compared to livers with a histopathologic score of 0. However, there was no significant difference in TAG content between the histopathologic scores 2 and 3. Best agreement with the quantitative TAG content could be achieved with ORO staining. This is in agreement with another study, performed with human and rat livers, which revealed good agreement between histopathologists based on H&E assessment but an inaccuracy compared to the true hepatic TAG concentrations [15]. The ORO stain was more accurate, because ORO identifies mainly TAG and less phospholipids and cholesterol [15]. A steriological analysis of fatty livers revealed a highly significant correlation with the hepatic TAG content in cattle [8]. Also Collins et al. revealed a high correlation between TAG content and volume fraction of fat droplets determined stereologically [23]. In our study we used a more practical technique, which is routinely used for the diagnosis of fatty liver and H&E stained specimens are often available for retrospective evaluation of cases.

In addition, one observer appeared to be stricter in scoring the specimens. A disagreement by one grade might also be due to the fact, that a categorical scoring system is used for a continuous disease process.

Formalin fixing and routine processing of the tissue samples can be an accurate method for histological assessment and grading. Even with the loss of fat or glycogen content a grading can be accomplished. The morphological appearance of cytoplasmic vacuoles,

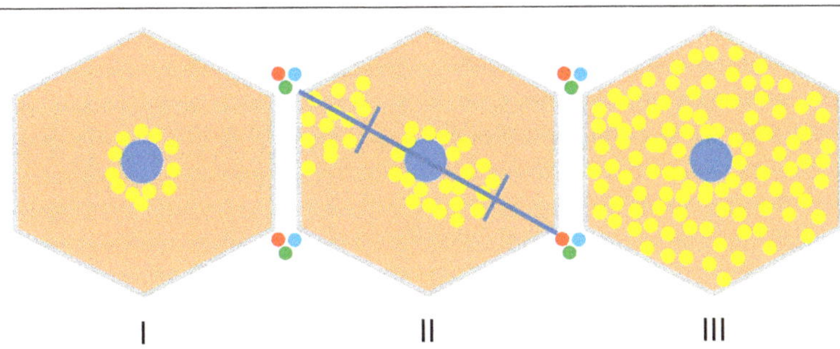

Fig. 8 Distribution of lipid infiltration within hepatic lobules, according to the severity of fatty liver. I = grade 1 (infiltration in one zone, around central vein); II = grade 2 (infiltration periportal to midzonal or zentrolobular to midzonal); III = grade 3 (infiltration panlobular)

Table 7 Number of whole livers with different grades in association to the localization of lipid infiltration within hepatic lobules

	Periportal	Zentrolobular	Zentro-panlobular	Panlobular
Grade 1	2	4	0	0
Grade 2	0	0	4	0
Grade 3	0	0	1	5

well demarcated versus foamy depositions/appearance, can be distinct enough to differentiate between fatty and glycogen depositions in H&E staining. However, special staining techniques improve the accuracy, especially when droplet size is quantified, although not being practical [15].

We suspected focal lipid infiltration macroscopically in one liver. However, the histopathological examination did not reveal obvious differences in the degree of lipid infiltration between the corresponding tissue samples. Focal fatty liver in cattle was rarely reported in the literature. Mohamed et al. [11] diagnosed focal fatty liver by ultrasonography in a heifer. In a study population of 106 Holstein cattle focal fatty liver was observed in 24% of the cows [10]. Focal lipid infiltration occurs adjacent to the insertion of serosal attachments [24] and is not related to the pathogenesis of fatty liver in cattle. In contrast to our results Raoofi et al. [10] and Mohamed et al. [11] observed a higher amount of fat vacuoles compared to the surrounding tissue.

Overall the distribution of TAG throughout the liver and the accuracy of human evaluation of liver biopsies may lead to acceptable diagnoses for clinical purposes. For the staging of lipid infiltration for research projects, some degree of variation needs to be considered.

A consensus on histopathologic diagnosing and staging of hepatic lipid infiltration should be established. In our study, we used a scoring system including the observation of lipid infiltration in the different lobular zones in order to divide into 4 categories (grade 0 to 3). We observed that mild infiltration was mainly located centrolobular and with increasing severity the distribution was panlobular. Although experienced pathologists evaluated the tissue, the assignment to the lobular zones may be somewhat subjective. An objective measurement of the lobules may have increased the accuracy. In addition, we used samples that were obtained at one point in time for every individual cow. In order to establish a reliable protocol for histopathological diagnostics, the development of fat infiltration over time should also be investigated.

Conclusion

In the present study, the difference in TAG content between any locations was less than 2%. Therefore, the TAG content of a liver biopsy taken at any side differs less than 2% from other locations throughout the liver. Based on the scoring system the TAG concentration was even distributed in 79.3% of the livers. The agreement between two pathologists in scoring histological specimens was moderate to fair. Overall the distribution of TAG throughout the liver and the accuracy of human evaluation of liver biopsies may lead to acceptable diagnoses for clinical purposes. For the staging of lipid infiltration for research projects, some degree of variation needs to be considered. A consensus on histopathologic diagnosing and staging of hepatic lipid infiltration should be established.

Abbreviations

H&E: Hematoxylin-eosin; ORO: Oil red O; PAS: Periodic acid-Schiff stain; TAG: Qtriacylglycerol

Acknowledgements

We would like to thank Dr. Clemens Bauer and Dr. Frank Götz (abattoir Zurich) for their assistance in collecting liver tissue, and Ms. Jeanne Peter for the preparation of the drawing (Fig. 1). Further we would like to thank Ms. Belinda Senn and Ms. Eliane Steiner for preparing and staining the tissue samples.

Funding

This research had no funding sources.

Authors' contributions

CG and MR initiated and organized the study. CG and RK were responsible for sample collection, SI and CG measured the TAG concentrations, MR and MH assessed the tissue samples, SH participated in statistical evaluation, CG drafted the manuscript, which was read and approved by all authors.

Competing interests

The authors declare that they have no potential conflicts of interest to disclose.

Consent for publication

Not applicable.

Author details

[1]Department of Farm Animals, Vetsuisse Faculty, University of Zurich, Winterthurerstrasse 260, Zurich, Switzerland. [2]Institute of Pharmacology and Toxicology, Vetsuisse Faculty, University of Zurich, Winterthurerstrasse 260, 8057 Zurich, Switzerland. [3]Institute of Veterinary Pathology, Vetsuisse Faculty, University of Zurich, Winterthurerstrasse 268, Zurich, Switzerland. [4]Section of Epidemiology, Vetsuisse Faculty, University of Zurich, Winterthurerstrasse 270, Zurich, Switzerland.

References

1. Bobe G, Young JW, Beitz DC. Invited review: pathology, etiology, prevention, and treatment of fatty liver in dairy cows. J Dairy Sci. 2004;87:3105–24.
2. Starke A, Haudum A, Busche R, Beyerbach M, Dänicke S, Rehage J. Technical note: analysis of total lipid and triacylglycerol content in small liver biopsy samples in cattle. J Anim Sci. 2010;88:2741–50.
3. Kemp SD, Zimmerman KL, Panciera DL, Monroe WE, Leib MS. Histopathologic variation between liver lobes in dogs. J Vet Intern Med. 2015;29:58–62.
4. Janiec DJ, Jacobson ER, Freeth A, Spaulding L, Blaszyk H. Histologic variation of grade and stage of non-alcoholic fatty liver disease in liver biopsies. Obes Surg. 2005;15:497–501.
5. Cole TL, Center SA, Flood SN, Rowland PH, Valentine BA, Warner KL, Erb HN. Diagnostic comparison of needle and wedge biopsy specimens of the liver in dogs and cats. J Am Vet Med Assoc. 2002;220:1483–90.

6. Maharaj B, Maharaj RJ, Leary WP, Cooppan RM, Naran AD, Pirie D, Pudifin DJ. Sampling variability and its influence on the diagnostic yield of percutaneous needle biopsy of the liver. Lancet. 1986;1:523–5.
7. Ratziu V, Charlotte F, Heurtier A, Gombert S, Giral P, Bruckert E, Grimaldi A, Capron F, Poynard T. Sampling variability of liver biopsy in nonalcoholic fatty liver disease. Gastroenterology. 2005;128:1898–906.
8. Gaal T, Reid IM, Collins RA, Roberts CJ, Pike BV. Comparison of biochemical and histological methods of estimating fat content of liver of dairy cows. Res Vet Sci. 1983;34:245–8.
9. O'Kelly JC. The concentrations of lipids in the plasma, liver and bile of genetically different types of cattle. Comp Biochem Physiol B. 1974;49:491–500.
10. Raoofi A, Bazargani TT, Tabatabayi AH. An abattoir survey on the frequency of fatty liver syndrome in dairy cows from the suburbs of Tehran. Trop Anim Health Prod. 2001;33:15–9.
11. Mohamed T, Oikawa S, Kurosawa T, Takehana K, Hosaka Y, Okada H, Koiwa M, Sato H. Focal fatty liver in a heifer: utility of Ultrasonography in diagnosis. J Vet Med Sci. 2004;66:341–4.
12. Theodossi A, Skene AM, Portmann B, Knill-Jones RP, Patrick RS, Tate RA, Kealey W, Jarvis KJ, O'Brian DJ, Williams R. Observer variation in assessment of liver biopsies including analysis by kappa statistics. Gastroenterology. 1980;79:232–41.
13. Fiorini RN, Kirtz J, Periyasamy B, Evans Z, Haines JK, Cheng G, Polito C, Rodwell D, Shafizadeh SF, Zhou X, Campbell C, Birsner J, Schmidt M, Lewin D, Chavin KD. Development of an unbiased method for the estimation of liver steatosis. Clin Transpl. 2004;18:700–6.
14. Franzén LE, Ekstedt M, Kechagias S, Bodin L. Semiquantitative evaluation overestimates the degree of steatosis in liver biopsies: a comparison to stereological point counting. Mod Pathol. 2005;18:912–6.
15. Levene AP, Kudo H, Armstrong MJ, Thursz MR, Gedroyc WM, Anstee QM, Goldin RD. Quantifying hepatic steatosis - more than meets the eye. Histopathology. 2012;60:971–81.
16. Norris AW, Chen L, Fisher SJ, Szanto I, Ristow M, Gonzalez FJ, Jozsi AC, Hirshman MF, Goodyear LJ, Kahn CR. Muscle-specific PPARgamma deficient mice develop insulin resistance but respond normally to thiazolidinediones. J Clin Invest. 2003;112:608–18.
17. Viera AJ, Garrett JM. Understanding interobserver agreement: the kappa statistic. Fam Med. 2005;37:360–3.
18. Bates D, Mächler M, Bolker B, Walker S. Fitting Linear Mixed-Effects Models using lme4. 2015:67. doi:10.18637/jss.v067.i01.
19. Reid IM, Collins RA. The pathology of post-parturient fatty liver in high-yielding dairy cows. Invest Cell Pathol. 1980;3:237–49.
20. Sejersen H, Sørensen MT, Larsen T, Bendixen E, Ingvartsen KL. Liver protein expression in dairy cows with high liver triglycerides in early lactation. J Dairy Sci. 2012;95:2409–21.
21. Haudum A, Starke A, Beyerbach M, Wohlsein P, Rehage J. Ultrasonographic assessment of liver dimensions in dairy cows with different hepatic triacylglycerol content. J Anim Sci. 2011;89:1392–400.
22. Gaál T, Husvéth F. Comparison of the liver biopsy sample and the "whole liver" in respect of lipid content and fatty acid composition of lipids. Acta Vet Hung. 1983;31:51–6.
23. Collins RA, Reid IM. A correlated biochemical and stereological study of periparturient fatty liver in the dairy cow. Res Vet Sci. 1980;28:373–6.
24. Stalker MJ, Haye MA. Liver and biliary system. In: Jubb, Kennedy & Palmer's Pathology of Domestic Animals. New York: Elsevier Saunders; 2007:297–298.

14

Evaluation of temporal surveillance system sensitivity and freedom from bovine viral diarrhea in Danish dairy herds using scenario tree modelling

Alessandro Foddai[1*], Anders Stockmarr[2] and Anette Boklund[1]

Abstract

Background: The temporal sensitivity of the surveillance system (TemSSe) for Bovine Viral Diarrhea (BVD) in Danish dairy herds was evaluated. Currently, the Danish antibody blocking ELISA is used to test quarterly bulk tank milk (BTM). To optimize the surveillance system as an early warning system, we considered the possibility of using the SVANOVIR ELISA, as this test has been shown to detect BVD-positive herds earlier than the blocking ELISA in BTM tests. Information from data (2010) and outputs from two published stochastic models were fed into a stochastic scenario tree to estimate the TemSSe. For that purpose we considered: the risk of BVD introduction into the dairy population, the ELISA used and the high risk period (HRP) from BVD introduction to testing (at 90 or 365 days). The effect of introducing one persistently infected (PI) calf or one transiently infected (TI) milking cow into 1 (or 8) dairy herd(s) was investigated. Additionally we estimated the confidence in low (PLow) herd prevalence (<8/4109 infected herds) and the confidence in complete freedom (PFree) from BVD (< 1/4109).

Results: The TemSSe, the PLow, and the PFree were higher, when tests were performed 365 days after BVD introduction, than after 90 days. Estimates were usually higher for the SVANOVIR than for the blocking ELISA, and when a PI rather than a TI was introduced into the herd(s). For instance, with the current system, the median TemSSe was 64.5 %, 90 days after a PI calf was introduced into eight dairy herds. The related median PLow was 72.5 %. When a PI calf was introduced into one herd the median TemSSe was 12.1 %, while the related PFree was 51.6 %. With the SVANOVIR ELISA these estimates were 99.0 %; 98.9 %, 43.7 % and 62.4 %, respectively.

Conclusions: The replacement of the blocking ELISA with the SVANOVIR could increase the TemSSe, the PLow and PFree remarkably. Those results could be used to optimize the Danish BVD surveillance system. Furthermore, the approach proposed in this study, for including the effect of the HRP within the scenario tree methodology, could be applied to optimize early warning surveillance systems of different animal diseases.

Keywords: Temporal surveillance system sensitivity, BVD, Scenario tree, Freedom from disease

Background

In Denmark, Bovine Viral Diarrhea (BVD) is considered an exotic disease [1]. An eradication program was initiated in 1994 [2, 3] and in the period from 2007–2011, only 3 out of approximately 4000 dairy herds were diagnosed with BVD by bulk tank milk (BTM) testing. All dairy herds were screened quarterly by BTM, while beef herds were screened at abattoir by blood sampling. If the BTM is classified as positive, individual animals are tested to find at least one antibody positive (sample size determined to have 95 % herd sensitivity, assuming 10 % within-herd prevalence). If the positive herd status is confirmed, all non-antibody positive animals are tested to detect the viremic animals and persistently infected (PI) cattle are eliminated [1].

* Correspondence: alessandrotula@hotmail.it
[1]Section of Epidemiology, National Veterinary Institute, Technical University of Denmark, Bülowsvej 27, DK-1870 Frederiksberg C, Denmark
Full list of author information is available at the end of the article

The BVD virus (BVDV) can cause uterine infections in pregnant cows, abortions, stillbirths or weak calves [4–6]. Cows exposed to BVDV in the first 120 days of pregnancy, can give birth to calves, which become PI [4, 5]. PI cattle shed the virus in large amounts throughout their lives, while other transiently infected (TI) animals shed the virus in small amounts for 2–3 weeks and become lifelong immune [4–6].

Because the Danish dairy herd size has increased remarkably since the eradication program started [1], an evaluation and eventual optimization of the Danish BVD surveillance system was considered necessary by the Danish Cattle Federation. An optimal early-warning system based on BTM testing should detect newly infected herds as soon as possible. Early-warning surveillance systems are those aimed to detect the "unexpected threats" in a timely way [7]. Hence an early-warning system should be based on a) the risk that some herds become infected and b) the time needed to detect antibodies in BTM. The latter is known to depend on the herd size and on the threshold prevalence of antibody positive milking cows, at which the BTM can be classified as positive with the antibody ELISA used [8, 9].

The time from the introduction of a pathogen into the country, until it is detected by the surveillance activities, can be defined as "high risk period" (HRP) [10], or "timeliness" [7]. In this study we use the former term, to emphasize the fact that the longer the time required for detection the higher is the risk the pathogen is spread from the first case(s) herd(s) to other Danish herds.

The probability of detecting a pathogen in a given time period has been called "temporal sensitivity" [11].

In two previous studies we estimated the detection time for BVD in Danish dairy herds [9] and the annual probability of BVD introduction into the Danish dairy population [12]. Here we considered Danish dairy herds and beef herds as two distinct populations in the country, as in Foddai et al. [12], and we evaluated the BVD surveillance system for the Danish dairy population (on national level).

Hence, the aims of the present study where (i) to evaluate the temporal sensitivity of the Danish BVD surveillance system (TemSSe) with fixed HRPs, according to different routes of BVDV introduction into Danish dairy herds (e.g. import of PI or TI animals), (ii) to investigate surveillance optimization, by taking into account the risk of BVDV introduction to the country, the herd size, and the antibody ELISA used on BTM; and (iii) to estimate (at country level) the confidence in low herd prevalence (PLow) and in complete freedom from BVD (PFree).

Methods

To estimate the TemSSe of the Danish surveillance system, we developed a stochastic scenario tree model (Section Stochastic scenario tree methodology used to estimate the TemSSe-Input parameters used in the stochastic scenario tree, Fig. 1) [13–15]. Inputs for the model were obtained by data analysis (Section Data analysis), risk assessment (Section Risk assessment per herd

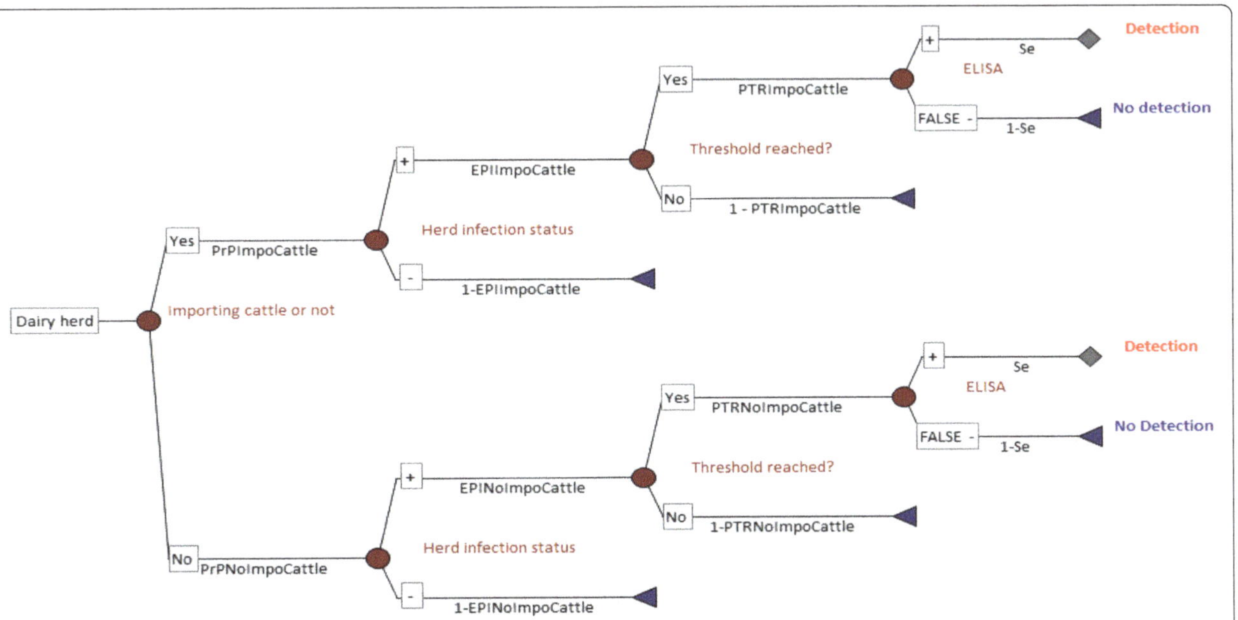

Fig. 1 Stochastic scenario tree representing the BVD surveillance system for Danish dairy herds. $PrP_{ImpoCattle}$ and $PrP_{NoImpoCattle}$ = proportion of dairy herds within the ImpoCattle and NoImpoCattle category, respectively. $EPI_{ImpoCattle}$ and $EPI_{NoImpoCattle}$ = effective probability of infection for the ImpoCattle and NoImpoCattle category. $PTR_{ImpoCattle}$ and $PTR_{NoImpoCattle}$ = probability that the threshold prevalence is reached within the milking paddock at 90 or 365 days from BVDV introduction within herds of the ImpoCattle and NoImpoCattle category. Se = Expected sensitivity of the antibody ELISA used (Danish blocking ELISA or SVANOVIR) on BTM, when the threshold prevalence of seropositive milking cows is reached

category) and modelling of within-herd BVD dynamics (Sections Modelling the within-herd BVD dynamics to estimate the PTR values-Threshold prevalence and sensitivity of the ELISAs used for BTM testing). Thereafter, the estimated TemSSe was used to calculate the negative predictive value (NPV) of the surveillance system (Section Negative predictive value of the surveillance system) [13], which represented the confidence in low herd prevalence (PLow) or in complete freedom from BVD (PFree) according to the design prevalence used. Finally, a sensitivity analysis (Section Sensitivity analyses) was carried out to investigate the impact of each input parameter on the output.

Stochastic scenario tree methodology used to estimate the TemSSe

The sensitivity of national veterinary surveillance systems (SSe) can be evaluated through use of stochastic scenario tree models [13–15]. Information from different surveillance sources can be combined into an overall SSe estimate, taking into account (I) the prevalence of infected herds in a country (between-herds design prevalence, P_H), (II) the prevalence of infected animals within a herd (within-herd design prevalence, P_U), (III) the relative risk (RR) of infection in the population's strata, and (IV) the sensitivity (Se) of the diagnostic test used [13].

Thus the SSe represents the probability that a population, infected with the assumed P_H and P_U, is correctly classified by the surveillance system [13–15]. Furthermore, in the methodology by Martin et al. [13] it is usually assumed that the specificity of the surveillance system is 100 %. Thus, if a positive sample is found, further confirmatory testing is made in the herd to avoid false positive results. In this study, we made the same assumption.

We developed a stochastic scenario tree (Fig. 1) in an Excel spreadsheet (Microsoft Office Excel 2007) using the software @Risk 6 (Palisade Corporation). The model was run using 10,000 iterations and Latin Hypercube sampling.

The P_H was set at 0.2 % (≈ 8 infected herds out of 4109 Danish dairy herds present in October 2010), or as 0.02 %, indicating that we aimed at detecting the first infected herd. The 0.2 % P_H is similar to the limit set by the World Animal Health Organization (OIE) to substantiate officially free status for infectious bovine rhinotracheitis (IBR) [16], which can be considered a disease similar to BVD, in terms of transmission routes. In fact, IBR is also caused by a virus, which can be transmitted between cattle herds and countries by moved animals, contaminated semen, embryos, and fomites [8, 17, 18]. Furthermore, these very low design prevalences used in this study give conservative TemSSe estimates, as the lower the disease prevalence between herds the more difficult it is to detect the disease.

P_U was here defined as the threshold prevalence of antibody positive milking cows, at which the BTM was classified positive by the ELISA used (see Section Threshold prevalence and sensitivity of the ELISAs used for BTM testing). With "fast" spreading pathogens, it can be assumed that P_U is quickly reached. In contrast, BVDV can be considered, in some cases, as a "slowly" spreading virus, especially if it is introduced into large herds through TI animals [9, 19]. Therefore, a long time could elapse before P_U is reached within a herd and before such a herd is detected by the surveillance system (long high risk period or HRP). For that reason, we estimated the temporal sensitivity of the surveillance system (TemSSe) for two different ELISAs, with a HRP of 90 or 365 days. Those two time intervals lead to TemSSe testing quarterly (current system) or testing one year after BVDV introduction into the Danish dairy population (as a worst case scenario), respectively. The effect of introducing one PI calf or one TI milking cow into naïve Danish dairy herd(s), on the final TemSSe was also investigated.

Moreover, dairy herds were divided into two levels of risk: herds with import of live cattle (ImpoCattle category) or without (NoImpoCattle category) (Fig. 1), based on a previous study [12].

Input parameters used in the stochastic scenario tree

We used data from the last trimester of 2010, when approximately 4109 herds (N) delivered milk and were tested in the BTM. Hence, we considered a single surveillance component, which includes the whole population of Danish dairy herds.

The steps needed for a BVD positive dairy herd, to give a positive BTM value are represented with nodes in the scenario tree in Fig. 1. The first node divided the population of tested dairy herds into the two risk categories: NoImpoCattle and ImpoCattle. Thus, the proportion of dairy herds in each risk category was represented by $PrP_{NoImpoCattle}$ and $PrP_{ImpoCattle}$, respectively (Fig. 1).

The probability that a herd was infected with BVD was represented by the effective probability of infection in the category ($EPI_{NoImpoCattle}$ and $EPI_{ImpoCattle}$) (Fig. 1). The EPI_j for each herd category "j" was obtained by multiplying the between-herds design prevalence (P_H) with the adjusted relative risk (ARRj) of infection [13, 14]. The relative risk estimates of each category ($RR_{NoImpoCattle}$ and $RR_{ImpoCattle}$) were adjusted to maintain the specified relativity (weight) and yet average to one over the whole population [15].

The NoImpoCattle category was used as risk reference category with $RR_{NoImpoCattle} = 1$, while for the

ImpoCattle category we set $RR_{ImpoCattle} = R_{ImpoCattle}/R_{NoImpoCattle}$. Where, the numerator and denominator represented the risk of BVD introduction for the ImpoCattle and NoImpocattle category, respectively (Sections Risk assessment per herd category and Risk estimates per herd category).

Thereafter, the ARR for the NoImpoCattle category was estimated as:

$$ARR_{NoImpoCattle} = \frac{1}{(PrPNoImpoCattle + RRImpoCattle \times PrPImpoCattle)} \quad (1)$$

While, the ARR for the ImpoCattle category was:

$$ARR_{ImpoCattle} = ARR_{NoImpoCattle} \times RR_{ImpoCattle} \quad (2)$$

Furthermore, the component unit sensitivity was calculated as the weighted sum of the probabilities with a positive outcome at the end of each limb of the stochastic scenario tree (Fig. 1). Therefore, at the assumed P_H, the overall probability (TemSSe) that at least one BTM positive herd was detected in the BTM surveillance system was:

$$TemSSe = 1\text{-}[1\text{-}(PrP_{NoImpoCattle} \times EPI_{NoImpoCattle} \times PTR_{NoImpoCattle} \times Se + PrP_{ImpoCattle} \times EPI_{ImpoCattle} \times PTR_{ImpoCattle} \times Se)]N \quad (3)$$

Where, $PTR_{NoImpoCattle}$ and $PTR_{ImpoCattle}$ represented the probability that the threshold prevalence of seropositive cows (P_U) was already reached within the milking group at the day of BTM testing (Sections Modelling the within-herd BVD dynamics to estimate the PTR values-Threshold prevalence and sensitivity of the ELISAs used for BTM testing), in NoImpoCattle or ImpoCattle herds, respectively.

The Se was the sensitivity of the test used on the BTM sample. We assumed that the test´s Se could be achieved, when the threshold prevalence was reached. A similar approach was used by Graat et al. [8] for IBR. When the threshold prevalence was not reached, we assumed that the BTM of an infected dairy herd was classified as negative.

Data analysis

Descriptive data analyses were carried out using the freeware R (R Development Core Team, 2012). Data from 2010, on herd size, milk deliverance, imports/exports of live cattle, imports of semen and embryos, were obtained from the Danish Cattle Federation for all Danish dairy cattle herds.

Moreover, for both herd categories, we estimated the percentage of imported goods from countries without or with endemic BVD (Table 1). We made this distinction on the BVD status of the sending countries, because it could affect the risk of introducing BVDV into the Danish cattle population [12].

Risk assessment per herd category

The annual risk of BVDV introduction in each risk category ($R_{ImpoCattle}$ and $R_{NoImpoCattle}$) was calculated by use of a stochastic model, which was previously developed by Foddai et al. [12]. Results of the data analysis (Table 1) were fed into such a model.

For the NoImpoCattle category, we included the following sources of infection: import of semen and embryos, visits by contaminated trucks used abroad, and visits by hoof trimmers practicing in cattle herds in Denmark and in other countries. For the ImpoCattle category, we included the same BVDV introduction routes plus imports of live cattle.

Information on truck visits and hoof trimmers was based on the previous study [12]. The overall annual number of truck visits, which could lead to introduction

Table 1 Results of data analysis for herds with import of cattle (ImpoCattle) and without (NoImpoCattle)

Herd Category	N	Imported cattle	Imported doses of semen	Imported embryos	Truck visits	Hoof trimmer visits[c]	PrP	EPI_j[d]
ImpoCattle	8	246 (10.6 %)	3776 (92.9 %)	5 (100 %)	5606 * 0.8 %[b] = 45	A*B* 0.2 %	0.2 %	0.025 %
NoImpoCattle	4101	0	301020 (46.9 %)	272 (99.3 %)	5606 * 99.2 %[b] = 5561	A*B* 99.8 %	99.8 %	0.024 %
Total	4109	246[a]	304796[a]	277 [a]	5606[a]		100 %	

N number of dairy herds delivering milk and tested in the fourth trimester of 2010 within each category, *PrP* proportion of herds within each category, *EPIj* effective probability of infection for each category "j". Between brackets is the percentage of cattle, doses of semen and embryos imported from countries where BVD is endemic. We assumed that trucks and hoof trimmers went only to countries with endemic BVD status [12]

[a], From Foddai et al. [12]

[b], According to Foddai et al. [12], in total 5606 truck visits at risk occur in Danish dairy herds during a one-year period. The estimated percentage of exports from the ImpoCattle and the NoImpoCattle category was 0.8 % and 99.2 %, respectively (Danish data 2010). We assumed that the number of trucks visits at risk in each category was proportional to the exports occurred from the category

[c] The number of hoof trimmers visiting cattle herds abroad (A) during a one-year period was a Pert distribution (5, 7, 18), while the number of times each hoof trimmer crosses the border (B) was Pert (1, 8, 30) (see Table 8 in [12]). The annual number of hoof trimmer visits, which could lead to BVDV introduction into each category, was assumed proportional to the herds present in the category and was estimated by: A * B * $PrP_{ImpoCattle}$ and A * B * $PrP_{NoImpoCattle}$

[d]The EPIj reported in the table was calculated using P_H 0.02 %. When we used P_H 0.2 % the EPIj was 0.20 for the ImpoCattle category and 0.19 % for the NoImpoCattle

of BVDV into Danish dairy herds, was previously estimated to 5606 [12]. For each of the two herd categories, we assumed that the number of truck visits at risk occurring in a year (or in a trimester) was proportional to the number of exports from the category (Table 1).

The annual number of hoof trimmer visits, which could lead to BVDV introduction into each category, was also estimated (Table 1) and was assumed to be proportional to the number of herds present in each category.

For other variables (e.g. the within-herd prevalence abroad, the probability of removing BVDV from contaminated trucks by disinfection, etc.), we used the same assumptions and input values described in Foddai et al. [12].

Hence, the risk of BVDV introduction in the ImpoCattle category ($R_{ImpoCattle}$) was estimated using five published stochastic scenario trees [12], which represented: live-animal imports (PAnim), semen imports (PSem), embryo imports (PEmb), truck visits (PTruck) and hoof trimmers practicing across borders (PTrim). Thus, information from the five trees was combined in Eq. 4, as:

$$R_{ImpoCattle} = 1\text{-}[(1\text{-PAnim}) \times (1\text{-PSem}) \times (1\text{-PEmb}) \times (1\text{-PTruck}) \times (1\text{-PTrim})] \quad (4)$$

The risk ($R_{NoImpoCattle}$) of BVDV introduction into the NoImpoCattle category was estimated using the same formula (Eq. 4) without the term (1-PAnim).

Modelling the within-herd BVD dynamics to estimate the PTR values

The prevalence of antibody positive milking cows was assumed to increase over time [9]. The probability (PTR) of reaching the threshold prevalence (or P_U within the milking group) needed for BVD detection by BTM testing within a fixed HRP, was estimated using a stochastic simulation model; which was previously developed to simulate within-herd BVD dynamics and to estimate the detection time in Danish dairy herds, according to the ELISA used on BTM [9].

The PTR is affected by the route of BVDV introduction to the herd (PI or TI animal), the herd size, the threshold prevalence of the antibody ELISA used, and the time elapsed between BVDV introduction and day of testing (HRP). Thus, the PTR was calculated as the number of model iterations out of 500, where the threshold prevalence was reached, at 90 or 365 days after the introduction of one PI calf or one TI milking cow, into a naïve Danish dairy herd. A total of 500 iterations appeared sufficient to simulate the within-herd disease dynamics in the study by Foddai et al. [9], where the model was validated using data from a recently infected Danish dairy herd.

The estimation of PTR was carried out for the minimum, most common, and maximum herd size within each herd category. Thereafter, the PTR values were used in the stochastic scenario tree (Fig. 1, Eq. 3) in a Pert distribution ($PTR_{NoImpoCattle}$ and $PTR_{ImpoCattle}$) to represent the variability, between herds of different sizes within each risk category.

In the NoImpoCattle category, the herd size was minimum 1, median 123 and maximum 1185 cows (source: Danish data, 2010). For the smallest herd we did not run the simulation model, because we considered the threshold prevalence as reached soon, after the first infectious animal was introduced into the herd. Therefore, for this herd size we used PTR 100 %. In the ImpoCattle category, the herd size was minimum 24, median 180 and maximum 1070 cows.

Furthermore, in Foddai et al. [9], it was assumed that on average Danish dairy herds have around 150 cows, 115 heifers and 8 calves. Here, we used the same herd structure with a similar ratio of heifers/cows and calves/cows (at day one in the model) (Table 2). All other parameters were kept unchanged from Foddai et al. [9], e.g. we used the same values for: the BVDV transmission rates within a cattle herd (βs), the inter-calving period, the probability of abortion, the culling rates, the dry period, the age at first calving, and the lactation length per individual cow (Table 2).

Threshold prevalence and sensitivity of the ELISAs used for BTM testing

In a previous study [9], the SVANOVIR®BVDV-Ab ELISA (Svanova Boehringer Ingelheim, Uppsala, Sweden) [20–23] was shown to detect BVD by BTM testing, significantly earlier than the Danish blocking ELISA [24, 25], which has been traditionally used in Denmark. As a part of this study, we therefore wanted to evaluate the surveillance system with the two different ELISAs, both used to test BTM samples for antibodies against BVDV.

The threshold prevalence has been estimated at 50 % for the Danish blocking ELISA [1] and at 6 % for the SVANOVIR ELISA [23]. Those thresholds were used in the stochastic simulation model by Foddai et al. [9], to estimate the PTR for each herd size (Table 3).

With a cut-off blocking % (bl%) of 50, the sensitivity (Se) of the Danish blocking ELISA on BTM is 100 % [26], while for the SVANOVIR ELISA, the Se has been estimated between 93.4 % and 99.6 % [27]. In the latter case, a Uniform distribution ranging between those two extremes was used in the Se input (Fig. 1, Eq. 3).

Negative predictive value of the surveillance system

The scenario tree methodology [13–15] is usually used to substantiate freedom from a disease or from

Table 2 Herd parameters used in the simulation model by Foddai et al. [9]

Parameter	Value
Small herds (cows, heifers, calves)	ImpoCattle (24, 18, 2)[a], NoImpoCattle (1, 0, 0)[a]
Medium herds (cows, heifers, calves)	ImpoCattle (180, 138, 10)[a], NoImpoCattle (123, 94, 6)[a]
Large herds (cows, heifers, calves)	ImpoCattle (1070, 820, 66)[a], NoImpoCattle (1185, 908, 64)[a]
Culling rate per year for cows	Pert distribution (min = 32 %, mode = 38 %, max = 43 %)[b]
Culling rate per year for heifers	Pert (4, 7, 12 %)[b]
Culling rate per year for calves	Pert (5, 12, 17 %)[b]
Parity distribution (1st, 2nd, 3rd and 4th)	(31, 27, 22, and 20 %)[b]
Percentage of dry cows	Pert (12, 17, 20 %)[b]
Age in the heifers group (in days)	(700; 768; 870)[b]
Days of inter-calving per cow between parity 1 and 2	(365, 399, 451)[b]
Days of lactation per cow between parity 1 and 2	(305, 339, 391)[b]
Days of inter-calving per cow after parity 2	(370, 391, 456)[b]
Days of lactation per cow after parity 2	(310, 331, 396)[b]

[a] The median number of cows was set according to Danish data (2010), for herds with imports of cattle (ImpoCattle) and without (NoImpoCattle). According to a previous study [9], usually, Danish dairy herds have around 150 cows. In that case the number of heifers (animals with age between 61 and approximately 900 days) and calves (age up to 60 days) present in the herd corresponded to 76.7 % (115/150) and 5.4 % (8/150) of the number of cows. We used similar (approximated) heifers/cows and calves/cows ratios in the herd sizes of the two risk categories
[b] Unchanged from Foddai et al. [9]

a pathogen (PFree) at country/area level. For that purpose, the negative predictive value (NPV or PFree) of the surveillance system is estimated to represent the confidence that a country, classified as free from a pathogen by the system, is truly free. In that case, the design prevalence (P_H and P_U) represents a hypothetical level of infection in the country. If a single positive unit is found, the country would lose the "free status".

In our study, we estimated the NPV of the surveillance system to show the PFree, but we also estimated the confidence (PLow) that the prevalence of infected herds is below the P_H. In the latter case, we did not exclude that in reality few positive herds could be present in the country.

Thus, to estimate PLow we used the TemSSe based on $P_H = 0.2$ %, while to estimate PFree the TemSSe was obtained using $P_H = 0.02$ %. The NPV of the surveillance system was:

Table 3 PTR values in herds of different size within each herd category (ImpoCattle or NoImpoCattle)

			PTR with 1 PI		PTR with 1 TI cow	
Herd category	Test	Herd size (in cows)	HRP = 90 days	HRP = 365 days	HRP = 90 days	HRP = 365 days
ImpoCattle	blocking ELISA	24	175 (35.0)	343 (68.6)	0 (0.0)	23 (4.6)
		180	0 (0.0)	206 (41.2)	0 (0.0)	2 (0.4)
		1070	0 (0.0)	14 (2.8)	0 (0.0)	0 (0.0)
	SVANOVIR	24	468 (93.6)	486 (97.2)	165 (33.0)	308 (61.6)
		180	240 (48.0)	376 (75.2)	0 (0.0)	38 (7.6)
		1070	3 (0.6)	281 (56.2)	0 (0.0)	9 (1.8)
NoImpoCattle	blocking ELISA	1	500 (100.0)	500 (100.0)	500 (100.0)	500 (100.0)
		123	0 (0.0)	248 (49.6)	0 (0.0)	3 (0.6)
		1185	0 (0.0)	8 (1.6)	0 (0.0)	0 (0.0)
	SVANOVIR	1	500 (100.0)	500 (100.0)	500 (100.0)	500 (100.0)
		123	314 (62.8)	397 (79.4)	2 (0.4)	47 (9.4)
		1185	1 (0.2)	278 (55.6)	0 (0.0)	12 (2.4)

PTR Iterations out of 500 (in %), where the threshold prevalence of antibody positive milking cows was reached in a herd, using the stochastic simulation model by Foddai et al. [9] and according to test used (blocking ELISA vs. SVANOVIR) sampling day (HRP of 90 or 365 days), and BVDV introduction route (PI calf or TI cow)

$$\text{PLow} = \frac{(1\text{-PriorPInf})}{(1\text{-PriorPInf}) + (\text{PriorPInf} \times (1\text{-TemSSe}))} \tag{5}$$

Where PLow (or PFree) is the confidence that the prevalence of infected herds was below the assumed design prevalence 0.2 % (or 0.02 %) at the beginning of the surveillance period, since no BVD cases have been detected with the investigated HRP.

PriorPInf is the probability that the country was "infected" with the assumed design prevalence, at the beginning of the surveillance period. This input was set to 50 %, which corresponds to a conservative uninformed prior [13].

Moreover, the PriorPInf was adjusted ($\text{PriorPInf}_{\text{Adj}}$) (Eq. 6), by taking into account the probability of BVDV introduction from abroad (PIntro):

$$\text{PriorPInfAdj} = \text{PriorPInf} + \text{PIntro} - (\text{PriorPInf} \times \text{PIntro}) \tag{6}$$

The annual median PIntro for Danish dairy herds has been estimated as 10.7 % (90 % prediction interval: 1.7 %; 36.6 %) [12]. This PIntro was used in Eq. 6 as a Pert distribution, for the scenarios with 365 days. With scenarios of 90 days the PIntro was divided by 4.

Thus, PLow and PFree were estimated for each infection scenario (introducing a PI or a TI animal into 1 or 8 dairy herds), HRP (90 or 365 days) and ELISA used.

Sensitivity analyses

The importance of each input parameter was investigated by using the regression mapped values in @Risk and the amount of change in the output due to a plus 1 standard deviation for each input was estimated.

The reference scenario was defined as: all dairy herds tested with the Danish blocking ELISA (current system) one year after introduction of a PI calf into a single dairy herd ($P_H = 0.02$ %), since PIs are the main source of BVDV spread between and within herds [28].

Results

Output of data analysis

The imports of live animals, doses of semen, embryos, and the visits of trucks and hoof trimmers are shown per herd category in Table 1. The percentage of goods imported from endemic countries is also shown for each category.

In 2010 cattle were imported to eight dairy herds, which represented the ImpoCattle category. On the other hand, the annual quantity of imported semen and embryos, as well as the estimated numbers of trucks and hoof trimmer visits were remarkably higher for the NoImpoCattle category than for the ImpoCattle category.

Risk estimates per herd category

The median annual risk of BVDV introduction in the NoImpoCattle category ($R_{\text{NoImpoCattle}}$) was 4.8 % (90 % prediction interval: 0.7 %, 21.8 %), while in the ImpoCattle category ($R_{\text{ImpoCattle}}$) it was 5.1 % (0.7 %, 22.4 %).

Based on those findings, the relative risk of BVDV introduction in the ImpoCattle category ($RR_{\text{ImpoCattle}}$) was calculated as the ratio between the two median risk estimates (and between their respective 90 % prediction intervals). Thus, the $RR_{\text{ImpoCattle}}$ was simulated from a Pert distribution with minimum 1, mode 1.03 and maximum 1.07, to calculate the $ARR_{\text{NoImpoCattle}}$ (Eq. 1) and the $ARR_{\text{ImpoCattle}}$ (Eq. 2).

PTR values according to infection scenario, herd size, HRP and ELISA

The PTR values were higher: i) in small herds than in large herds, ii) for BVDV introductions through one PI calf than through one TI milking cow, and iii) with HRP of 365 days than with HRP of 90 days. Moreover, the SVANOVIR ELISA had higher PTR values than the Danish blocking ELISA (Table 3).

In the ImpoCattle category, the PTR ranged from 0 %, e.g. when a PI calf or a TI cow was introduced into the largest herd (with 1070 cows), and the BTM was tested 90 days later with the blocking ELISA; to 97.2 % when one PI calf was introduced into the smallest herd (24 cows), and the BTM was tested one year later with the SVANOVIR (Table 3).

In the NoImpoCattle category the PTR ranged from 0 %, e.g. when a PI calf or a TI cow was introduced into the largest herd (1185 cows), and the BTM was tested 90 days later with the blocking ELISA; to 100 % when one infectious animal was introduced into a herd of a single cow (Table 3).

Temporal sensitivity and negative predictive value of the surveillance system with $P_H = 0.2$ %

If BVDV was introduced into at least 8 dairy herds ($P_H = 0.2$ %), with the Danish blocking ELISA (current surveillance system), the median TemSSe ranged from 64.4 % to 98.2 %. The related median PLow estimates were 72.5 % and 97.6 %, respectively (Table 4).

Using the same P_H, with the SVANOVIR ELISA, the median TemSSe ranged from 64.0 % to 99.8 %. The related median PLow estimates were 72.1 % and 99.7 %, respectively (Table 4).

Usually, the blocking ELISA gave lower TemSSe and PLow estimates than the SVANOVIR. For instance, when one PI calf was introduced into at least eight dairy herds and a HRP of 90 days was used, the median

Table 4 Temporal surveillance sensitivity with related confidence in low herd prevalence and in freedom from disease

$P_H = 0.2$ %	1 PI introduced		1 TI introduced	
ELISA_HRP	TemSSe	PLow	TemSSe	PLow
blocking_90	64.5 (7.9; 97.3)	72.5 (50.4; 97.2)	64.4 (7.8; 97.3)	72.5 (50.4; 97.2)
SVANOVIR_90	99.0 (86.9; 99.9)	98.9 (87.7; 99.9)	64.0 (8.0; 97.0)	72.1 (50.4; 96.9)
blocking_365	98.2 (79.2; 99.8)	97.6 (78.5; 99.8)	65.6 (8.5; 97.4)	68.9 (44.9; 96.7)
SVANOVIR_365	99.8 (99.3; 99.9)	99.7 (99.1; 99.9)	78.9 (30.5; 98.3)	78.3 (52.0; 97.8)
$P_H = 0.02$ %	1 PI introduced		1 TI introduced	
ELISA_HRP	TemSSe	PFree	TemSSe	PFree
blocking_90	12.1 (1.0; 36.2)	51.6 (48.2; 59.5)	12.1 (1.0; 36.2)	51.6 (48.2; 59.5)
SVANOVIR_90	43.7 (22.6; 56.9)	62.4 (54.6; 68.6)	12.0 (1.0; 35.5)	51.5 (48.2; 59.2)
blocking_365	39.3 (17.8; 55.4)	55.7 (46.3; 64.5)	12.5 (1.1; 36.6)	47.3 (40.0; 55.8)
SVANOVIR_365	53.5 (46.7; 59.2)	62.3 (55.5; 67.3)	17.7 (4.4; 40.0)	48.6 (41.0; 57.2)

TemSSe Median temporal surveillance system sensitivity, *PLow* Confidence in low herd prevalence, *PFree* Confidence in complete freedom from BVD. Estimates are reported as median % (90 % prediction intervals), according to ELISA used (blocking = Danish blocking ELISA, SVANOVIR = SVANOVIR ELISA), and HRP (high risk period of 90 or 365 days) after the introduction of 1 persistently infected (PI) calf or 1 transiently infected (TI) milking cow into 1 (between-herds design prevalence $P_H = 0.02$ %) or 8 ($P_H = 0.2$ %) Danish dairy herd(s)

TemSSe and the PLow were 34.5 and 26.4 percentage points (respectively) higher for the SVANOVIR ELISA (Table 4).

Only in the scenario where one TI cow was introduced to the herds and a HRP of 90 days was used, the median TemSSe and the PLow were slightly higher (around +0.4 %) for the blocking ELISA. With a HRP of 365 days the opposite situation was observed, and the TemSSe and the PLow were remarkably higher for the SVANOVIR (+13.3 and +9.4 %, respectively) than for the blocking ELISA (Table 4).

Temporal sensitivity and negative predictive value of the surveillance system with $P_H = 0.02$ %

If BVDV was introduced into a single dairy herd (P_H = 0.02 %), and all Danish dairy herds were tested with the blocking ELISA, the median TemSSe ranged from 12.1 % to 39.3 %. The related median PFree estimates were 51.6 % and 55.7 %, respectively (Table 4).

Using the same P_H, with the SVANOVIR, the median TemSSe ranged from 12.0 % to 53.5 %, while the related PFree estimates were 51.5 % and 62.3 %, respectively (Table 4).

Also in this case, usually, the TemSSe and the related PFree were higher in the surveillance system based on the SVANOVIR ELISA. For instance, when one PI calf was introduced into a dairy herd and a HRP of 90 days was used, the median TemSSe and the PFree were 31.6 and 10.8 percentage points (respectively) higher for the SVANOVIR ELISA than for the blocking ELISA (Table 4).

Output of sensitivity analysis

According to the sensitivity analysis, the input with the highest impact on the estimated TemSSe was the PTR distribution used in the NoImpoCattle category ($PTR_{NoImpoCattle}$). When this input was increased with 1 standard deviation the TemSSe increased between 0 and 11.4 %. The second input in order of importance was the PTR distribution used in the ImpoCattle category ($PTR_{ImpoCattle}$). In that case, the increase caused on the TemSSe ranged between 0 and 0.03 %. The other inputs caused a change lower than 0.03 %.

For the PFree, the most important input was still the $PTR_{NoImpoCattle}$. Increasing such an input with one standard deviation caused an increase in the PFree between 0 and 4.5 %. The second most important input was the annual (overall) probability of BVDV introduction (PIntro) into the Danish dairy population. Increasing the PIntro of 1 standard deviation caused a decrease in the PFree between 0 and 3.1 %. All the other inputs caused a change lower than 0.006 %.

Discussion

A new approach – Including the high risk period in the stochastic scenario tree methodology

To evaluate the BVD surveillance system in Danish dairy herds, we followed the concepts from Martin et al. [13]. Additionally, we estimated the temporal surveillance system sensitivity (namely the TemSSe) and the related negative predictive value (Eq. 5), to substantiate the confidence (PFree) in complete freedom from BVD (< 1 infected herd), and the confidence (PLow) in low herd prevalence (< 8 infected herds).

The way we adapted the scenario tree model allowed us to estimate the TemSSe. Thurmond [11] suggested that the sensitivity for an assay should not be considered as constant, since it is affected by the different disease states. This means that the sensitivity of the test is affected by the time elapsed since a herd (or an animal) became infected. In our case, the transition state herd sensitivity of the ELISA used on BTM, was conditioned upon the immune status of the milking herd. That status varied in time according to HRP, test used, BVDV introduction route (with PI or TI animals) and herd size. Uncertainty, due to all these variables was included in our TemSSe estimates (Table 4), by using the PTR parameter (Table 3) in the scenario tree (Fig. 1), between the infection node "Herd infection status" and the detection node "ELISA".

In this way, we could evaluate if the surveillance system can actually function as an early-warning system, or if optimization was needed, to increase the probability of detecting infected herds (TemSSe) within the aimed time period.

To our knowledge, this is the first study, where the impact of the HRP is included in the evaluation of a surveillance system using stochastic scenario trees. We believe that this approach should be considered, especially when early-warning surveillance systems are established for slowly spreading diseases, as is the case of BVD in large dairy herds after introduction of TI animals [9, 19].

Temporal sensitivity and confidence in low herd prevalence (TemSSe and PLow)

Conclusions on surveillance sensitivity and disease status at national level should be related to a specific time period, when the pathogen could have been introduced into the country. Hence, in our case, the TemSSe and PLow/PFree should be related to the period when the BVDV could have been introduced into the Danish dairy herd(s).

Between December 2011 and December 2012, no dairy herds have been found positive in Denmark. Therefore, we can assume that very few (≤ 8) or no dairy herds were infected in the country at the beginning of 2012 and became BTM positive. Currently, there is no regulated between-herds design prevalence to substantiate BVD status at country level [29]. Thus we considered $P_H = 0.2$ % as a reasonable cut-off level. Such design prevalence has been set up by the OIE to substantiate officially free status from IBR [16], which has been eradicated from Denmark [17].

If BVDV was introduced by a PI calf in at least 8 dairy herds, the median probability of detecting at least one of these herds after one year by BTM testing, would be > 95 % with both ELISAs. According to the European Food Safety Authority (EFSA) this could be considered as an acceptable level of confidence (e.g. when no value is indicated in the legislation) [30]. The confidence in low herd prevalence (PLow) would be high as well (Table 4). This means that if the aim of the surveillance system was to substantiate on annual basis that the prevalence of herds infected by at least one PI animal is <0.2 %, there is no need to replace the Danish blocking ELISA with the SVANOVIR ELISA.

On the other hand, if the objective of the surveillance system was to detect BVD by 90 days after introduction of a PI calf in 0.2 % dairy herds, then the SVANOVIR ELISA could be preferred, because only that test showed median TemSSe and PLow higher than 95 % (Table 4). PI animals are usually considered to be the major sources of BVDV spread [28], between and within cattle herds, and considering BVDV introductions by those animals could therefore be sufficient.

As we showed with the PTR values (Table 3), outbreaks due to TI cows can occur with low probabilities, and if the objective of the surveillance system is to detect BVD after introduction of a TI cow in 0.2 % dairy herds, the SVANOVIR ELISA could be used, although in that case, the median TemSSe and the PLow would be < 95 % (Table 4).

Confidence in complete freedom from BVD (PFree)

With $P_H = 0.02$ % (corresponding to 1/4109 infected herds), the TemSSe and the related PFree were <95 %, with both ELISAs (Table 4).

Hence, if we apply those findings to the BTM testing made in the fourth trimester of 2010, with the Danish blocking ELISA, it can be concluded that the probability of detection and the confidence in complete freedom from BVD, one year after one single herd was infected by a PI calf were very low (Table 4).

However, if the SVANOVIR ELISA had been used, under the same infection and HRP scenarios, the TemSSe and the related PFree would have been higher than with the Danish blocking ELISA (Table 4).

Impact of herd size, HRP and ELISA on the TemSSe and its related PLow/PFree

In this study, we found that, if one TI cow was introduced to the herd(s) and a HRP of 90 days was used, the TemSSe and the related PLow/PFree were similar or slightly higher for the blocking ELISA than for the SVANOVIR (Table 4). Under the same infection scenario, when a HRP of 365 days was used, the contrary was observed. This was due to the fact that, with an HRP of 90 days, detection occurred with both tests in the NoImpoCattle herd with one cow (where the PTR was 100 %). With the SVANOVIR, the PTR was >0 % also in the ImpoCattle herd with 24 cows and in the NoImpoCattle herd with 123 cows (Table 3). However, the Se of the SVANOVIR ELISA on BTM was assumed to be lower [27] than the Se of the Danish blocking ELISA [26].

Thus, when 1) the herd size is very small (e.g. <50 cows), 2) the threshold prevalence of the test used is low, and consequently 3) the PTR is around 100 %, the TemSSe would become more dependent on the Se of the test used on BTM, than on the PTR. Then there is less need to consider the threshold prevalence of the test used and the PTR parameter (the node "Threshold reached?" in Fig. 1) could be removed. In fact, in very small herds, even a high threshold prevalence of 50 % can be reached in a short HRP with very high probability (high PTR).

When the Danish BVD eradication program was launched in 1994, the average herd size was 42 cows [2], while currently it has increased to approximately 150 cows [9]. Hence, in the ′90s detection could occur with the Danish blocking ELISA, even testing quarterly and especially if a PI was introduced to the herd(s). In the current situation, this is more difficult, since the size of Danish dairy herds is continuously increasing [1] and the HRP (with the PTR value) have higher importance than in the past. Thus, the PTR needs to be used in the evaluation of the current surveillance system. With larger herds, tests that can detect a lower prevalence of seropositive animals, in a short HRP and with higher PTR should be preferred. This is the case of the SVANOVIR compared to the blocking ELISA. Using the former, a higher TemSSe would be achieved.

With those points in mind, it can also be noted that, once the threshold prevalence has been reached, increasing the BTM testing frequency would increase the costs related to the higher number of samples tested, but also the probability of detection (TemSSe). In contrast, using a higher BTM testing frequency, before the threshold prevalence is reached in the milking group, would be inefficient. In the latter case, changing the test (rather than increasing the testing frequency), could increase the TemSSe without increasing the costs of the surveillance system (if we assume that tests have similar commercial price from the manufacturer).

Importance of disease epidemiology and infection scenario

In this study, we showed that the TemSSe was higher for HRP of 365 days than for HRP of 90. This was due to two main reasons: a) the longer the time an infectious animal is kept in the herd, the higher the probability that such an animal causes an outbreak with seroconversion of several milking cows, and b) within the first 90 days from BVDV introduction, no new PI calves could be born in the herd from recently infected cows (TI).

The first observation is also valid for other diseases, while the second is peculiar to the epidemiology of BVD. In fact, PI calves are born from PI cows or from susceptible cows, which become infected within the first four months of pregnancy [4, 5]. In the latter case, PI calves will be born in the herd at least 5 months after introduction of the first infectious animal (one PI calf or one TI cow in our infection scenarios), because the cattle pregnancy lasts around 280 days.

When PIs are present in a herd, the immunization of other herd mates occurs quicker than when only TI animals are present [9]. In fact in simulation studies, it is usually assumed that the within group transmission rate of TIs is approximately 17 times lower than in PIs, and that only the latter are able to spread BVDV between animals groups (e.g. from calves to milking cows) [9, 31, 32]. Thus, detection of BVD infected herds by BTM testing becomes more likely (higher PTR) in the presence of PI cattle, since the number of newly infected animals per unit of time is very high.

These observations are in accordance with our results. In most of the iterations where we introduced a TI and the within herd BVDV spread was simulated, the outbreak died out, before the threshold prevalence was reached. Thus, in those scenarios, the TemSSe was low (Table 4) due to the low PTR values (Table 3).

Therefore, the approach proposed in this study, allowed us to evaluate the Danish BVD surveillance system taking into account all the main epidemiological characteristics of the disease.

Information from sensitivity analysis

In the sensitivity analysis, we confirmed that the PTR is an important parameter to consider, when the temporal sensitivity of the surveillance system is estimated.

Moreover, we showed how an increase in the probability of BVDV introduction into the Danish dairy population (PIntro) could cause a decrease in the PFree. Testing imported animals at the border, could reduce the PIntro and could increase the confidence in freedom (PFree), as previously argued for bovine tuberculosis [33].

Limitations of the study

Our estimates (TemSSe, PLow and PFree) could be considered as conservative, since we assumed that herds became infected by introduction of one BVDV positive animal only (a PI or a TI). In reality, more infected animals could be introduced to one herd at the same time leading to higher values of PTR and TemSSe.

Moreover, the PTRs were estimated for three herd sizes within each herd category. A more precise modeling for all herd sizes would have required to run the simulation at least 1185 times, to determine the PTR for each dairy herd size (from 1 to 1185 cows). Because this was not feasible, to include uncertainty, we set the PTR values as Pert distributions within each risk category (Fig. 1).

In the model used for the risk assessment [12], we did not include veterinarians visiting infected cattle herds abroad. This was based on interviews with farmers and vets, who stated that the veterinary equipment and medicines used in herds outside Denmark were not used in Danish herds [12].

Finally, in line with Foddai et al. [9] we assumed that detection by BTM testing could occur when a fixed threshold prevalence of antibody positive milking cows was reached within a herd, and that, these cows had similar milk production and antibody levels in milk. In reality, this is not always the case, and thus, we used a simplification. Further studies could investigate how the sensitivity of the ELISA used on BTM samples changes per day (after the introduction of the infectious animal(s)), according to prevalence of seroconverted milking cows, their individual milk production and antibody titer.

Conclusions

Using the SVANOVIR ELISA on BTM, would increase the temporal sensitivity and the related confidence in BVD freedom (and in low herd prevalence), compared to the current situation, where the Danish blocking ELISA is used. Those results could be considered to optimize the BVD surveillance system in Danish dairy herds and to substantiate freedom from disease in the Danish dairy population. Moreover, in this study, we showed a novel idea on how to include the effect of the high risk period within the stochastic scenario tree methodology. By using this approach surveillance systems of different animal diseases could be evaluated and optimized as early warning systems.

Abbreviations

$ARR_{ImpoCattle}$, adjusted relative risk of infection for the ImpoCattle category; ARRj, adjusted relative risk of infection for the risk category "j"; $ARR_{NoImpoCattle}$, adjusted relative risk of infection for the NoImpoCattle category; BTM, bulk tank milk; BVD, bovine viral diarrhea; BVDV, bovine viral diarrhea virus; EFSA, European food safety authority; ELISA, enzyme-linked immunosorbent assay; $EPI_{ImpoCattle}$, effective probability of infection for the ImpoCattle category; EPIj, effective probability of infection for the category "j"; $EPI_{NoImpoCattle}$, effective probability of infection for the NoImpoCattle category; HRP, high risk period; IBR, infectious bovine rhinotracheitis; ImpoCattle, dairy herds which import live cattle; NoImpoCattle, dairy herds which do not import live cattle; NPV, negative predictive value of the surveillance system; PAnim, annual risk of BVDV introduction due to import of live cattle (estimated only for the ImpoCattle category, Eq. 4); PEmb, annual risk of BVDV introduction due to import of embryos (estimated for each category); PFree, confidence in complete freedom from BVD (P_H < 0.02 % or <1/4109 infected herds); P_H, between-herds design prevalence; PI, persistently infected cattle; PIntro, probability of BVDV introduction into the Danish dairy population; PLow, confidence in low herd prevalence (P_H < 0.2 % or < 8/4109 herds); PriorPInf, prior probability that the country is infected at the assumed P_H at the beginning of the surveillance period; $PriorPinf_{Adj}$, prior probability that the country is infected at the assumed P_H at the beginning of the surveillance period adjusted for the PIntro; $PrP_{ImpoCattle}$, proportion of dairy herds which import live cattle; $PrP_{NoImpoCattle}$, proportion of dairy herds which do not import live cattle; PSem, annual risk of BVDV introduction due to import of semen (estimated for each category); PTrim, annual risk of BVDV introduction due to hoof trimmers practicing abroad (estimated for each category); $PTR_{ImpoCattle}$, probability that the threshold prevalence is reached within the ImpoCattle herds on the day of testing; $PTR_{NoImpoCattle}$, probability that the threshold prevalence is reached within the NoImpoCattle herds on the day of testing; PTruck, annual risk of BVDV introduction due to trucks used abroad (estimated for each category); P_U, within-herd threshold/design prevalence of antibody positive milking cows; $R_{ImpoCattle}$, risk of BVDV introduction for the ImpoCattle category; $R_{NoImpoCattle}$, risk of BVDV introduction for the NoImpoCattle category; $RR_{ImpoCattle}$, risk of BVDV introduction for the ImpoCattle category relative to the $R_{NoImpoCattle}$; $RR_{NoImpoCattle}$, relative risk of BVDV introduction for the NoImpoCattle category (equal to 1); Se, test sensitivity; SSe, surveillance system sensitivity; TemSSe, temporal surveillance system sensitivity; TI, transiently infected cattle

Acknowledgements

This study was supported by the Ministry of Food, Agriculture and Fisheries of Denmark (Grant number 3412-09-02603) and by the Danish Cattle Federation (SEGES). We would like to thank Dr. Peter Lind for his critical reading of the manuscript.

Funding

This work was funded by the Ministry of Food, Agriculture and Fisheries of Denmark (NaturErhvervstyrelsen, via Innovationsloven J.nr. 3412-09-02603) and by the Danish Cattle Federation (SEGES).

Authors´ contributions

AF carried out the literature review, developed the stochastic simulation model, made the calculations and drafted the manuscript. AS contributed to the statistical methodology and critically revised the manuscript. AB participated in the study design, in the development of the simulation model, and helped to draft the manuscript. All authors read and approved the final manuscript.

Competing interests

The authors declare that they have no competing interests.

Consent for publication

Not applicable.

Author details

[1]Section of Epidemiology, National Veterinary Institute, Technical University of Denmark, Bülowsvej 27, DK-1870 Frederiksberg C, Denmark. [2]Statistics and Data Analysis Section, Department of Applied Mathematics and Computer Science, Technical University of Denmark, Matematiktorvet, DK-2800 Lyngby, Denmark.

References

1. Foddai A, Enøe C, Stockmarr A, Krogh K, Uttenthal Å. Challenges for bovine viral diarrhoea virus antibody detection in bulk milk by antibody enzyme-linked immunosorbent assays due to changes in milk production levels. Acta Vet Scand. 2015. doi:10.1186/s13028-015-0125-z.
2. Bitsch V, Rønsholt L. Control of bovine viral diarrhea virus infection without vaccines. Vet Clin Noth Am Food Anim Pract. 1995;11:627–40.
3. Bitsch V, Hansen K-EL, Rønsholt L. Experiences from the Danish programme for eradication of bovine virus diarrhoea (BVD) 1994-1998 with special reference to legislation and causes of infection. Vet Microbiol. 2000;77:137–43.

4. McClurkin AW, Littledike ET, Cutlip RC, Frank GH, Coria MF, Bolin SR. Production of cattle immunotolerant to bovine viral diarrhea virus. Can J Comp Med. 1984;48:156–61.
5. Brownlie J, Clarke MC, Howard CJ, Pocock DH. Pathogenesis and epidemiology of bovine virus diarrhoea virus infection of cattle. Ann Rech Vét. 1987;18:157–66.
6. Baker JC. Clinical aspects of bovine virus diarrhoea virus infection. Rev sci tech Off int Epiz. 1990;9:25–41.
7. Hoinville LJ, Alban L, Drewe JA, Gibbens JC, Gustafson L, Häsler B, et al. Proposed terms and concepts for describing and evaluating animal-health surveillance systems. Prev Vet Med. 2013;112:1–12.
8. Graat EAM, de Jong MCM, Frankena K, Franken P. Modelling the effect of surveillance programmes on spread of bovine herpesvirus 1 between certified cattle herds. Vet Microbiol. 2001;79:193–208.
9. Foddai A, Enøe C, Krogh K, Stockmarr A, Halasa T. Stochastic simulation modeling to determine time to detect bovine viral diarrhea antibodies in bulk tank milk. Prev Vet Med. 2014;117:149–59.
10. Horst HS, Huirne RBM, Dijkhuizen AA. Risks and economic consequences of introducing classical swine fever into The Netherlands by feeding swill to swine. Rev sci tech Off int Epiz. 1997;16:207–14.
11. Thurmond MC. Special article. Conceptual foundations for infectious disease surveillance. J Vet Diagn Invest. 2003;15:501–14.
12. Foddai A, Boklund A, Stockmarr A, Krogh K, Enøe C. Quantitative assessment of the risk of introduction of bovine viral diarrhea virus in Danish dairy herds. Prev Vet Med. 2014;116:75–88.
13. Martin PAJ, Cameron AR, Greiner M. Demonstrating freedom from disease using multiple complex data sources 1: A new methodology based on scenario trees. Prev Vet Med. 2007;79:71–97.
14. Martin PAJ, Cameron AR, Barfod K, Sergeant ESG, Greiner M. Demonstrating freedom from disease using multiple complex data sources 2: Case study-Classical swine fever in Denmark. Prev Vet Med. 2007;79:98–115.
15. Martin PAJ. Current value of historical and ongoing surveillance for disease freedom: Surveillance for bovine Johne's disease in Western Australia. Prev Vet Med. 2008;84:291–309.
16. OIE. Infectious bovine rhinotracheitis/infectious pustular vulvovaginitis. Terrestrial Animal Health Code. 2010. Chapter 11.11. Article 11.11.2. http://web.oie.int/eng/normes/mcode/en_chapitre_1.11.11.pdf Accessed 30 May 2016.
17. OIE. Infectious bovine rhinotracheitis/infectious pustular vulvovaginitis. OIE Terrestrial Manual; 2010. Chapter 2.4.13, 1-17. http://www.oie.int/fileadmin/Home/eng/Health_standards/tahm/2.04.13_IBR_IPV.pdf Accessed 30 May 2016.
18. Yates WDG. A review of infectious bovine rhinotracheitis, shipping fever pneumonia and viral-bacterial synergism in respiratory disease of cattle. Can J Comp Med. 1982;46:225–63.
19. Moerman A, Straver PJ, de Jong MCM, Quak J, Baanvinger T, van Oirschot JT. A long term epidemiological study of bovine viral diarrhoea infections in a large herd of dairy cattle. Vet Rec. 1993;132:622–6.
20. Juntti N, Larsson B, Fossum C. The use of monoclonal antibodies in enzyme linked immunosorbent assays for detection of antibodies to bovine viral diarrhoea virus. J Vet Med (B). 1987;34:356–63.
21. Niskanen R, Alenius S, Larsson B, Juntti N. Evaluation of an enzyme-linked immunosorbent assay for detection of antibodies to bovine virus diarrhoea virus in milk. J Vet Med (B). 1989;36:113–8.
22. Niskanen R, Alenius S, Larsson B, Jacobsson S-O. Determination of level of antibodies to bovine virus diarrhoea virus (BVDV) in bulk tank milk as a tool in the diagnosis and prophylaxis of BVDV infections in dairy herds. Arch Virol. 1991;Suppl 3:245–51.
23. Niskanen R. Relationship between the levels of antibodies to bovine viral diarrhoea virus in bulk tank milk and the prevalence of cows exposed to the virus. Vet Rec. 1993;133:341–4.
24. Rønsholt L, Nylin B, Bitsch V. A BVDV antigen- and antibody blocking ELISA (DVIV) system used in a Danish voluntary eradication program. In: Edwards S, Paton DJ, Wensvoort G, editors. Proceedings of the third ESVV Symposium on Pestivirus Infections: 19-20 September 1996; Lelystad, The Netherlands. Weybridge: Central Veterinary Laboratory, European Society for Veterinary Virology; 1997. p. 150–3.
25. Bitsch V, Houe H, Nylin B, Rønsholt L. Examination of blood and bulk tank milk samples to monitor the bovine virus diarrhoea infection status of cattle herds. In: Edwards S, Paton DJ, Wensvoort G, editors. Proceedings of the third ESVV Symposium on Pestivirus Infections: 19-20 September 1996; Lelystad, The Netherlands. Weybridge: Central Veterinary Laboratory, European Society for Veterinary Virology; 1997. p. 158–61.
26. Houe H. Epidemiological features and economical importance of bovine virus diarrhoea virus (BVDV) infections. Vet Microbiol. 1999;64:89–107.
27. Lindberg A. Validation of an antibody ELISA used on bulk milk to screen dairy herds for bovine viral diarrhoea virus infection. In: Proceedings of the 9th International Symposium on Veterinary Epidemiology and Economics. 2000.
28. Niskanen R, Lindberg A, Larsson B, Alenius S. Lack of virus transmission from bovine viral diarrhoea virus infected calves to susceptible peers. Acta Vet Scand. 2000;41:93–9.
29. OIE.http://www.oie.int/en/animal-health-in-the-world/the-world-animal-health-information-system/old-classification-of-diseases-notifiable-to-the-oie-list-a/Accessed 8 July 2015.
30. European Food Safety Authority (EFSA). Modelling the impact of a change in meat inspection sensitivity on the surveillance of bovine tuberculosis (bTB) at the country level. Supporting Publications; 2013. EN-450, 1-40. www.efsa.europa.eu/publications Accessed 30 May 2016
31. Viet A-F, Fourichon C, Seegers H, Jacob C, Guihenneuc-Jouyaux C. A model of the spread of the bovine viral-diarrhoea virus within a dairy herd. Prev Vet Med. 2004;63:211–36.
32. Ezanno P, Fourichon C, Viet A-F, Seegers H. Sensitivity analysis to identify key-parameters in modelling the spread of bovine viral diarrhoea virus in a dairy herd. Prev Vet Med. 2007;80:49–64.
33. Foddai A, Nielsen LR, Willeberg P, Alban L. Comparison of output-based approaches used to substantiate bovine tuberculosis free status in Danish cattle herds. Prev Vet Med. 2015;121:21–9.
34. Dyreforsøgsloven, LBK nr. 474 af 15/05/2014. Bekendtgørelse af lov om dyreforsøg. In Danish. https://www.retsinformation.dk/pdfPrint.aspx?id=162938.
35. Directive 2010/63/EU, of the European Parliament and of the Council of 22 September 2010 on protection of animals used for scientific purposes. O J L 2010; 276/33-79. http://eur-lex.europa.eu/LexUriServ/LexUriServ.do?uri=OJ:L:2010:276:0033:0079:en:PDF.

15

Characterization of the resistance class 1 integrons in *Staphylococcus aureus* isolates from milk of lactating dairy cattle in Northwestern China

Longping Li[1,2,3] and Xin Zhao[1,4*]

Abstract

Background: Integrons are mobile DNA elements and they have an important role in acquisition and dissemination of antimicrobial resistance genes. However, there are limited data available on integrons of *Staphylococcus aureus* (*S. aureus*) from bovine mastitis, especially from Chinese dairy cows. To address this knowledge gap, bovine mastitis-inducing *S. aureus* isolates were investigated for the presence of integrons as well as characterization of gene cassettes. Integrons were detected using PCR reactions and then further characterized by a restriction fragment-length polymorphism analysis and amplicon sequencing.

Results: All 121 *S. aureus* isolates carried the class 1 integrase gene *intI*1, with no *intI*2 and *intI*3 genes detected. One hundred and three isolates were positive for the presence of 12 resistance genes, either alone or in combination with other gene cassettes. These resistance genes encoded resistance to trimethoprim (*dhfrV, dfrA1, dfrA12*), aminoglycosides (*aadA1, aadA5, aadA4, aadA24, aacA4, aadA2, aadB*), chloramphenicol (*cmlA6*) and quaternary ammonium compound (*qacH*) and were organized into 11 different gene cassettes arrangements (A-K). The gene cassette arrays *dfrA1-aadA1* (D, 44.6%), *aadA2* (K, 31.4%), *dfrA12-orfX2-aadA2* (G, 27.3%) and *aadA1* (A, 25.6%) were most prevalent. Furthermore, 74 isolates contained combinations of 2 to 4 gene cassette arrays. Finally, all of the integron/cassettes-positive isolates were resistant to aminoglycoside antibiotics.

Conclusions: This is the first study on the integrons and gene cassette arrays in *S. aureus* isolates from milk of mastitic cows from Northwestern China and provide the evidence for class 1 integron as possible antibiotic resistance determinants on dairy farms.

Keywords: Class 1 integron, Gene cassettes, Antibiotic resistance, Bovine milk, *Staphylococcus aureus*

Background

Staphylococcus aureus (S. aureus) induced bovine mastitis causes heavy losses for the dairy industry worldwide. In addition, acquisition and spread of antibiotic resistance among staphylococcal strains is a major concern in treatment of staphylococcal infections in humans and animals. The evolution of multidrug resistance is relatively fast due to horizontal or lateral gene transfer, which is influenced by a wide range of mobile genetic elements [1]. In particular, integrons comprise a substantial proportion of these elements and are often found in plasmids and/or transposons that enhance spread of resistance genes [2].

Integrons are mobile DNA elements that are composed of three DNA segments, including a 5′ conserved segment (5'CS), an internal variable region with one or more resistance gene cassettes of different lengths and sequences, and a 3′ conserved segment (3'CS). The 5'CS encodes a recombinase (integrase) which allows for site-specific insertion of resistance-gene cassettes between two highly conserved adjacent nucleotide sequences

* Correspondence: zhaoxin@nwsuaf.edu.cn
[1]College of Animal Science and Technology, Northwest A&F University, YangLing, Shaanxi, People's Republic of China
[4]Department of Animal Science, McGill University, 21,111 Lakeshore, Ste. Anne de Bellevue, Quebec H9X 3V9, Canada
Full list of author information is available at the end of the article

(5'CS and 3'CS) and a promoter that ensures expression of the integrated gene cassettes. The 3'CS contains a quaternary ammonium compound resistance gene *qacΔE*-1, the sulfonamide resistance gene *sul*1 and an open reading frame of an unknown function [3]. At present, at last 5 classes of mobile integrons have been distinguished by their highly conserved class-specific integrase genes [3] and gene cassettes are widely disseminated in diverse bacteria [2]. Mobile integrons have been a major driver in the spread of antibiotic resistance and ongoing use of antibiotics has swelled their numbers in recent years [2].

The role of class 1 integrons is well known in the spread of antibiotic resistance genes in gram-negative bacteria [4, 5]. A previous study has also shown that gram-positive bacteria were a major reservoir of class 1 antibiotic resistance integrons in poultry litter [6]. Furthermore, nosocomial infection caused by resistance class 1 integron-carrying *S. aureus* [7] or methicillin-resistant *S. aureus* [8, 9] were also reported in Southern China. However, investigation of integrons in bovine milk-associated *S. aureus* is still lacking. Therefore, the aims of this work were (i) to determine the class 1 integron-associated genes in *S. aureus* isolates obtained from dairy farms in Shaanxi province in Northwestern China and (ii) to amplify and characterize their gene cassette contents and antibiotic resistance gene cassette arrays by sequencing of *S. aureus* isolates.

Methods

Bacterial strains and susceptibility testing

Milk samples were collected from 4 dairy farms in three cities (BaoJi, XiAn and XianYang) in the Shaanxi Province. These 4 farms had 24 to 68 lactating cows and samples were collected between September 2011 and October 2012. Milk samples were taken from cows with clinical mastitis. Cows were selected by veterinarians based on clinical mastitis which was manifested with decreased milk production, color change of the milk and inflammation of udders. Milk samples were collected after cleaning the teats, discarding a few streams of milk and scrubbing the teat ends with cotton balls moistened with 70% alcohol. In all, 121 *S. aureus* isolates were isolated from 214 individual quarter milk samples which were obtained from 161 cows. Conventional antimicrobials including penicillin, kanamycin, gentamicin and erythromycin were used alone or combination for the treatment of clinical mastitis on these farms.

One hundred and twenty one *S. aureus* isolates from bovine milk were identified by biochemical methods, including a test for catalase, a tube coagulase test and anaerobic fermentation of mannitol. The identification was also confirmed by PCR amplification and sequencing of species-specific region of the gene encoding 16S rRNA and by amplification of thermonuclease gene (*nuc*) [10].

Antibiotic susceptibility testing was performed by the standard disc diffusion method on Mueller-Hinton agar (Oxoid, UK) plates according to Clinical and Laboratory Standards Institute (CLSI) guidelines [11, 12]. The antimicrobial agents (Oxoid, England) tested were (the disk concentration of the antibiotics indicated in parentheses): penicillin G (10 U), ampicillin (10 μg), gentamicin (10 μg), kanamycin (30 μg), tetracycline (30 μg), erythromycin (15 μg), clindamycin (2 μg) and chloramphenicol (30 μg) as in our previous studies [10, 13]. In addition, the susceptibility of antibiotics trimethoprim/sulfamethoxazole (SXT, 1.25/23.75, 25 μg) was conducted in the present study. *S. aureus* ATCC 25923 was used as a quality control. Results of susceptibility testing were recorded according to the CLSI [11, 12] guidelines following incubation at 35 °C for 18 h.

Detection of three classes of integrases, gene cassettes and 3' conserved region

Plasmid and chromosome DNA of bacterial isolates were extracted using a commercial DNAout kit (Tiandz Inc., Beijing, China) as described previously [13]. The presence of class 1, 2, and 3 integrons was analyzed in all isolates in the present study. The primers were used to detect the *intI*1 [6], *intI*2 [6] and *intI*3 [14] genes as in previous studies using a touchdown program: 94 °C for 5 min, ten touchdown cycles of 94 °C for 30 s, 68 °C for 30 s (decreasing by 1 °C /cycle) and 72 °C for 1 min, followed by 30 cycles of 30 s at 94 °C, 30 s at 58 °C and 1 min at 72 °C, with a final extension for 10 min at 72 °C (C1000™ Thermal Cycler, BIO-RAD, Beijing, China). The PCR products of the class 1 integrase gene (*intI*1) were randomly selected for sequencing to ensure specificity and accuracy. Amplicons of *intI*1 were purified with a PCR purification kit (TransGen Biotech, Beijing, China), ligated into the *pEASY*®-T5 Zero Cloning Vector (TransGen Biotech, Beijing, China), and then sequenced at Nanjing GenScript Ltd. (Nanjing, China). Sequence comparisons were performed using the BLAST program (http://www.ncbi.nlm.nih.gov/BLAST/). In addition, 2 × *TransTaq*®-T PCR SuperMix (+dye) (TransGen Biotech, Beijing, China) was employed to amplify the variable region (gene cassettes) between the *intI*1 gene and the 3'CS of class 1 integron using primers sets 5'CS and 3'CS as described previously [14] and following a three-step program [15] with minor modifications: 1 min of denaturation at 94 °C, 1 min of annealing at 52 °C and 2 min of extension at 72 °C for a total of 35 cycles; 5 s were added to the extension time at each cycle. Furthermore, two genes strongly associated with Class 1 integrons, the non-mobile cassettes conferring resistance to sulfadiazine, *sul*1, and to quaternary ammonium compounds, *qacΔE*-1, were also detected [6].

Restriction fragment length polymorphism (RFLP) and DNA sequencing analysis of cassette regions

To determine whether each cassette gene PCR amplicon with the same size had the same content, cassette PCR products were digested with *Alu*I or *Rsa*I (New England Biolabs, Beijing, China). The PCR products with the same RFLP pattern were considered to contain same gene cassettes. To analyze the sequences of the gene cassettes, 1 or 2 cassette amplicons representative of each restriction profile were excised, purified and ligated into *pEASY*®-T5 Zero Cloning Vector (TransGen Biotech, Beijing, China). The ligation mixture was transformed into *Trans*-T1 Phage Resistant Chemically Competent Cell (TransGen Biotech, Beijing, China) and recombinants were selected on Luria agar containing ampicillin (100 mg/L). Recombinant plasmid DNA was purified by standard methods and subjected to sequencing at Nanjing GenScript Ltd. (Nanjing, China). The nucleotide sequences were analyzed and compared with available sequences in GenBank using the BLAST program by the National Center for Biotechnology Information (http://www.ncbi.nlm.nih.gov/blast/).

Results

Antimicrobial resistance profiles

High levels of resistance to antibiotics had been found in the bovine milk-associated *S. aureus* isolates, including to penicillin G (97/121, 80.2%), ampicillin (97/121, 80.2%), kanamycin (83/121, 68.6%), gentamicin (82/121, 67.8%), tetracycline (52/121, 43%), erythromycin (40/121, 33.1%; 38 isolates exhibited the inducible macrolide-lincosamide-streptogramin resistance phenotype and 2 isolates expressed the constitutive macrolide-lincosamide-streptogramin resistance phenotype and chloramphenicol (36/121, 29.8%) in our previous reports using the same isolates [10, 13]. Furthermore, 32.2% (39/121) of the isolates were resistant to SXT in the present study (Table 1). In addition, 89.3% (108/121) of the isolates were resistant to at least one antimicrobial, 47.1% (57/121) to three or more different classes of antimicrobials.

Detection of class 1, 2 and 3 integrons

One hundred and twenty-one bovine milk associated *S. aureus* isolates were assayed by PCR amplification for the three classes of integrase genes. All isolates yielded a

Table 1 Antimicrobial resistance phenotypes[a] and genotypes

Antimicrobial subclass	Antimicrobial agent	Phenotypes		Genotypes (No. of isolates contained different resistance gene cassettes or type)				
		No. of resistant isolates (%)	No. of sensitive isolates	*aadA1, aadA2, aadA4, aadA5, aadA24, aacA4 or aadB*	*dhfrV, dfrA1 or dfrA12*	*qacH*	*cmlA6*	type H integron only
Aminoglycosides	gentamicin	82(67.8)	–	61	/	/	/	21
		–	39	24	/	/	/	6
	kanamycin	83(68.6)	–	62	/	/	/	21
		–	38	25	/	/	/	6
Folate pathway inhibitors	trimethoprim/ sulfamethoxazole (SXT)	39(32.2)	–	/	28	/	/	5
		–	82	/	46	/	/	20
Phenicols	chloramphenicol	36(29.8)	–	/	/	/	5	4
		–	85	/	/	/	13	23
Pencillins	Penicillin G	97(80.2)	/	/	/	/	/	25
		–	24	/	/	/	/	3
	ampicillin	97(80.2)	/	/	/	/	/	25
		–	24	/	/	/	/	3
Macrolides	erythromycin	40(33.1)	/	/	/	/	/	11
		–	81	/	/	/	/	16
Lincosamides	clindamycin	2(1.7)	/	/	/	/	/	1
		–	119	/	/	/	/	26
Tetracyclines	tetracyclin	52(43)	/	/	/	/	/	11
		–	69	/	/	/	/	12

[a]Part of phenotypes has been published in our previous studies [10, 13]

280 bp PCR products, none of the isolates yielded 233 and 717 bp products. These suggested that all tested isolates had the class 1 integrase gene (*intI*1), with no class 2 and 3 integrase genes detected. Furthermore, all isolates were positive for the presence of *qacΔE-1*1 and *sul*1 genes, which are also the markers for class 1 integrons.

Characterization of antibiotic resistance gene cassettes in class 1 integrons

Antibiotic resistance genes were detected using a CS-based PCR assay in which the primers flanked the variable region of the integron. Among 121 *intI*1 gene positive *S. aureus*, 85.1% (103/121) of isolates carried different gene cassettes.

Sequence analyses of PCR products representing each of the PCR-RFLP profiles identified a single gene cassette in the amplicons with sizes of 781 bp (A), 1048 bp (B), 874 bp (I) and 1009 bp (K). Two gene cassettes present in tandem were detected in the 1360 bp (C), 1586 bp (D), 1608 bp (E) and 1648 bp (F) amplicons, and three gene cassettes were present in the 1913 bp (G) and 3149 bp (J) amplicons. The 153 bp (H) amplicon was found to be a class 1 integron variable region without any gene cassette but contained partial sequences of 5′ and 3′ conserved segments of integrons. The resistance gene cassettes of those isolates are shown in Table 2, while the schematic representation of the inserted gene cassettes is shown in Fig. 1. In the present study, we found 2 open reading frames with unknown functions (*orfX1* and *orfX2*) and one 153 bp amplicon (H). The remaining genes encoded resistance to trimethoprim (*dhfrV*, *dfrA1* and *dfrA12*), aminoglycosides (*aadA1*, *aadA5*, *aadA4*, *aadA24*, *aacA4*, *aadA2* and *aadB*), chloramphenicol (*cmlA6*) and quaternary ammonium compound (*qacH*) (Table 2). Furthermore, 11 distinct kinds of gene cassette arrays (A-K) were identified (Table 2, Fig. 1). Of them, the gene cassette arrays *dfrA1-aadA1* (D, 44.6%), *aadA2* (K, 31.4%), *dfrA12*-hypothetical protein-*aadA2* (G, 27.3%) and *aadA1* (A, 25.6%) were found to be most prevalent. In addition, among 103 gene cassettes-positive isolates, 74 isolates contained 2 to 4 gene cassettes combinations, 26 isolates with H only and 3 isolates with G only (Table 3). The most frequent combinations were D + K + I + H (22 isolates) and J + G + D (18 isolates). To our knowledge, those gene cassette arrays have not been reported in bovine milk-associated *S. aureus* isolates previously, although all of these gene cassette arrays exist in other types of bacteria.

The association between integron/cassettes and antimicrobial resistance

All of the integron/cassettes-positive isolates were resistant to aminoglycoside antibiotics. However, no gene cassette conferring resistance to erythromycin, clindamycin, tetracycline, penicillin and ampicillin was identified in the present study, thus the gene cassette arrays observed in our study did not account for those antimicrobial resistance phenotypes.

Table 2 Different gene cassette amplicon and numbers of isolates containing the gene cassette among 121 bovine milk associated *S. aureus*

Cassette types	Amplicon size (bp)	Resistance gene (s)	No. of isolates (%)
A	781	*aadA1*	31 (25.6)
B	1048	*aadA5*	3 (2.5)
C	1360	*dhfrV*	4 (3.3)
		aadA4	
D	1586	*dfrA1*	54 (44.6)
		aadA1	
E	1608	hypothetical protein (*orfX1*)	3 (2.5)
		aadA24	
F	1648	*aacA4*	4 (3.3)
		aadA1	
G	1913	*dfrA12*	33 (27.3)
		hypothetical protein (*orfX2*)	
		aadA2	
H	153	–	71 (58.7)
I	874	*qacH*	2 (1.65)
J	3149	*aadB*	18 (14.9)
		aadA1	
		cmlA6	
K	1009	*aadA2*	38 (31.4)

Among 39 SXT resistant and 36 chloramphenicol resistant *S. aureus* isolates, 28 and 5 of them contained gene cassettes *dhfrV*, *dfrA1* or *dfrA12* and *cmlA6*, 6 and 7 isolates were negative for any gene cassettes, 5 and 4 isolates were positive for the type H integron (153 bp), respectively. On the other hand, among 82 SXT susceptible and 85 chloramphenicol susceptible *S. aureus* isolates, 45 and 13 *S. aureus* isolates contained gene cassettes *dhfrV*, *dfrA1* or *dfrA12* and *cmlA6*, respectively. The rest of susceptible isolates harbored other cassettes (Table 1).

Discussion

To the best of our knowledge, this is the first report of the presence of a class 1 integron and the resistance gene cassettes in bovine milk associated *S. aureus* isolates in China. The emergence of different resistance gene cassettes/arrays may reflect antibiotic selective pressure in our region, and provide useful antimicrobial

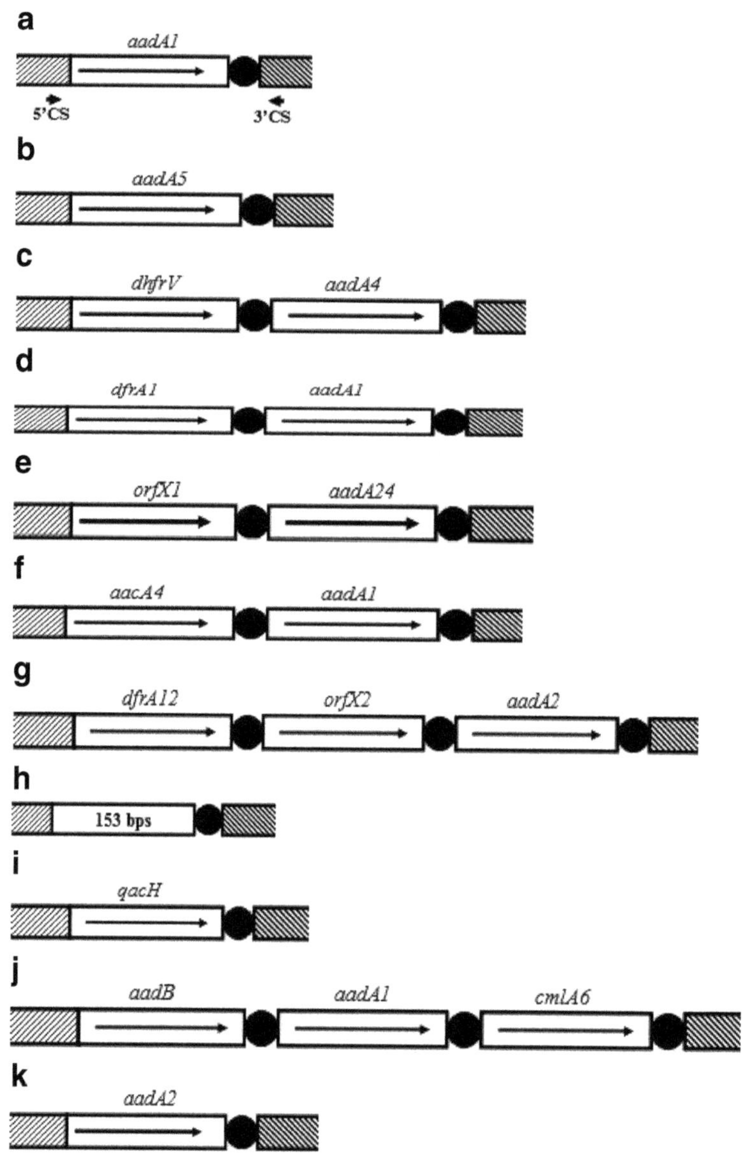

Fig. 1 Schematic representation of the variable regions identified in bovine milk associated isolates. Cassette types (**a**-**k**) are arranged as identified in Table 2. Gene cassettes are shown as boxes, with arrows for the orientation of transcription and black circles for the 59-base elements. Hatched rectangles designate 5'- and 3'- conserved segments (CSs). The 5'CS and 3'CS amplification primers are indicated by arrowheads

surveillance information for the rational and effective use of antimicrobials. Different types of class 1 integrons were characterized which carried a variety of antibiotic resistance gene cassettes/arrays. All bovine milk-associated *S. aureus* isolates were carried by class 1 integrons. However, previous studies have documented a lower proportion of class 1 integrons among *S. aureus* isolates from hospital or farm animals, such as 53% (16/30) of clinical and environmental hospital-associated *S. aureus* isolates [7], 42.5% (76/179) of nosocomial methicillin-resistant *S. aureus* strains in China [8]. The differences between our *S. aureus* isolates and others' reported ones may reflect the difference in selection pressure due to antibiotics use in our studied region. The class 1 integrons play a major role in the acquisition, expression, and dissemination of antibiotic resistance genes and may potentially have the capability to integrate nearly all the antimicrobial resistant genes such as β-lactamases, aminoglycosides, chloramphenicol, sulfonamides, macrolides, trimethoprim, fosfomycin, quinolone and quaternary ammonium compound [16] to disseminate multidrug resistance in clinical bacteria isolates. In our study, we found that aminoglycosides resistance determinants (*aadA1, aadA5, aadA4, aadA24, aacA4, aadA2* and *aadB*), trimethoprim resistance genes (*dhfrV, dfrA1* and *dfrA12*) and chloramphenicol

Table 3 Different combination of gene cassette arrays among 121 bovine milk associated *S. aureus* isolates

Gene cassette arrays	No. of isolates (%)
H	26 (21.5%)
D+ K + A + H	22 (18.2%)
J + G + D	18 (14.9%)
G + H	8 (6.6%)
D + K + H	8 (6.6%)
D + A	5 (4.1%)
F + C + B + A	3 (2.5%)
G	3 (2.5%)
E + K + H	2 (1.7%)
G + K + H	2 (1.7%)
G + K + I	2 (1.7%)
D + H	1 (0.83%)
F + C + H	1 (0.83%)
K + A	1 (0.83%)
G + E + H	1 (0.83%)

resistance determinants (*cmlA6*) were prevalent among pathogenic *S. aureus* isolated from milk of mastitic cows. Those antimicrobial-resistance genes organized into 11 different gene cassettes arrangements (Fig. 1) and 74 isolates contained different combinations of gene cassette arrays, such as 22 isolates with the combination D+ K + A + H and 18 isolates with the combination J + G + D (Table 3). This may be due to the fact that aminoglycosides, chloramphenicol and trimethoprim were often widely used for the prevention or treatment of bovine mastitis in the past years in our region.

A BLAST search for our resistance gene cassettes in GenBank revealed high similarities were found with the published sequence in *Salmonella* or *Escherichia coli* isolates from different geographical locations or sources. This data suggests that class 1 integron/cassettes may contribute significantly to the horizontal transfer of antimicrobial resistance genes among different bacterial species from different sources or geographical locations [16]. However, the gene cassettes observed in class 1 integrons examined in our study did not account for the total resistance phenotypes observed (erythromycin, tetracyclines, clindamycin and pencillins antibiotics) among the pathogenic *S. aureus* isolates (Table 1). This suggests bovine milk associated *S. aureus* resistance to erythromycin, tetracyclines and pencillins may be not associated with integron-mediated antimicrobial resistance. And this could possibly reflect the presence of other mobile DNA elements, such as plasmid and transposon, or other resistance genes that we could not detected in this study.

The occurrence of resistance gene cassettes associated aminoglycosides resistance genes including *aadA1*, *aadA5*, *aadA4*, *aadA24*, *aacA4*, *aadA2* and *aadB* were relatively high among the *S. aureus* isolates. Similarly, a large proportion was reported in the case of clinical isolates of *Escherichia coli* in China hospital [14] and *Salmonella enterica* isolates from animal and human in Iran [4]. Furthermore, extensive existence of a variety of *aadA* alleles among *S. aureus* isolates indicates frequent recombination events in integrons. Similar allelic diversity has also been reported in *Escherichia coli* [17]. Therefore, class 1 integron/cassettes mediated capture of gene cassettes conferring resistance to aminoglycosides agents may play an important role among the antibiotic-resistant *S. aureus* isolates obtained from bovine milk. In addition, aminoglycoside antibiotic (neomycin, kanamycin or gentamicin) alone or combined with an antibiotic from another family (pencillins, ampicillin, macrolides or tetracyclines) has been widely used in the treatment of *S. aureus* intramammary infection on the local dairy farms. The class 1 integron exhibits a number of characteristics that might have adopted to antimicrobial agents by obtaining diverse gene cassettes with resistant genes. Human activities promote the diversification and dissemination of class 1 integrons [18].

In the present study, 45 and 13 isolates had the *dhfrV*, *dfrA1* or *dfrA12* and *cmlA6* gene cassettes while being phenotypically susceptible to SXT and chloramphenicol, respectively. The reasons for this phenomenon may be due to the presence of the weak promoter (strong promoter Pc variant) or repressors led to gene cassette expression transcriptional inactivation [19, 20]. The resistance gene cassette expression also can be affected by differences in the length and the sequence of *attC* sites, since the *attC* sites are imperfect inverted repeats that may form stem-loop structures similar to ρ-independent transcription terminators once they have been transcribed [19, 21]. From the clinical standpoint, those susceptible *S. aureus* isolates may be the genetic source for the evolution of resistance to clinically relevant antibiotics through mutation of regulatory sequences or integrase-mediated gene cassette rearrangements [19].

Conclusions

In summary, all 121 tested bovine milk-associated *S. aureus* isolates carried class 1 integrons and 103 isolates contained different gene cassettes and we have presented evidence for the first time for class 1 integron as possible antibiotic resistance determinants on Chinese dairy farms.

Abbreviations

3′CS: 3′ conserved segment; 5′CS: 5′ conserved segment; *S. aureus*: *Staphylococcus aureus*; SXT: Trimethoprim/sulfamethoxazole

Acknowledgments

LPL and XZ thank the farmers for providing milk samples.

Funding
This study was financially supported by a grant from the National Science Foundation of China (31372282) and the China Thousand Talents Program. This work was also supported By Young Talent fund of University Association for Science and Technology in Shaanxi, China (20160236). The funders had no role in study design, data collection and analysis, decision to publish, or preparation of the manuscript.

Authors' contributions
LPL and XZ conceived and designed the experiments; LPL conducted the laboratory experiments; LPL and XZ performed the data analysis; LPL and XZ wrote the manuscript. Both authors read and approved the final manuscript.

Consent for publication
Not applicable.

Competing interests
The authors declare that they have no competing interests.

Author details
[1]College of Animal Science and Technology, Northwest A&F University, YangLing, Shaanxi, People's Republic of China. [2]Life Science Research Center, Yulin University, Yulin 719000, People's Republic of China. [3]Shaanxi Provincial Engineering and Technology Research Center of Cashmere Goat, Yulin University, Yulin 719000, People's Republic of China. [4]Department of Animal Science, McGill University, 21,111 Lakeshore, Ste. Anne de Bellevue, Quebec H9X 3V9, Canada.

References

1. Malachowa N, DeLeo FR. Mobile genetic elements of *Staphylococcus aureus*. Cell Mol Life Sci. 2010;67(18):3057–71.
2. Gillings MR. Integrons: past, present, and future. Microbiol Mol Biol Rev. 2014;78(2):257–77.
3. Mazel D. Integrons: agents of bacterial evolution. Nat Rev Microbiol. 2006; 4(8):608–20.
4. Firoozeh F, Zahraei-Salehi T, Shahcheraghi F. Molecular clonality and detection of class 1 integron in multidrug-resistant *Salmonella enterica* isolates from animal and human in Iran. Microb Drug Resist. 2014;20(6):517–24.
5. Wu K, Wang F, Sun J, Wang Q, Chen Q, Yu S, Rui Y. Class 1 integron gene cassettes in multidrug-resistant gram-negative bacteria in southern China. Int J Antimicrob Agents. 2012;40(3):264–7.
6. Nandi S, Maurer JJ, Hofacre C, Summers AO. Gram-positive bacteria are a major reservoir of class 1 antibiotic resistance integrons in poultry litter. Proc Natl Acad Sci. 2004;101(18):7118–22.
7. Xu Z, Shi L, Zhang C, Zhang L, Li X, Cao Y, Li L, Yamasaki S. Nosocomial infection caused by class 1 integron-carrying *Staphylococcus aureus* in a hospital in South China. Clin Microbiol Infec. 2007;13(10):980–4.
8. Xu Z, Li L, Shirtliff ME, Peters BM, Li B, Peng Y, Alam MJ, Yamasaki S, Shi L. Resistance class 1 integron in clinical methicillin-resistant *Staphylococcus aureus* strains in southern China, 2001-2006. Clin Microbiol Infec. 2011;17(5):714–8.
9. Xu Z, Li L, Alam MJ, Zhang L, Yamasaki S, Shi L. First confirmation of integron-bearing methicillin-resistant *Staphylococcus aureus*. Curr Microbiol. 2008;57(3):264–8.
10. Li L, Zhou L, Wang L, Xue H, Zhao X. Characterization of methicillin-resistant and -susceptible staphylococcal isolates from bovine milk in northwestern china. PLoS One. 2015;10(3):e0116699.
11. Clinical and Laboratory Standards Institute. Performance standards for antimicrobial disk and dilution susceptibility tests for bacteria isolated from animals: informational supplement. Wayne: CLSI document; 2004. p. M31–S1.
12. Clinical and Laboratory Standards Institute. Performance standards for antimicrobial susceptibility testing; twenty-first informational supplement. Wayne: CLSI document; 2011. p. M100–S21.
13. Li L, Feng W, Zhang Z, Xue H, Zhao X. Macrolide-lincosamide-streptogramin resistance phenotypes and genotypes of coagulase-positive *Staphylococcus aureus* and coagulase-negative staphylococcal isolates from bovine mastitis. BMC Vet Res. 2015;11:168.
14. Su J, Shi L, Yang L, Xiao Z, Li X, Yamasaki S. Analysis of integrons in clinical isolates of *Escherichia coli* in China during the last six years. FEMS Microbiol Lett. 2006;254(1):75–80.
15. Levesque C, Piche L, Larose C, Roy PH. PCR mapping of integrons reveals several novel combinations of resistance genes. Antimicrob Agents Chemother. 1995;39(1):185–91.
16. Partridge SR, Tsafnat G, Coiera E, Iredell JR. Gene cassettes and cassette arrays in mobile resistance integrons. FEMS Microbiol Rev. 2009;33(4):757–84.
17. Blahna MT, Zalewski CA, Reuer J, Kahlmeter G, Foxman B, Marrs CF. The role of horizontal gene transfer in the spread of trimethoprim-sulfamethoxazole resistance among uropathogenic *Escherichia coli* in Europe and Canada. J Antimicrob Chemother. 2006;57(4):666–72.
18. Gillings MR. Class 1 integrons as invasive species. Curr Opin Microbiol. 2017;38:10–5.
19. Rowe-Magnus DA, Mazel D. The role of integrons in antibiotic resistance gene capture. Int J Med Microbiol. 2002;292(2):115–25.
20. Rowe-Magnus DA, Mazel D. Resistance gene capture. Curr Opin Microbiol. 1999;2(5):483–8.
21. Martinez-Freijo P, Fluit AC, Schmitz FJ, Grek VS, Verhoef J, Jones ME. Class I integrons in gram-negative isolates from different European hospitals and association with decreased susceptibility to multiple antibiotic compounds. J Antimicrob Chemother. 1998;42(6):689–96.

Rumen-derived lipopolysaccharide enhances the expression of lingual antimicrobial peptide in mammary glands of dairy cows fed a high-concentrate diet

Di Jin, Guangjun Chang, Kai Zhang, Junfei Guo, Tianle Xu and Xiangzhen Shen*

Abstract

Background: Long-term high-concentrate diet (HCD) feeding can cause subacute ruminal acidosis in cows and subsequently trigger systemic inflammatory and immune responses. Therefore, we conducted the present study in which twelve lactating cows installed with ruminal fistula were randomly assigned to the HCD group (forage: concentrate = 4:6, $n = 6$) or the low-concentrate diet (LCD) group (forage:concentrate = 6:4, $n = 6$) and were fed for 20 weeks. Ruminal fluid, plasma and mammary gland tissue samples were collected at week 20 for analysing lipopolysaccharide (LPS), pro-inflammatory cytokines, and immune relevant gene expression. The aim of this study was to investigate the effect of rumen-derived LPS on lingual antimicrobial peptide (LAP) synthesis and immune responses in mammary glands of lactating cows fed a HCD.

Results: Compared with the LCD group, the ruminal pH was lower in the HCD group, while LPS concentrations in the rumen, lacteal artery and vein were higher. The expression of LAP, BNBD5, IL-1β, IL-6, IL-8, and TNF–α was enhanced in the HCD group. LAP protein expression was higher in the HCD group than that in the LCD group. The expression of nuclear factor kappa B (NF-κB) did not change, but was activated, as the amounts of phosphorylated NF-kB and phosphorylated inhibitory kBα increased in the HCD group compared with that in the LCD group.

Conclusions: After long-term HCD feeding, rumen-derived LPS translocated to the blood stream, triggered inflammatory and immune responses and enhanced LAP synthesis via the NF-kB signalling pathway in mammary glands of lactating cows.

Keywords: High-concentrate diet, Lipopolysaccharide, Lingual antimicrobial peptide, Mammary gland, Cow

Background

Long-term consumption of a high-concentrate diet (HCD), which occurs to meet the energy requirements for high milk yields, is hazardous to the health of ruminants. The consumption of a HCD causes the accumulation of organic acid and a decrease in pH values in the rumen [1]. When the ruminal pH decreases below 5.6 for more than 3 h per day, subacute ruminal acidosis (SARA) occurs [2], and free LPS is released from Gram-negative bacteria and is translocated into the circulatory system across the epithelial barrier of the gastrointestinal tract [3, 4]. Subsequently, LPS initiates systemic inflammatory responses [3, 5, 6] and host innate immune responses [7] because the circulating LPS stimulates the release of pro-inflammatory cytokines such as interleukin (IL)-1β, IL-6, and tumour necrosis factor-α (TNF-α) by monocytes, macrophages and vascular cells [8]. The release of LPS and cytokines results in enhanced secretion of several antimicrobial peptides, such as lingual antimicrobial peptide (LAP), in mammary epithelial cells (MECs) [9].

LAP, which was first found in inflamed bovine tongue epithelium [10], is a type of β-defensin that often exists in epithelial cells when inflammation occurs. Defensins contribute to innate immunity and are effective against a

* Correspondence: xzshen@njau.edu.cn
College of Veterinary Medicine, Nanjing Agricultural University, Nanjing, People's Republic of China

broad spectrum of microorganisms [11]. Other previously identified bovine β-defensins include tracheal antimicrobial peptide, bovine neutrophil β-defensin (BNBD), enteric β-defensin, and many other types of bovine β-defensins [12–15]. The localization of LAP in MECs has been reported [9]; its expression is constitutive [16], such as in healthy [17], involuted [18], *E.coli*-challenged [19] or *Streptococcus uberis*-challenged [20] mammary gland tissues. LAP is also secreted from the alveolar epithelium into milk in healthy [21] and LPS-challenged [22] udders. One study demonstrated that the level and kinetics of the mRNA concentrations of LAP in *E. coli*-challenged MECs were associated with induction time but were independent of the pathogen dose [23]. Furthermore, Liu et al. [24] demonstrated that the p65 subunit of nuclear factor-kB (NF-kB) plays a stimulatory role while C/EBPβ have a repressive function in LAP expression, and mutations of the binding sites for these two transcription factors abrogated LAP expression in MECs. Some reports also found widespread LAP mRNA expression in infected bovine digestive tract mucosal and respiratory tissues [25, 26]. These results strongly indicated that LAP plays a crucial role in the innate immune response.

The effects of a HCD on gastrointestinal tract or mammary gland health have received increasing attention [27, 28]. The relationship between LAP synthesis and immune responses in mammary glands of dairy cows intramammary stimulated by LPS or pathogens has been described previously. However, the effect of rumen-derived LPS caused by a HCD on LAP synthesis and immune responses in mammary glands of cows remains unknown. Furthermore, the mechanism of this effect remains to be clarified. Therefore, the objective of this study was to investigate the effects of rumen-derived LPS on LAP synthesis and immune responses in mammary glands of cows fed a HCD.

Methods

Animals, diet and experimental design

Twelve multiparous mid-lactating Holstein cows (Purchased from the Experimental Farm of Nanjing Agricultural University, average body weight, 455 ± 28 kg) fitted with ruminal fistula were randomly assigned into two groups: the high-concentrate diet (HCD) group (forage:-concentrate = 4:6, $n = 6$) as the treatment group and the low concentrate diet (LCD) group (forage:concentrate = 6:4, $n = 6$) as the control group (fed for 20 weeks). Dietary ingredients and nutrient compositions, which were published previously [29], are presented in Table 1. The experimental cows were housed in individual tie-stalls at the Dairy Farm of Nanjing Agricultural University (Nanjing, China). Cows were fed three times daily at 04:00, 12:00, and 20:00 h and were allowed free access to fresh water throughout the experiment. All of the cows received LCD for 4 weeks (as an adaption period) before the formal experiment to ensure the similarity of their metabolic status.

Table 1 Ingredients and nutrient composition of the experimental diets

Ingredients, % of DM	Percentage (%)of ingredients (dry matter)	
	LCD[1]	HCD[1]
Corn silage	30.00	20.00
Alfalfa hay	30.00	20.00
Maize	22.78	33.6
Wheat bran	5.15	15
Soybean meal	9.81	9
Calcium phosphate dibasic	0.92	0.53
Powder	0	0.52
Salt	0.35	0.35
Premix[2]	1	1
Total	100	100
Forage:concentrate	6:4	4:6
Nutritional composition[3]		
Net energy, MJ/kg	6.32	6.74
CP, %	16	16.2
EE, %	3.96	4.15
NDF, %	37.71	31.92
ADF, %	22.75	17.55
NFC, %	32.43	40.31
Ca, %	0.9	0.8
P, %	0.45	0.45

[1] LCD, low-concentrate diet; HCD, high-concentrate diet
[2] The premix contained VA,1,900kU/kg; VD, 250kU/kg; VE, 3000 mg/kg; niacin, 4000 mg/kg; Cu, 1200 mg/kg; Fe, 525 mg/kg; Zn, 13,000 mg/kg; Mn, 5500 mg/kg; I, 170 mg/kg; Co, 50 mg/kg; Se, 27 mg/kg
[3] The calculated nutritional composition values

Sample collection

Cows were milked at 05:00, 13:00, and 21:00 h, and milk yield was recorded daily. Milk samples were collected to measure the somatic cell count (SCC) (MilkoScan FT1, FOSS, Denmark) once a week. Ruminal fluid samples were collected via the ruminal fistula for 3 consecutive days during the 18th week, at 2-h intervals starting at 04:00 h (after the morning feeding) for 12 h. The collected ruminal fluid samples were filtered through 2 layers of gauze. Rumen fluid samples were centrifuged at 10,000 × g for 45 min, and then the supernatant was aspirated and passaged through a disposable 0.22-μm filter. The filtrate was collected in a sterile, depyrogenated glass tube (previously heated at 250 °C for 2 h) and heated at 100 °C for 30 min. Samples were cooled at room temperature for 15 min and stored at—20 °C for subsequent LPS measurements. Blood samples were collected at 4 h after feeding on the sampling day of the 18th week. The samples were obtained via the lacteal artery and vein with 5-mL vacuum tubes containing sodium heparin. Samples were kept on ice until transported to the laboratory. Plasma was isolated from blood samples by centrifugation at 1469× g at 4 °C for 15 min and then stored at −20 °C for LPS, enzyme, and cytokine analyses. Mammary gland tissue samples were obtained by biopsy on the sampling day from the same quarter of the mammary gland [19]. Samples were rinsed with 0.9 % saline, fixed in 4 % (w/v) paraformaldehyde in PBS for immunohistochemistry, or snap-frozen in liquid nitrogen and then stored at −70 °C for RNA and protein extraction. Incisions were sutured, and antibiotics were administered intramuscularly to avoid infection.

Ruminal pH, ruminal and plasma LPS

The ruminal pH values were measured with a pH metre immediately after collecting ruminal fluid. The concentration of LPS in ruminal fluid (CE64406) and plasma (CE80545) samples was determined by a chromogenic endpoint assay (Chinese Horseshoe Crab Reagent Manufactory Co., Ltd., Xiamen, China) with a minimum detection limit of 0.05 EU/mL (ruminal liquid) or 0.01 EU/mL (plasma) according to the manufacturer's instructions.

Pro-inflammatory enzyme activity and cytokine concentrations

The analyses of the myeloperoxidase (MPO) and β-N-acetyl glucosaminidase (NAG) concentrations were performed using commercial kits (Jiancheng Bioengineering Institute of Nanjing, Nanjing, China) that used an enzymatic colorimetric detection method read by a microplate reader (Epoch, BioTek, USA) [29]. Plasma IL-1β, IL-6, and TNF-α concentrations were determined using commercial radioimmunoassay kits (Beijing North Institute of Biological Technology, Beijing, China) and a gamma radioimmunoassay counter (SN-6105, Hesuo Rihuan Photoelectric Instrument Co., Ltd., Shanghai, China). All of the procedures were performed according to the manufacturer's instructions.

Reverse transcription and quantitative PCR

Total RNA was extracted from 100 mg of mammary gland tissue samples using RNAiso Plusreagent (TaKaRa Co., Otsu, Japan) according to the manufacturer's protocol. The quantity and quality of the total RNA were assessed by a spectrophotometer (Eppendorf Biotechnology, Hamburg, Germany). The first-strand cDNA was synthesized using 250 ng of the total RNA template and a PrimeScript RT Master Mix Perfect Real Time Kit (TaKaRa Co., Otsu, Japan). The primers for the target genes were designed based on known cattle sequences using Primer Premier Software 5.0 (Premier Biosoft International, USA);these primers are shown in Table 2. The qPCR was performed on an ABI 7300 system (Applied Biosystems, Foster City, CA, USA) using a SYBR Premix Ex TaqKit (TaKaRa Co., Otsu, Japan) according to the manufacturer's instructions. The $2^{-\Delta\Delta Ct}$ method was used to analyse the qPCR data, and glyceraldehyde-3-phosphate dehydrogenase (GAPDH) was used as an internal reference.

Immunohistochemical analysis of LAP

Mammary gland tissue samples for immunohistochemical analysis were fixed in 4 % (w/v) paraformaldehyde in PBS, dehydrated and processed for paraffin sectioning (4 μm thick). Oven-dried sections were deparaffinised, followed by antigen retrieval by microwaving in EDTA buffer, and treated with 3 % H_2O_2 for 10 min to block the endogenous peroxidase. The sections were washed in PBS after each step. Then, the air-dried sections were blocked within 5 % BSA for 20 min followed by incubation with rabbit anti-LAP antibody (kindly provided by Isobe N [17], 2 mg/mL) in PBS with 0.5 % bovine serum albumin (BSA) overnight at 4 °C. After the sections were washed in PBS, they were incubated with HRP-conjugated secondary antibody using a REAL EnVision +/HRP Rabbit/Mouse Kit (Dako, Denmark) for 50 min at 4 °C, followed by washing in PBS. The products were visualized by incubating the sections with a diaminobenzidine (DAB) reaction mixture. The sections were then counterstained with haematoxylin, dehydrated and covered. Images were recorded by Nikon H550Smicroscope and NIS Elements F 3.0 software.

Western blot analysis

Total protein from 100 mg of mammary gland tissue was extracted using RIPA lysis buffer (Beyotime, Shanghai, China), and the concentration was determined using a bicinchoninic acid protein assay (BCA) kit (Pierce, Rockford, IL, USA). The proteins were fractionatedby sodium dodecyl sulphate-polyacrylamide gel electrophoresis (SDS-PAGE) and then transferred onto nitrocellulose membranes (Bio Trace, Pall Co., USA). After the membranes were blocked with 5 % BSA at room temperature for 2 h, they were incubated with anti-LAP (1 μg/mL), anti-phospho-NF-kB p65 (Ser536) (Beyotime,Shanghai, China, 1:1000), or anti-phospho-IkB-α (Ser36/Ser32) (Signalway Antibody, LLC., Maryland, USA, 1:1000) at 4 °C overnight. Then, the membranes were incubated with horseradish peroxidase (HRP)-conjugated secondary antibody (E030130-01, Earth Ox, CA, 1:10000) at room temperature for 1 h. Antibodies directed against β-tubulin (Santa Cruz Biotechnology, Inc., 1:200) and GAPDH (Beyotime, Shanghai, China, 1:500) were used as internal controls

Table 2 The primer sequences of target and internal reference genes used in qPCR

Gene	Forward primer	Reverse primer	GenBank accession	PCR products
LAP	AGGCTCCATCACCTGCTCCTT	CCTGCAGCATTTTACTTGGGCT	AY911374.1	183 bp
BNBD5	CTGGATCACCTGCTCCTCGT	GGTGCCAATCTGTCTCATGTTG	NM_001130761.1	150 bp
IL-1β	GGCCAAAGTCCCTGACCTCT	CTGCCACCATCACCACATTC	NM_174093.1	167 bp
IL-6	GGAGGAAAAGGACGGATGCT	GGTCAGTGTTTGTGGCTGGA	NM_173923.2	227 bp
IL-8	CCTCTTGTTCAATATGACTTCCA	GGCCCACTCAATAACTCTC	NM_173925.2	170 bp
TNF-α	CTTCTGCCTGCTGCACTTCG	GAGTTGATGTCGGCTACAACG	AF_348421.1	156 bp
NF-kB	ATACGTCGGCCGTGTCTAT	GGAACTGTGATCCGTGTAG	NM_001045868.1	129 bp
GAPDH	GGGTCATCATCTCTGCACCT	GGTCATAAGTCCCTCCACGA	XM_001252479	177 bp

for normalization. The membrane was washed, and the target proteins were visualized by enhanced chemiluminescence (ECL) using a BeyoECL Plus Kit (Beyotime, Shanghai, China). ECL signals were recorded by an imaging system (Fujifilm, Japan) and analysed with Quantity One software (Bio-Rad, USA). The membranes containing phospho-NF-kB p65 and phospho-IkBα were stripped by stripping buffer (Boster Biotechnology, Inc., Wuhan, China) according to the manufacturer's instructions and incubated with antibodies directed against NF-kB p65 and IkB-α (Beyotime,Shanghai, China, 1:500), respectively. Then, the steps were followed as described above.

Statistical analysis

The data of LPS concentrations of the rumen and lacteal artery and vein, as well as the concentrations of pro-inflammatory enzymes and cytokines in the lacteal vein plasma, were analysed using a MIXED model with repeated measures using SPSS Statistics (IBM, Inc., New York, USA). The data of the rumen pH values, the mRNA and protein expressions were analysed using the paired *t*-test. The correlations between the LAP mRNA expression and the protein expression were analysed using Pearson's model. The data are expressed as the mean ± SEM. Differences were considered statistically significant if $p < 0.05$ or highly significant if $p < 0.01$, and a tendency was considered if $0.05 < p < 0.10$.

Results

Ruminal pH and LPS content

The mean pH value of ruminal fluid in the HCD group was lower than that in the LCD group ($p < 0.05$) and was lower than 5.6 for more than 3 h per day (Table 3).

The LPS concentration in rumen samples was higher in the HCD group than that in the LCD group at 4 h after feeding ($p < 0.01$). The plasma LPS concentrations in the lacteal artery and vein increased in the HCD group compared with that in the LCD group ($p < 0.01$ for both, Table 3).

Evaluation of MPO activity, NAG activity and pro-inflammatory cytokine concentrations

The MPO activity in the mammary tissue and the lacteal vein plasma in the HCD group was up-regulated compared with the LCD group as shown in Fig. 1a and b ($p < 0.05$ and $p < 0.01$, respectively).

The NAG activity in the lacteal vein plasma in the HCD group was higher than that in the LCD group (Fig. 1c, $p < 0.01$).

The concentrations of pro-inflammatory cytokines (IL-1β, IL-6, and TNF-α) in plasma from the lacteal vein increased in the HCD group compared with the LCD group (Fig. 1d, e, f, $p < 0.05$ for IL-1β and IL-6, $p < 0.01$ for TNF-α).

Expression of immune-related genes in the mammary gland

The mRNA analysis showed that the expression of LAP, IL-1β, IL-6, IL-8, and TNF-α increased in the HCD group compared with the LCD group ($p < 0.05$, Fig. 2). However, no significant difference but a up-regulated tendency for BNBD5 expression was observed (Fig. 2). The expression of the transcription factor NF-kB, which is implicated in regulating inflammatory and immune responses, did not differ between the HCD group and the LCD group (Fig. 2).

Table 3 Ruminal pH and LPS content in rumen samples and in plasma samples from the lacteal artery and vein

Item	Treatment		SEM	*p* value
	LCD	HCD		
Mean pH value	6.02	5.90	0.03	0.03
Time < pH 5.6, h/d	1.65	3.72	0.50	0.01
LPS in rumen, kEU/mL	47.17	79.04	7.94	<0.01
LPS in plasma of lacteal artery, EU/mL	0.47	0.86	0.06	<0.01
LPS in plasma of lacteal vein, EU/mL	0.12	0.27	0.03	<0.01

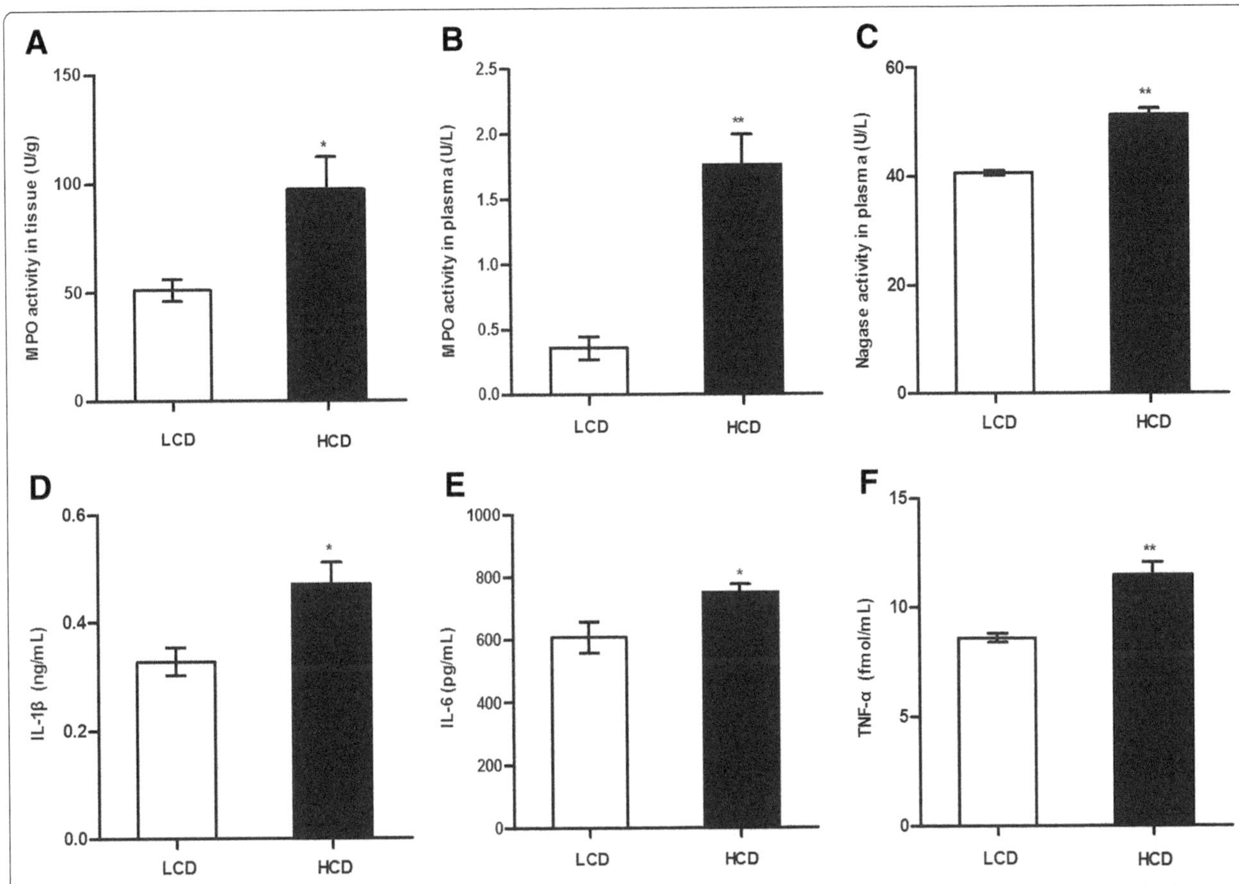

Fig. 1 Effect of a HCD on MPO activity, NAG activity, and pro-inflammatory cytokine concentrations. **a** MPO activity in the mammary gland tissue. **b** MPO activity in the lacteal vein plasma. **c** NAG activity in the lacteal vein plasma. **d** IL-1β concentration in the lacteal vein plasma. **e** IL-6 concentration in the lacteal vein plasma. **f** TNF-α concentration in the lacteal vein plasma. Significant difference is indicated (*$p < 0.05$, **$p < 0.01$)

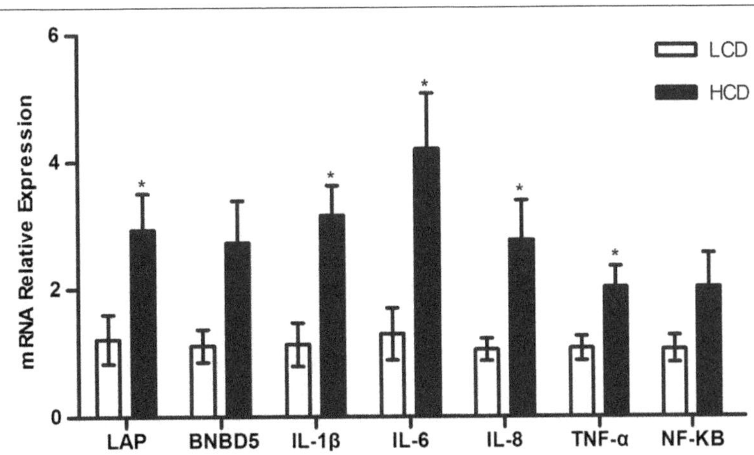

Fig. 2 Quantitative real-time PCR. Quantification of the mRNA expression involved in immune and inflammatory responses in mammary glands of dairy cows fed a LCD or HCD. Single asterisk (*) indicates $p < 0.05$

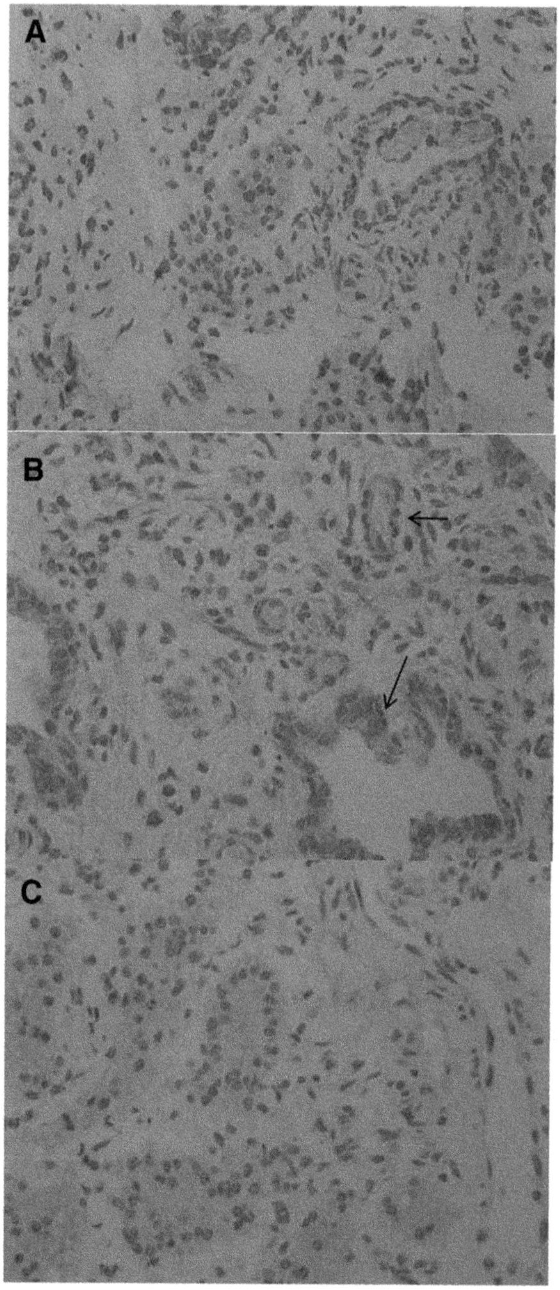

Fig. 3 Sections of mammary gland immunostained for LAP (×400). **a** Alveolar epithelial cells from the LCD group that are immunonegative for LAP. **b** Alveolar epithelial cells from the HCD group that are immunopositive for LAP. Arrows show positive immunoreactions for LAP. **c** Alveoli cultured with normal rabbit IgG as controls for LAP

Immunohistochemical analysis of LAP

Positive immunoreactions of LAP in the HCD group were observed in the cytoplasm of mammary gland alveolar epithelium cells (Fig. 3b). Sections from the LCD group and the control sections incubated with normal rabbit IgG, which was substituted for LAP antibody, did not show positive reactions (Fig. 3a and c).

LAP protein expression

LAP protein expression in mammary gland was upregulated in the HCD group compared with that in the LCD group (Fig. 4a and b). In addition, individual relative LAP protein expression with its corresponding mRNA relative copy numbers showed a strong correlation ($r^2 = 0.85$, $p < 0.01$, Fig. 4c).

NF-kB signalling pathway analysis

Although the expression of NF-kB and IkBα proteins did not differ between the HCD group and LCD group, the amounts of phosphorylated NF-kB and phosphorylated IkBα increased in the HCD group compared with that in the LCD group ($p < 0.01$, $p < 0.05$, respectively, Fig. 5).

Discussion

The present study demonstrated that rumen-derived LPS caused enhanced inflammatory responses, immune responses and LAP synthesis in mammary glands of cows fed a HCD.

Ruminants are often fed a relatively high proportion of grains or easily fermentable carbohydrates to support high milk yields [30]. Although this type of diet can promote cost efficiency over the short term, it would do harm to ruminant health and welfare in the long term [1]. In our present study, two different diets (HCD and LCD) were designed and used based on practical experience. The high incidence of SARA, a common digestive disorder in many dairy herds, is associated with feeding ruminants a HCD [1, 3, 5, 30] and also the physical form of total mixed ratio (both forage and grain) [31, 32], because a HCD can cause a decrease in ruminal pH due to the accumulation of short-chain fatty acids including organic acids and volatile fatty acids (VFA) [1, 33]. As shown in Table 3, the mean pH value of ruminal fluid in the HCD group was lower than that in the LCD group ($p < 0.05$) and was lower than 5.6 for more than 3 h per day, which is consistent with previous studies and which indicated that SARA was successfully induced by the HCD [2]. Alterations in the ruminal environment appear to lead to Gram-negative bacteria lysis and subsequent release of large amounts of LPS into the ruminal fluid [3, 5]. In the present study, the concentrations of LPS in the ruminal fluid and in the plasma from the lacteal artery and the lacteal vein of the HCD group were significantly higher than were those in the LCD group ($p < 0.01$ for all three of them) as shown in Table 3. These results are consistent with previous reports [1, 3] and indicated that a higher level of LPS did translocate into the circulatory system across the epithelial barrier of the gastrointestinal tract in the HCD group.

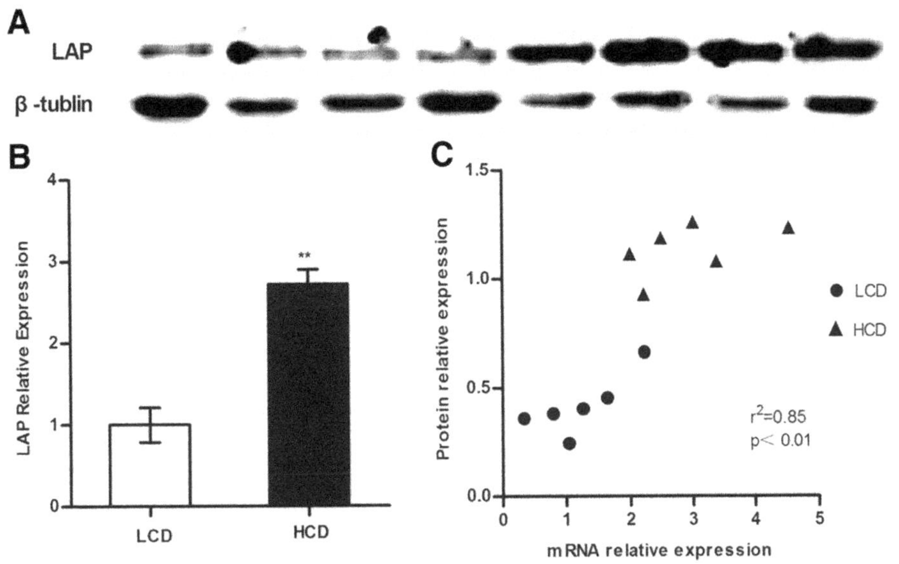

Fig. 4 Effect of a HCD on LAP protein expression in mammary gland. **a** Western blot results of LAP and β- tubulin proteins (Bands 1-4 represent LCD group samples, while bands 5–8 represent HCD group samples). **b** The mean relative protein expression of LAP against the reference β-tubulin protein as quantified by band density. Significant difference is indicated (*$p < 0.05$). **c** Correlation between LAP mRNA and protein expression levels (*r*, coefficient of correlation (Pearson); *p*, significance of correlation)

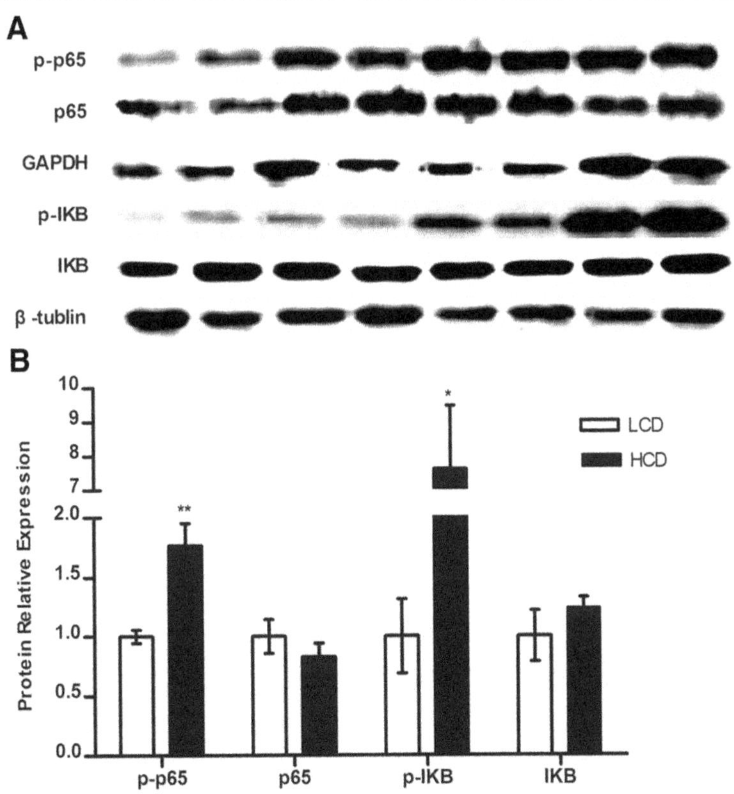

Fig. 5 Effect of a HCD on the NF-kB signalling pathway in mammary gland. **a** Western blot results of phospho-NF-kB p65, p65 (GAPDH as an internal control), phospho-IkB, and IkB (β- tubulin as an internal control) proteins. Bands 1–4 represent LCD group samples, while bands 5–8 represent HCD group samples. **b** The mean relative expression of the target protein against the reference protein as quantified by band density. Significant difference is indicated (*$p < 0.05$, **$p < 0.01$)

LPS in the peripheral blood initiates inflammatory responses [3, 5] by activating some inflammation-related enzymes and by secreting several pro-inflammatory cytokines. MPO, which is stored in azurophilic granules of neutrophils, is a neutrophil bactericidal protein. Its activity reflects the state of inflammation and oxidative stress [34]. NAG, one of the enzymes responsible for the metabolism of proteoglycans, is a good marker for estimating the degree of infection in bovine mammary glands [35]. Pro-inflammatory cytokines (IL-1β, IL-6, and TNF-α) play crucial role in inflammation [36]. In the current study, MPO activity, NAG activity, and pro-inflammatory cytokine (IL-1β, IL-6, and TNF-α) concentration in plasma from the lacteal vein in the HCD group increased compared with those in the LCD group, which provided evidence for activation of systemic inflammatory responses.

A high concentration of LPS enhanced the expression of various cytokines and β-defensins in the MECs of dairy cows [37]. MECs are located at the interface between the body and the environment. These cells not only function as a physical barrier for pathogens but also participate in the innate immune defence via producing inflammatory mediators such as cytokines, chemokines, and host defence peptides [38]. In the present study, the mRNA analysis of genes expressed in the mammary gland that are associated with inflammatory responses and immune responses, including LAP, IL-1β, IL-6, IL-8, and TNF-α, were all enhanced in the HCD group compared with that in the LCD group. However, BNBD5 mRNA expression was not significantly different but had an increasing tendency probably owing to the high variability between individual values. The LPS concentration in the lacteal vein in the HCD group might explain why the inflammatory and subsequent immune responses were activated in the mammary gland.

Not only the mRNA analysis but also the protein analysis of LAP showed a difference between the HCD group and the LCD group. LAP secretion in the cytoplasm of the MECs was demonstrated previously by immunohistochemistry [17]. In the present study, positive immunoreactions of LAP were observed in the cytoplasm of MECs in sections obtained from the HCD group, while the sections from the LCD group and the control sections did not show positive reactions. Furthermore, the western blot results confirmed that the LAP protein concentration in the HCD group was higher than that in the LCD group and had a strong correlation with its corresponding relative copy numbers of mRNA. Additionally, the finding that the SCC and the LAP protein concentration both increased in the HCD group compared with those in the LCD group is consistent with Kawai et al [39] and Isobe et al [40], who found that the SCC is associated with the LAP protein concentration in the milk. Recently, Kawai et al [41] also found that LF concentration decreased with decrease in SCC after antibiotics treatment, but not LAP concentration and LPO activity, which differed depending on the severity of mastitis.

The p65 subunit of NF-kB is known to play a stimulatory role in LAP expression in MECs. NF-kB can be activated by LPS and is implicated in regulating immune responses [42]. Mature NF-kB, which is sequestered in the cytoplasm in a quiescent state by interaction with IkBα, is a heterodimer of two subunits, p50 and p65, and the major pathway leading to their activation is based on IkBα phosphorylation and degradation [43]. The p50/p65 heterodimer translocates into the nucleus and binds to the kB sites on the promoter region of genes when the nuclear localization signal is exposed [44]. As shown previously, although NF-kB mRNA expression and NF-kB and IkBα protein expression did not differ between the HCD group and LCD group, phosphorylated NF-kB and phosphorylated IkBα protein expression significantly increased in the HCD group compared with that in the LCD group, which indicated that the NF-kB signalling pathway was activated after HCD administration. This result was consistent with previously reported results [24], thus we supposed that activation of the NF-kB signalling pathway is one of the mechanism that enhances LAP synthesis and immune responses.

Conclusions

In summary, our study clearly documented that after long-term HCD feeding, rumen-derived LPS translocated to the blood stream, triggers inflammatory and immune responses and enhances LAP synthesis via the NF-kB signalling pathway in mammary glands of lactating cows. Our findings may shed light on the mechanism involved in rumen-derived LPS on LAP synthesis and immune responses in mammary glands of dairy cows fed a HCD.

Abbreviations

BCA, bicinchoninic acid protein assay; BNBD, bovine neutrophil β-defensin; BSA, bovine serum albumin; DAB, diaminobenzidine; ECL, chemiluminescence; GAPDH, glyceraldehyde-3-phosphate dehydrogenase; HCD, high-concentrate diet; HRP, horseradish peroxidase; IkB, inhibitory kappa B α; IL, interleukin; LAP, lingual antimicrobial peptide; LCD, low-concentrate diet; LPS, lipopolysaccharide; MECs, mammary epithelial cells; MPO, myeloperoxidase; NAG, β-N-acetyl glucosaminidase; NF-kB, nuclear factor-kappaB; SARA, subacute ruminal acidosis; SCC, somatic cell count; TNF-α, tumour necrosis factor-α

Acknowledgements

The authors gratefully appreciate Dr Naoki Isobe (Graduate School of Biosphere Science, Hiroshima University, Japan) for kindly providing us anti-LAP antibody.

Funding

This study was supported by grants from the National Natural Science Foundation of China (31172371), the National Basic Research Programme of China (2011CB100802), and the Priority Academic Programme Development (PAPD) of Jiangsu Higher Education Institutions.

Authors' contributions

DJ conducted most of the measurements of parameters in this study, performed the statistical analyses and drafted the manuscript. GC, KZ, TX and JG contributed to animal experiment and sample collection. XS conceived the idea, designed the experiments and finalized the manuscript. All authors read and approved the final manuscript.

Competing interests

None of the authors has any financial or personal competing interests that would have influenced the content of the paper or interfered with their objective assessment of the manuscript.

Consent for publication

Not applicable.

References

1. Plaizier JC, Krause DO, Gozho GN, McBride BW. Subacute ruminal acidosis in dairy cows: the physiological causes, incidence and consequences. Vet J. 2008;176:21–31.
2. Kleen JL, Hooijer GA, Rehage J, Noordhuizen JP. Subacute ruminal acidosis (SARA): a review. J Vet Med A Physiol Pathol Clin Med. 2003;50:406–14.
3. Khafipour E, Krause DO, Plaizier JC. A grain-based subacute ruminal acidosis challenge causes translocation of lipopolysaccharide and triggers inflammation. J Dairy Sci. 2009;92:1060–70.
4. Gozho GN, Plaizier JC, Krause DO, Kennedy AD, Wittenberg KM. Subacute ruminal acidosis induces ruminal lipopolysaccharide endotoxin release and triggers an inflammatory response. J Dairy Sci. 2005;88:1399–403.
5. Emmanuel DGV, Dunn SM, Ametaj BN. Feeding high proportions of barley grain stimulates an inflammatory response in dairy cows. J Dairy Sci. 2008; 91:606–14.
6. Ametaj BN, Koenig KM, Dunn SM, Yang WZ, Zebeli Q, Beauchemin KA. Backgrounding and finishing diets are associated with inflammatory responses in feedlot steers. J Anim Sci. 2009;87:1314–20.
7. Beutler B, Rietschel ET. Innate immune sensing and its roots: the story of endotoxin. Nat Rev Immunol. 2003;3:169–76.
8. Heine H, Rietschel ET, Ulmer AJ. The biology of endotoxin. Mol Biotechnol. 2001;19:279–96.
9. Swanson K, Gorodetsky S, Good L, Davis S, Musgrave D, Stelwagen K, et al. Expression of a beta-defensin mRNA, lingual antimicrobial peptide, in bovine mammary epithelial tissue is induced by mastitis. Infect Immun. 2004;72:7311–4.
10. Schonwetter BS, Stolzenberg ED, Zasloff MA. Epithelial antibiotics induced at sites of inflammation. Science. 1995;267:1645–8.
11. Kagan BL, Ganz T, Lehrer RI. Defensins: a family of antimicrobial and cytotoxic peptides. Toxicology. 1994;87:131–49.
12. Diamond G, Zasloff M, Eck H, Brasseur M, Maloy WL, Bevins CL. Tracheal antimicrobial peptide, a cysteine-rich peptide from mammalian tracheal mucosa: peptide isolation and cloning of a cDNA. Proc Natl Acad Sci U S A. 1991;88:3952–6.
13. Tarver AP, Clark DP, Diamond G, Russell JP, Erdjument-Bromage H, Tempst P, et al. Enteric beta-defensin: molecular cloning and characterization of a gene with inducible intestinal epithelial cell expression associated with Cryptosporidium parvum infection. Infect Immunity. 1998;66:1045–56.
14. Cormican P, Meade KG, Cahalane S, Narciandi F, Chapwanya A, Lloyd AT, et al. Evolution, expression and effectiveness in a cluster of novel bovine beta-defensins. Immunogenetics. 2008;60:147–56.
15. Goldammer T, Zerbe H, Molenaar A, Schuberth HJ, Brunner RM, Kata SR, et al. Mastitis increases mammary mRNA abundance of beta-defensin 5, toll-like-receptor 2 (TLR2), and TLR4 but not TLR9 in cattle. Clin Diagn Lab Immunol. 2004;11:174–85.
16. Roosen S, Exner K, Paul S, Schroder JM, Kalm E, Looft C. Bovine beta-defensins: identification and characterization of novel bovine beta-defensin genes and their expression in mammary gland tissue. Mamm Genome. 2004;15:834–42.
17. Isobe N, Hosoda K, Yoshimura Y. Immunolocalization of lingual antimicrobial peptide (LAP) in the bovine mammary gland. Anim Sci J. 2009;80:446–50.
18. Singh K, Davis SR, Dobson JM, Molenaar AJ, Wheeler TT, Prosser CG, et al. cDNA microarray analysis reveals that antioxidant and immune genes are upregulated during involution of the bovine mammary gland. J Dairy Sci. 2008;91:2236–46.
19. Petzl W, Zerbe H, Gunther J, Yang W, Seyfert HM, Nurnberg G, et al. Escherichia coli, but not Staphylococcus aureus triggers an early increased expression of factors contributing to the innate immune defense in the udder of the cow. Vet Res. 2008;39:18.
20. Swanson KM, Stelwagen K, Dobson J, Henderson HV, Davis SR, Farr VC, et al. Transcriptome profiling of Streptococcus uberis-induced mastitis reveals fundamental differences between immune gene expression in the mammary gland and in a primary cell culture model. J Dairy Sci. 2009;92:117–29.
21. Isobe N, Nakamura J, Nakano H, Yoshimura Y. Existence of functional lingual antimicrobial peptide in bovine milk. J Dairy Sci. 2009;92:2691–5.
22. Isobe N, Morimoto K, Nakamura J, Yamasaki A, Yoshimura Y. Intramammary challenge of lipopolysaccharide stimulates secretion of lingual antimicrobial peptide into milk of dairy cows. J Dairy Sci. 2009;92:6046–51.
23. Gunther J, Liu S, Esch K, Schuberth HJ, Seyfert HM. Stimulated expression of TNF-alpha and IL-8, but not of lingual antimicrobial peptide reflects the concentration of pathogens contacting bovine mammary epithelial cells. Vet Immunol Immunop. 2010;135:152–7.
24. Liu S, Shi X, Bauer I, Gunther J, Seyfert HM. Lingual antimicrobial peptide and IL-8 expression are oppositely regulated by the antagonistic effects of NF-kappaB p65 and C/EBPbeta in mammary epithelial cells. Mol Immunol. 2011;48:895–908.
25. Stolzenberg ED, Anderson GM, Ackermann MR, Whitlock RH, Zasloff M. Epithelial antibiotic induced in states of disease. Proc Natl Acad Sci U S A. 1997;94:8686–90.
26. Isobe N, Sugino T, Taniguchi K, Moriya N, Hosoda K, Yoshimura Y. Differential localization of lingual antimicrobial peptide in the digestive tract mucosal epithelium of calves. Vet Immunol Immunop. 2011;142:87–94.
27. Liu JH, Xu TT, Zhu WY, Mao SY. A high-grain diet alters the omasal epithelial structure and expression of tight junction proteins in a goat model. Vet J. 2014;201:95–100.
28. Zhou J, Dong G, Ao C, Zhang S, Qiu M, Wang X, et al. Feeding a high-concentrate corn straw diet increased the release of endotoxin in the rumen and pro-inflammatory cytokines in the mammary gland of dairy cows. BMC Vet Res. 2014;10:172.
29. Ershun Z, Yunhe F, Zhengkai W, Yongguo C, Naisheng Z, Zhengtao Y. Cepharanthine attenuates lipopolysaccharide-induced mice mastitis by suppressing the NF-kB signaling pathway. Inflammation. 2014;37:331–7.
30. Morgante M, Stelletta C, Berzaghi P, Gianesella M, Andrighetto I. Subacute rumen acidosis in lactating cows: an investigation in intensive Italian dairy herds. J Anim Physiol An N. 2007;91:226–34.
31. De Nardi R, Marchesini G, Stefani AL, Barberio A, Andrighetto I, Segato S. Effect of feeding fine maize particles on the reticular pH, milk yield and composition of dairy cows. J Anim Physiol An N. 2014;98:504–10.
32. Beauchemin KA, Yang WZ. Effects of physically effective fiber on intake, chewing activity, and ruminal acidosis for dairy cows fed diets based on corn silage. J Dairy Sci. 2005;88:2117–29.
33. Xu T, Tao H, Chang G, Zhang K, Xu L, Shen X. Lipopolysaccharide derived from the rumen down-regulates stearoyl-CoA desaturase 1 expression and alters fatty acid composition in the liver of dairy cows fed a high-concentrate diet. BMC Vet Res. 2015;11:52.
34. Nussbaum C, Klinke A, Adam M, Baldus S, Sperandio M. Myeloperoxidase: a leukocyte-derived protagonist of inflammation and cardiovascular disease. Antioxid Redox Sign. 2013;18:692–713.
35. Kankofer M, Wiercinski J, Zerbe H. Activity of Placental β-N-acetyl-glucosaminidase in cows with and without retained fetal membranes. Reprod Domest Anim. 2000;35(3-4):97–100.
36. Notebaert S, Demon D, Vanden Berghe T, Vandenabeele P, Meyer E. Inflammatory mediators in Escherichia coli-induced mastitis in mice. Comp Immunol Microbiol Infect Dis. 2008;31:551–65.
37. Strandberg Y, Gray C, Vuocolo T, Donaldson L, Broadway M, Tellam R. Lipopolysaccharide and lipoteichoic acid induce different innate immune responses in bovine mammary epithelial cells. Cytokine. 2005;31:72–86.
38. Rainard P, Riollet C. Innate immunity of the bovine mammary gland. Vet Res. 2006;37:369–400.
39. Kawai K, Akamatsu H, Obayashi T, Nagahata H, Higuchi H, Iwano H, et al. Relationship between concentration of lingual antimicrobial peptide and somatic cell count in milk of dairy cows. Vet Immunol Immunop. 2013;153:298–301.

40. Isobe N, Shibata A, Kubota H, Yoshimura Y. Lingual antimicrobial peptide and lactoferrin concentrations and lactoperoxidase activity in bovine colostrum are associated with subsequent somatic cell count. Anim Sci J. 2013;84:751–6.
41. Kawai K, Korematsu K, Akiyama K, Okita M, Yoshimura Y, Isobe N. Dynamics of lingual antimicrobial peptide, lactoferrin concentrations and lactoperoxidase activity in the milk of cows treated for clinical mastitis. An Sci J. 2015;86:153–8.
42. Vallabhapurapu S, Karin M. Regulation and function of NF-kappaB transcription factors in the immune system. Annu Rev Immunol. 2009;27:693–733.
43. Scheidereit C. IkappaB kinase complexes: gateways to NF-kappaB activation and transcription. Oncogene. 2006;25:6685–705.
44. Pahl HL. Activators and target genes of Rel/NF-kappaB transcription factors. Oncogene. 1999;18:6853–66.

17

Effect of pre-analytical treatments on bovine milk acute phase proteins

Funmilola C. Thomas[1,5], Andre M. Santana[2], Mary Waterston[3], Hayley Haining[4] and Peter David Eckersall[1*]

Abstract

Background: Samples for diagnostic procedures often require some form of pre-analytical preparation for preservation or safe handling during transportation prior to analysis in the laboratory. This is particularly important for milk samples which frequently need preservatives to retain milk composition as close to that found in freshly collected samples as possible.

Methods: Milk samples were treated by heating at 56 °C for 30 min or preserved by addition of either potassium dichromate or bronopol respectively. Haptoglobin (Hp), mammary associated serum amyloid A3 (M-SAA3) and C-reactive protein (CRP) were measured in the various treatment groups and in control samples which were not treated, using enzyme linked immunoassays. The concentrations of each APP were compared between treated and non-treated groups using the Wilcoxon signed ranks tests.

Results: Heat treatment of samples was found to have a significant lowering effect on milk M-SAA3 and CRP but not Hp. The use of potassium dichromate and bronopol as preservatives in milk had no significant effects on milk Hp and M-SAA3 concentration but lowered milk CRP values compared to controls.

Conclusions: The observed effects of heating and preservative use on milk APP should be taken into consideration when assaying samples which have undergone heat treatment as a result of international transfer regulations involving biological samples or samples needing chemical preservation prior to transport to laboratory.

Keywords: C-reactive protein, Haptoglobin, Thermal stability, Mammary associated serum amyloid A3, Milk preservatives

Background

Acute phase proteins are continuing to gain importance particularly as diagnostic tools in veterinary medicine [1]. With the discovery of the major APP in bovine milk and its correlation with mastitis [2], there has been a growing interest of their exploitation as biomarkers of bovine mastitis which can be adapted to a rapid, on farm measuring format. Given that bovine mastitis is one of the most prevalent and costly conditions affecting dairy cows which greatly impacts on the economics, welfare and public health aspects of dairy farming, the substantiation of APP for mastitis diagnosis is on-going and methods to apply their measurement to a rapid format has been explored [3, 4].

* Correspondence: David.Eckersall@glasgow.ac.uk
[1]Institute of Biodiversity Animal Health and Comparative Medicine, University of Glasgow, Bearsden Road, Glasgow, UK
Full list of author information is available at the end of the article

Acute phase proteins are proteins that are synthesized in the liver and other organs in response to inflammatory stimuli and released into body fluids such as serum, milk [1], ovarian fluid [5] among others. These samples are therefore collected and assayed for APP in order to make clinical diagnosis of inflammatory and other disease conditions.

Samples for diagnostic procedures often require some form of preparation, preservation or safe handling for transportation prior to analysis in the laboratory. This is especially true for milk samples which frequently need the use of preservatives to retain milk composition as close to that found in freshly collected samples as possible. Chemical preservation is used to avoid cellular degradation prior to somatic cell counts (SCC) or progesterone analysis (pregnancy test). Storage at freezing temperatures of between −20 °C and −80 °C is also a common way of preserving milk samples.

In addition, as a requirement for the safe transfer of biological samples across international borders, some national regulations demand heat treatment of liquid samples in order to inactivate harmful pathogens in the samples have been put in place (for example; Council directive 82/894/EEC of 21 December 1982, under the European Communities Act 1972).

However, high temperature is known to denature proteins and to influence the concentration and immunologic activities of some serum proteins [6]. Heat treatment has also been shown to result in irreversible changes in the structure of milk proteins [7]. In addition, heating of milk has been shown to affect various characteristics of milk ranging from allergenicity of milk proteins [8, 9] to the concentration of water soluble vitamins, as well as concentration and activity of some milk proteins [10–12]. Furthermore, it was observed that heating colostrum to 60 °C for 60 min resulted in inactivation of pathogens and loss (although minimal) in colostral IgG [11]. A number of other studies have also focused on the effect of heat on residues of drugs (quninolones, aminoglycosides) in milk [13, 14].

In the handling of milk samples for analysis, the use of preservatives has become a requirement for some tests due to distances of laboratories from the farms where samples are obtained. Potassium dichromate ($K_2Cr_2O_7$) as in Lactab Marks III® is a corrosive and toxic biocide which is often used in the preservation of milk samples in order to maintain milk composition from time of collection to analysis. It is known to be effective as a milk preservative [15], and is used to maintain progesterone levels in milk for pregnancy diagnosis. Another broad spectrum milk preservative, bronopol ($C_3H_6BrNO_4$; 2-bromo-2-nitro-1, 3 propanediol), which unlike potassium dichromate is reported to have no toxic effects, is often used particularly for preserving milk somatic cells prior to analysis through its action as an antimicrobial agent [16, 17]. Other preservatives that have been used in milk samples include Mercury chloride ($HgCl_2$), sodium azide [18], Azidiol [19] and formalin [20].

Although the efficacy of these milk preservatives in maintaining milks' natural composition including protein, fat and SCC has been proven [21–23], there has also been early reports [24–26] of alterations in analyte concentrations following the use of these preservatives. For example potassium dichromate was shown to lead to a decrease in protein concentration within 24 h after milk collection [27]. In addition, bronopol based preservatives were shown to alter the mid infrared measurements of protein contents of milk samples [15]. In another study however, the use of the preservatives potassium dichromate, bronopol and sodium azide did not significantly change fats, proteins, lactose and milk solids content of milk when samples were analyzed within 7 days of preservative use, but were reported to lead to a decline in these components when measured in samples stored for over 7 days after preservation [17]. Use of bronopol has also been reported to yield higher values of % proteins and fat compared to values measured in aliquots of the samples following the use of potassium dichromate [25]. Differences in effects on milk microflora between use of different preservatives have also been reported [28]. Other parameters that have been reported to be affected by chemical preservative use include the electrical conductivity (EC) of milk by potassium dichromate [29].

From these reports, it can be inferred that there is a possibility that the use of chemical preservatives may affect analytes in milk. Indeed it has been recommended that where the preservative used in milk interferes with the analyte being tested; a suitable correction factor should be used [30].

It is not known if the use of chemical preservatives in milk meant for analysis or heat treatment of milk samples in order to inactivate pathogens before international transit, has any effect on the concentration of milk APP. It is important to be aware of such effects if they exist, if such treated milk samples are to be used for analysis of APP.

This study was carried out to evaluate the effects of heat treatment and chemical preservative use on milk APP concentration. This consisted of the assessment of milk Hp, M-SAA3 and CRP following the use of the milk preservatives; potassium dichromate (Lactab Marks III®) and bronopol (broad spectrum Microtab® II) for milk preservation after 24 h and heating samples to pathogen-inactivation temperature (56 °C) for 30 min.

Methods

Samples

Milk samples were residual samples from previously described investigations [31] and also the residue of milk from commercial dairy farms in the West of Scotland submitted to Veterinary Diagnostic Services, University of Glasgow, for evaluation for mastitis. The samples covered a wide concentration range of APP; encompassing low values in non-mastitic milk and high values as expected in mastitic milk samples.

After mastitis diagnosis the residual of each sample was made into aliquots, stored at −20 °C before being subjected to various pre-analytical treatments before APP determination.

Sample treatments

Aliquots (~500 μl) were treated in one of the following ways described below, while control sample aliquots were not treated.

Heat treatment was carried out by heating milk samples at 56 °C for 30 min in a water bath after which they were analyzed for Hp ($n = 59$), M-SAA3 ($n = 44$) and CRP ($n = 53$) alongside controls (unheated).

To evaluate the effect of preservatives, another set of milk aliquots had 1 mg of potassium dichromate (Lactabs Mark III®, Thompson and Capper Ltd, Cheshire, UK) per 10 ml of milk sample added and allowed to gently mix and act for 24 h at 4–8 °C on a shaker. Subsequently samples were analyzed for Hp ($n = 35$), M-SAA3 ($n = 18$) and CRP ($n = 20$). A different set of milk aliquots had 8 mg (1 tablet) of bronopol (Broad spectrum Microtabs®, Advanced instruments, *Inc.*, Massachusetts, USA) added to 40 ml of milk sample and gently mixed for 24 h at 4–8 °C. Hp ($n = 54$), M-SAA3 ($n = 21$) and CRP ($n = 21$) were also measured in these samples. Control samples (without any form of treatment) were also held at 4–8 °C for 24 h before analysis.

Acute phase protein assays

Haptoglobin

Samples were assayed for Hp using an in-house ELISA developed for measuring milk Hp as described in Thomas et al.[31]. Dilutions were varied (ranging from 1:200 to 1:3200) from sample to sample depending on its Hp concentration, in order for the optical density value to fall within the interpolating range of the standard curve.

Mammary associated serum amyloid A

Tridelta Development Ltd supplied the Phase™ range SAA ELISA kit. Assay was performed as described in Thomas et al. [31]. Sample dilutions were also modified to accommodate samples with very high M-SAA3 concentration samples (range of dilutions were from 1:125 to 1: 1000).

C-reactive protein

Cow C-reactive protein (CRP) ELISA kits for assay of bovine milk CRP were supplied by the Life Diagnostics *Inc.* (West Chester, USA). The assay was performed as described in Thomas et al.[31]. Sample dilutions ranged from 1: 5 to 1:50.

Statistical analyses

Tests for normality of each APP data were carried out using the Kolmogorov–Smirnov and Shapiro–Wilk test along with normal probability plots and quantile-quantile (Q-Q) plots.

To assess the effects on each milk APP of heat treatment and use of preservatives: potassium dichromate and bronopol, the Wilcoxon signed ranks test (non-parametric) was used to compare treated and untreated groups. The Spearman's rho correlation test was also carried out to assess the correlation between each APP in the different treatment groups. A P value of <0.05 was considered significant. All statistical analyses were carried out using the IBM statistical package for social sciences (SPSS) software; version 21 (IBM Corporation USA, 2012).

Results

Table 1 shows the statistical observations from the comparison of three acute phase proteins values between various treated and untreated milk samples.

Table 1 Effect of heat treatment and preservative use on Hp, M-SAA3 and CRP values in bovine milk

Protein	Treatment	Significant difference P	Regression equation in the form of $y = c + mx$[a]	Correlation coefficient R^2/P
Hp	Heat $n = 59$	0.066	Heated Hp = 13.05 + 0.76 x unheated Hp	0.85/$P < 0.001$
SAA	Heat $n = 44$	0.000	Heated SAA = 38.6 + 0.28 unheated SAA	0.925/ $P < 0.001$
CRP	Heat $n = 53$	0.007	Heated CRP = −57.69 + 0.61 unheated CRP	0.932/$P < 0.001$
Hp	Potassium dichromate $n = 35$	0.080	Treated Hp = −270 + 1.04 non treated Hp	0.982/$P < 0.001$
SAA	Potassium dichromate $n = 18$	0.363	Treated SAA = −0.30 + 0.99 non treated SAA	0.642/$P < 0.05$
CRP	Potassium dichromate $n = 20$	0.003	Treated CRP = 5.41 + 1.47 non treated	0.913/$P < 0.001$
Hp	Bronopol $n = 54$	0.155	Treated Hp = 52.4 + 0.95 non treated Hp	0.882/ $P < 0.001$
SAA	Bronopol $n = 21$	0.088	Treated SAA = 1.10 + 0.0005 non treated	0.262/$P > 0.05$
CRP	Bronopol $n = 21$	0.028	Treated CRP = 2.15 + 0.96 non treated	0.765/$P < 0.001$

[a]y = treated group, x = non treated group, c = y value when x = 0, m = gradient

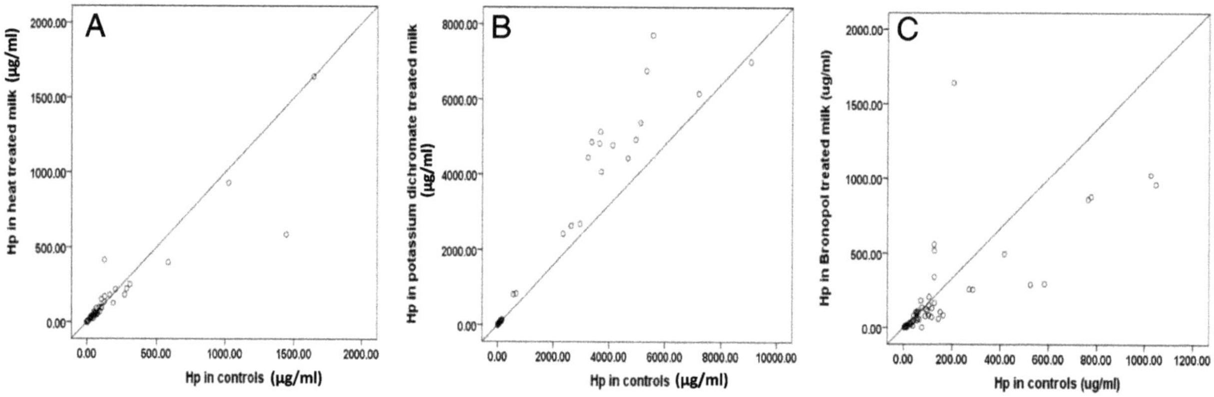

Fig. 1 a Regression plot of Hp concentrations in heated treated versus unheated milk samples ($n = 59$); (**b**) Regression plot of Hp concentration in potassium dichromate treated versus untreated milk samples ($n = 35$); (**c**): Regression plot of Hp concentrations in bronopol treated versus untreated milk samples ($n = 54$)

Haptoglobin

The intra-assay and inter-assay precision of the Hp ELISA were 6 % and 27 % correspondingly while the limit of detection (LOD) of the assay was 0.4 µg/ml as reported in Thomas et al.[31].

No significant difference was observed between the milk Hp concentration of heated and unheated samples ($P = 0.06$) and a significant correlation between heated and unheated samples' Hp was observed. The use of potassium dichromate and bronopol as milk preservatives for a duration of 24 h also did not result in a significant difference in milk Hp from controls (Table 1). Regression curves of concentrations of Hp in treated versus untreated milk samples are shown Fig. 1a-c. Table 2 shows the means and standard deviations of the concentrations in control and treated samples of Hp MSAA3 and CRP in the different treatment protocols.

Mammary associated serum amyloid A

Intra-assay and inter-assay CV of the MSAA3 ELISA was 7 % and 33 % respectively and a LOD of 0.6 µg/ml was established for the assay [31]. The M-SAA3 concentrations in heated samples were significantly different (lower) than those recorded from the unheated samples. However use of either potassium dichromate or bronopol preservatives did not indicate a significant difference in the values of M-SAA3 in milk from control values (Table 1). Regression curves of concentrations of M-SAA3 in treated versus untreated milk samples are shown Fig. 1b.

C-reactive protein

Intra-assay and inter-assay CV of the CRP assay was 4 % and 7 % respectively and the LOD was 1.8 ng/ml [31]. Heating milk samples was also observed to have a significant lowering effect on milk CRP values ($P = 0.007$). A significant correlation was observed between CRP values in heated and unheated milk samples ($r = 0.97$). The use of potassium dichromate and bronopol in milk was shown to produce a significant increase in milk CRP values compared to controls ($P = 0.003$, and 0.028 respectively) (Table 1). Regression curves of concentrations of CRP in treated versus untreated milk samples are shown Fig. 1c.

Table 2 Means and standard deviations of APP concentrations in controls versus treatments with heat, potassium dichromate and bronopol

Protein	Treatment	Total (n)	Mean ± SD of controls	Mean ± SD of treated
Hp (µg/ml)	Heat	59	130.50 ± 311.86	112.88 ± 257.95
Hp (µg/ml)	Potassium dichromate	35	1533.94 ± 2144.03	1791.14 ± 2483.71
Hp (µg/ml)	Bronopol	54	160.29 ± 245.48	204. ± 321.11
MSAA3 (µg/ml)	Heat	44	153.72 ± 270.84	94.39 ± 178.48
MSAA3 (µg/ml)	Potassium dichromate	18	17.16 ± 47.40	16.76 ± 47.20
MSAA3 (µg/ml)	Bronopol	21	1.94 ± 1.80	1.10 ± 2.05
CRP (ng/ml)	Heat	53	939.57 ± 3.69	514.86 ± 2.28
CRP (ng/ml)	Potassium dichromate	20	25.88 ± 47.68	43.56 ± 77.20
CRP (ng/ml)	Bronopol	21	11.01 ± 7.04	12.74 ± 9.03

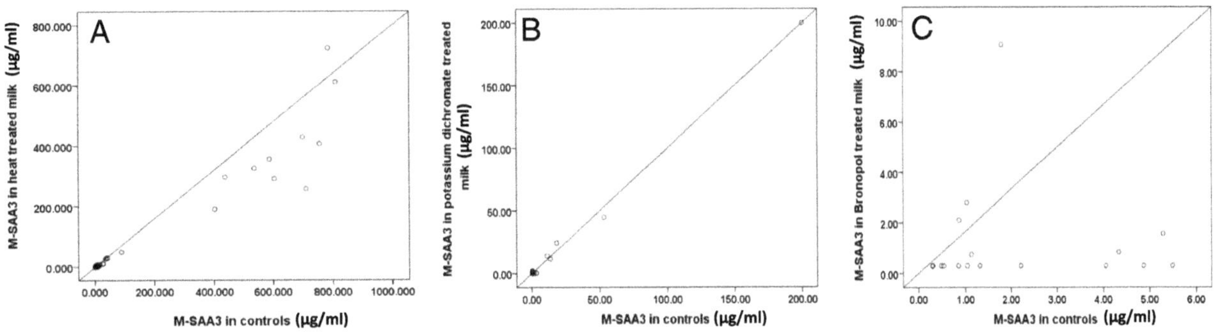

Fig. 2 a Regression plot of MSAA3 in heated versus unheated milk samples ($n = 44$); (**b**) Regression plot of MSAA3 concentrations in potassium dichromate treated versus untreated milk samples ($n = 18$); (**c**) Regression plot of MSAA3 concentrations in bronopol treated versus untreated milk samples ($n = 21$)

Discussion

In this study, three pre-analytical treatments regimes were carried out on milk samples with the aim of comparing the APP values in treated and untreated samples.

Studies have shown that some protocols of heat treatment could have significant reduction effects on milk (colostral) proteins especially IgG [32, 33] and on lactoferrin [34]. The observed lowering of M-SAA3 and CRP concentrations in heated samples and changes in CRP values of chemically preserved milk in the present study, lends further credence to these observations. It is generally known that high temperatures can result in the irreversible change in milk proteins structure [7], moreover it has been shown that aggregation of whey proteins can result upon heating of milk [35], this may explain some of the changes observed in the concentration of protein analytes in milk (which are mainly contained in the whey fraction).

It has been shown that high temperatures, can result in dissociation of CRP pentameres into its monomeric forms in blood [36], and although this may not affect the total concentration of serum CRP, it signals possible protein structural changes that are possible upon heating. There has been no report of the effect of heating on the structure of bovine milk CRP or M-SAA3. It can be inferred that a denaturation effect of heat on the protein structure or moiety of these compounds as reported by Hausen et al. [6] causes a loss of functional ability of these proteins to bind antibodies.

This finding should therefore be taken into consideration when assaying for M-SAA3 and CRP in samples which have undergone heating as a result of international transfer regulations involving heat treatment of biological samples. In such studies, control samples should be treated in the same manner as experimental samples or a correction factor based on the reduction in concentration should be applied. Structural changes in Haptoglobin have been associated with increased body temeprature (heat shock conditions) [37]. However, no significant change in milk Hp concentration due to heating was observed in this study. Studies have shown good correlation between milk Hp and SCC [31, 38], and in the study of Hachana et al. [39], heating milk samples to 60 °C prior to analysis did not alter the Fossomatic SCC, although the duration for which samples were held at that temperature before analysis was not ascertained.

Only CRP concentrations were affected by the use of chemical preservatives in milk, with the concentration showing a significant increase following treatment with

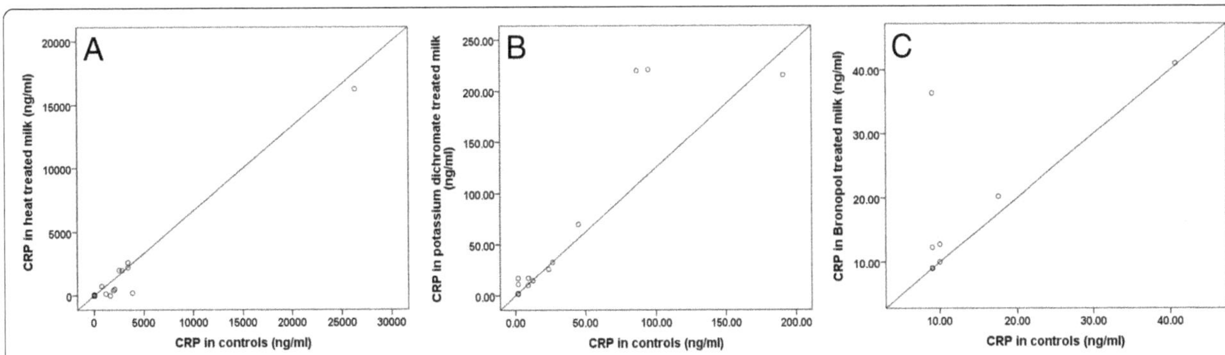

Fig. 3 a : Regression plot of CRP concentrations in heated versus unheated milk samples ($n = 53$); (**b**) Regression plot of CRP in potassium dichromate treated versus untreated milk samples ($n = 20$); (**c**) Regression curve of CRP in bronopol treated versus untreated milk samples ($n = 21$)

both potassium dichromate and Bronopol but the reason for this is not clear. Therefore samples being collected for SCC or for progesterone analysis can also be used for determination of Hp and M-SAA3 concentrations but should be cautiously used for CRP determination.

In this study, a minimum time of 24 h was used to assess the possible preservative effects of potassium dichromate and Bronopol and it is possible that samples preserved for longer (beyond 24 h) may develop changes that could significantly alter milk concentration of Hp and M-SAA3 as observed for some other proteins by Chalermsan et all [17]. The milk used in this study were residual samples collected for either clinical investigation or prior experimental study and selected to give a range of APP concentrations. The effect of chemical preservatives was assessed after 24 h mixing and incubation at 4–8 °C in order replicate to some degree the likely treatment of samples between collection addition of preservative and analysis. While it would be ideal to assess the effect of heat and chemical preservation on fresh samples treated immediately, this was beyond the scope of the present investigation. In future investigation of APP in milk, where preservatives are being used to allow SCC or progesterone analysis, it would be of value to assess the effects on APP under a variety of conditions. Nevertheless, it is recommended that samples needing to be transported over long distances without any means of cold-chain storage, can be preserved using either potassium dichromate (Lactabs Mark III) or Bronopol (Microtabs), but should be analyzed as soon as possible and a correction factor should be applied in cases of CRP assay. It would be interesting to examine the effect of prolonged storage of milk with preservatives on milk APP.

Conclusions

Heating milk samples and preservative use in milk can affect the measurement of APP concentration. However, if this is recognized and taken into consideration when assaying milk samples which have undergone heat treatment, as maybe required to conform with international transport regulations, or when assaying samples needing chemical preservation, then APP analysis will give valid results especially when control samples are treated with identical procedures.

Abbreviations

APP, acute phase proteins; APR, acute phase response; CRP, C-reactive protein; CV, coefficient of variation; ELISA, enzyme linked immunosorbent assay; Hp, haptoglobin; IMI, intramammary infection/inflammation; LOD, limit of detection; M-SAA3, mammary associated serum amyloid A3; SAA, serum amyloid A; SCC, somatic cell counts; SPSS, statistical package for social sciences; x, times

Acknowledgements

The authors wish to acknowledge the management and staff of Cochno Research Farm for assistance in collection of milk samples.

Funding

The study was supported with funds from the Federal University of Agriculture Abeokuta/ TETFund, Nigeria and the BBSRC, UK (grant number BB/M022021).

Author's contribution

All authors shared the planning of the experiment. FCT, AMS and MW undertook the experimental procedures and conversion of assay output into APP concentration. Interpretation of data was performed by FCT under supervision of HH and PDE with PDE being the corresponding author. All authors contributed to the preparation and have approved the final version of the manuscript.

Competing interest

The authors declare that they have no competing interest.

Consent for publication

Not applicable.

Author details

[1]Institute of Biodiversity Animal Health and Comparative Medicine, University of Glasgow, Bearsden Road, Glasgow, UK. [2]Department of Veterinary Clinics and Surgery, FCAV, UNESP, Sao Paulo, Brazil. [3]Institute of Infection, Immunity and Inflammation, GBRC University Place, University of Glasgow, Glasgow, UK. [4]Veterinary Diagnostic Services, School of Veterinary Medicine, University of Glasgow, Bearsden Road, Glasgow, UK. [5]Present Address: Department of Veterinary Physiology and Pharmacology, College of Veterinary Medicine, Federal University of Agriculture, Abeokuta, Nigeria.

References

1. Eckersall PD, Bell R. Acute phase proteins: Biomarkers of infection and inflammation in veterinary medicine. Vet J. 2010;185:23–7.
2. Eckersall PD, Young FJ, McComb C, Hogarth CJ, Safi S, Weber A, McDonald T, Nolan AM, Fitzpatrick JL. Acute phase proteins in serum and milk from dairy cows with clinical mastitis. Vet Rec. 2001;148:35–41.
3. Åkerstedt M, Björck L, Waller KP, Stemesjö Å. Biosensor assay for determination of haptoglobin in bovine milk. J Dairy Res. 2006;73(March ?August2006):299–305.
4. Tan X, Ding SQ, Hu YX, Li JJ, Zhou JY. Development of an immunosensor assay for detection of haptoglobin in mastitic milk. Vet Clin Pathol. 2012; 41(4):575–81.
5. Lavery K, Gabler C, Day J, Killian G. Expression of haptoglobin mRNA in the liver and oviduct during the oestrous cycle of cows (Bos taurus). Anim Reprod Sci. 2004;84:13–26.
6. Hausen BS, Signor C, Kober H, Tatsch E, Pereira RS, Duarte T, Hermes CL, Duarte MMMF, Moresco RN. Effect of temperature on albumin cobalt binding and its influence on ischemia-modified albumin levels in patients with suspected acute coronary syndrome. Clin Lab. 2012;169–172.
7. Raikos V. Effect of heat treatment on milk protein functionality at emulsion interfaces. A review. Food Hydrocoll. 2010;259–265.
8. Tiwari R, Wolber FM. Effect of heat treatment on stability and allergenicity of cow milk and goat milk proteins. J Allergy Clin Immunol. 2010;125:53.
9. Bloom KA, Huang FR, Bencharitiwong R, Bardina L, Ross A, Sampson HA, Nowak-Wegrzyn A. Effect of heat treatment on milk and egg proteins allergenicity. Pediatr Allergy Immunol. 2014;25:740–6.
10. Conesa C, Rota C, Castillo E, Pérez MD, CALVO M, SÁNCHEZ L. Effect of heat treatment on the antibacterial activity of bovine lactoferrin against three foodborne pathogens. Int J Dairy Technol. 2010;63:209–15.
11. Elizondo-Salazar JA, Jayarao BM, Heinrichs AJ. Effect of heat treatment of bovine colostrum on bacterial counts, viscosity, and Immunoglobulin G concentration. J Dairy Sci. 2010;93:961–7.
12. Morisawa Y, Kitamura A, Ujihara T, Zushi N, Kuzume K, Shimanouchi Y, Tamura S, Wakiguchi H, Saito H, Matsumoto K. Effect of heat treatment and enzymatic digestion on the B cell epitopes of cow's milk proteins. Clin Exp Allergy. 2009;39:918–25.
13. Roca M, Castillo M, Marti P, Althaus RL, Molina MP. Effect of heating on the stability of quinolones in milk. J Agric Food Chem. 2010;58:5427–31.
14. Zorraquino MA, Althaus RL, Roca M, Molina MP. Effect of heat treatments on aminoglycosides in milk. J Food Prot. 2009;72:1338–41.

15. Barbano D, Wojciechowski K, Lynch J. Effect of preservatives on the accuracy of mid-infrared milk component testing. J Dairy Sci. 2010; 93(12):6000–11.
16. Ardoe Y. Bronopol as a preservative in milk samples for the determination of cell content using Fossomatic. Milchwissenschaft. 1982;37:139–42.
17. Chalermsan N, Vijchulata P, Chirattanayuth P, Sintuwanit S, Surapat S, Engkagul A. Effects of preservatives on raw milk components analyzed by infrared spectrophotometry. Kasetsart Journal, Natural Sciences. 2004;38:38–43
18. Luck H. Determination of the fat protein and lactose content of 10 day preserved milk samples by means of the Ir milk analyzer. S Afr J Dairy Technol. 1975:75–78
19. Barcina Y, Zorraquino MA, Pedauye J, Ro GRF. Azidiol as a preservative for milk samples. An Vet. 1987;3:65–9.
20. Upadhyay N, Goyal A, Kumar A, Ghai DL, Singh R. Preservation of milk and milk products for analytical purposes. Food Rev Int. 2014;30:203–24.
21. Sánchez A, Sierra D, Luengo C, Corrales JC, Morales CT, Contreras AGC. Influence of storage and preservation on fossomatic cell count and composition of goat milk. J Dairy Sci. 2005;88:3095–100.
22. Sešķēna RJL. Influence of chemical preservatives on the quality and composition indices of raw milk samples. Acta Univ Latv. 2007;723:171–80.
23. Butler GSS. Suitability of bronopol preservative treated milk for fatty acid determination. J Dairy Res. 2011;78:220–5.
24. DA Coleman MB. Effects of several factors on quantification of fat, protein, and somatic cells in milk. J Dairy Sci. 1989;72:3295–303.
25. Monardes HG, Moore RK, Corrigan BRY. Preservation and storage mechanisms for raw milk samples for use in milk-recording schemes. J Food Prot. 1996;59:151–4.
26. Bertrand JA. Influence of shipping container, preservative, and breed on analysis of milk components of shipped samples. J Dairy Sci. 1996;79:145–8.
27. Ng-Kwai-Hang KF, Hayes JF. Effects of potassium dichromate and sample storage time on fat and protein by Milko-Scan and on protein and casein by a modified Pro-Milk Mk II method. J Dairy Sci. 1982;65:1895–9.
28. Benda P. The effect of some preservatives on natural microflora in milk samples. Vet Med (Praha). 1995;40:359–64.
29. Lien C, Young C, Chang HWY. Study on the effect of detection of preservative doses in raw milk samples. J Agric Assoc Taiwan. 2011;12:287–302.
30. The Effect of Bronopol on the Freezing Point and Impedance of Milk Samples. [www.aicompanies.com/wp-content/uploads/2015/12/EffectBronopol_Brochure_WEB.pdf]. verified 19 July 2016.
31. Thomas FC, Waterston M, Hastie P, Parkin T, Haining H, Eckersall PD. The major acute phase proteins of bovine milk in a commercial dairy herd. BMC Vet Res. 2015;11:207.
32. Johnson JL, Godden SM, Molitor T, Ames THD. Effects of feeding heat-treated colostrum on passive transfer of immune and nutritional parameters in neonatal dairy calves. J Dairy Sci. 2007;90:5189–98.
33. Godden SM, Smith S, Feirtag JM, Green LR, Wells SJFJ. Effect of on-farm commercial batch pasteurization of colostrum on colostrum and serum immunoglobulin concentrations in dairy calves. J Dairy Sci. 2003;86:1503–12.
34. Teixeira AGV, Bicalho MLS, Machado VS, Oikonomou G, Kacar C, Foditsch C, Young R, Knauer WA, Nydam DV, Bicalho RC. Heat and ultraviolet light treatment of colostrum and hospital milk: effects on colostrum and hospital milk characteristics and calf health and growth parameters. Vet J. 2013;197:175–81.
35. Wijayanti HB, Bansal N, Deeth HC. Stability of Whey proteins during thermal processing: a review. Compr Rev Food Sci Food Saf. 2014;13:1235–51.
36. Taylor KE, van den Berg CW. Structural and functional comparison of native pentameric, denatured monomeric and biotinylated C-reactive protein. Immunology. 2007;120:404–11.
37. Ettrich R, Brandt W, Kopecký V, Baumruk V, Hofbauerová K, Pavlícek Z. Study of chaperone-like activity of human haptoglobin: conformational changes under heat shock conditions and localization of interaction sites. Biol Chem. 2002;383:1667–76.
38. Pyörälä S, Hovinen M, Simojoki H, Fitzpatrick J, Eckersall PD, Orro T. Acute phase proteins in milk in naturally acquired bovine mastitis caused by different pathogens. Vet Rec. 2011;168:535.
39. Hachana Y, Kraiem K, Hani K, Haddad BPM. Effects of time storage and analytical temperature on Fossomatic cell counts in cows milk. Milchwissenschaft-Milk Sci Int. 2008;63:267–9.

18

Low prevalence of antibodies against *Toxoplasma gondii* in dairy cattle from China's central region

Hui Dong[1†], Yao Yao Lu[1†], Rui Jing Su[1], Ying Hua Wang[2], Meng Yao Wang[1], Yi Bao Jiang[1] and Yu Rong Yang[1*]

Abstract

Background: *Toxoplasma gondii* is an intracellular protozoan that can infect humans and other animals, including cattle. Cattle are one of the world's main sources of meat, and people who consume raw or undercooked meat and milk of cattle infected with *T. gondii* can become infected. In this study, a total of 5292 dairy cattle serum samples, collected from 17 cities (Henan Province, China) from January 2015 to September 2017, were screened for antibodies against *T. gondii*.

Results: Antibodies to *T. gondii* were found in 1.93% (102/5292) (95% CI, 1.56–2.30) of dairy cattle using a modified agglutination test (cut-off 1:100). The results showed that geographic location and season may be risk factors for *T. gondii* infection of cattle ($P < 0.05$), and the seroprevalence of *T. gondii* in cattle along the Yellow River is higher than other areas.

Conclusions: This is the first large-scale investigation on the seroprevalence of *T. gondii* infection in cattle from Central China. This survey shows that the *T. gondii* infection rate of dairy cattle is low; however, these findings provide additional information on the epidemiology of Chinese *T. gondii*. The possibility of dairy cattle exposure to *T. gondii* in Central China can not be ignored, and the consumption of raw or undercooked beef or milk may pose a risk to human health.

Keywords: *Toxoplasma gondii*, Cattle, Seroprevalence, Modified agglutination test, Risk factor, Henan Province, China

Background

Toxoplasma gondii is an intracellular protozoan parasite that can infect a wide variety of host species, including cattle [1]. The seroprevalence of *T. gondii* infection varies among host species. Although cattle appear to be poor hosts for *T. gondii, T. gondii* can still infect cattle and viable *T. gondii* strain had been isolated from the intestines of naturally infected cows [2, 3]. Cattle infected with *T. gondii* pose a risk for toxoplasmosis in people who consume raw or undercooked meat and unpasteurized milk [4]. In humans, *T. gondii* can cause encephalitis, retinitis, newborn hydrocephalus [1, 5], and even death [6].

Approximately 211 million cattle are raised in China, accounting for 14.4% of the world's cattle population (National Bureau of Statistics of China, 2015 update). Surveys regarding *T. gondii* infection in cattle have been reported in some parts of China. However, only three reports on cattle *T. gondii* infections in the central region of China have been published to date, which indicated that the prevalence of *T. gondii* was 20.10% (79/393) in 2011–2013 by IHA (95% CI 16.14–24.06) [7], 5.38% (43/800) in 2011–2012 by IgG test paper (95% CI 3.81–6.94) [8], and 0 (0/102) in 2013–2014 by IHA [9], and there requires more reports as a support to summarize and analyze the epidemiological situation in the region. Large quantities of milk and meat are consumed each year in China, and the safety of the cattle products with respect to *T. gondii* infection is unknown. The objective of this investigation was to estimate the seroprevalence and risk of *T. gondii* infection in dairy cattle from Central China. To our knowledge, the present study is the most extensive investigation of *T. gondii* infections in dairy cattle from Central China.

* Correspondence: yangyu7712@sina.com
[†]Hui Dong and Yao Yao Lu contributed equally to this work.
[1]College of Animal Science and Veterinary Medicine, Henan Agricultural University, Zhengzhou, People's Republic of China
Full list of author information is available at the end of the article

Methods

Investigation site and serum samples

Henan Province is located in the central region of China. Henan Province (latitude 34.90°N, longitude 113.50°E) has a humid and subtropical climate. From east to west, the plains transition into the hilly mountains. The average annual temperature is 15.7 °C to 12.1 °C, and the average annual precipitation is 1380.6 to 532.5 mm. The sera of 5292 dairy cattle from 49 farms in 17 cities were collected by local veterinarians from January 2015 to September 2017 (Table 1, Fig. 1). Cattle feed consists of silage, hay, and fresh grass. The cattle were females, and their ages ranged from 2 to 15 years. The farm names and sample collection dates were recorded. The sera were used for *Brucellosis* screening, which in turn also allowed us to survey for *T. gondii* infection. The cattle sera were separated from jugular vein blood and transported to the Laboratory of Veterinary Pathology, Henan Agricultural University (Zhengzhou, Henan, China) in cooler boxes. The samples were stored at 4 °C and tested for *T. gondii* antibodies within one week.

Assessment for *T. gondii* antibodies

The serum samples were tested for antibodies against *T. gondii* by modified agglutination test (MAT) [10]. Sera with MAT titers of 1:100 or higher were considered positive for *T. gondii* [1]. Whole formalin-treated *T. gondii* tachyzoite antigens were obtained from the University of Tennessee Research Foundation (Knoxville, TN, USA; https://utrf.tennessee.edu/). *T. gondii*-positive mouse sera were provided by Dr. J. P. Dubey (Beltsville, ARS, USDA) as reference sera. All serum samples were tested at 1:100, then the dilution was doubled to the maximum titer, and positive controls and negative controls were run on each plate.

Statistical analysis

Statistical analysis was performed using GraphPad Prism 4.0 software (GraphPad Software Inc., San Diego, CA, USA). The data were analyzed using the chi-square test or Fisher's exact test to assess the association between seropositivity and risk factors based on region (17 cities), season (spring, summer, autumn, and winter), and geographic location (north of the Yellow River, south of the Yellow River).

Results

Our survey indicated that 1.93% (102/,5292) (95% CI, 1.56–2.30) of the examined dairy cattle were seropositive for *T. gondii* infection by MAT, with titers of 1:100 in 102 cattle, 1:200 in 40, 1:400 in 21, 1:800 in 17, 1:1600 in 12, and 1:3200 in 7 (Table 1). The seroprevalence of *T. gondii* infection in cattle from 17 cities ranged from 0 to 10%. No information on the geographical locations from 756 cattle samples was available. The seroprevalence rates of *T. gondii* varied with regions. The differences in *T. gondii* seroprevalence among different regions are shown in Table 2. A high prevalence was observed in LuoYang and ZhouKou compared to the other regions ($P < 0.05$), and no seropositive sera from cattle were observed in SanMenXia and XinYang.

Risk factors in relation to geographic location and season were analyzed. The prevalence of *T. gondii* in dairy cattle along the Yellow River was higher than in the other areas (Fig. 1). The seroprevalence of *T. gondii* in cattle from south of the Yellow River (3.67%, 57/2116) was higher than that in the north of the Yellow River (1.69%, 41/2420), with a statistically significant odds ratio of 1.606 (95% CI, 1.071–2.410) ($P = 0.027$). In the summer, no positive sera for *T. gondii* were detected in dairy cattle by MAT ($n = 416$). In spring, *T. gondii* antibodies were identified in 3 (0.71%) of 424 samples. In winter, 14 (1.52%) of 921 serum samples tested positive for *T. gondii* antibodies. Out of the 3531 samples collected in autumn, 85 (2.41%) samples were determined to be positive. Seroprevalence was higher in the autumn in contrast to the spring and winter, and the difference between autumn and spring was statistically significant (OR 3.462; $P = 0.039$) (Table 3).

Discussion

It has been estimated that about one third of the world population has been infected with *T. gondii* [5, 11]. Infection is often most common in areas that have hot, humid climates and lower altitudes [1]. Cattle can be readily infected with *T. gondii*, but they are considered poor hosts because these have developed a more effective immune response to *T. gondii* infection than sheep that possibly facilitates in *T. gondii* elimination from tissues, as well as transient antibody responses [12–14]. However, *T. gondii* has been isolated from cattle tissues or unpasteurized milk [13], indicating that meat and milk may be sources of *T. gondii* infections. *T. gondii* has also been detected in cattle semen [1], suggesting it may be transmitted by venereal contact or artificial insemination.

Several test methods have been used in the diagnosis of *T. gondii* infection in humans. However, these could not be applied to cattle based on its incompatibility with the bovine immunoglobulin G system [15]. Current understanding of the specificity and sensitivity of serological diagnosis of *T. gondii* infection in cattle is limited. Furthermore, cattle may not readily acquire persistent *T. gondii* infections, and the actual prevalence rates based on serum antibodies to *T. gondii* are difficult to ascertain. Previous reports have screened for *T. gondii* antibodies in cattle sera at a 1:100 dilution or higher by MAT [1, 16], prompting us to use this as a cut-off in the present study.

Table 1 Seroprevalence of *T. gondii* in cattle in Henan Province

Location/ City	Samples obtained date	Tested No.	No. of seropositive samples /(titer)						% (Positive No. /Test No.)	95% CI
			1:100	1:200	1:400	1:800	1:1600	1:3200		
I ZhengZhou	24 Sep 2015[b]	209	2	–	–	–	–	–	0.54% (2/369)	0.02–2.09
	21 Apr 2017	30	–	–	–	–	–	–		
	01 Aug 2017	130	–	–	–	–	–	–		
II KaiFeng	19 Oct 2015	42	3	–	–	–	–	–	3.49% (3/86)	0.77–10.18
	24 Sep 2015	44	–	–	–	–	–	–		
III PingDingShan	25 Sep 2015	300	7	3	2	2	1	–	1.98% (7/353)	0.88–4.12
	27 Jun 2017	4	–	–	–	–	–	–		
	07 Jun 2017	49	–	–	–	–	–	–		
IV LuoYang	22 Sep 2015	280	31	23	12	9	6	6	10.00% (31/310)	7.10–13.88
	22 Mar 2017	30	–	–	–	–	–	–		
V AnYang	15 Jan 2015	369	2	–	–	–	–	–	0.41% (2/488)	0.01–1.58
	05 Jul 2016	4	–	–	–	–	–	–		
	25 Apr 2017	32	–	–	–	–	–	–		
	07 Jun 2017	31	–	–	–	–	–	–		
	28 Jun 2017	22	–	–	–	–	–	–		
	29 Jun 2017	30	–	–	–	–	–	–		
VI JiaoZuo	16 Oct 2015	200	7	–	–	–	–	–	2.29% (7 /306)	1.02–4.74
	22 Apr 2017	29	–	–	–	–	–	–		
	26 Jun 2017	29	–	–	–	–	–	–		
	05 Sep 2017	48	–	–	–	–	–	–		
VII HeBi	15 Jan 2015	552	12	2	2	2	2	–	2.07% (12/581)	1.14–3.62
	15 Jun 2017	29	–	–	–	–	–	–		
VII XinXiang	16 Oct 2015	452	11	2	1	1	–	–	2.28% (11/482)	1.23–4.09
	22 Mar 2017	30	–	–	–	–	–	–		
IX PuYang	15 Oct 2015	250	2	–	–	–	–	–	0.71% (2/283)	0.02–2.71
	05 Sep 2017	33	–	–	–	–	–	–		
X XuChang	25 Sep 2015	200	5	2	2	2	2	–	1.77% (5/282)	0.64–4.20
	26 Apr 2017	52	–	–	–	–	–	–		
	27 Apr 2017	30	–	–	–	–	–	–		
XI LuoHe	25 Sep 2015	50	1	–	–	–	–	–	2.00% (1/50)	0.01–11.47
XII SanMenXia	23 Sep 2015	81	–	–	–	–	–	–	0 (0/150)	–
	21 Apr 2017	23	–	–	–	–	–	–		
	07 Sep 2017	46	–	–	–	–	–	–		
XIII NanYang	14 Oct 2015	170	2	–	–	–	–	–	1.06% (3/283)	0.21–3.22
	21 Mar 2017	16	–	–	–	–	–	–		
	22 Mar 2017	39	–	–	–	–	–	–		
	27 Jun 2017	29	–	–	–	–	–	–		
	07 Sep 2017	29	1	–	–	–	–	–		
XIV XinYang	18 Apr 2017	31	–	–	–	–	–	–	0 (0/31)	–
XV ZhouKou	19 Oct 2015	50	2	1	–	–	–	–	5.63% (4/71)	1.80–14.03
	21 Mar 2017	21	2	2	2	1	1	1		
XVI ZhuMaDian	25 Apr 2017	31	–	–	–	–	–	–	0.76% (1/131)	0.01–4.62
	29 Jun 2017	10	–	–	–	–	–	–		

Table 1 Seroprevalence of *T. gondii* in cattle in Henan Province *(Continued)*

Location/ City	Samples obtained date	Tested No.	No. of seropositive samples /(titer)						% (Positive No. /Test No.)	95% CI
			1:100	1:200	1:400	1:800	1:1600	1:3200		
	06 Sep 2017	90	1	–	–	–	–	–		
XVII JiYuan	24 Sep 2015	250	6	1	–	–	–	–	2.50% (7/280)	1.11–5.17
	22 Mar 2017	30	1	1	–	–	–	–		
XVII NK[a]	14 Oct 2015	657	4	3	–	–	–	–	0.53% (4/756)	0.15–1.40
	16 Oct 2015	50	–	–	–	–	–	–		
	04 Jul 2017	49	–	–	–	–	–	–		
Total		5292	102	40	21	17	12	7	1.93% (102/5292)	1.59–2.34

[a]NK means the city information of samples were not known
[b]The samples sampled on same date from a single farm

Dubey and Jokelainen et al. have conducted *T. gondii* serological testing of cattle [13, 17], and emphasize that the results of *T. gondii* serological screening of cattle should be interpreted with caution. Serology is an indirect method, and the positive results indicate that the host has been exposed to the parasite and produced measurable humoral immune responses. An increase in seroprevalence in cattle up to the age of five years has been reported [17]. The details of the age of the cattle included in the current study were unfortunately unknown, but the age range was wide and did not include only old cattle.

The overall estimated global prevalence of *T. gondii* in cattle using various detection methods is 14.96% (8286/55,377, 1.40–91.80%) [1]. Furthermore, the overall estimated seroprevalence of *T. gondii* in other larger ruminants, specifically water buffalo, horse, and camel was

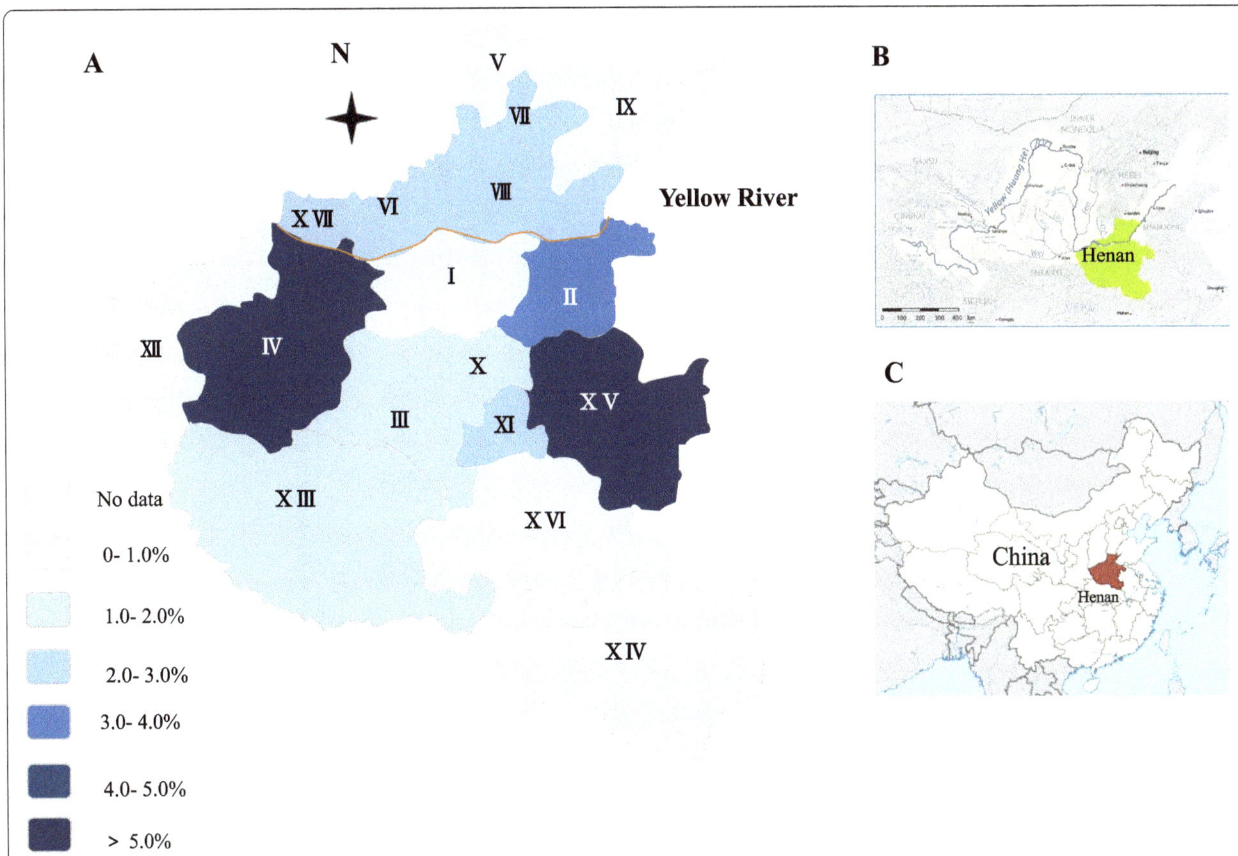

Fig. 1 Location of samples received from Henan Province of China. Figures were adapted from Wikipedia. **a** Cities in Henan Province I: ZhengZhou, II: KaiFeng, III: PingDingShan, IV: LuoYang, V: AnYang, VI: JiaoZuo, VII: HeBi, VIII: XinXiang, IX: PuYang, X: XuChang, XI: LuoHe, XII: SanMenXia, XIII: NanYang, XIV: XinYang, XV: ZhouKou, XVI: ZhuMaDian, XVII: JiYuan. Yellow line is the Yellow River. **b** The Yellow River area in China (adapted from Wikipedia). **c** Henan Province in China (adapted from Wikipedia)

Table 2 *P* values of comparison of seroprevalence of *T. gondii* infection in cattle from different cities by Fisher's exact test

Location	ZhengZhou	KaiFeng	PingDingShan	LuoYang	AnYang	JiaoZuo	HeBi	XinXiang	PuYang	XuChang	LuoHe	SanMenXia	NanYang	XinYang	ZhouKou	ZhuMaDian
KaiFeng	0.0489*	–	–	–	–	–	–	–	–	–	–	–	–	–	–	–
PingDingShan	0.1003	0.4192	–	–	–	–	–	–	–	–	–	–	–	–	–	–
LuoYang	0.0001*	0.0790	0.0001*	–	–	–	–	–	–	–	–	–	–	–	–	–
AnYang	1.0000	0.0259*	0.0398*	0.0001*	–	–	–	–	–	–	–	–	–	–	–	–
JiaoZuo	0.0867	0.4624	0.7940	0.0001*	0.0316*	–	–	–	–	–	–	–	–	–	–	–
HeBi	0.0937	0.4263	1.0000	0.0001*	0.0271*	0.8113	–	–	–	–	–	–	–	–	–	–
XinXiang	0.0484*	0.4553	0.8146	0.0001*	0.0119*	1.0000	0.8350	–	–	–	–	–	–	–	–	–
PuYang	1.0000	0.0851	0.3114	0.0001*	0.6271	0.1794	0.1625	0.1476	–	–	–	–	–	–	–	–
XuChang	0.2488	0.3963	1.0000	0.0001*	0.1067	0.7744	1.0000	0.7955	0.2856	–	–	–	–	–	–	–
LuoHe	0.3176	1.0000	1.0000	0.1022	0.2541	1.0000	1.0000	1.0000	0.3872	1.0000	–	–	–	–	–	–
SanMenXia	1.0000	0.0473*	0.1095	0.0001*	1.0000	0.1015	0.1400	0.0751	0.5461	0.1685	0.2500	–	–	–	–	–
NanYang	0.6572	0.1419	0.5244	0.0001*	0.3625	0.3433	0.4082	0.2742	1.0000	0.5043	0.4800	0.5545	–	–	–	–
XinYang	1.0000	0.5644	1.0000	0.0938	1.0000	1.0000	1.0000	1.0000	1.0000	1.0000	1.0000	–	1.0000	–	–	–
ZhouKou	0.0073*	0.7020	0.0942	0.3615	0.0030*	0.1332	0.0855	0.1134	0.0163*	0.0845	0.2746	0.0100*	0.0322*	0.3111	–	–
ZhuMaDian	1.0000	0.3032	0.6889	0.0002*	0.5106	0.4450	0.4805	0.4767	1.0000	0.6695	0.4773	0.4662	1.0000	1.0000	0.1256	–
JiYuan	0.0440*	0.7051	0.7873	0.0002*	0.0139*	1.0000	0.8048	0.8105	0.1050	0.5758	1.0000	0.1017	0.2203	1.0000	0.2430	0.4450

*The difference was considered significant when *P* value less than 0.05

Table 3 Seroprevalence and risk factors of *T. gondii* in dairy cattle in Henan Province

Factor	Category	No. of tested	Number of seropositive samples /(cut-off)						Prevalence (%)	95% CI
			1:100	1:200	1:400	1:800	1:1600	1:3200		
Geography[a]	North of the Yellow River	2420	41	6	3	3	2	0	1.69%	1.25–2.30
	South of the Yellow River	2116	57	31	18	14	10	7	3.67%	2.08–3.48
Season[b]	Summer	416	0	0	0	0	0	0	–	–
	Spring[a]	424	3	3	2	1	1	1	0.71%	0.14–2.16
	Winter	921	14	2	2	2	2	2	1.52%	0.88–2.56
	Autumn[a]	3531	85	19	17	14	9	6	2.41%	1.95–2.97

[a]*P*-value < 0.05 by two-tailed chi-square tests for geography factor
[b]Spring includes March, April and May. Summer includes June, July, and August. Autumn includes September, October, and November. Winter includes December, January and February

13.49%, 9.34%, and 35.92%, respectively [1]. The seroprevalence of *T. gondii* based on screening cattle from China is 9.08% (1560/17168, 1.73–46.40%) [18]. In the present study, the seroprevalence of *T. gondii* in cattle was 1.93% (102/5292), which is lower than the rest of the world and China's average infection rate. It is also lower than the seroprevalence of *T. gondii* in free-range chickens (18.86%, 132/700) [19], ostrich (10.20%, 20/197) [20], sheep (29.33%, 83/283) [21], swine (13.08%, 304/2325) [22], domestic cats (50%, 21/42) [23], and large cats (88.9%, 8/9) [24] in Henan Province. The maximum titer against *T. gondii* antibodies in dairy cattle was 3200 in this study. Considering the effective immune response to *T. gondii* infection in cattle, cattle with high antibody titers may be in the post-acute phase of toxoplasmosis, whereas those with low antibody titers may be in the chronic phase and may contain viable cysts of the parasite in their tissues [1, 25, 26]. However, seronegativity does not guarantee that the meat or milk from cattle will be free of *T. gondii.*

The seroprevalence of 1.93% in this survey indicates that cattle from central China are widely exposed to *T. gondii*. The route of *T. gondii* infection in cattle is probably by ingestion of *T. gondii* oocysts shed by infected felids. This finding suggests that cattle from Henan Province come into contact with *T. gondii* oocysts from cats or from soil, water, or feed. A single *T. gondii* felid could shed millions of oocysts after ingestion of raw meat containing *T. gondii* cysts, and the oocysts can survive in the environment for several years [27]. Furthermore, 100%, 10%, and 71% of cats shed oocysts after primary, secondary, and tertiary infection *T. gondii* [1, 28]. The high seroprevalence of *T. gondii* in domestic cats (50%, 21/42) and wild captured large cats (88.9%, 8/9) from Henan Province [23, 24] suggest that the risk of environmental contamination from felids should be given more attention.

Cattle can acquire *T. gondii* oocysts from the environment by direct contact with soil, feed, or water. The seroprevalence was highest in samples collected in autumn, suggesting that fresh grass used as cattle food may harbor *T. gondii* oocysts, and a humid environment favors the survival of oocysts in this season. The results of the present study agree with those of previous reports [29–31]. However, some studies have shown that season is not a risk factor for *T. gondii* infection [32, 33].

Henan Province is located in the downstream area of the Yellow River. LuoYang and Zhou Kou are much closer to the Yellow River than other cities, and *T. gondii* prevalence in the two regions is relatively higher ($P < 0.05$, Fig. 1). *T. gondii* oocysts can be transported via freshwater runoff into the ocean, thereby posing a threat to humans and other animals residing close to the river [34]. In the present study, risk factor analysis showed that geographic location is associated with *T. gondii* seroprevalence. The seroprevalence of *T. gondii* infection in the cattle from south of the Yellow River was higher than that of the north of the Yellow River ($P < 0.05$), which may be attributable to the mountainous geographic environment and higher temperature.

To our knowledge, neither natural cases of toxoplasmosis nor isolated strains of *T. gondii* in Chinese cattle have been reported to date. Cattle are considered poor hosts for *T. gondii* and good hosts for *Neospora caninum*. *N. caninum* infection in cattle has been found to be widespread in central China [35, 36], and this may cause reproductive losses in the cattle [37]. The dominant *T. gondii* genotype in China is ToxoDB#9 [38]. Several studies have identified *T. gondii* ToxoDB#9 in swine, domestic cats, and sheep in Henan Province [21, 23, 24, 39, 40]. However, only ToxoDB#225 *T. gondii* has been identified in Chinese cattle [40]. Additional investigations on cattle toxoplasmosis are thus warranted.

Conclusions

This is the first large-scale investigation on the seroprevalence of *T. gondii* infection in cattle from central China. The rate of *T. gondii* infection in dairy cattle is relatively low, and this information may be integrated into the Chinese *T. gondii* epidemiology database. Additionally, geographic location appeared as a risk factor for *T. gondii* seropositivity, with prevalence of *T. gondii* in cattle along the Yellow River higher than those of other areas.

Abbreviations
MAT: Modified agglutination test; PCR: Polymerase chain reaction

Acknowledgments
We thank Jian Li and Zifu Zhu (Henan Agricultural University, Zhengzhou, China) for performing some of the laboratory detections.

Funding
The China Postdoctoral Science Foundation (2016 M600577) and the Program for Science and Technology Innovation Talents in Universities of Henan Province (17HASTIT038) supported this study.

Authors' contributions
HD and YYL performed the laboratory tests, data analysis, and wrote the manuscript; RJS, YHW, MYW, and YBJ participated in sample collection and laboratory testing; and YRY designed the study protocol, analyzed the results, and assisted in writing the manuscript. All authors read and approved the final manuscript.

Consent for publication
Not applicable.

Competing interests
The authors declare that they have no competing interests. None of the authors of this report have financial or personal relationships with other people or organizations that could inappropriately influence its content.

Author details
[1]College of Animal Science and Veterinary Medicine, Henan Agricultural University, Zhengzhou, People's Republic of China. [2]Center for Animal Disease Control and Prevention of Henan Province, Zhengzhou, China.

References
1. Dubey JP. Toxoplasmosis of animals and humans. 2nd ed. Boca Raton: CRC Press; 2010.
2. Dubey JP. Isolation of *Toxoplasma gondii* from a naturally infected beef cow. J Parasitol. 1992;78:151–3.
3. Dubey JP, Thulliez P. Persistence of tissue cysts in edible tissues of cattle fed *Toxoplasma gondii* oocysts. Am J Vet Res. 1993;54:270–3.
4. Dubey JP. Toxoplasmosis- a waterborne zoonosis. Vet Parasitol. 2004;126:57–72.
5. Weiss LM, Dubey JP. Toxoplasmosis: a history of clinical observations. Int J Parasitol. 2009;39:895–901.
6. Wang ZD, Liu HH, Ma ZX, Ma HY, Li ZY, Yang ZB, Zhu XQ, Xu B, Wei F, Liu Q. *Toxoplasma gondii* infection in immunocompromised patients: a systematic review and meta-analysis. Front Microbiol. 2017;8:389.
7. Li JN. Analysis of toxoplasma gondii infection and risk factors in Xinxiang area. Xinxiang Med. Col., 2015; master's thesis, in Chinese.
8. Wang J, Shi DM, Cheng HC, Huo J. Serological survey of toxoplasmosis of cattle in Zhengzhou City. Chin J Vet Med. 2013;49(9):39–40 in Chinese.
9. Zhou XX, Zhou H, Ning XD, Li J, Jian FC, Zhang LX, Zhao QY, Ning CS. Seroprevalence of *Toxoplasma gondii* in cattle, sheep and goats in partial area of China. Chin Herb Sci. 2014;34(5):43–6 in Chinese.
10. Dubey JP, Desmonts G. Serological responses of equids fed *Toxoplasma gondii* oocysts. Equine Vet J. 1987;19(4):337–9.
11. Montoya JG, Liesenfeld O. Toxoplasmosis. Lancet. 2004;363(9425):1965–76.
12. Esteban-Redondo I, Innes EA. *Toxoplasma gondii* infection in sheep and cattle. Comp Immunol Microbiol Infect Dis. 1997;20(2):191–6.
13. Dubey JP. A review of toxoplasmosis in cattle. Vet Parasitol. 1986;22:177–202.
14. Munday BL, Corbould A. Serological responses of sheep and cattle exposed to natural *Toxoplasma* infection. Aust J Exp Biol Med Sci. 1979;57(2):141–5.
15. Dubey JP, Desmonts G, McDonald C, Walls KW. Serologic evaluation of cattle inoculated with *Toxoplasma gondii:* comparison of Sabin-Feldman dye test and other agglutination tests. Am J Vet Res. 1985;46:1085–8.
16. Qin SY, Zhou DH, Cong W, Zhang XX, Lou ZL, Yin MY, Tan QD, Zhu XQ. Seroprevalence, risk factors and genetic characterization of *Toxoplasma gondii* in free-range white yaks *(Bos grunniens)* in China. Vet Parasitol. 2015;211:300–2.
17. Jokelainen P, Tagel M, Motus K, Viltrop A, Lassen B. *Toxoplasma gondii* seroprevalence in dairy and beef cattle: large-scale epidemiological study in Estonia. Vet Parasitol. 2017;236:137–43.
18. Dong H, Su RJ, Lu YY, Wang MY, Liu J, Jian FC, Yang YR. Prevalence, risk factors, and genotypes of *Toxoplasma gondii* in food animals and humans (2000-2017) from ChinaJ. Front Microbiol. 2018;9:2108.
19. Feng YJ, Lu YY, Wang YH, Liu J, Zhang LX, Yang YR. *Toxoplasma gondii* and *Neospora caninum* in free-range chickens in Henan Province of China. Biomed Res Int. 2016;2016:8290536.
20. Feng YJ, Lu YY, Wang YH, Zhang LX, Yang YR. *Toxoplasma gondii* and *Neospora caninum* in farm-reared ostriches *(Struthio camelus)* in China. BMC Vet Res. 2017;13(1):301.
21. Yang Y, Feng Y, Yao Q, Wang Y, Lu Y, Liang H, Zhu X, Zhang L. Seroprevalence, isolation, genotyping, and pathogenicity of *Toxoplasma gondii* strains from sheep in China. Front Microbiol. 2017;8:136.
22. Wang HY, Ren QE, Jian FC, Zhou Y, Zhang LX, Ning CS. Investigation on the swine *Toxoplasma gondii* infection in some districts of Henan province and the inoculation experiment in mice infected with isolates from pigs. Chin J Zoonoses. 2008;24(10):930–2 in Chinese.
23. Yang Y, Ying Y, Verma SK, Cassinelli AB, Kwok OC, Liang H, Pradhan AK, Zhu XQSC, Dubey JP. Isolation and genetic characterization of viable *Toxoplasma gondii* from tissues and feces of cats from the central region of China. Vet Parasitol. 2015;211:283–8.
24. Yang YR, Feng YJ, Lu YY, Dong H, Li TY, Jiang YB, Zhu XQ, Zhang LX. Antibody detection, isolation, genotyping, and virulence of *Toxoplasma gondii* in captive felids from China. Front Microbiol. 2017;8:1414.
25. Carmo ELD, Morais RDAPB, Lima MS, Moraes CCG, Albuquerque GR, Silva AVD, Povoa MM. *Anti-Toxoplasma gondii* antibodies in beef cattle slaughtered in the metropolitan region of Belem. Brazilian Amazon Rev Bras Parasitol Vet. 2017;26(2):226–30.
26. Opsteegh M, Teunis P, Zuchner L, Koets A, Langelaar M, van der Giessen J. Low predictive value of seroprevalence of *Toxoplasma gondii* in cattle for detection of parasite DNA. Int J Parasitol. 2011;41(3–4):343–54.
27. Afonso E, Germain E, Poulle ML, Ruette S, Devillard S, Say L, Villena I, Aubert D, Gilot-Fromont E. Environmental determinants of spatial and temporal variations in the transmission of *Toxoplasma gondii* in its definitive hosts. Int J Parasitol Parasites Wildl. 2013;2:278–85.
28. Zulpo DL, Sammi AS, Dantos JRD, Sasse JP, Martins TA, Minutti AF, Cardim ST, Barros LDD, Navarro IT, Garcia JL. Toxoplasma gondii: a study of oocyst re-shedding in domestic cats. Vet Parasitol. 2018;249:17–20.
29. Dubey JP. *Toxoplasma gondii* oocyst survival under defined temperatures. J Parasitol. 1998;84:862–5.
30. Boughattas S, Behnke J, Sharma A, Abu-Madi M. Seroprevalence of toxoplasma *gondii* infection in feral cats in Qatar. BMC Vet Res. 2017;13(1):26.
31. Liu XC, He Y, Han DG, Zhang ZC, Li K, Wang S, Xu LX, Yan RF, Li XR. Detection of *Toxoplasma gondii* in chicken and soil of chicken farms in Nanjing region. China Infect Dis Poverty. 2017;6(1):62.
32. Zhang XX, Shi W, Zhang NZ, Shi K, Li JM, Xu P, Zhao Q, Prevalence DR. Genetic characterization of *Toxoplasma gondii* in donkeys in northeastern China. Infect Genet Evol. 2017;54:455–7.
33. Elfahal AM, Elhassan AM, Hussien MO, Enan KA, Musa AB, El Hussein AM. Seroprevalence of *Toxoplasma gondii* in dairy cattle with reproductive problems in Sudan. ISRN Vet Sci. 2013;2013:895165.
34. Fayer R, Dubey JP, Lindsay DS. Zoonotic protozoa: from land to sea. Trends Parasitol. 2004;20:531–6.

35. Yang YR, Zhang QF, Kong YG, Ying YQ, Kwok OCH, Liang H, Dubey JP. Low prevalence of *Neospora caninum* and *Toxoplasma gondii* antibodies in dogs in Jilin, Henan and Anhui provinces of the People's Republic of China. BMC Vet Res. 2014;10:295.
36. Qian W, Wang T, Yan W, Zhang M, Han L, Xue R, Song S, Lv C. Seroprevalence and first multilocus microsatellite genotyping of *Neospora caninum* in dairy cattle in Henan. central China Vet parasitol. 2017;244:81–4.
37. Dubey JP, Hemphill A, Calero-Bernal R, Schares G. Neosporosis in animals. Boca Raton: CRC Press; 2017.
38. Pan M, Lyu C, Zhao J, Shen B. Sixty years (1957-2017) of research on toxoplasmosis in China-an overview. Front Microbiol. 2017;8:1825.
39. Wang H, Zhang L, Ren Q, Yu F, Yang Y. Diagnosis of swine toxoplasmosis by PCR and genotyping of *Toxoplasma gondii* from pigs in Henan. Central China BMC Vet Res. 2017;13:152.
40. Qian WF, Yan WC, Wang TQ, Shao XD, Zhai K, Han LF, Lv CC. Genetic characterization of *Toxoplasma gondii* from domestic animals in Central China. Trop Biomed. 2015;32:540–4.

Brucellosis in the Addis Ababa dairy cattle: the myths and the realities

Bedaso Mammo Edao[1,2*], Gizachew Hailegebreal[3], Stefan Berg[4], Aboma Zewude[6], Yemiserach Zeleke[6], Teshale Sori[2], Gizat Almaw[5], Adrian M. Whatmore[4], Gobena Ameni[6] and James L. N. Wood[1]

Abstract

Background: Bovine brucellosis is considered as an important disease among livestock and people in sub-Saharan African countries including Ethiopia. A cross-sectional study was conducted from November 2016 to May 2017 to estimate the prevalence and associated risk factors, and to assess knowledge-attitude and practices (KAP) of farm workers about bovine brucellosis in Addis Ababa dairy farms.

Results: A total of 1550 cattle from 127 dairy farms were serially tested using the Rose Bengal Plate Test (RBPT), Competitive Enzyme-Linked Immunosorbant Assay (c-ELISA) and Complement Fixation Test (CFT). Forty-three (2.77%) of the collected sera were positive by the RBPT and only one of these was positive by c-ELISA (0.06%) and none was positive by CFT. The knowledge of farm workers towards the disease was very low and risk factors associated with *Brucella* infection were apparent in the study area.

Conclusion: Seropositivity for *Brucella* spp. was found in only a very small percentage by c-ELISA test, although risk factors for transmitting *Brucella* infection were present. The results suggest that bovine brucellosis is currently not a generalized problem in dairy cattle of Addis Ababa. Since this favorable disease situation is not the result of informed policy, there is no guarantee that it will continue unchanged. Setting clear policy in control of the disease and implementing "One Health" are the most constructive approaches recommended.

Keywords: Addis Ababa, Brucella, Dairy cattle, Ethiopia, Risk factors, Seroprevalence

Background

Ethiopia has one of the largest cattle populations in Africa [1] despite gaining minimum return from this resource as a result of various technical and non-technical factors, including infectious diseases. Bovine brucellosis is one of the infectious diseases hampering productivity of cattle and has been reported from several parts of the country [2] Bovine brucellosis is a zoonotic disease with economic and public health impact, particularly for human and animal populations in developing countries that rely mainly on livestock production [3]. The disease can generally cause significant loss of productivity through abortion, stillbirth, low herd fertility and low milk production [4].

* Correspondence: edaobedaso@gmail.com; bme27@cam.ac.uk
[1]Disease Dynamics Unit, Department of Veterinary Medicine, University of Cambridge, Madingley Road, Cambridge CB3 0ES, UK
[2]College of Veterinary Medicine, Addis Ababa University, P.O.Box 34, Bishoftu, Ethiopia
Full list of author information is available at the end of the article

Bovine brucellosis is considered to be predominantly caused by *Brucella abortus*; and, to a much lesser extent by *B. melitensis*, where cattle are kept together with infected goat or sheep. It is characterized clinically by abortion at first gestation, retained fetal membrane (RFM), metritis, orchitis and epididymitis [5, 6]. Sources of infection for the transmission of the bovine brucellosis are aborted fetuses, retained fetal membranes, and vaginal discharges and milk from infected animals. Direct contact with an aborting cow and the aborted fetus and indirect contact with contaminated fomites are the most common means of transmission of the disease in cattle. Ingestion of contaminated feed, fodder, water and grazing on contaminated pasture, may also play a secondary role in disease transmission [5, 7]. The disease has been eradicated from many developed countries; it however, remains a major public and animal health problem in many developing countries, where rural income relies mainly on livestock and dairy products [8]. Brucellosis in human often originates from domestic animal reservoir and associated with various risk factors and behavioral practices such as consumption of raw milk and milk products and close contact with infected animals [9, 10].

The risk factors that influence the spread and maintenance of brucellosis are related to management systems, the genetic content of susceptible animal population, biology of agents causing the disease and environmental factors. These include the size and composition of the herd, age of the animals, frequent contact between infected and susceptible herds, poor farm biosecurity and climate change [5, 11].

In Ethiopia there is no documented information on how and when bovine brucellosis was introduced and established. However, in the last two decades several serological surveys have showed that it is endemic and widespread [2, 12–26]. These studies in animals and humans were largely confined to serological surveys and commonly targeted cattle, occasionally sheep and goats, and rarely camels. So far, there was only one attempt to identify *Brucella* species in the country [27]; the distribution and proportion of their natural hosts was also not studied exhaustively [19]. According to [28], *Brucella* seroprevalence in dairy cattle of Ethiopia revealed highest prevalence in central Ethiopia followed by southern part whereas lowest prevalence was revealed in western part of the country.

Market oriented new dairy industries are emerging in Ethiopia so as to contribute hugely towards filling the gap between an increasing national demand and supply for milk and milk products. Cow dairy development roadmap of the country is aimed at increasing the productivity of indigenous cattle through improvements in genetics, health and feeding to satisfy consumption demand and start export of cow milk and milk products. It is important that public and private dairy industries and cooperatives require up to date and consistent scientific data on bovine brucellosis. This would in turn assist in developing baseline information to establish nationwide bovine brucellosis intervention policies aimed at controlling and eradicating the disease.

Thus, this study was designed to estimate the seroprevalence of brucellosis in dairy cattle, to identify potential risk factors and assess knowledge, attitude and practices (KAP) of the farm workers towards the disease.

Results

Knowledge-attitude and practices (KAP) of the farm workers about brucellosis

Demographic characteristics

Out of 127 surveyed farms, 130 farm workers in 59 farms were willing to participate in the KAP study. Of the 130 farm workers responsible for the management of the farm, the majority (88%) was male, and more than 50% were between ages 25 to 60 years. Half of the participants were not married (51.5%) and most of them had family size below 10 people. Eighty percent of the respondents had attended only primary school and more than 5 % were illiterate (Table 1).

Risk factors

Most of the study participants reported several risk factors for acquiring bovine brucellosis. The majority of participants, 96% of farm workers in small scale, 100% in medium size and 92.6% in large herd sized farms were not aware of bovine brucellosis. Most respondents, 83.6% in small scale, 60% in medium size and 81.5% in large farms disposed of dead fetus/after birth to open dump in the environment and more than 5 % of participants in small herd sized farms fed aborted materials to dogs. Almost all participants in small scale and medium sized farms and 77.8% in large-scale farms practiced assisted parturition without wearing protective gloves or masks. More than 60 and 80% percent of farm workers in all farm sizes consume raw milk and meat, respectively (Table 2).

Farm characteristics

Of the 127 farms, 103, 17 and 7 were small, medium and large herd sized farms, respectively. Of the farms assessed by a questionnaire survey, more than 70% of medium and large farms, as well as 42.7% of small sized farms, had reproductive problems (abortion, stillbirth, retained fetal membrane and repeat breeding) in their farms. The majority of farms were using artificial

Table 1 Demographic characteristics of farm workers in the study area ($n = 130$)

Demographic characteristics	Category	number	Percent
Gender	Female	16	12.3
	Male	114	87.7
Age	12–24	54	41.5
	25–60	72	55.4
	> 60	4	3.1
Marital status	Single	67	51.5
	Married	63	48.5
Level of education	Illiterate	9	6.9
	Primary	104	80
	Secondary	13	10
	Technical/Diploma	3	2.3
	Degree	1	0.8
No of people in the household[a]	1–5	61	46.9
	6–10	67	51.5
	> 10	2	1.6

[a]Household defined as family members regularly living together and sharing meals, *n* number

Table 2 Knowledge-attitudes and practices of farm workers about *Brucella* infection in the study area

Variables	Proportion of respondents (n)		
	Herd size		
	Small (*n* = 73) n (%)	Medium (*n* = 30) n (%)	Large (*n* = 27) n (%)
Awareness about brucellosis			
Yes	3 (4)	0 (0)	2 (7.4)
No	70 (96)	30(100)	25 (92.6)
Dead fetus/After birth disposal			
Burning/Burying	7 (9.6)	12 (40)	5 (18.5)
Open dump	61 (83.6)	18 (60)	22 (81.5)
Give to dog	5 (6.8)	0 (0)	0 (0)
Assist parturition			
Yes	72 (98.6)	30(100)	21 (77.8)
No	1 (1.4)	0 (0)	6 (22.2)
Consume raw milk			
Yes	50 (68.5)	18 (60)	20 (74)
No	23 (31.5)	12 (40)	7 (26)
Consume raw meat			
Yes	60 (82.2)	25(83.3)	24 (88.9)
No	13 (17.8)	5 (16.7)	3 (11.1.)

n number

insemination (AI) for breeding purposes and 71.4% of large sized farms raised their own replacement animals whereas most of small and medium sized farms used both (raised their own stock and purchase). The practices of provision of separate pens for parturition and aborted animals were 28.6 and 14.3% in large sized farms, respectively, whereas there were no such practice in small and medium sized farms and almost all farms used flushing with tap water to clean pens after parturition. The majority (85.7%) of the large herd sized farms, and more than 40% of medium and small-scale farms, used separate feed and water supply for each animal. Reproductive problems and age were prominent culling criteria in all farms and the majority of farms (> 70%) in the study area did not report frequent contact of dairy animals with other species (sheep and goat) (Table 3).

Seroprevalence of bovine brucellosis in dairy cattle

A total of 1550 dairy cattle were tested with RBT and 43 (2.77%) of them were positive in this test. The RBT positive sera samples were further tested using c-ELISA and CFT. Only one animal was confirmed seropositive for bovine brucellosis in the study area based on c-ELISA and no sero reactor animal was found by CFT. In addition to RBT positive sera samples, equal number of randomly selected RBT negative sera were shipped to APHA, Weybridge, UK and further tested using RBT and C-ELISA. However, the result was the same. The seroprevalence of bovine brucellosis in Addis Ababa dairy farms was thus 0.06% (1/1550) based on c-ELISA test. The herd level seroprevalence of brucellosis based on c-ELISA was 0.8% (1/127). The prevalence of antibodies to *Brucella* spp. in small, medium and large sized cattle farms was 0, 5.8% (1/17) and 0%, respectively (Table 4).

Discussion

Improvement of knowledge-attitudes and practices among urban dairy farm workers could have a significant impact on the reduction of many zoonotic infections, including brucellosis. The results of the KAP study show that the majority of farm workers in the studied dairy farms were not aware of bovine brucellosis (96.1%). Farm workers with a primary and lower level of education were less likely to have heard of brucellosis when compared to those with secondary and higher level of education. Animal attendants with a primary and lower level of education are hence likely at a higher risk of exposure to the disease. Similar findings were reported by a study conducted in Tajikistan [29]. Low awareness of the disease in the study area might be explained by the low prevalence of brucellosis in dairy cows. The majority (88%) of farm workers were male. This could be due to the fact that farm works in urban and peri-urban intensive diary is labor demanding, as a result of which most farm owners prefer to employ male farm workers.

Practices posing a high risk of *Brucella* transmission are very common: most participants reported assisting in animal parturition, disposing aborted fetuses/after birth in open environment without protective gloves or masks and consumption of raw meat and milk. The reason could be poor knowledge of the disease and risks of transmission but also lack of resources used for personal protection such as gloves, aprons and antiseptics. Creation of awareness of the farm owners and attendants is important to control brucellosis in the area even though the prevalence in animals was low in this serological survey.

In the present study, bovine brucellosis at individual animal level was 0.06% (1/1550) and herd level prevalence was 0.8% (1/127) using c-ELISA whereas no seropositive animal was found on CFT in dairy farms of Addis Ababa. This observation is consistent with previous reports made by [30] in Eastern and Western Showa zones of central Ethiopia using Rose Bengal Plate test (RBPT), serum agglutination test (SAT) and complement fixation test (CFT) (*n* = 564). This report is also consistent with [31], who reported no positive reactors in intensive dairy farms of the Addis Ababa area (*n* = 747). Similarly, a study by [23] could not find positive reactors in Adama, central Ethiopia (*n* = 52) and northern Ethiopia (Mekele and Gondar) (*n* = 252). A study

Table 3 Summary of characteristics of dairy farms in the study area

Variables	Herd size		
	Small ($n = 103$) Frequency (%)	Medium ($n = 17$) Frequency (%)	Large (n = 7) Frequency (%)
Reproductive problems			
Yes	44 (42.7)	12 (70.6)	5 (71.4)
No	59 (57.3)	5 (29.4)	2 (28.6)
Service type			
AI	71 (69)	9 (53)	5 (71.4)
Bull Both	9 (8.7) 23 (22.3)	4 (23.5) 4 (23.5)	0 (0) 2 (28.6)
Replacement stock			
Raised own	31 (30.1)	8 (47)	5 (71.4)
Purchased Both	5 (4.9) 67 (65)	1 (6) 8 (47)	0 (0) 2 (28.6)
Parturition pen			
Yes	0 (0)	0 (0)	2 (28.6)
No	103 (100)	17 (100)	5 (71.4)
Cleaning pen after parturition			
Flushed with tap water	101 (97)	17 (100)	7 (100)
Flushed with water and disinfect	2 (3)	0(0)	0 (0)
Separate pen for aborted cow			
Yes	1 (1)	0 (0)	1 (14.3)
No	102 (99)	17 (100)	6 (85.7)
Feed and water supply			
Own	50 (48.5)	7 (41.2)	6 (85.7)
Communal	5 (4.9)	4 (23.5)	0 (0)
Both	48 (46.6)	6 (35.3)	1 (14.3)
Culling criteria			
Reproductive problem	57 (55.3)	6 (35.3)	4 (57.1)
Logistics	7 (6.8)	3 (17.7)	0 (0)
Age	39 (37.9)	8 (47)	3 (42.9)
Contact with other spp. [a]			
Yes	12 (11.6)	4 (23.5)	2 (28.6)
No	91 (88.4)	13 (76.5)	5 (71.4)

Contact with other spp. [a] Sheep and goat

Table 4 Individual animal level and herd level prevalence of bovine brucellosis in dairy farms of Addis Ababa

Farm type	Individual farm				Herd level			
	No tested	RBT Positive n^a (%)	C-ELISA Positive n^a (%)	CFT Positive n^a (%)	No Tested	RBT Positive n^a (%)	C-ELISA Positive n^a (%)	CFT Positive n^a (%)
Small	821	5(0.6)	0(0)	0(0)	103	5(4.8)	0(0)	0(0)
Medium	363	9(2.5)	1(0.27)	0(0)	17	6(35.3)	1(5.8)	0(0)
Large	366	29(7.9)	0(0)	0(0)	7	3(42.9)	0(0)	0(0)
Total	1550	43(2.77)	1(0.06)	0(0)	127	14(11)	1(0.8)	0(0)

n^a number positive

conducted in Debrebirhan and Ambo towns by [32] reported that there was only one sero-reactor animal to *Brucella* infection using CFT($n = 415$).

In contrast, there are reports of a higher prevalence of antibodies to *Brucella* spp. in Addis Ababa dairy farms, 1.5% [11] and 2.2% [23]. A similar study by [26] reported 1.4% in Asella and Bishoftu towns using the card test (CT), RBPT, indirect Enzyme-Linked Immuno Sorbant Assay (i-ELISA) and Complement Fixation Test (CFT). In Ethiopia, brucellosis in animals has been reported from different localities of the country; with prevalence ranging from 0.4 to 8% particularly associated with cattle in both intensive and extensive management systems [2, 11–15, 18–25]. A high seroprevalence of brucellosis (38%) in cattle in western Ethiopia has been reported [33], while most of the studies suggested a low seroprevalence (below 5%) in cattle.

In the present study only one sero reactor animal to *Brucella* spp. was found in cattle populations of Addis Ababa dairy farms. According to the individual animal record in the farm, this seroreactor animal in the farm was purchased from outside the capital and had late abortion history at first calving. This is remarkable, as bovine brucellosis is considered the world's most common bacterial zoonosis [34] and listed among top five zoonotic diseases in Ethiopia [35]. The hypothesis that brucellosis is endemic in the investigated dairy farms of Addis Ababa could thereby not be confirmed in the present study. However, the presence of one or more positive reactors in the herd is a reliable predictor of the presence of infection [5]. Seropositivity for *Brucella* spp. was found in only a very small percentage by c-ELISA test, although risk factors for obtaining *Brucella* infection such as, age variety, origin of animals, different level of parity, history of abortion, and herd size composition were present as revealed by farm characteristics analysis.

In epidemiological studies, the use of two or more tests applied serially is recommended to maximize the accuracy of test results. RBPT is highly sensitive test and c-ELISA and CFT are highly specific and also sensitive and usually used as confirmatory tests [36]. The combination of these tests in this study could therefore maximize the accuracy of the findings. False positive serological reactors in RBT are due to cross-reactions with Smooth Lipopolysaccharide (S-LPS) antigens of other bacteria. As there has never been history of vaccination, seropositivity in this case is due to natural infection.

The difference in test results of C-ELISA and CFT is due to the variation in sensitivity of the tests. The C-ELISA test is more sensitive as compared to CFT for the diagnosis of brucellosis even though both tests have similar specificities of 99.9% [36]. Moreover, CFT is prone to prozone effect (low dilutions of some titrated sera from infected animals do not fix complement) that could lead to a false negative result [37, 38].

The very low prevalence could be explained by the cross sectional study design if informal culling practices suggested by [17] had been instituted. These include culling of cows with abortion history of at least two times and above for any reason and removing seropositive reactors from the herd for economic reasons. The other possible explanation might be absence of infectious foci, such as *Brucella*-infected dairy farms or ranches in the surrounding areas, which could spread the disease among contact herds. Movement of infected animals to susceptible herds is a common route of transmission [5]. The random selection used in our study design should have detected clustered infection if it were common [39].

Conclusions

Urban and peri-urban dairy farming offers an important opportunity to improve the livelihood of people in low-income countries. There was only one seropositive animal found for bovine brucellosis in the present study by the c-ELISA test. This study showed poor knowledge of brucellosis and abundant high-risk behaviors among the farm workers. Poor knowledge and high-risk practices strengthens the logic for including health education as part of control programs. At present, there is no officially coordinated program for control of bovine brucellosis in Ethiopia. The disease is unlikely to be a significant limiter of dairy production in the Addis Ababa area due to the low prevalence, but it may be present in other animal types. Since this favorable disease situation is not the result of informed policy, there is no guarantee that it will continue unchanged.

The current study warrants the need for constant surveillance program in case the prevalence rates do change. Implementing a *"test and slaughter"* program in the zero grazing system to eliminate the existing low risk of brucellosis could avert the cost-related limitation of brucellosis control. Modernization of husbandry systems and testing of new animals before introducing to dairy farms should be encouraged. A multi-sectorial framework should be promoted involving all stakeholders working in public and animal health in the context of a "One Health" approach. Since the current study was limited to cross sectional study design, future studies in the dairy farms should follow longitudinal study types to ascertain actual burden of the disease in the study area.

Methods

Description of the study area

The study was conducted in Addis Ababa dairy farms. Addis Ababa, the capital city of Ethiopia, lies at an elevation of 2300 m (7,500 ft) above sea level and is featured by a grassland biome. It is geographically located at 9°1′

48″N latitude and 38°44′24″E longitude. It has a typical highland climate with temperature ranging from 11 °C - 24 °C. Addis Ababa has a mean annual rainfall of 1300 mm with bimodal distribution. The city is divided into 10 boroughs (Fig. 1), called sub cities, and 99 wards (kebeles) [40].

Dairy cattle production systems in Ethiopia are classified into commercial dairy systems, urban and peri-urban smallholder dairy, rural smallholder (mixed crop and livestock production), and pastoral and agro pastoral [23]. The dairy systems hold mainly exotic breeds or crosses with the local zebu breeds, while the rural husbandry systems stock mainly zebus. This study focused on urban and peri-urban smallholder farms, which produce milk for home use and sale, and commercial dairy systems, which are producing milk as a full-time business. These dairies constitute the main dairy source for the capital and produce milk for sale [26]. Within these systems dairy farms were classified based on size of the herd and herd management into large scale farms, with more than 50 animals, medium scale farms (20 to 50 animals) and smallholder farms (< 20 animals).

Variables collected

Explanatory variables that were hypothesized to be risk factors for the disease were assessed at both individual animal and farm level. Information related to herd structure was extracted from individual herd records and when this was not the case from farm owners or managers interview using a pre tested structured questionnaires. Herd level parameters studied include herd size as described above, the presence of reproductive problems such as abortion, retained fetal membranes and still birth in the farm, separate calving pen, brucellosis testing in the farm, frequent contact with animals in other herds and species, which were categorized as yes or no variables. The major reasons for culling were coded either as reproductive problems, old age or logistics. Breeding strategy was artificial insemination, bull or both. The afterbirth (aborted fetuses and fetal membranes) disposal method was also categorized into burying, open dump or feeding to dogs.

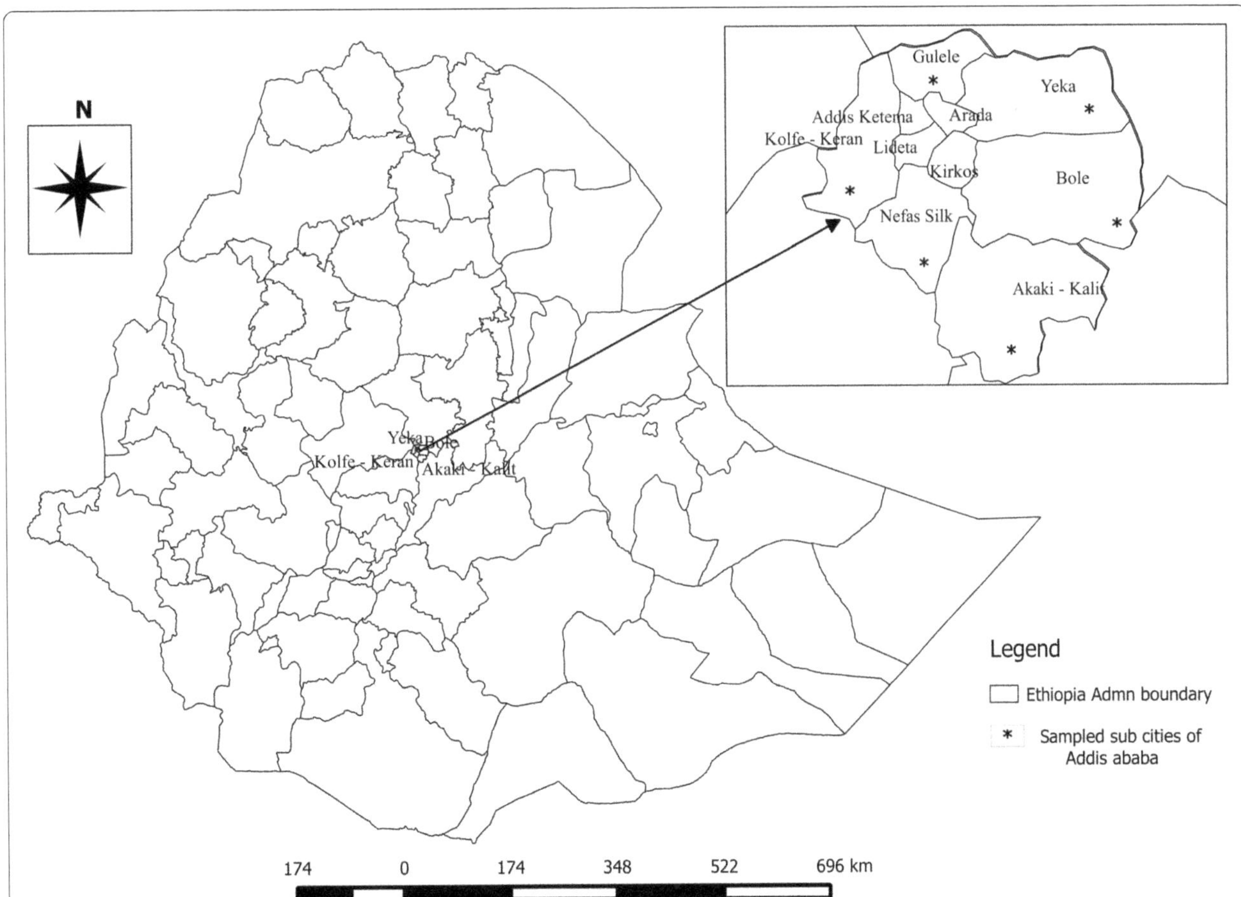

Fig. 1 Map of Addis Ababa showing studied/sampled sub cities. The map depicted in Fig. 1 is our own developed from Ethiopian shape files using QGIS software

The feeding and water supply strategies were classified into communal and own.

Individual animals were categorized as young 6–17 months or adult (≥17 months), breed as Holstein-Friesian (HF), cross, or local. The origin of each individual animal was defined as either own stock or purchased. The clinical reproductive disorder experience, such as abortion, stillbirth and retained fetal membrane were categorized as either yes or no variables. Parity number and frequency of some of the aforementioned disorders were also recorded. Physiological status of an animal was categorized as pregnant, non-pregnant, lactating and lactating-pregnant. Repeat breeding was also assessed based on the animal owner's general observation. Accordingly, cows that demanded 3 or more services per pregnancy were categorized as a repeat breeder otherwise they were categorized as a non-repeat breeder.

Study population

The target study populations were dairy cattle above six months of age, which consist of breeding females, replacement heifers, and available bulls. The breeds of these animals were crosses of local breed with HF. None of the animals tested were vaccinated against brucellosis. For the Knowledge, Attitude and Practice (KAP) study, occupationally associated farm workers, willing to be interviewed, were included.

Study design and sample size

A cross-sectional study design was employed from November 2016 to May 2017 to address the objectives of the study. According to Addis Ababa city urban agriculture bureau there are about 880 registered smallholder, medium scale and large commercial dairy farms in Addis Ababa. Individual farm was considered as a herd and the primary sampling unit. The sample size for dairy farms was calculated using a 9.1% herd level seroprevalence of bovine brucellosis [23], 95% confidence interval (CI) and 5% required precision [41]. Hence, a total of 127 dairy farms were considered for this study and proportional allocation was made for each sub city based on the number of farms. In each sub city, herd-sampling frame was established in collaboration with sub city veterinary department and farms were selected randomly using computer generated random numbers. Before data collection, consent was made with the identified farms owners requesting their farms to be included in the study. Farms where the owners were not willing to participate in the study were replaced by other farms. All cattle above six months of age in the selected dairy farms were included and a total of 1550 animals were sampled for serological screening. For the KAP study farm workers from sero surveyed farms, who agreed to be interviewed, were included. Hence, 130 farm workers from 59 farms participated in the study.

Data collection

Data concerning farm workers KAPs towards the disease were collected by interviewing individuals using a pre-tested structured questionnaire. Verbal consent was obtained from the respondents and the objective of the survey explained to them before start of the interview. The interviews were conducted in local languages (Afaan Oromo or Amharic). The questionnaire focused on demographic characteristic of the interviewee, knowledge-attitude about the disease, handling and afterbirth/aborted fetus disposal practices, habit of raw animal product consumption and animal feeding and housing practices.

Blood samples (10 ml) from the jugular vein of each animal were collected, using sterile needles and plain vacutainer tubes labelled with individual animal identification number. The blood samples were centrifuged at 3000×g for 10 min to obtain the serum within 12 h of collection. Sera were decanted into cryo-vials, identified and stored at −20 °C until screened for antibodies against natural *Brucella* exposure using serological analysis.

Rose Bengal plate test (RBPT)

All sera samples collected were initially screened by RBPT using RBPT antigen (Animal and Plant Health Agency, New Haw, Addlestone, Surrey, KT15 3NB, United Kingdom) according to OIE (2016) procedures. Sera and antigen were taken from refrigerator and left at room temperature for half an hour before the test to reach room temperature. Briefly, RBT antigen (30 µl) was added onto a glass slide next to an equal amount of cattle sera. The antigen and test serum were mixed thoroughly in a plastic applicator, shaken for 4 min, and agglutination was read immediately. Any observed agglutination by the naked eye was considered to be a positive reaction.

Competitive ELISA

All RBPT positive sera were further tested using the COMPELISA 160 and 400, a competitive ELISA kit for the detection of antibodies against *Brucella* in serum samples (Animal and Plant Health Agency, New Haw, Addlestone, Surrey, KT15 3NB, United Kingdom) at Addis Ababa University, Aklilu Lemma Institute of Pathobiology (AAU-ALIPB). The test was performed according to the manufacturer's instructions. The test was conducted in 96-well polystyrene plates that are pre-coated with *Brucella* species lipopolysaccharide (LPS) antigen. 20 µl of each test serum was added to each well followed by 100 µl of prepared conjugate solution. The plates were then shaken vigorously for two minutes and incubated at room temperature for 30 min on rotary shaker, at 160 revs/min. Plates were washed 5 times and dried. Hydrogen peroxidase substrate and chromogen solution was developed for

10 min. 100 µl of o-Phenylenediamine dihydrochloride (OPD) solution was added to all wells and the plates were incubated at room temperature for 10 to 20 min. Micro plate reader was switched on and the units allowed stabilizing for 10 min. The reaction was then being stopped using stopping solution. Optical densities (OD) were read at 450 nm using micro plate reader. The lack of color development indicated that the sample tested was positive. A positive/negative cut-off was calculated as 60% of the mean of the OD of the 4 conjugate control wells. Any test sample giving an OD equal to or below this value was regarded as being positive. An animal was considered positive if it tested seropositive on both RBPT and c-ELISA in serial interpretation.

Complement fixation test (CFT)

All samples that were RBPT-positive were further subjected to complement fixation test as a confirmatory test at the National Veterinary Institute (NVI), Bishoftu, Ethiopia. The *Brucella* antigen and control sera (positive and negative) used during the test were produced by Animal and Plant Health Agency, UK. The standardization of the antigen was made at 1:20 working dilution (strength). The *Brucella* antigen, complement and 3% sensitized sheep red blood cells were added after the test sera were serially diluted (1:5, 1:10, 1:20, and 1:40) in microtitre plates. Then the plates were incubated at 37 °C for 30 min. The test was considered positive when the reading was as partial fixation (50% haemolysis) or complete fixation (no haemolysis) at 1:10 dilution. The validity of the test was considered when there was complete hemolysis in negative control serum and the positive control shows inhibition of hemolysis.

Case definition

Animals were considered positive to brucellosis when they tested positive on either RBPT/CFT or RBPT/c-ELISA tests in parallel interpretation. Similarly, a herd or farm was considered seropositive when at least one animal in a herd or farm tested positive. Since there is no history of vaccination against brucellosis in Ethiopia, seropositivity observed in this study was considered to be due to natural infection.

Data analysis

Data generated from the questionnaire survey and laboratory investigations were recorded and coded using a Microsoft Excel spreadsheet (Microsoft Corporation) and analyzed using STATA version 13.0 for Windows (Stata Corp. College Station, TX, USA). Descriptive statistics were used for demographic and farm characteristics as well as KAPs relating to bovine brucellosis. The seroprevalence was calculated as the number of seropositive samples divided by the total number of samples tested.

Abbreviations

AAU-ALIPB: Addis Ababa University, Aklilu Lemma Institute of Pathobiology; AW: Adrian Whatmore; AZ: Aboma Zewde; BM: Bedaso Mammo; cELISA: Competitive Enzyme Linked Immunosorbent Assay; CFT: Complement Fixation Test; CI: Confidence Interval; GA: Gobena Ameni; GH: Gizachew Hailegebreal; GZA: Gizat Almaw; HF: Holstein Friesian; iELISA: Indirect Enzyme Linked Immunosorbent Assay; JW: James Wood; KAP: Knowledge Attitude Practice; LPS: Lipopolysaccharide; NVI: National Veterinary Institute; RBPT: Rosbengal Plate Test; RFM: Retained Fetal Membranes; SAT: Serum Agglutination Tests; SB: Stefan Berg; TS: Teshale Sori; YZ: Yemiserach Zeleke; ZELS: Zoonoses and Emerging Livestock Systems

Acknowledgments

We are grateful to the UK Department for International Development, the Biotechnology and Biological Sciences Research Council, the Economic & Social Research Council, the Medical Research Council, the Natural Environment Research Council and the Defence Science & Technology Laboratory, for funding the study under the Zoonoses and Emerging Livestock Systems (ZELS) programme, ref.: BB/L018977/1.The authors also express special thanks to University of Cambridge, Department of Veterinary Medicine, Animal and Plant Health Agency, UK, Addis Ababa University Aklilu Lemma Institute of Pathobiology (AAU-ALIPB) and National Animal Health Diagnostic and Investigation Centre, Ethiopia for facilitating the study and owners or managers of the dairy farms included in the study for their collaboration during the study period. We thank Andrew Taylor, APHA, for his support and advice on testing for Brucellosis. Members of the veterinary departments in the surveyed sub cities and districts are also very much appreciated for their cooperation in the fieldwork.

Funding

This work was funded by the Biotechnology and Biological Sciences Research Council, the Department for International Development, the Economic & Social Research Council, the Medical Research Council, the Natural Environment Research Council and the Defence Science & Technology Laboratory, under the Zoonoses and Emerging Livestock Systems (ZELS) programme, ref.: BB/L018977/1. The funders were not engaged in the design of the study, collection, analysis and interpretation of the data.

Authors' contributions

BM and GH collected all the required data. GZA contributed to the collection of the data. BM and GH performed the laboratory work. AZ and YZ contributed to the lab work. JW, GA, SB, AW, BM and GH designed the study. BM and GH analyzed and interpreted the data. BM and GH drafted the manuscript. JW, GA, SB AW and TS critically and substantially revised the manuscript. All authors read and approved the final manuscript.

Ethics approval and consent to participate

Ethical clearance for this study was obtained from University of Cambridge, Department of Veterinary Medicine, Addis Ababa University, Aklilu Lema Institute of Pathobiology, Addis Ababa University, College of Veterinary Medicine and Agriculture (AAU-CVMA), Minutes of Animal Research Ethics and Review committee. Before conducting the research, participants were informed with the objectives of the study and written and signed consent was obtained from the farm owners or managers of the dairy farms to take samples from their cattle and farm workers to be included in the KAP survey.

Consent for publication

Not applicable.

Competing interests

The authors declare that they have no competing interests.

Author details

[1]Disease Dynamics Unit, Department of Veterinary Medicine, University of Cambridge, Madingley Road, Cambridge CB3 0ES, UK. [2]College of Veterinary Medicine, Addis Ababa University, P.O.Box 34, Bishoftu, Ethiopia. [3]South Agricultural Research Institute, Worabe Agricultural Research Center, P.O.Box

21, Worabe, Ethiopia. [4]Animal and Plant Health Agency, Woodham Lane, Addlestone KT15 3NB, UK. [5]National Animal Health Diagnostic and Investigation Centre, Sebeta, Ethiopia. [6]Aklilu Lemma Institute of Pathobiology, Addis Ababa University, P.O.Box 1176, Addis Ababa, Ethiopia.

References

1. CSA. Federal Democratic Republic of Ethiopia. Central Statistical Agency, Agricultural Sample Survey Report on Livestock and Livestock Characteristics. 2015; Volume II, 2014/15. Addis Ababa, Ethiopia.
2. Ibrahim N, Belihu K, Lobago F, Bekana M. Seroprevalence of bovine brucellosis and its risk factors in Jimma zone of Oromia region, South-Western Ethiopia. Trop Anim Health Prod. 2010;42:34–40.
3. Schelling E, Diguimbaye C, Daoud S, Nicolet J, Boerlin P, Tanner M, Zinsstag J. Brucellosis and Q-fever sero prevalence of nomadic pastoralists and their livestock in Chad. Prev Vet Med. 2003;61:279–93.
4. Diaz Aparicio E. Epidemiology of brucellosis in domestic animals caused by *Brucella melitensis*, *Brucella suis* and *Brucella abortus*. Rev Sci Tech. 2013;32(1):53–60.
5. Radostits ED, Gay CC, Inchcliff KW. Veterinary Medicine Textbook of the disease of cattle, sheep, pigs, goats, and horses. 10th ed. Newyork: W. B. Sounders Company ltd; 2006. p. 963–94.
6. OIE. BRUCELLOSIS. In: Manual of diagnostic tests and vaccines for terrestrial animals; 2016. Available at: http://www.oie.int/fileadmin/Home/eng/Health standards/tahm/ 2.01.04_BRUCELLOSIS.pdf. Accessed 2 Feb 2017.
7. Debassa G, Tefera M, Addis M. Small ruminant brucellosis: serological survey in Yabello District, Ethiopia. Asia J Anim Sci. 2013;7(1):14–21.
8. Roth F, Zinsstag J, Orkhon D, Chimed-Ochir G, Hutton G, Cosivi O, et al. Human health benefits from livestock vaccination for brucellosis: case study. Bull World Health Org. 2003;81:867–76.
9. Musa MT, Eisa MZ, El Sanousi MA, Wahab EM, Perrett L. Brucellosis in camels (*Camelus dromedarius*) in Darfur, western Sudan. J Comp Pathol. 2008;13(8):151–5.
10. Bechtol D, Carpenter LR, Mosites E, Smalley D, Dunn JR. *Brucella melitensis* infection following military duty in Iraq. Zoonoses Public Health. 2011;58(2):489–92.
11. McDermott JJ, Arimi SM. Brucellosis in sub-Saharan Africa: epidemiology, control and impact. Vet Microbiol. 2002;90(1–4):111–34.
12. Berhe G, Belihu K, Asfaw Y. Seroepidemiological investigation of bovine brucellosis in the extensive cattle production system of Tigray region of Ethiopia. Intern J Appl Res Vet Med. 2007;5:2.
13. Jergefa T, Kelay B, Bekana M, Teshale S, Gustafson H, Kindahl H. Epidemiological study of bovine brucellosis in three agro-ecological areas of Central Oromiya, Ethiopia. Rev sci tech Off Int Epiz. 2009;28(3):933–43.
14. Asmare K, Asfaw Y, Gelaye E, Ayelet G. Brucellosis in extensive management system of zebu cattle in Sidama zone, Southern Ethiopia. Afr J Agric Res. 2010;5:257–63.
15. Degefu H, Mohamud M, Hailemelekot M, Yohannes M. Seroprevalence of bovine brucellosis in agro pastoral areas of Jijjiga zone of Somali National Regional State, eastern Ethiopia. Ethiop Vet J. 2011;15(1):37–47.
16. Megersa B, Biffa D, Niguse F, Rufael T, Asmare K, Skjerve E. Cattle brucellosis in traditional livestock husbandry practice in southern and eastern Ethiopia, and its zoonotic implication. Acta Vet Scandinavica. 2011;53:24.
17. Tesfaye G, Tsegaye W, Chanie M, Abinet F. Seroprevalence and associated risk factors of bovine brucellosis in Addis Ababa dairy farms. Trop Anim Health Prod. 2011;43:1001–5.
18. Megersa B, Biffaf D, Abunna F, Regassa A, Godfroid J, Skjerve E. Seroepidemiological study of livestock brucellosis in a pastoral region. Epidemiol Infect. 2012;140:887–96. https://doi.org/10.1017/S0950268811001178.
19. Yohannes M, Mersha T, Degefu H, Tolosa T, Woyesa MBB. Serological survey in Guto-Gida District, east Wollega zone, Ethiopia. Global Veterinaria. 2012;8(2):139–43.
20. Adugna KE, Agga GE, Zewde G. Seroepidemiological survey of bovine brucellosis in cattle under a traditional production system in western Ethiopia. Rev Sci Tech. 2013;32(3):765–73.
21. Gumi B, Firdessa R, Yamuah L, Sori T, Tolosa T, Aseffa A, Zinsstag J, Schelling E. Seroprevalence of brucellosis and Q-fever in southeast Ethiopian pastoral livestock. J Vet Sci Med Diagn. 2013;2:1.
22. Tschopp R, Abera B, Sourou SY, Guerne-Bleich E, Aseffa A, Wubete A, Zinsstag J, Young D. Bovine tuberculosis and brucellosis prevalence in cattle from selected milk cooperatives in Arsi zone, Oromia region, Ethiopia. BMC Vet Res. 2013;9:163.
23. Asmare K, Sibhat B, Molla B, Ayelet G, Shiferaw J, Martin AD, Skjerve E, Godfroid J. The status of bovine brucellosis in Ethiopia with special emphasis on exotic and cross bred cattle in dairy and breeding farms. Acta Trop. 2013;126:186–92.
24. Alemu F, Admasu P, Feyera T, Niguse A. Seroprevalence of bovine brucellosis in eastern Showa, Ethiopia. Acad J Anim Dis. 2014;3(3):27–32.
25. Asgedom H, Damena D, Duguma R. Seroprevalence of bovine brucellosis and associated risk factors in and around Alage district, Ethiopia. Springer Plus. 2016;5:851. https://doi.org/10.1186/s40064-016-2547-0.
26. Geresu MA, Ameni G, Kassa T, Tuli G, Arenas A, Gezahegne MK. Seropositivity and risk factors for *Brucella* in dairy cows in Asella and Bishoftu towns, Oromia regional state, Ethiopia. Afr J Microbiol Res. 2016; 10(7):203–13. https://doi.org/10.5897/AJMR2015.7707.
27. Geresu MA, Ameni G, Wubete A, Arenas-Gamboa AM, Gezahegne MK. Isolation and Identification of *Brucella* Species from Dairy Cattle by Biochemical Tests: The First Report from Ethiopia. World Vet J. 2016;6(2):80–8.
28. Asmare K, Krontveit IR, Ayelet G, Sibhat B, Godfroid J, Skjerve E. Meta-analysis of *Brucella* seroprevalence in dairy cattle of Ethiopia. Trop Anim Health Prod. 2014;46:1341–50. https://doi.org/10.1007/s11250-014-0669-3.
29. Lindahl E, Sattorov N, Boqvist S, Magnusson UA. Study of knowledge, attitudes and practices relating to brucellosis among small-scale dairy farmers in an urban and Peri- urban area of Tajikistan. PLoS One. 2015;10(2):e0117318.
30. Alem W, Solomon G. A retrospective sero-epidemiology study of bovine brucellosis in different production systems in Ethiopia. In: Proceeding of 16th annual conference; 2002. p. 53–7. June 5–6, Addis Ababa, Ethiopia.
31. Belihu K. Analysis of dairy cattle breeding practices in selected areas of Ethiopia. PhD Thesis. Berlin: Humboldt University; 2002.
32. Bashitu L, Afera B, Tuli G, Aklilu F. SeroPrevalence study of bovine brucellosis and its associated risk factors in Debrebirhan and ambo towns. J Adv Dairy Res. 2015;3:131. https://doi.org/10.4172/2329-888X.1000131.
33. Rashid M. Reproductive wastage in cattle due to bovine brucellosis. In: Proceeding of the fourth livestock improvement conference. Addis Ababa: Ethiopian Institute of Agricultural Research; 1993. p. 270–2.
34. Pappas G, Papadimitriou P, Akritidis N, Christou L, Tsianos E. The new global map of human brucellosis. Lancet Infect Dis. 2006;6:91–9.
35. Pieracci EG, Hall AJ, Gharpure R, Haile A, Walelign E, Deressa A, Bahiru G, Kibebe M, Walke H, Belay E. Prioritizing zoonotic diseases in Ethiopia using a one health approach. One Health. 2016;2:131–5.
36. Godfroid J, Nielsen K, Saegerman C. Diagnosis of brucellosis in livestock and wildlife. Croat Med J. 2010;51:296–305. https://doi.org/10.3325/cmj.2010.51.296.
37. Plackett P, Alton GG. A mechanism for prozone formation in the complement fixation test for bovine brucellosis. Aust Vet J. 1975;51:374.
38. Corbel MJ. Brucellosis in humans and animals. Geneva: Produced by the, WHO in collaboration with the, FAO and OIE; 2006.
39. Fosgate GT, Carpenter TE, Chomel BB, Case JT, Debess EE, Reilly KF. Time-space clustering of human brucellosis, California, 1973–1992. Emerg Infect Dis. 2002; 8(7):672–8. https://doi.org/10.3201/eid0807.010351. PMID: 12095433.
40. National Metrology Institute of Ethiopia. 2016. http://www.most.gov.et/ national-metrology-institute-of-ethiopia. Accessed Nov 9 2016.
41. Thrusfield M. Sampling in veterinary epidemiology. 3rd ed. London: Black well Science Ltd.; 2007. p. 214–56.

Isolation of *Streptococcus agalactiae* in a female llama (*Lama glama*) in South Tyrol (Italy)

Alexander Tavella[1*], Astrid Bettini[1], Monia Cocchi[1], Ilda Idrizi[1], Stefano Colorio[1], Laura Viel[1], Claudia Zanardello[1] and Patrik Zanolari[2]

Abstract

Background: *Streptococcus agalactiae* is pathogenic for both animals and humans. In dairy cattle it commonly causes mastitis, with great economic losses, and there is scientific evidence of mastitis, caseous lymphadenitis, contagious skin necrosis and purulent infections associated with *S. agalactiae* in camels (*Camelus dromedarius*) as well. In humans, it is a common component of the respiratory and gastrointestinal microflora, but it can also act as a pathogen, especially in elderly people and immunocompromised patients, as well as in pregnant women and newborns.

Case presentation: A 10-year old non-pregnant female llama (*Lama glama*) was conferred to the Institute for Animal Health Control, in Bolzano for necropsy after sudden death. The animal had not shown unusual behaviour and had a low to normal nutritional condition (body condition score 2/5). The breeder had reported a chronic suppurative subcutaneous infection in the intermandibular area, resistant to therapy (therapy unknown). After necropsy, several samples were processed for histological, bacteriological and parasitological examinations.

Conclusions: This report describes, to the best of our knowledge, the first isolation of *S. agalactiae* in llamas (*Lama glama*). The animal came from a herd that counts approximately 200 South American camelids (llamas, alpacas) along with several horses, chicken, rabbits, cats and dogs; this farm offers services, such as trekking and pet therapy activities.

Keywords: *Streptococcus agalactiae*, *Lama glama*, South American camelids, Lancefield group B *Streptococcus*

Background

Llamas and Alpacas have gained increasing interest in the last 10 years and are now frequently held as farm animals and for hobby purposes (trekking, pet therapy) in the alpine regions of Northern Italy. In the Autonomous Province of Bolzano – South Tyrol (Italy), the South American camelids (SACs) population has grown lately and currently counts 800 individuals (583 llamas and 217 alpacas – informations from the local Veterinary Service 2017-databank), representing an important *niche product* in local livestock breeding. In fact, SACs are often held in multispecies farming systems with sheep and horses, and could come into close contact with humans for trekking and pet therapy reasons.

Streptococcus agalactiae (S. agalactiae), a Lancefield Group B *Streptococcus* (GBS), is an important pathogen affecting both animals and humans [1]. In dairy cattle, it is a major cause of mastitis and an important source of economic loss [2]. GBS have also been found in many other animals, such as camels, dogs, cats, crocodiles, seals, fish and dolphins [3–5]. More in detail, *S. agalactiae* has been frequently observed in camels (*Camelus dromedarius*), as a causative agent of mastitis, caseous lymphadenitis, contagious skin necrosis and purulent infections [6–8]. Moreover, authors described that the camel strains are different from the bovine strains, and that they resemble much more the human strains [3]. To our knowledge, no *S. agalactiae* strains have been isolated from llamas (*Lama glama*).

S. agalactiae has also been observed in humans. It is a common component of the microflora of the respiratory and gastrointestinal tracts, approximately isolated in

* Correspondence: atavella@izsvenezie.it
[1]Istituto Zooprofilattico Sperimentale delle Venezie, Viale dell'Università 10, 35020 Legnaro, Italy
Full list of author information is available at the end of the article

30% of healthy humans. In the elderly and in immunocompromised patients, this bacterium can be associated with urinary tract and skin and soft-tissue infections, bacteremia, osteomyelitis, meningitis and endocarditis [9, 10]. Moreover, in pregnant women, invasive maternal infection is associated with abortion, preterm delivery [11], sepsis and meningitis in newborns [12]. In humans, invasive infections caused by *S. agalactiae* are reported increasingly worldwide [13]. In the pathogenetic mechanism, the adherence to host epithelial cells is the first critical step of the infectious process, leading to the formation of microbial biofilms. These consist of microcolonies encased in extracellular polysaccharide material. Bacteria inside biofilms have increased resistance to antimicrobial agents and disinfectants [14]. Furthermore, the production of biofilms is correlated with both pathogenicity and virulence of the bacteria.

In this report, we describe the case of a 10-year-old non-pregnant female llama (*Lama glama*), conferred for necropsy to the Institute for Animal Health Control (*Istituto Zooprofilattico Sperimentale delle Venezie*, IZSVe) in Bolzano (Italy) after sudden death.

Case presentation

Clinical history

The animal had been kept all summer with other six llamas on an alpine pasture, but it originated from a larger llama breeding farm where no *S. agalactiae*-cases were ever recorded. The affected llama had not shown an unusual behavior. The breeder reported a chronic suppurative subcutaneous infection in the intermandibular area, that had persisted for several months. The nutritional condition of the individual was low to normal (body condition score 2/5) and was evaluated through adspection and palpation of the thorax, the abdomen and the back [15]. The animal was found dead by the breeder on the alpine pasture.

Necropsy

The physical exam revealed a 2–3 cm fistulating lump in the intermandibular area. After removing both hair and skin surfaces, a delimited suppurative infection site was observed in the subcutaneous tissue, in form of an abscess. Inspection of the anatomic area did not reveal a connection to the animal's oral cavity. A diffused subcutaneous oedema was noted in the intermandibular area, as well as in the neck and the caudal portion of the head. The retropharyngeal lymph nodes were enlarged and oedematous. After opening the lump, a yellowish creamy content was observed.

The exam of the abdominal cavity showed a severe sero-fibrinous ascites. The C3 intestinal compartment showed a catarrhal-haemorrhagic inflammation of the mucous membrane, while the proximal part of the bowel portion was hyperaemic; furthermore, a catarrhal duodenojejunitis was recognized. A severe nematode larvae infestation was identified in the intestinal content.

The liver had a light brown colour and a brittle consistency and showed a diffused necrosis.

In the thoracic cavity, a high-grade of sero-fibrinogenous pleural effusion was identified. The lung showed a pronounced oedema, as well as emphysematous areas; furthermore, an interstitial pneumonia was observed.

The pericardium and heart base were oedematous. No other pathological lesions were observed in any other organ.

Selected samples from liver and lung tissues were routinely processed for histopathological examination. Moreover, bacteriological examinations from the intermandibular abscess, the liver and the lungs were performed by routine laboratory tests, using blood agar (in-house protocol) at 37 ± 1 °C in aerobiosis for 24 h. The flotation procedure was performed, using a solution with specific gravity of 1.3.

Histology

Samples of lung and liver parenchyma were first collected and fixed in 10% neutral buffered formalin, then processed, paraffin embedded, stained with Haematoxylin and Eosin (HE) and observed by standard light microscopy for histological examination. A diffuse mild to moderate neutrophilic and lymphocytic interstitial infiltration with alveolar emphysema was observed in the lung parenchyma (Fig. 1).

While examining the liver parenchyma, a multifocal necrosis of hepatocytes with foci of neutrophilic and lymphocytic infiltration and diffuse hydropic degeneration of hepatocytes were detected (Fig. 2).

Bacterial examination

After incubation, a pure culture of translucid grey colonies, with a complete ß hemolysis was submitted for identification. Gram staining revealed Gram positive cocci in linear chains, catalase negative, and esculin negative as well.

To confirm the presence of ß hemolytic *Streptococci*, the CAMP test was performed, indicating a positive reaction, the serological assay (Streptococcal^{EN} Grouping Kit, Oixoid, Wade road, Basingstoke, Hampshire, UK) was performed as well, showing group B antigens. Biochemical identification of isolates (api® 32 Strep, bioMéerieux, Marcy-L'Etoile, France), revealed *S. agalactiae* (profile: 16122051110; %ID = 81.2; T = 0.77), with an excellent level of genus identification.

S. agalactiae was isolated from the purulent material of the intermandibular abscess and from the pulmonary parenchyma.

Biofilm formation was evaluated using micro titer plates as described by Stepanovic and coworkers [14].

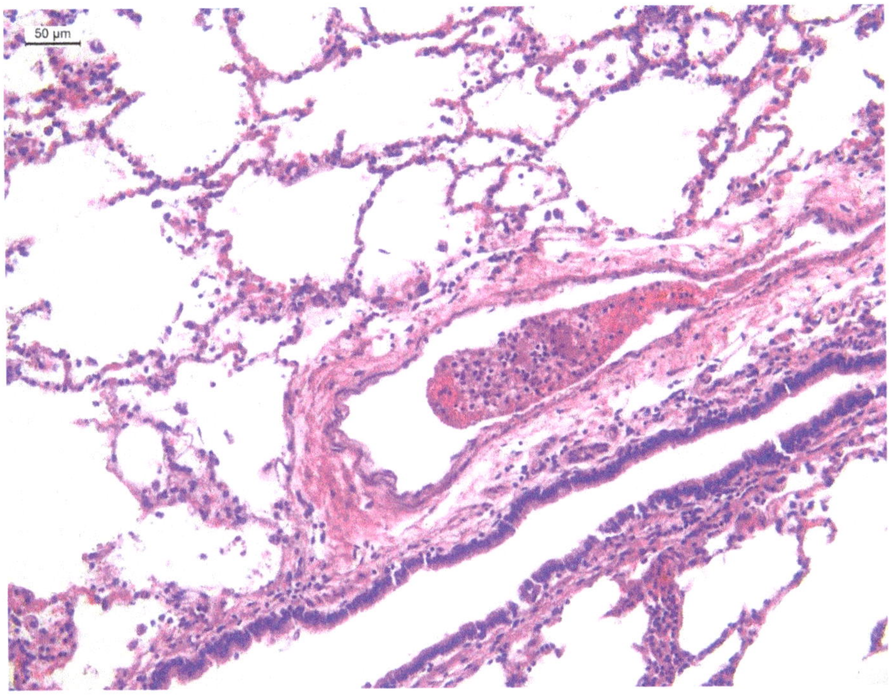

Fig. 1 Lung - Mild to moderate neutrophilic and lymphocytic interstitial infiltration with alveolar emphysema. Presence of a thrombus in the blood vessel. HE, 20X

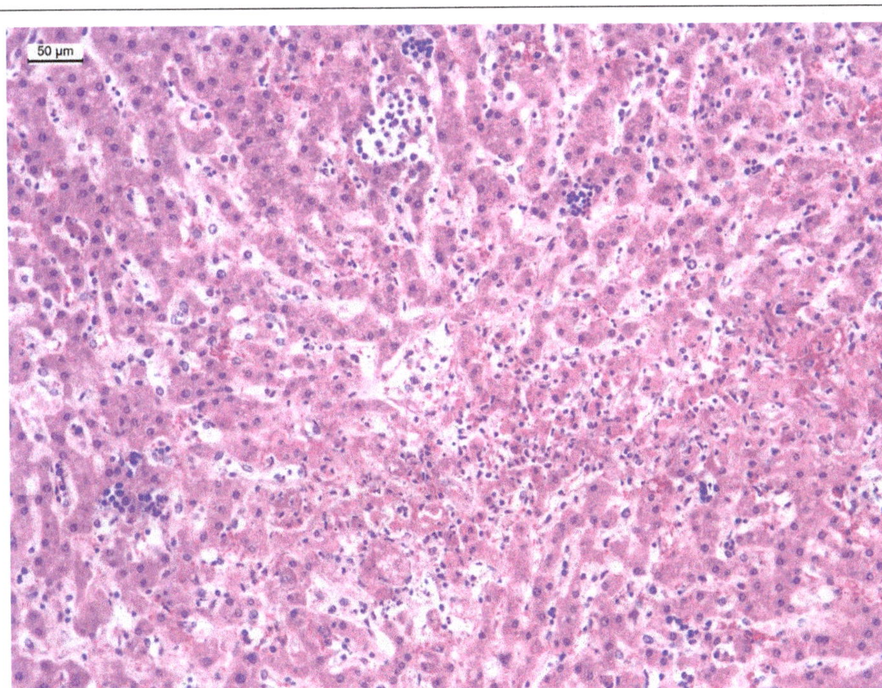

Fig. 2 Liver – Foci of necrosis of hepatocytes with neutrophilic infiltration. HE, 20X

In order to confirm the identification of the isolated strains obtained from the abscess and from the lung, Matrix-Assisted Laser Desorption/Ionization Time-Of-Flight Mass-Spectrometry (MALDI TOF MS) (Maldi Biotyper, Bruker Daltonics) was performed. This technique allows the identification of a bacterium by determining the molecular mass of fingerprint peptides (mainly ribosomal proteins), and comparing the mass fingerprint of the unknown strains to a database containing reference mass fingerprints. As specified by the manufacturer, a score value of < 1.7 indicates that identification is not reliable, scores between 1.7 and 2.0 demonstrate that identification is reliable at the genus level, scores between 2.0 and 2.3 evidence that identification is reliable at the genus level and probable at the species level. Scores higher than 2.3 indicate highly probable species identification.

S. agalactiae strains were recognized with a score of 2.454 (abscess isolate), and a score of 2.465 (lung isolate), indicating a highly probable species identification.

The strains formed biofilms under in vitro conditions, and were identified as moderate biofilm producers.

Determination of the minimum inhibitory concentration (MIC)

Minimum Inhibitory Concentration (MIC) of the lung and abscess isolates was evaluated in accordance with the guidelines of the Clinical and Laboratory Standards Institute [16] for microwell dilution testing, using a commercial plate (Micronaut-S, Merlin, Kleinstrasse 14, Bornheim, Hersel, Germany). The antimicrobials tested were ampicillin, ceftiofur, enrofloxacin, florphenycol, spectinomycin, tetracycline, tilmicosin, trimethoprim/sulphamethoxazole. The MIC results are shown in Table 1.

Parasitology

Parasitology of the feces reported an elevated number of oval eggs with a thin single wall, measuring approximately 45x30μm, attributable to *Strongyloides spp.*. Several eggs had transformed to the larval stage. Furthermore, yellow-brownish eggs (approximately 70 × 30 μm) showing a "lemon" shape with evident stoppers at the poles referred to *Trichuris spp.* were observed. The elevated number of eggs observed at the microscope and the evidence of the severe enteral parasitaemia allowed the indirect evaluation of the infestation intensity.

Molecular characterization

Both isolates were submitted for 16S sequencing. The alignment using the NCBI GenBank BLAST function (Additional files 1 and 2) and the Clustal Omega application (Additional file 3) confirmed the identification of *S. agalactiae.*

Discussion and conclusions

S. agalactiae has already been isolated from skin lesions, periarticular abscesses and mastitic udder in camelids (*Camelus dromedarius*) [3, 6, 17]. Furthermore, it has been described as an opportunistic pathogen of the upper respiratory tract [18]. *S. agalactiae* has been isolated from caseus lymphadenitis, contagious skin necrosis and purulent infections in camels [3]; it has also been observed in wounds caused by *Hyalomma* spp. in dromedaries in Kenya [19]. So far, no *S. agalactiae* have been described in llamas, instead *S. zooepidemicus, S. pyogenes, S. faecalis* and *S. uberis* have been isolated from alpacas in Peru [20]. Therefore, to the best of our knowledge, this represents the first isolation of *S. agalactiae* from llamas (*Lama glama*).

In this report, the identification of *S. agalactiae* was performed in an abscess and in the lung parenchima in an animal presenting a parasitic infestation.

Parasitism in SACs has become a major health concern, as several parasites, in particular gastrointestinal nematodes, are believed to cause mild-to-severe clinical diseases and economic losses. Although death from parasitic gastroenteritis associated with gastrointestinal nematodes may occur, infections tend to be more insidious, often presenting nonspecific clinical signs (i.e. diarrhoea, anorexia and poor growth) or asymptomatic conditions [21]. Domesticated SACs, comprising llamas, often share pasture with other livestock species and are often farmed under more intensive grazing conditions as in their native countries, factors which may significantly increase the risk of nematode infections [22]. In the present case, an elevated number of eggs attributable to *Strongyloides spp.* and *Trichuris spp.* were observed: this has been identified in the large intestine of SACs in previous studies as well [21–23]. Infections with Trichuris spp. can lead to severe anemia [15], even though many clinical signs are often absent at the early stages of infection [24]. The abdominal cavity lesions observed can be ascribed to the intense infestation by gastrointesintal

Table 1 Inhibitory concentrations of the isolated *S. agalactiae* strain

Strain	Ampicillin μg/ml	Ceftiofur μg/ml	Enrofloxacin μg/ml	Florfenicol μg/ml	Spectinomycin μg/ml	Tetracycline μg/ml	Tilmicosin μg/ml	Trimethprim/Sulfamethoxazole μg/ml
Abscess	0.125	0.125	0.5	1	128	0.25	4	0.125/2.375
Lung	0.125	0.125	0.5	1	128	0.25	4	0.125/2.375

nematodes [23, 25]. Moreover, this could act as a predisposing factor for septicemic streptococcosis.

S. agalactiae is a component of the microflora of the respiratory, genital and gastrointestinal tracts in both humans and ruminants. In humans, GBS is also an important cause of morbidity and mortality in newborns, in the elderly and in immunocompromised adults [26]. Primary manifestations include bacteremia, skin and soft tissue infections, while other signs are referred to pneumonia, osteomyelitis and urinary tract infections [27, 28]. Furthermore, invasive infections are increasingly reported worldwide: Takahashi and coworkers [13] have highlighted significant differences in clinical aspects, including prognosis, between disease-entities caused by *S. agalactiae* and other Streptococci in human diseases. In veterinary medicine, *S. agalactiae* is the main cause of subclinical mastitis in cattle. To our knowledge, no report of septicemic conditions have been observed in SACs. In other species, the bacterial colonization and infection requires the capacity of the bacterium to adhere and to persist. The formation of biofilm-like communities could facilitate microbial survival and proliferation by enhancing resistance to host defenses and nutrient deprivation [29]. In our report, the isolates were indeed biofilm producers. The production of biofilm was correlated with pathogenicity and virulence of some bacteria [30]. In fact, microorganisms inside biofilms have increased resistance to antimicrobial agents and disinfectants, indicating the pathogenicity potential in the isolated strains. The analysis of virulence factors production was not performed in this report, though previous reports have highlighted the potential of virulence factors production by *S. agalactiae* in bovine mastitis [31]. Furthermore, biofilm formation appears to be a prerequisite for colonization of the bovine mammary gland. *S. agalactiae* isolates recovered from bovine subclinical mastitis produced different pH-dependent biofilm levels, suggesting that biofilm production is modulated by environmental factors [28, 32]. No data are available for isolates from llamas.

Identification of the bacteria was performed by a commerical kit and by MALDI TOF MS. This last procedure is routinely used in many laboratories, and has shown to be a rapid and reliable technique for the identification of veterinary bacteria [33]. Furthermore, Lartigue and coworkers [34] have identified over 100 *S. agalactiae* isolates, further charachterized by serotyping and multilocus sequence typing, with the use of MALDI TOF MS. Based on the previously described evaluation scores, a score higher than 2.3 allows the highly probable species identification. In this case report, *S. agalactiae* strains were identified with a score of 2.454 (abscess isolate), and a score of 2.465 (lung isolate).

The determination of MIC was performed in order to evaluate potential antimicrobial resistances. Since no CLSI official breakpoints are available for llamas, and in order to provide a laboratory result, the MIC results (Table 1) were evaluated using the bovine clinical breakpoints, even though several authors have observed that llamas, but also other SACs, and bovines present different pharmacokinetics and pharmacodynamics [35, 36]. The evaluation, with bovine breakpoints, highlights sensitivity to all antimicrobials tested. While no official breakpoints are available for SACs, non official breakpoints have been proposed for ampicillin, ceftiofur and enrofloxacin [15]. Based on these data, the MIC results would suggest ampicillin and ceftiofur sensitivity by IV and IM administration, and an intermediate resistance to enrofloxacin by IV administration.

The breeding of SACs in the Autonomous Province of Bolzano – South Tyrol (Italy) has grown considerably in the last 10 years, conveying our Province a leading position in this field. These animals have gained great fame especially as hobby animals in zoos, private hotel gardens, as trekking animals and for pet therapy purposes, leading to a close contact between animals and humans, pointing out the possible risk of the transmission to humans. In fact, transmission of *S. agalactiae* from animals to humans has been briefly described in cattle [37, 38]; the authors describe a probable linkage between cattle exposure and detection of GBS in humans, stating that *S. agalactiae* can be transmitted between bovines and humans in a farm environment and that increased cattle exposure is associated with higher risk of infection. To this regard, no data are available for SACs, and molecular characterization of the isolates should be performed in order to establish the zoonotic potential of *S. agalactiae*.

Abbreviations

CAMP test: Christie Atkins Munch Petersen test; CLSI: Clinical and Laboratory Standards Institute; GBS: Lancefield Group B *Streptococcus*; HE: Haematoxylin and Eosin; IM: Intramuscolar inoculation; IV: Intravenous inoculation; IZSVe: Istituto Zooprofilattico delle Venezie; m/z: Mass to charge ratio; MALDI TOF MS: Matrix- Assisted Laser Desorption/Ionization Time-Of-Flight Mass Spectrometry; MIC: Minimum Inhibitory Concentration; *S. agalactiae*: *Streptococcus agalactiae*; SACs: South American camelids

Acknowledgements

N/A

Funding

N/A

Authors' contributions

AT performed necropsy and developed the manuscript, AB co-authored the manuscript, MC performed bacteriological examinations, II performed necropsy and parasitology, SC co-authored the manuscript, LV executed Maldi-Tof and MIC examinations, CZ performed histopathological examinations, PZ provided scientific and technical support. All authors read and approved the final manuscript.

Consent for publication

No personal data was included in the manuscript; no consent for publication was applicable.

Competing interests

The authors declare that they have no competing interests.

Author details

[1]Istituto Zooprofilattico Sperimentale delle Venezie, Viale dell'Università 10, 35020 Legnaro, Italy. [2]Clinic for Ruminants, Vetsuisse-Faculty, University of Bern, Bremgartenstrasse 109A, 3012 Bern, Switzerland.

References

1. Fischer A, Liljander A, Kaspar H, Muriuki C, Fuxelius HH, Bongcam-Rudloff E, de Villiers EP, Huber CA, Frey J, Daubenberg C, Bishop R, Younan M, Jores J. Camel *Streptococcus agalactiae* populations are associated with specific disease complexes and acquired the tetracycline resistance gene *tetM* via a *Tn916*-like element. Vet Res. 2013;44:86.
2. Lyhs U, Kulkas L, Katholm J, Waller KP, Saha K, Tomusk RJ, Zadoks RN. Streptococcus agalactiae serotype IV in humans and cattle. Northern Europe Emerg Infect Dis. 2016;22:2097–103.
3. Younan M, Bornstein S. Lancefield group B and C streptococci in east African camels (Camelus dromedarius). Vet Rec. 2007a;160:330–5.
4. Delannoy CMK, Crumlish M, Fontaine MC, Pollock J, Foster G, Dagleish MO, Turnbull JF, Zadocks RN. Human Streptococcus agalactiae strains in acquatic mammal and fish. BMC Microbiol. 2013;13:41.
5. Ma Y, Hao L, Ke H, Liang Z, Ma J, Liu Z, Li Y. *LuxS*/AI-2 in *Streptococcus agalactiae* reveals a key role in acid tolerance and virulence. Res Vet Sci. 2017;115:501–7.
6. Younan M, Müller W, Bornstein S. Streptococcus agalactiae infections in camels (Camelus dromedarius) in Kenya. Rev Élev Méd Vét Pays Trop. 2001a; 53:169–71.
7. Younan M, Ali Z, Mueller W, Bornstein S. Application of the California mastitis test in intramammary Streptococcus agalactiae and Staphylococcus aureus infections in camels (Camelus dromedarius) in Kenja. Prev Vet Med. 2001b;51:307–16.
8. Zubair S, de Villiers EP, Younan M, Andersson G, Tettelin H, Riley DR, Jores J, Bongcam-Rudloff E, Bishop RP. Genome sequences of two pathogenic Streptococcus agalactiae isolated from the one-humped camel Camelus dromedaries. Gen Ann. 2013;1:4.
9. Edwards MS, Baker CJ. Group B streptococcal infections in elderly adults. Clin Inf Dis. 2005;41:839–47.
10. Farley MM. Group B streptococcal disease in non pregnant adults. Clin Inf Dis. 2001;33:556–61.
11. Tevdorashvili G, Tevdorashvili D, Andqhuladze M, Tevdoarshvili M. Prevention and treatment strategy in pregnant women with group B streptococcal infection. Georg Med News. 2015;241:15–23.
12. Manning SD, Springman AC, Million AD, Milton NR, McNamara SE, Somsel PA, Bartlett P, Davies HD. Association of Group B Streptococcus Colonization and Bovine Exposure: a prospective cross-sectional cohort study. PLoS One. 2010;5:1. https://doi.org/10.1371/journal.pone.0008795.
13. Takahashi T, Ubukata K, Watanabe H. Invasive infection caused by Streptococcus dysgalactiae subsp. equisimilis: characteristics of strains and clinical features. J Infect Chemot. 2011;17(1):1–10.
14. Stepanović S, Cirković I, Ranin L, Svabić-Vlahović M. Biofilm formation by Salmonella spp. and Lysteria monocytogenes on plastic surface. Let Appl Microbiol. 2004;38:428–32.
15. Cebra C, Anderson DE, Tibary A, Van Saun RJ, Johnson LW. Nutritional Assessment. In: Lama and alpaca care: medicine, surgery, reproduction, nutrition and herd health. 1st ed. Amsterdam: Elvesier; 2014. p. 112–3.
16. Clinical and Laboratory Standards Institute (CLSI). Performance Standards for Antimicrobial Disk and Dilution Susceptibility Tests for Bacteria Isolated from Animals; Approved Standard. 4th ed. Wayne: VET01-A4; 2013.
17. Tibary A, Fite C, Anouassi A, Sghiri A. Infectious cases of reproductive loss in camelids. Theriogen. 2006;66:633–47.
18. Younan M, Bornstein S, Glück IV. Peri-arthricular abscesses in camel calves in North Kenya. J Cam Prac Res. 2007b;14:161–4.
19. Wernery U, Kaaden OR. Infestations with Ectoparasites. In: Infectious diseases in camelids. 2nd ed. Vienna: Blackwell Science Berlin; 2002. p. 328–9.
20. Fowler ME. Infectious diseases. In: Medicine and surgery of south American camelids. 3rd ed. Ames: Blackwell Publishing; 2010. p. 148–94.
21. Ballweber LR. Ecto- and endoparasites of New World camelids. Vet Clin North Am Food Anim Pract. 2009;25(2):295–310. https://doi.org/10.1016/j.cvfa.2009.02.003.
22. Jabbar A, Campbell AJD, Charles JA, Gasser RB. First report of anthelmintic resistance in *Haemoncus contortus* in alpacas in Australia. Parasit Vectors. 2013;6:243. https://doi.org/10.1186/1756-3305-6-243.
23. Rohbeck S. Parasite infections of the digestive and respiratory systemms of South American camelids: Investigations on their epidemiology and control measures in a herd in sourthern Hessen, Germany and the biology of *Eimeria macusaniensis*. Giessen: Thesis doctor of veterinary medicine, Justus Liebig University; 2006.
24. Franz S, Wittek T, Joachim A, Hinney B, Dadak AM. Llamas and alpacas in Europe: Endoparasites of the digestive tract and their pharmacotherapeutic control. Vet J. 2015;204(3):255–62. https://doi.org/10.1016/j.tvjl.2015.04.019.
25. Sprenger LK, Yoshitani UY, Buzatti A, Molento MB. Occurrence of gastrointestinal parasites in wild animals in state of Paranà, Brazil. An Acad Bras Cienc. 2018;90(1):231–8. https://doi.org/10.1590/0001-3765201720150030.
26. Schuchat A. Epidemiology of group B streptococcal disease in the United States: shifting paradigms. Clin Microbiol Rev. 1998;11(3):497–513.
27. Krzysciak W, Pluskwa K, Jurczak A, Koscielniak D. The pathogenicity of the Streptococcus genus. Eur J Clin Microbiol Infect Dis. 2013;32:1361–76.
28. Rosini R, Immaculada M. Biofilm formation by Streptococcus agalactiae: influence of environmental conditions and implicated virulence factors. Front Cell Infect Microbiol. 2015;5:1–4. https://doi.org/10.3389/fcimb.2015.00006.
29. Ebrahimi A, Moatamedi A, Lotfalian S, Mirshokraei P. Biofilm formation, hemolysin production and antimicrobial susceptibilities of Streptococcus agalactiae isolated from the mastitis milk of dairy cows in Shahrekord district. Iran Vet Res Forum. 2013;4:269–72.
30. Costerton JW, Lewandowski Z, Caldwell DE, Korber DR, Lappin-Scott HM. Microbial biofilms. Annu Rev Microbiol. 1995;49:711–45.
31. Kaczorek E, Malaczewska J, Wòjcik R, Siwicki AK. Biofilm production and other virulence factors in Streptococcus spp. isolated from clinical cases of bovine mastitis in Poland. BMC Vet Res. 2017;13:398. https://doi.org/10.1186/s12917-017-1322-y.
32. Miranda PSD, Lannes-Costa PS, Pimentel BAS, Silva LG, Ferreira-Carvalho BT, Menezes GC, Mattos-Guaraldi AL, Hirata RJ, Mota RA, Nagao PE. Biofilm formation on different pH conditions by Streptococcus agalactiae isolated from bovine mastitic milk. Lett Appl Microbiol. 2018. https://doi.org/10.1111/lam.13015.
33. Randall LP, Lemma F, Koylass M, Rogers J, Ayling RD, Worth D, Klita M, Steventon A, Line K, Wragg P, Muchowski J, Kostrzewa M, Whatmore AM. Evaluation of MALDI-ToF as a method for the identification of bacteria in the veterinary diagnostic laboratory. Res Vet Sci. 2015;101:42–9. https://doi.org/10.1016/j.rvsc.2015.05.018.
34. Lartigue MF, Héry-Arnaud G, Haguenoer E, Domelier AS, Schmit PO, van der Mee-Marquet N, Lanotte P, Mereghetti L, Kostrzewa M, Quentin R. Identification of Streptococcus agalactiae isolates from varioius phylogenetic lineages by matrix-assisted laser desorption ionization-time of fligt mass spectrometry. J Clin Microbiol. 2009;47(7):2284–7. https://doi.org/10.1128/JCM.00175-09.
35. Rubio-Langre S, De Lucas JJ, Litterio NJ, Aguilar S, Boggio JC, San Andrès MI. Pharmacokinetic behavior of marbofloxacin after intravenuous, subcutaneous and intramuscular administration in llamas (*Lama glama*). Small Ruminant Res. 2012;106(1):64–9.
36. Kreil V, Ambros L, Prados AP, Tarragona L, Monfrinotti A, Bramuglia G, Rebuelto M. Pharmacokinetics of immediate and sustained release cephalexin administered by different routes to llamas (*Lama glama*). Adv Pharmacol Sci. 2016;2016:1–6. https://doi.org/10.1155/2016/4621039.
37. Keefe GP. Streptococcus agalactiae mastitis: a review. Can Vet J. 1997;38: 429–37.
38. Carvalho-Castro GA, Silva JR, Paiva LV, Custòdio DAC, Moreira RO, Mian GF, Prado IA, Chalfun-Junior A, Costa GM. Molecular epidemiology of *Streptococcus agalactiae* isolated from mastitis in Brazilian dairy herds. Braz J Microbiol. 2017;48:551–9. https://doi.org/10.1016/j.bmj.2017.02.004.

Gastrointestinal nematodes of dairy goats, anthelmintic resistance and practices of parasite control in Northern Italy

Sergio Aurelio Zanzani[1], Alessia Libera Gazzonis[1], Annarita Di Cerbo[1], Marian Varady[2] and Maria Teresa Manfredi[1*]

Abstract

Background: Gastrointestinal nematodes (GINs) are one of the main constraints to ruminant production worldwide. Anthelmintic resistance (AR) has been reported in goats throughout Europe, yet little is known about the AR status in Italy. The aims of the study were: i) determine the frequency of AR in GINs in goat flocks in Northern Italy, Italy, ii) survey goat farmers on the current practices of parasite control, iii) update the species composition of the gastrointestinal helminthofauna. Thirty three flocks were enrolled and 1288 individual fecal samples were collected. Based on the egg per gram (EPG), 15 flocks were selected to evaluate the presence of AR in GINs with the Fecal Egg Count Reduction Test (FECRT). A questionnaire surveyed 110 dairy goat farmers to acquire information about farm management and drenching practices against GINs. Further, the gastrointestinal tracts of 42 goats were analyzed.

Results: The FECRs indicated that five of the 15 flocks had problems of AR, which was identified in all two of the anthelmintic classes tested. Resistance and suspected resistance was found in 40% of the flocks selected for AR testing that were treated with benzimidazoles while 20% of the flocks treated with eprinomectin had resistant GINs. *Teladorsagia/Trichostrongylus* L3 were isolated from the post-treatment coprocultures of all flocks with resistance but not from the flock with suspected oxfendazole resistance. Treatments against helminths were performed once annually in 73.63% of the flocks, but 20.00% of farmers declared not regularly treating their goats every year. Annual treatments usually occurred in autumn or winter at dose rate for sheep. *Te. circumcincta, H. contortus, Tr. colubriformis, Skrjabinema caprae* and *Oesophagostomum venulosum* were the most abundant and prevalent species of the gastrointestinal tract.

Conclusions: Strategies to prevent the development of AR should be widely adopted in Northern Italy. Further, farmers and practitioners should be educated about the importance of using the correct dose rates in goats. In addition, the presence of highly pathogenic GINs coupled with high worm burden in all sector of gastrointestinal tract and the prevalence values further suggest that improved diagnosis and active surveillance of GINs infection is needed.

Keywords: Parasites, Gastrointestinal nematodes, Anthelmintic resistance, Goat, Italy

Background

Gastrointestinal nematodes (GINs) remain one of the main constraints to ruminant production, since they can causing reduction in skeletal growth, live-weight gain and in milk yield [1,2]. GINs infection is of economic importance also in goat farming worldwide and the peculiarities of goat-nematode interactions could affect the control strategies of GINs. In particular, goats can be significantly more heavily infected than sheep and both the acquisition and expression of immune responses against GINs are less efficient in goats than sheep [3]. Further, the control of GINs infection in goats is still largely based on use of drugs at regular intervals and currently is made more complicated by the presence of resistant nematodes to one or more type of drugs. The phenomenon of anthelmintic resistance (AR) is spread in many countries with differences in prevalence and some data showed that GINs develop anthelmintic resistance more rapidly in goats [4,5]. A few recent studies have examined the gastrointestinal parasites of goats [6,7], but none have described the control practices used on dairy-goat farms in Northern Italy and especially in Lombardy. Anthelmintic

* Correspondence: mariateresa.manfredi@unimi.it
[1]Department of Veterinary Science and Public Health, Università degli Studi di Milano, Milan, Italy
Full list of author information is available at the end of the article

resistance (AR) against benzimidazoles has previously been reported in this area but is considered rare [8]. Knowledge of the presence of this phenomenon and of the control practices that are normally employed by farmers could be useful for improving both the management of gastrointestinal parasites and the milk production of grazing goats in marginal or mountain areas. Nevertheless, knowledge of the specific composition of the gastrointestinal helminthofauna can also supply valuable information for parasite control. The main aim of this study was to evaluate the presence of AR in the gastrointestinal nematodes of various breeds of dairy goats in Northern Italy by the fecal egg count reduction test (FECRT) and to determine which species could be implicated in resistance using larval cultures. A secondary aim was to survey goat farmers on the current practices of parasite control on farms in Northern Italy. Another goal was to update the species composition of the gastrointestinal helminthofauna by identifying the adult nematodes recovered from gastrointestinal tracts.

Methods

A survey was performed with a questionnaire from November 2010 to June 2011 in Lombardy, a region of Northern Italy (45°40'N, 9°30'E). Members of SATA, a regional association offering zootechnical and veterinarian support to breeders, were involved in the study. The majority of surveyed farms were in the areas of Northern Italy most suitable for goat breeding.

All animal procedures used in this study were approved by the Milan University Institutional Animal Care and Use Committee.

Evaluation of AR

Fecal egg count reduction test (FECRT)

Individual fecal samples were collected from the rectums of 1288 dairy goats from 33 flocks. All animals were older than two years of age. The parasitological analyses used a modified McMaster method [9,10], with 40.5 ml of a sodium nitrate and sucrose-based flotation solution (specific gravity = 1.30) and 4.5 g of feces. The numbers of eggs per gram (EPG) were calculated. Based on the EPGs, 15 flocks were selected to evaluate in vivo the presence of AR in GINs with the FECRT, in accordance with the guidelines of the World Association for the Advancement of Veterinary Parasitology (WAAVP) [11,12].

Benzimidazoles and probenzimidazoles (BZs) and macrocyclic lactones (MLs) were the classes of anthelmintic selected for evaluation in this survey based on the information obtained from the questionnaire. Albendazole (ABZ) (7.5 mg/kg body weight (bw)), oxfendazole (OFZ) (10 mg/kg bw), febantel (10 mg/kg bw) and netobimin (15 mg/kg bw) were tested in ten flocks. Ivermectin (IVM) and moxidectin (MOX) at 0.4 mg/kg were administrated to goats of one and two flocks, respectively. Finally, eprinomectin (EPM) at an extra-label dose of 1 mg/kg bw was tested in two flocks. All drugs were orally dosed except EPM, which was applied with a pour-on formulation. The dose rates of the anthelmintics were twice as high as those for sheep or cattle. The body weight of the visually heaviest goat of each flock was estimated to determine the anthelmintic dose rate. The number of dosed animals in each flock, based on the recommendations provided by Coles et al. [11], varied from 20 to 32 goats depending on flock size.

Individual fecal samples were collected on the day of treatment (T1). T1 mean EPG and standard deviation were calculated for each flock. FECRTs were performed only on goats with EPGs ≥ 150 at T1. Individual post-treatment fecal samples (T2) were collected after 8–10 days from goats treated with BZs and after 14–17 days from goats treated with MLs. The fecal egg count reduction (FECR) was calculated with the formula proposed by Cabaret and Berrag [13], in which each host serves as its own control:

$$\text{FECR} = (1/n)\Sigma(100\text{x}(1-[T_{i2}-T_{i1}]))$$

where T_{i2} is post-treatment and T_{i1} is pre-treatment EPG in host i from a total of n hosts. The tested population of nematodes was classified as resistant when the FECR was <95% and the lower limit of the 95% confidence interval (CI) was <90% or as "suspected resistant" when the FECR was <95% or the lower limit of the 95% CI was <90% [11].

Larval cultures

Pre- and post-treatment larval cultures were performed from pooled fecal samples for each flock; pools were composed with feces collected from each animal of the group. The feces sampled at T1 and T2 were incubated for 10 days at 25°C in a large vessel, ensuring adequate moisture (50-80%), and third-stage larvae (L3) were recovered from the coprocultures by a Baermann technique [9]. The first 100 randomly selected L3 of each pool were then examined with a light microscope (Axioscope 2, Zeiss) and were identified to the generic level as *Teladorsagia/Trichostrongylus*, *Oesophagostomum/Chabertia*, *Haemonchus*, *Nematodirus*, *Bunostomum* and *Cooperia*, according to MAFF [9] and van Wyk et al. [14]. When fewer than 100 L3 were isolated from a pool, the percentage of larval type was calculated on the basis of the counted L3.

Farm questionnaire

A questionnaire surveyed 110 dairy goat farmers from 51, 43, 10 and six farms from eastern (Bergamo (BG) and Brescia (BS) provinces), western (Como (CO), Lecco (LC) and Varese (VA) provinces), northern (Sondrio (SO) province) and southern (Pavia (PV) province) areas of

Lombardy, respectively. Each farmer was asked about their farm management, including the presence and dimensions of pastures, demography and breed of goats, the presence of other species on the farm (e.g. cattle) and drenching practices against GINs, such as treatment times, frequency, products and dosages of anthelmintic drugs. Farm veterinarians contributed to answering these last questions.

Composition of gastrointestinal helminthofaunas

The gastrointestinal tracts of two or three goats from the 15 flocks previously selected to evaluate the presence of AR were collected at abattoirs. The species composition and abundance of abomasal and intestinal nematodes of 42 goats were analyzed. Abomasa and small and large intestines were examined separately; they were opened and the contents were filtered through metallic sieves to collect adult helminths [9]. Ten percent of the content was examined with a stereomicroscope, and the extracted worms were counted. Cleared male worms were identified according to established descriptions [15-20]. The nomenclature used for Ostertagiinae was that proposed by Drózdz [15] and Durette-Desset [21]. The numbers of nematodes in the content fractions were adjusted to the total volumes from the abomasa or intestines. The total number of nematodes (males and females) and the number of individuals of each species based on the taxonomy of adult male worms were recorded. Epidemiological indexes, mean abundance (A) and prevalence (P), were calculated for each species of helminth when possible [22]. The helminth community structure was examined by the importance value, I, that was calculated for each of the helminth species according to Thul et al. [23].

Results

Evaluation of AR

Adult goats of the 15 flocks selected for testing AR had a mean pre-treatment EPG of 426.93 (±249.11) (Table 1). Baermannized L3 isolated from pre-treatment pools of individual fecal samples showed that *Teladorsagia/Trichostrongylus* were the most common GINs in the study area: *Teladorsagia/Trichostrongylus* L3 were recovered from 13 of the 15 flocks (86.6%) and were the most prevalent GINs in eight of the 13 flocks (61.5%) infected with these worms. *Oesophagostomum/Chabertia* and *Haemonchus* L3 were recovered from 11 (73.3%) and nine (60%) flocks, respectively, and *Haemonchus* was the most prevalent GIN in six of the nine flocks (66.6%) infected with this species. *Bunostomum*, isolated from five flocks (33.3%), was the most prevalent GIN in flock 11. According to Table 1 *Nematodirus* L3 were isolated only from flock 3, and *Cooperia* L3 were not found in any flock.

The percentages of the FECRs and the 95% confidence intervals of the fifteen investigated flocks are shown in Table 2. The FECRs after treatment indicated that 4 cases and 1 doubt of the 15 flocks (33.33%) had problems of AR, which was identified in the tested anthelmintic classes. One of the three flocks dewormed with ABZ had a post-treatment FECR fitting the criteria for resistance proposed by Coles [11]: the FECR was 81.11% and the lower 95% CI limit was 69%. One of the two flocks dewormed with OFZ was classified as "suspected resistant" (FECR = 95.17, lower 95% CI limit = 88%). The FECR was 100% in all flocks treated with netobimin, demonstrating the complete efficacy of this drug. Both flocks treated with febantel had resistant worms, with very low FECRs (40.52 and 76.01%, respectively) and lower 95% CI limits (58% and -3%, respectively). ML treatments were effective in four of five flocks; one of the two flocks treated with EPM was classified as resistant, having a FECR of 87.43% and a lower 95% CI limit of 76%. All areas of Northern Italy had flocks with FECRs less than 100%. The flocks with resistance or suspected resistance were all located in the eastern provinces. *Teladorsagia/Trichostrongylus* L3 were isolated from the post-treatment coprocultures of all flocks with resistance but not from the flock with "suspected" OFZ resistance. A low percentage of *Haemonchus* larvae (25%) were isolated from the coproculture of the flock with EPM resistance (Table 2).

Farm questionnaire

The questionnaire survey indicated that 42.73% (47/110) of the goat farms practiced grazing, and the majority of farms with pastures (60.5%) were in the western area. Pastures had a mean area of 11.63 ha (range, 1–50 ha) and altitudes from 200 to 1500 m a.s.l. Goats usually grazed from March or May to October or November during the day (or at night in the hottest months) and were kept in the fold at night (or day in the hottest months), depending on the region. Intensive breeding was the most common system in farms from southern (100%) or eastern (90.2%) areas, whereas only 55.8% or 50% of farms from western or northern areas, respectively, had this type of farming. Farms in the eastern and northern areas had the largest mean flock sizes (84.65 and 100.85, respectively); these farms also had the highest number of does (Table 3). Only one breed of goat was farmed in 89.10% (98/110) of the flocks. Eleven of 110 (10%) farms had does belonging to autochthonous breeds (Orobica, Nera di Verzasca, Bionda dell'Adamello and Frisa Valtellinese), but only five farms raised only one these breeds. The autochthonous breeds were raised on farms of northern (3/11), eastern (5/11) and western (3/11) Lombardy according to the geographical area from which they originated. A large number of farmers (49.10%) confirmed maintaining other species (cattle or sheep) on their farms, but generally in low numbers. Goat farms were mainly specialized for cheese production. The percentage of farms producing milk for cheese varied from 70% in northern areas to 100% in southern areas (Table 3).

Table 1 Mean eggs per gram (EPG) from fecal egg counts (FEC) and third-stage larvae identified in the goats of flocks selected for testing for anthelmintic resistance

Flock no.	Province*	Area	Pre-treatment FEC		Pre-treatment coproculture				
			Mean EPG	±SD°	*Telad/Trich*	*Oesop/Chab*	*Haem*	*Nemat*	*Bunost*
1	BS	E	248	100	96	4	0	0	0
2	BS	E	1317	1040	0	3	97	0	0
3	BS	E	250	62	81	0	0	9	10
4	BS	E	424	294	82	5	13	0	0
5	BG	E	307	397	78	17	5	0	0
6	BG	E	720	484	30	27	43	0	0
7	BG	E	261	110	81	19	0	0	0
8	BG	E	367	245	84	0	0	0	16
9	BG	E	468	264	32	29	39	0	0
10	BG	E	289	225	92	4	0	0	4
11	VA	W	312	144	29	0	0	0	71
12	VA	W	494	350	0	0	100	0	0
13	VA	W	400	397	24	7	69	0	0
14	VA	W	218	62	20	4	76	0	0
15	SO	N	329	85	77	5	12	0	6

**BS* = Brescia, *BG* = Bergamo, *VA* = Varese, *SO* = Sondrio, ° = standard deviation, *Telad/Trich* = *Teladorsagia/Trichostrongylus*, *Oesop/Chab* = *Oesophagostomum/Chabertia*, *Haem* = *Haemonchus*, *Nemat* = *Nematodirus*, *Bunost* = *Bunostomum*.

Table 2 Fecal egg count reduction (FECR), 95% confidence intervals (CI) and third-stage larvae (L3) identified in post-treatment coprocultures in selected goat flocks

Anthelmintic class	Drug	Flock no.	FECR (%)	95% CI	Status	Post-treatment coproculture	
						L3 type (%)	No. L3 seen
BZs	ABZ	1	100	°	S		0
	OFZ	3	100	°	S		0
	OFZ	4	95.17	88-100	SR**		0
	ABZ	5	81.11	69 -94	R*	*Teladorsagia/Trichostrongylus* (100)	23
	ABZ	6	100	°	S		0
	Febantel	7	76.01	58-94	R*	*Teladorsagia/Trichostrongylus* (100)	15
	Febantel	8	40.52	−3 to 84	R*	*Teladorsagia/Trichostrongylus* (100)	38
	Netobimin	11	100	°	S		0
	Netobimin	12	100	°	S		0
	Netobimin	13	100	°	S		0
MLs	IVM	10	100	°	S		0
	MOX	14	98.82	97-100	S		0
	MOX	15	98.57	96-100	S		0
	EPM	2	99.32	98-100	S		0
	EPM	9	87.43	76-99	R*	*Teladorsagia/Trichostrongylus* (75), *Haemonchus* (25)	16

BZs, Benzimidazoles and probenzimidazoles; *MLs*, Macrocyclic lactones.

ABZ, Albendazole; *OFZ*, Oxfendazole; *IVM*, Ivermectin; *MOX*, Moxidectin; *EPM*, Eprinomectin.

S, susceptible.

*resistant (R) flock: FECR <95% AND lower 95% CI limit <90%.

**suspected resistant (SR) flock: FECR <95% OR lower 95% CI limit <90%.

°95% CI not calculated because the reduction was 100%.

Table 3 Characteristics of goat farms and flock demographics

Location of farm	No. farms surveyed	Breeding system (% of farms)			Pasture (% of farms)	Mean flock size (Does, range)	Main breeds	Production of cheese (% of farms)
		E*	SE°	I§				
Eastern area	51	5.8	1.9	90.2	31.4	84.65 (18–840)	Alpine Saanen Orobica	82.35
Northern area	10	0	30	50	40	100.85 (21–190)	Alpine Frisa Valtellinese	70
Western area	43	34.8	9.3	55.8	60.5	40.97 (4–156)	Alpine Saanen Nera di Verzasca	86
Southern area	6	0	0	100	16.6	39.20 (7–93)	Alpine	100

* = extensive, goats are free to graze and browse all over the year.
° = semi-extensive, goats are kept in the fold during winter before the kidding period (January to March) and from March to November they are free to graze and browse.
§ = intensive, goats are reared indoor all over the year.

Treatments against helminths were performed once annually in 73.63% (81/110) of the flocks, but 20.00% (22/110) of farmers declared not regularly treating their goats every year. No differences in frequency of treatments were detected depending on the breeding system and reared breeds. Annual treatments usually occurred in autumn or winter (November to January), depending on the farm and particularly on the dry status of does. A few flocks received an additional treatment in summer when necessary, and a low percentage of farms (2.72%) regularly administered two treatments. The choice of anthelmintic varied among the farms but was dominated by BZs (87.27%); only 2.73% (3/110) of farmers treated their flocks with MLs. Anthelmintics were not rotated on any of the farms. In 89.09% (98/110) of the flocks, anthelmintic dose rates were calculated on the basis of the visually estimated body weight of the goat that appeared heaviest. The dose rate for sheep was used to treat the goats in all these flocks. Eleven of the 110 farmers (10%) did not indicate the dose rate used in their flocks. Annual treatments were usually done without any parasitological analyses, which were only performed on particular occasions (Table 4).

Table 4 Responses of 110 goat farmers to a questionnaire about practices of gastrointestinal-nematode control in dairy goats

Factor	Percentage (n)		
Control of gastrointestinal parasites			
Pasture management	0		
Anthelmintic treatment	100%		
Coprological analysis			
Before treatment	0		
Post treatment	0		
Anthelmintic used			
Benzimidazoles and probenzimidazoles	87.27% (96)	Albendazole	25.45% (28)
		Febendazole	25.45% (28)
		Oxfendazole	8.18% (9)
		Netobimin	23.64% (26)
		Febantel	4.55% (5)
Macrocyclic lactones	2.73%(3)	Ivermectin	0.91% (1)
		Moxidectin	0.91% (1)
		Eprinomectin	0.91% (1)
Nd*	10.00% (11)	-	-
Frequency of treatment/year			
0	20.00%		
1	73.63%		
2	2.72%		
Nd	3.63%		
Time of treatment			
Autumn/winter	82%		
Summer	9%		
Autumn/winter + summer	9%		

*not done.

Composition of gastrointestinal helminthofauna

All necropsied goats were infected by GINs (P = 100%). *T. circumcincta* was the most abundant (mean A = 558.62) and most frequent (P = 73%) abomasal parasite in all studied goats, followed by *H. contortus* (mean A = 69, P = 46%). Other abomasal species belonging to the subfamily Ostertagiinae (*Ostertagia ostertagi, Os. leptospicularis, Spiculopteragia spiculoptera* morph *spiculoptera, Te. pinnata* and *Te. trifurcata*) were also present but with low abundances and prevalences (Table 5). *Tr. colubriformis* and *N. lanceolatus* (mean A = 444.31 and 41.21, respectively) were the dominant species in the small intestine, whereas

Table 5 Mean abundance (A), prevalence (P), importance value (I) and location of gastrointestinal nematodes recovered in dairy goats

Nematode species	A (SD°)	Min-Max	P (%)	I	
				Values	Categories§
Abomasum					
Teladorsagia circumcincta	558.62 (896.06)	0-3500	73	92.47	D
Haemonchus contortus	69 (288.29)	0-1900	46	7.2	D
Spiculopteragia spiculoptera morph *spiculoptera*	2.44 (8.23)	0-40	11	0.06	CO-D
Teladorsagia pinnata	0.91 (3.54)	0-21	9	0.018	CO-D
Teladorsagia trifurcata	0.33 (2.23)	0-15	2	0.016	CO-D
Trichostrongylus axei	0.44 (1.44)	0-5	8	0.008	S
Ostertagia ostertagi	0.55 (2.19)	0-10	6	0.008	S
Ostertagia leptospicularis	0.44 (1.79)	0-10	6	0.006	S
Spiculopteragia spiculoptera morph *mathevossiani*	0.33 (1.65)	0-10	4	0.003	S
Ostertagia lyrata	0.11 (0.74)	0-5	2	0.0006	S
Small intestine					
Trichostrongylus colubriformis	444.31 (1431.91)	0-7200	34	95.99	D
Nematodirus lanceolatus	41.21 (185.87)	0-1000	14	3.56	D
Nematodirus spathiger	4.65 (24.12)	0-130	7	0.2	CO-D
Bunostomum trigonocephalum	1.83 (5.94)	0-30	13	0.16	CO-D
Trichostrongylus capricola	0.68 (3.71)	0-20	3	0.014	CO-D
Nematodirus sp	1.03 (4.09)	0-20	7	0.04	CO-D
Nematodirus filicollis	0.17 (0.93)	0-5	7	0.007	S
Trichostrongylus vitrinus	0.17 (0.93)	0-5	3	0.003	S
Cooperia pectinata	0.17 (0.93)	0-5	3	0.003	S
Strongyloides papillosus	0.17 (0.93)	0-5	3	0.003	S
Large intestine					
Skrjabinema caprae	1059.34 (1321.72)	0-6244	95	89.03	D
Oesophagostomum venulosum	50.92 (73.22)	0-275	58	2.61	D
Chabertia ovina	2.42 (5.88)	0-20	21	0.04	CO-D
Trichuris ovis	4.18 (20.16)	0-123	11	0.039	CO-D
Total adult nematodes					
Abomasum	1370.08 (1999.53)	0-8165	80		
Small intestine	1870.17 (4146.71)	0-19580	78		
Large intestine	2862.68 (3409.54)	0-16337	97		

°SD = standard deviation.
§Based on importance values helminth species were classified into categories; Dominant (D), species strongly characteristic of the community (I > 1.0), Codominant (CO-D) species contributing significantly to the community (0.01 < I < 1.0), Subordinate species occurring infrequently, they do not contribute significantly to the community (0 < I < 0.01).

Skrjabinema caprae and *Oesophagostomum venulosum* had the highest epidemiological parameters in the large intestine (Table 5). All sections of the gastrointestinal tract had high mean worm burdens, varying from 1370.08 to 2862.68 nematodes.

Discussion

The questionnaire indicated that the anthelmintic classes BZ and ML were used against GINs in Northern Italy and the FECRTs performed in accordance with WAAVP guidelines demonstrated that AR involved all two anthelmintic classes, with an overall prevalence of 33.33% (5/15 flocks selected for AR testing). Resistance and suspected resistance was found in 40% (4/10) of the flocks selected for AR testing that were treated with BZs, while 20% (1/5) of the flocks treated with MLs had resistant GINs. The absence of L3 in the post-treatment coproculture from the flock with suspected resistance to OFZ, however, did not permit an accurate assessment of sensitivity to this drug. Excluding this flock would lower the percentage of flocks presenting AR to BZs to 30% (3/10) and the overall prevalence of AR resistance to 26.67% (4/15).

Very little data is available on AR on Italian goat farms. Resistance to BZs in *Tr. colubriformis* was recently demonstrated in southern Italy [24]. A generic resistance to BZs was reported by Genchi et al. [8], and AR has been reported in Italian sheep [8,25,26]. The presence of AR in goats throughout Europe is extremely variable, probably because several factors affect the development of resistance in GINs. Chartier et al. [27] found that AR to BZs was present in 100% (15/15) of tested flocks in western France, and a recent Norwegian survey detected AR in 7.69% (1/13) of tested flocks, suggesting that AR in Norway was only beginning [28]. AR has been detected at low frequencies in other European countries such as Sweden [29], Germany [30] and Slovakia [31]. BZs are the class of anthelmintic involved in most cases of AR [27,28,32].

AR to BZs in Northern Italy also appears to be more widespread than AR to MLs; this difference is probably due to the long-term use of BZs and the lack of rotation with other drugs on dairy-goat farms in this region. The isolation of L3 from post-treatment coprocultures from flocks with resistance to BZs indicated that only the *Teladorsagia/Trichostrongylus* group was resistant to this anthelmintic class. Reports of AR to ML on European goat farms are rare. AR in *Teladorsagia*, *Trichostrongylus* and *Haemonchus* has been reported in Denmark, Switzerland, Germany and the UK [30,33,34]. The L3 isolated from the T2 coproculture of flock 15 of our study, which were classed as resistant to EPM, were *Teladorsagia/Trichostrongylus* and *Haemonchus.* The T1 and T2 coprocultures from this flock showed that AR probably involved primarily *Teladorsagia/Trichostrongylus. Haemonchus* L3 were the most abundant (39%) in T1 coprocultures, followed by *Teladorsagia/Trichostrongylus* L3 (32%), while the proportion was inverted in T2 coprocultures (75% *Teladorsagia/Trichostrongylus* L3 and 25% *Haemonchus* L3). Nevertheless, AR to EPM in flock 15 appeared to involve both *Teladorsagia/Trichostrongylus* and *Haemonchus*. Even though AR to EPM in goats had already been described in Switzerland and Germany, the results for this flock were quite unexpected. EPM is only registered for use in cattle in Italy, and the questionnaire indicated that its extra-label use occurred in only 0.91% (1/110) of Northern Italy farms. Flock 15 had never been treated with EPM, but the goats had been dewormed once a year for four years with IVM at the dose rate indicated for sheep. This practice in flock 15 may have caused the appearance of AR to the subclass of avermectins or to the entire ML anthelmintic class, despite the low frequency of treatment. The development of AR to the entire ML anthelmintic class (so also to MOX) due to the repeated use of ivermectin is reasonably still uncommon in goats; a treatment with MOX in goats that failed against IVM-resistant *Ostertagia* (*Teladorsagia*) species was described by Leathwick [35] in New Zealand, and the GINs of a goat flock described by Scheuerle et al. [30] were resistant to EPM but still sensitive to MOX. The efficacy of MOX against GINs should be evaluated in flock 15, and generally on goat farms with verified AR to avermectins, because resistance to milbemycins differs from that to avermectins [36]. MOX, if effective, could be used to prevent the development of AR to milbemycins.

The questionnaire submitted to goat farmers of Northern Italy indicated that the annual number of treatments is very low respect to other countries [4,37]. In fact, the farmer answers showed that treatments against GINs were performed once annually in 73.63% of the flocks and less than once annually in 20.00% of the flocks. The low frequencies of treatments in Lombardy probably did not prevent the appearance of AR in GINs, as also suggested by Chartier et al. [38]. Moreover, the questionnaire revealed that errors in dose rates were common. First of all, the majority of farmers (89.81%) declared to treat their goats with a dose rates for sheep and in all farms visual weight estimate was used. As a consequence, underdosing could result and selection of anthelminthic resistant worms was made easy. According to several authors, this event occurs more frequently in goats drenched with BZs as the bioavailability of the drugs is low than in sheep [4,39,40]. In addition, BZs, resulting in this study implicated in AR, were the most used drugs in Northern Italy and the alternation of anthelminthic families was not performed in any farms. Otherwise, very few drugs against GINs are registered in Italy for goats (6) at the same dose rate for sheep and they are not allowed in lactating goats (2 MLs, 1 levamisole/oxyclozanide) or a withdrawal time (min 3-max 9 days) is required (1 BZs, 1 proBZs, 1 morantel). It follows that the alternation of anthelminthic families to prevent AR as suggested is very difficult in Northern Italy [39]. Finally, the questionnaire survey showed poor drench practices widespread among goat farmers that probably generated the AR detected in GINs. However, it should be further considered that the control practice against GINs in goat flocks adopted by farmers in Northern Italy are not effective enough to improve the productivity in milking goats [39]. In fact, in this area the treatments are usually done in autumn-early winter; at this time treatments are less useful to prevent both the development of the disease and the contamination of pasture by nematode eggs being previously demonstrated that highest fecal egg counts in goats from Northern Italy occurred mainly in summer [6].

The survey on the composition of gastrointestinal helminthofaunas by the necroscopy of goats showed that *Te. circumcincta*, *H. contortus*, *Tr. colubriformis*, *Sk. caprae* and *Oesophagostomum venulosum.* were the most abundant and prevalent species of the gastrointestinal tract. Most of the species were previously reported in naturally infected goats in Northern Italy but *N. lanceolatus* is described for the first time in Italian goats [41,42]. According to Rossi [19], *N. lanceolatus*, placed in synonymy with *N. oiratianus* by Samson [43], showed distinctive features of spicul tips, bursal rays and lateral lobes of the caudal bursa. Some of the recorded helminths can also infect wild ruminants. For example, *Sp. spiculoptera* morph *spiculoptera*, *Sp. spiculoptera* morph *mathevossiani* and *Os. leptospicularis* are common parasites in *Cervus elaphus* and *Capreolus capreolus* [44], and *Te. circumcincta* is one of the most dominant species in the Alpine chamois, *Rupicapra rupicapra* [41,44]. The host distributions of these parasites suggest possible interactions between goats and wild ruminants. Similar interactions probably also occur in domestic species: goats and sheep share numerous gastrointestinal parasites (e.g. *H. contortus*), and the presence of *Os. ostertagi* in goats is probably due to its presence in cattle in the same pasture zones. Further, the analysis of helminthic community structure shows that these species, particularly *Sp. spiculoptera* morph *spiculoptera*, give an important contribution to composition of the community even if they have been classified into the categories of codominant or subordinate species by the importance values. In addition, the detection of a few species typical of cervids emphasizes that the goats may incur other parasitic risks resulting from interaction with these hosts on pastures [45]. Overall, a high number of helminth species is found than previous studies [46]. In this respect, it should be noted that the helminthofauna of a few goats from semi-extensive farms sampled in this survey shows a scarce number of helminth species and appears dominated by *Te. circumcincta*. It may result from the few ecological niches offered by that environment to parasites life cycle rather the frequent use of anthelminthics [33,47]. By quantitative analysis, *Te. circumcincta* and *Tr. colubriformis* showed both high prevalence and worm burden in agreement especially with data of Chartier and Reche [46] but disagree with those of Domke et al. [48] that in goats from Norway found very low prevalence and abundance of *Tr. colubriformis*. As suggested by the authors, it may depend on the scarce ability to overwintering on pastures by this parasite. Further, *H. contortus* showed higher prevalence rates than has been previously found in goats from the same area or from different areas with temperate climate [41,46,49]. This parasite is especially frequent in dry areas where higher worm burden are also recorded; however, *H. contortus* has a wide distribution being also found in regions characterized by a very severe climate due its ability to survive inside the hosts [42,49-51]. At last, the presence of high worm burden of *Sk. caprae* in the large intestine is remarkable; this parasite is generally considered with a very low pathogenicity but it can cause in dairy goats restlessness, itching and lesions in the perianal region [52]. In addition, all sector of gastrointestinal tract showed high worm burdens similar to those recorded in previous surveys as in goats from dry areas of Central Spain or Central and northern Greece [40,49]. These finding coupled with the high prevalence values ranging from 80% to 97% of goats with abomasal and intestinal infection respectively demonstrate that the GINs influence is very important in the surveyed area of Northern Italy including a few pastures with alpine features where despite the climatic conditions could be hard the GINs life cycle don't seem undergo any interruption. The present study provides sufficient data to make a picture of GINs infection in goat farms in Northern Italy by qualitative and quantitative point of view and based on these finding the risk posed by GINs infection on dairy goats raised in Northern Italy seems quite high.

Conclusions

The established presence of AR in 30% of the tested flocks in Northern Italy suggests that strategies to prevent the development of AR should be widely adopted in this region. The answers to questionnaire show that farmers and practitioners should be firstly educated about the importance of using the correct dose rates in goats and about the rotation of different anthelmintic classes, because AR can develop even at low frequencies of treatment. The management of resistance on the farms presenting AR should include the use of targeted selective treatments to sustain susceptible GINs population in refugia [53]. The questionnaire survey further provide some information on the practices adopted to control GINs in dairy goats that could be useful to improve the treatments against GINs in Northern Italy. However, the results obtained during the survey showed the presence of highly pathogenic GINs, high worm burden in all sector of gastrointestinal tract and high prevalence values. These data also coupled with a few published reports demonstrating some changes on abundance, seasonality and spatial spread of GINs in the last years probably related to climate change [54] suggested that improved diagnosis and active surveillance of GINs infection should be wanted to allow a suitable control of GINs and improve host production. Furthermore, the knowledge of the composition of gastrointestinal helminthofauna and of the epidemiological parameters is also essential to reach this important objective and control the AR.

Competing interests
The authors declare that they have no competing interests.

Authors' contributions
The study idea was conceived by SAZ and MTM. ADC and MV participated in the design of the study; ADC helped administering the questionnaire. SAZ and AG participated in the acquisition of the laboratory data. SAZ and MTM carried out the statistical analysis. Data interpretation was done by all authors. SAZ, MTM and MV drafted the manuscript. All authors contributed to the critical revision of the manuscript for important intellectual content and have seen and approved the final draft.

Acknowledgements
The authors thank the COST Action FA0805 (CAPARA) and all veterinary practitioners and SATA technicians of regional breeder association of Lombardy who have helped in our on-going research project.

Author details
[1]Department of Veterinary Science and Public Health, Università degli Studi di Milano, Milan, Italy. [2]Parasitological Institute of the Slovak Academy of Science, Košice, Slovak Republic.

References

1. van Houtert MF, Sykes AR: **Implications of nutrition for the ability of ruminants to withstand gastrointestinal nematode infections.** *Int J Parasitol* 2010, **26:**1151–1167.
2. Waller PJ: **Sustainable helminth control of ruminants in developing countries.** *Vet Parasitol* 1997, **71:**195–207.
3. Hoste H, Sotiraki S, Landau SY, Jackson F, Beveridge I: **Goat–Nematode interactions: think differently.** *Trends Parasitol* 2010, **26:**376–381.
4. Domke AVM, Chartier C, Gjerde B, Leine N, Vatn S, Østeras O, Stuen S: **Worm control practice in Norwegian sheep and goat flocks.** *Acta Vet Scand* 2011, **53:**29.
5. Waller PJ: **The development of anthelminthic resistance in ruminant livestock.** *Acta Trop* 1994, **56:**233–243.
6. Di Cerbo AR, Manfredi MT, Zanzani S, Stradiotto K: **Gastrointestinal infection in goat farms in Lombardy (Northern Italy): analysis on community and spatial distribution of parasites.** *Small Rum Res* 2010, **88:**102–112.
7. Manfredi MT, Di Cerbo AR, Zanzani S, Stradiotto K: **Breeding management in goat farms of Lombardy, Northern Italy: riskfactors connected to gastrointestinal parasites.** *Small Rum Res* 2010, **88:**113–118.
8. Genchi C, Traldi G, Di Sacco B, Tassi B: **Anthelmintic Resistant Nematodes in Farms Animals in Italy.** In *Anthelmintic Resistance in Nematodes of Farm Animals.* Edited by Coles GC, Borgsteede FHM, Geerts S. Brussels: European Commission, Agriculture; 1993:107–113.
9. Ministry of Agriculture, Fisheries and Food: *Manual of Veterinary Parasitological Laboratory Techniques.* 3rd edition. London: Her Majesty's Stationary Office (HMSO); 1986.
10. Raynaud JP: **Étude de l'efficacité d'une technique de coproscopie quantitative pour le diagnostic de routine et le contrôle des infestations parasitaires des bovins, ovins, équins et porcins.** *Ann Parasit Hum Comp* 1970, **45:**321–342.
11. Coles GC, Bauer C, Borgsteede FHM, Geerts S, Klei TR, Taylor MA, Waller PJ: **World association for the Advancement of Veterinary Parasitology (W.A.A.V. P.): methods for the detection of anthelmintic resistance in nematodes of veterinary importance.** *Vet Parasitol* 1992, **44:**35–44.
12. Coles GC, Jackson F, Pomroy WE, Prichard RK, Von Samson-Himmelstjerna G, Silvestre A, Taylor MA, Vercruysse J: **The detection of anthelmintic resistance in nematodes of veterinary importance.** *Vet Parasitol* 2006, **136:**167–185.
13. Cabaret J, Berrag B: **Faecal egg count reduction test for assessing anthelmintic efficacy: average versus individually based estimations.** *Vet Parasitol* 2004, **121:**105–113.
14. van Wyk JA, Cabaret J, Michael LM: **Morphological identification of nematode larvae of small ruminants and cattle simplified.** *Vet Parasitol* 2004, **119:**277–306.
15. Drozdz J: **Studies oh helminths and helminthiases in Cervidae. I. Revision of the subfamily Ostertaginae (Sarwar, 1956).** *Acta Parasitol* 1965, **13:**445–481.
16. Durette-Desset MC: **Les Nematodirinae (Nematoda) chez les ruminants et chez les lagomorphes.** *Ann Parasit Hum Comp* 1981, **54:**313–329.
17. Gibbons LM: **Revision of the African species of the genus *Cooperia* Ransom, 1907 (Nematoda, Trichostrongylidae).** *Syst Parasitol* 1981, **2:**219–252.
18. Lichtenfels JR, Pilitt PA, Lancaster MB: **Systematics of the nematodes that cause ostertagiasis in cattle, sheep and goats in North America.** *Vet Parasitol* 1988, **27:**3–12.
19. Rossi P: **Sur le genre *Nematodirus* Ransom, 1907. (Nematoda: Trichostrongyloidea).** *Ann Parasit Hum Comp* 1983, **58:**557–581.
20. Skrjabin KI: **Key to Parasitic Nematodes.** In *Academy of Sciences of the URSS, 1952.* 3rd edition. Jerusalem: Israel Program for Scientific Translation; 1961.
21. Durette-Desset MC: **Sur la divisions generiques des Nematodes Ostertaginae (Trichostrongylidae).** *Ann Parasit Hum Comp* 1982, **64:**356–373.
22. Bush AO, Lafferty KD, Jeffrey ML, Shostak AW: **Parasitology meets ecology on its own terms: Margolis et al. revisited.** *J Parasitol* 1997, **83:**575–583.
23. Thul JE, Forrester DJ, Abercrombie CL: **Ecology of parasitic helminth of wood ducks, *Air sponsa*, in the Atlantic flyway.** *Proc Helminthol Soc Wash* 1985, **52:**297–310.
24. Cringoli G, Veneziano V, Rinaldi L, Sauvé C, Rubino R, Fedele V, Cabaret J: **Resistance of trichostrongyles to benzimidazoles in Italy: a first report in a goat farm with multiple and repeated introductions.** *Parasitol Res* 2007, **101:**577–581.
25. Ambrosini F: **The resistance of *Trichostrongylids* to anthelminthics in sheep.** *JAEID* 2000, **94:**105–126.
26. Traversa D, Paoletti B, Otranto D, Miller J: **First report of multiple drug resistance in trichostrongyles affecting sheep under field conditions in Italy.** *Parasitol Res* 2007, **101:**1713–1716.
27. Chartier C, Pors I, Hubert J, Rocheteau D, Benoit C, Bernard N: **Prevalence of anthelmintic resistant nematodes in sheep and goats in western France.** *Small Ruminant Res* 1998, **29:**33–41.
28. Domke AVM, Chartier C, Gjerde B, Höglund J, Leine N, Vatn S, Stuen S: **Prevalence of anthelmintic resistance in gastrointestinal nematodes of sheep and goats in Norway.** *Parasitol Res* 2012, **111:**185–193.
29. Hőglund J, Gustafsson K, Ljungstrőm BL, Engstrőm A, Donnan A, Skuce P: **Anthelmintic resistance in Swedish sheep flocks based on a comparison of the results from the fecal egg count reduction test and resistant allele frequencies of the beta-tubulin gene.** *Vet Parasitol* 2009, **161:**60–68.
30. Scheuerle MC, Mahling M, Pfister K: **Anthelmintic resistance of *Haemonchus contortus* in small ruminants in Switzerland and southern Germany.** *Wien Klin Wochenschr* 2009, **121:**46–49.
31. Varady M, Corba J, Cernaská D: **A survey on anthelminthic resistance in nematode parasites of sheep in the Slovak Republic.** *Vet Parasitol* 2006, **135:**39–45.
32. Hong C, Hunt KR, Coles GC: **Occurrence of anthelmintic resistant nematodes on sheep farms in England and goat farms in England and Wales.** *Vet Parasitol* 1996, **139:**83–86.
33. Maingi N, Bjørn H, Thamsborg SM, Bøgh HO, Nansen P: **A survey of anthelmintic resistance in nematode parasites of goats in Denmark.** *Vet Parasitol* 1996, **66:**53–66.
34. VLA Surveillance Report: **Multiple anthelmintic resistance in goat nematodes.** *Vet Rec* 2006, **158:**357–36.
35. Leathwick DM: **A case of moxidectin failing to control ivermectin resistant *Ostertagia* species in goats.** *Vet Rec* 1995, **136:**443–444.
36. Prichard R, Ménez C, Lespine A: **Moxidectin and the avermectins: consanguinity but not identity.** *Int J Parasitol Drugs Drug Res* 2012, **2:**134–153.
37. Maingi N, Bjørn H, Thamsborg SM, Dangolla A, Kyvsgaard NC: **Worm control practices on sheep fams in Denmark and implications for the develpment of anthelmintc resistance.** *Vet Parasitol* 1996, **66:**39–52.
38. Chartier C, Soubirac F, Pors I, Silvestre A, Hubert J, Couquet C, Cabaret J: **Prevalence of anthelmintic resistance in gastrointestinal nematodes of dairy goats under extensive management conditions in southwestern France.** *J Helminthol* 2001, **75:**325–330.
39. Hoste H, Chartier C, Etter E, Goudeau C, Soubirac F, Lefrileux Y: **A questionnaire survey on the practices adopted to control gastrointestinal nematode parasitism in dairy goat farms in France.** *Vet Res Comm* 2000, **2000**(24):459–469.
40. Papadopoulus E, Arsenos G, Sotiraki S, Deligiannis C, Lainas T, Zygoyiannis D: **The epizootiology of gastrointestinal nematode parasite in Greek dairy breeds of sheep and goats.** *Small Rum Res* 2003, **47:**193–202.
41. Genchi C, Manfredi MT, Bossi A: **Les infestations naturelles par les strongles digestifs sur les paturages de haute montagne: interaction entre la chèvre et le chamois.** *Les Colloques de l'INRA* 1984, **28:**502–505. Niort, France.
42. Torina A, Dara S, Marino AMF, Sparagano OAE, Vitale F, Reale S, Caracappa S: **Study of gastrointestinal nematodes in Sicilian sheep and goats.** *Ann N Y Acad Sci* 2004, **1026:**187–194.
43. Samson KS: **The sinonimy of the ruminant parasites *Nematodirus oiratianus* Raevskaia, 1920 and *Nematodirus lanceolatus* Ault, 1944.** *Proc Helminth Soc Wash* 1968, **35:**52–53.
44. Zaffaroni E, Manfredi MT, Citterio C, Sala M, Piccolo G, Lanfranchi P: **Host specificity of abomasal nematodes in free ranging alpine ruminants.** *Vet Parasitol* 2000, **90:**221–230.

45. Alberti EG, Gioia G, Sironi G, Zanzani S, Riccaboni P, Magrini M, Manfredi MT: ***Elaphostrongylus cervi*** **in a population of red deer (*Cervus elaphus*) and evidence of cerebrospinal nematodiasis in small ruminants in the province of Varese, Italy.** *J Helminthol* 2011, **85:**313–318.
46. Chartier C, Reche B: **Gastrointestinal helminthes and lungworms of French dairy goats: prevalence and geographical distribution in Poitou-Charentes.** *Vet Res Comm* 1992, **16:**327–355.
47. Manfredi MT: **Biologia dei nematodi gastrointestinali dei ruminanti.** *Parassitologia* 2006, **48:**397–401.
48. Domke AVM, Chartier C, Gjerde B, Leine N, Vatn S, Stuen S: **Prevalence of gastrointestinal helminths, lungworms and liver fluke in sheep and goats in Norway.** *Vet Parasitol* 2013, **194:**40–48.
49. Valcárcel F, Garcia Romero C: **Prevalence and seasonal pattern of caprine trichostrongyles in a dry area of central Spain.** *J Vet Med B* 1999, **46:**673–681.
50. Lindqvist A, Ljungström BL, Nilsson O, Waller PJ: **The dynamics, prevalence and impact of nematode infections in organically raised sheep in Sweden.** *Acta Vet Scand* 2001, **42:**377–389.
51. Pino LA, Morales G, Aldana E, Perdomo L, Molina E: **Caracterización microecológica de los nematodos parásitos de ovinos de zonas áridas de Venezuela. (Un nuevo criterio para el control).** *Rev Ibér Parsitol* 1986, **46:**395–401.
52. Borgsteede FH, Dercksen DP: **Coccidial and helminth infection in goats keept indoors in the Netherlands.** *Vet Parasitol* 1996, **65:**321–326.
53. Jackson F, Varady M, Bartley DJ: **Managing anthelmintic resistance in goats. can we learn lessons from sheep?** *Small Ruminant Res* 2012, **103:**3–9.
54. van Dijk J, Sargison ND, Kenyon F, Skuce PJ: **Climate change and infectious disease: helminthological challenges to farmed ruminants in temperate regions.** *Animal* 2010, **4:**377–392.

Feeding mastitis milk to organic dairy calves: effect on health and performance during suckling and on udder health at first calving

Katharina Abb-Schwedler[1*], Ariane Maeschli[1], Renate Boss[2], Hans U Graber[2], Adrian Steiner[3] and Peter Klocke[1,4]

Abstract

Background: Infection pathways of *S. aureus* udder infections in heifers are still not well understood. One hypothesis is that calves become infected with *S. aureus* via feeding mastitis milk. Especially on small-scale farms, pasteurisers are not economic. The purpose of this randomised comparative study was to investigate the influence of feeding milk containing *S. aureus* genotype B (SAGTB) on the health and development of calves and udder health of the respective heifers. Additionally, a method reducing the bacterial load to obtain safer feeding milk was tested. Thirty-four calves were fed mastitis milk from cows with subclinical SAGTB mastitis. One group was fed untreated milk (UMG). For the other group, milk was thermised at 61°C for one minute (heat treated milk group = HMG). After weaning, calves were followed up until first calving. A milk sample of these heifers was taken at first milking to compare udder health of both groups.

Results: Thermisation of milk led to an effective reduction of *S. aureus* in the feeding milk. 78% of the analysed pools were free of *S. aureus*, a reduction of at least one log was obtained in the other pools.
Quarter milk samples revealed that two heifers had a *S. aureus* intramammary infection, but caused by a genotype different from genotype B.
During the suckling period, the UMG had a significantly higher incidence rate of 1.09 diarrhoea cases per 100 calf days at risk compared to 0.26 cases per 100 calf days in the HMG ($p < 0.05$).

Conclusions: Under the conditions of this study, no effects of feeding milk containing SAGTB on udder health after first calving were observed. But a power analysis indicated that the sample size in the current setup is insufficient to allow for assessment on mastitis risk after SAGTB exposition, as a minimal number of 4 calves infected (vs. 0 in the HMG) would have shown significant effects. High bacterial load, however, was associated with an increased incidence rate of diarrhoea. Thus, thermisation as a minimal preventive measure before feeding mastitis milk to calves might be beneficial for maintaining calf health.

Keywords: Dairy cattle, Heifer mastitis, Feeding milk, *S. aureus*, Calf health

Background

Feeding whole milk to dairy calves during the first three months of life is mandatory for organic farms in Switzerland [1]. This feeding regimen is supposed to meet the physiological needs of young calves. Godden et al. [2] found lower morbidity and mortality rates in whole milk-fed calves compared to those receiving milk replacer. Whole milk has a higher energy content and a more balanced composition of nutrients than milk replacer [3]. Maternal hormones and growth factors are included in whole milk but not in milk replacer [4].

Due to economic considerations, milk that is not suitable for sale is frequently fed to dairy calves. This milk often originates from diseased cows and thus is contaminated with antibiotics or pathogenic microorganisms. Selim and Cullor [5] found high concentrations of bacteria, including pathogens, in milk fed to calves. Feeding untreated mastitis milk can facilitate the transmission of infectious pathogens and provoke disease in calves [2].

* Correspondence: abb-schwedler@gmx.net
[1]Research Institute of Organic Agriculture, Ackerstrasse 21, 5070 Frick, Switzerland
Full list of author information is available at the end of the article

The mastitis prevalence levels of heifers range from 12.3 to 57% of infected quarters at first parturition [6]. In some studies, environmental mastitis pathogens such as *Escherichia coli* (*E. coli*) and *Streptococcus uberis* (*S. uberis*) have been found to be predominant [7,8], while in others coagulase-negative Staphylococci (CNS) had the highest incidence rate [9,10]. *Staphylococcus aureus* (*S. aureus*) as a major pathogen [11] was also frequently isolated in cases of clinical mastitis around parturition: Waage et al. [12] reported 44.3% of clinically affected heifer quarters infected with *S. aureus*. In a Swedish study, *S. aureus* was reported to be the most commonly isolated udder pathogen in veterinary treated mastitis of primiparous cows [13]. In Switzerland, *S. aureus* genotype B (SAGTB) in particular reaches within-herd prevalence of up to 87.5%, because it is contagious and pathogenic [14].

Epidemiological studies in Switzerland and New Zealand indicated that feeding mastitis milk to calves is a risk factor for heifer mastitis [15,16]. This hypothesis was already tested in different studies with conflicting results [17-20]. Schalm [17] concluded from his trials that udder infections with *Streptococcus agalactiae* (*S. agalactiae*) were transmitted via suckling among pen mates after consuming infectious milk. In another study, milk containing haemolytic Staphylococci was fed to 12 calves and inter-suckling was avoided. Nevertheless, 5 animals suffered from mastitis with haemolytic Staphylococci at calving [18]. Roberson et al. [19] tested herds with a high prevalence of coagulase-positive staphylococcal mastitis and where waste milk was fed to calves for mastitis prevalence of heifers when calving and compared them to low prevalence herds. No statistically significant difference was found. Another study was performed on feeding milk inoculated with *S. aureus*. After first parturition, two out of 29 heifers from the treatment group and 6 out of 35 control heifers were affected with *S. aureus* mastitis [20].

One recommended method to decrease the risk of milk with a high bacterial load for calves is pasteurisation before feeding [21]. It was concluded, however, that this strategy is not economically feasible for small-scale farms. Godden et al. [2] considered pasteurisation unfeasible for less than 23 calves fed per day, while Jamaluddin et al. [22] found it to be unprofitable even for farms with less than 315 calves fed per day. Thermisation is a sub-pasteurisation method that uses temperatures of 57 to 68°C for at least 15 s such that after heating the milk shows a positive reaction to the phosphatase test [23]. With this technique, the number of spoilage bacteria in milk can be reduced markedly with minimal heat damage to milk proteins [24]. Especially for coagulase-positive Staphylococci, a time temperature combination of 60°C for 30 s was reported to reduce the bacterial load by 3.3 log colony-forming units (CFU)/ml [25].

The objective of this study was to compare the effect of feeding untreated versus heat-treated whole milk originating from cows affected with subclinical *S. aureus* mastitis on the health of dairy calves during a 3-month suckling period and on their udder health at first calving. An additional objective was to determine the ability of thermisation to reduce the *S. aureus* load in whole milk from cows with subclinical *S. aureus* mastitis. Therefore, a non-blinded randomised comparative study was performed.

Methods

Animals and animal keeping

The experimental protocol used in this study was approved by the Cantonal Veterinary Office Aargau (Switzerland) before onset of the study.

A total of 34 calves of 3 breeds (17 Simmental, 16 Holstein, 1 Brown Swiss) born between January and November 2009 were recruited from five different dairy farms (all running free stall housing systems, 40 to 110 cows, within a distance of 10 km from the experimental unit). Four of these calves were twins.

As no data was available concerning the effect size (according to Cohen [26]) of feeding *S. aureus* mastitis milk to calves, it was estimated from the data of an epidemiological study [15]. There, a mean logarithmically transformed Somatic Cell Count (logSCC) of 2.12 was found in first calving heifers from farms practising mastitis milk feeding to calves. Farms not practising this management had a mean logSCC of 1.36. Using WinEpiscope program (WinEpiscope 2.0), assuming a variation coefficient of ≤0.33 corresponding to effect size $s \sim 0.7$, for two-sided hypothesis-testing, a minimum sample size of 16 animals per group was calculated. In order to have two spare calves available, it was decided to recruit 34 calves.

Until introduced into the experiment, the calves were housed in individual calf igloos and fed with the milk of their respective dams that were udder health monitored per monthly SCC-checks and yearly quarter milk sampling for bacteriological culture in the previous and current lactation. In cases of clinical mastitis or elevated SCC (>150,000 cells/ml) additional quarter milk samples were analysed. At the average age of 11 days (SD ±3 days), calves were transferred from their farm of origin to the experimental unit. Calves from every farm were, after initial random selection for the first calf, alternatingly assigned to one of two groups (see Figure 1): 1) Untreated milk group fed with untreated pooled whole milk from cows with subclinical *S. aureus* mastitis (UMG) confirmed by weekly repeated quarter milk sample investigation; 2) Heat-treated milk group fed with heat-treated milk from the same cows (HMG). The groups were housed in two separate small group pens on deep litter straw. These were cleaned twice a day and

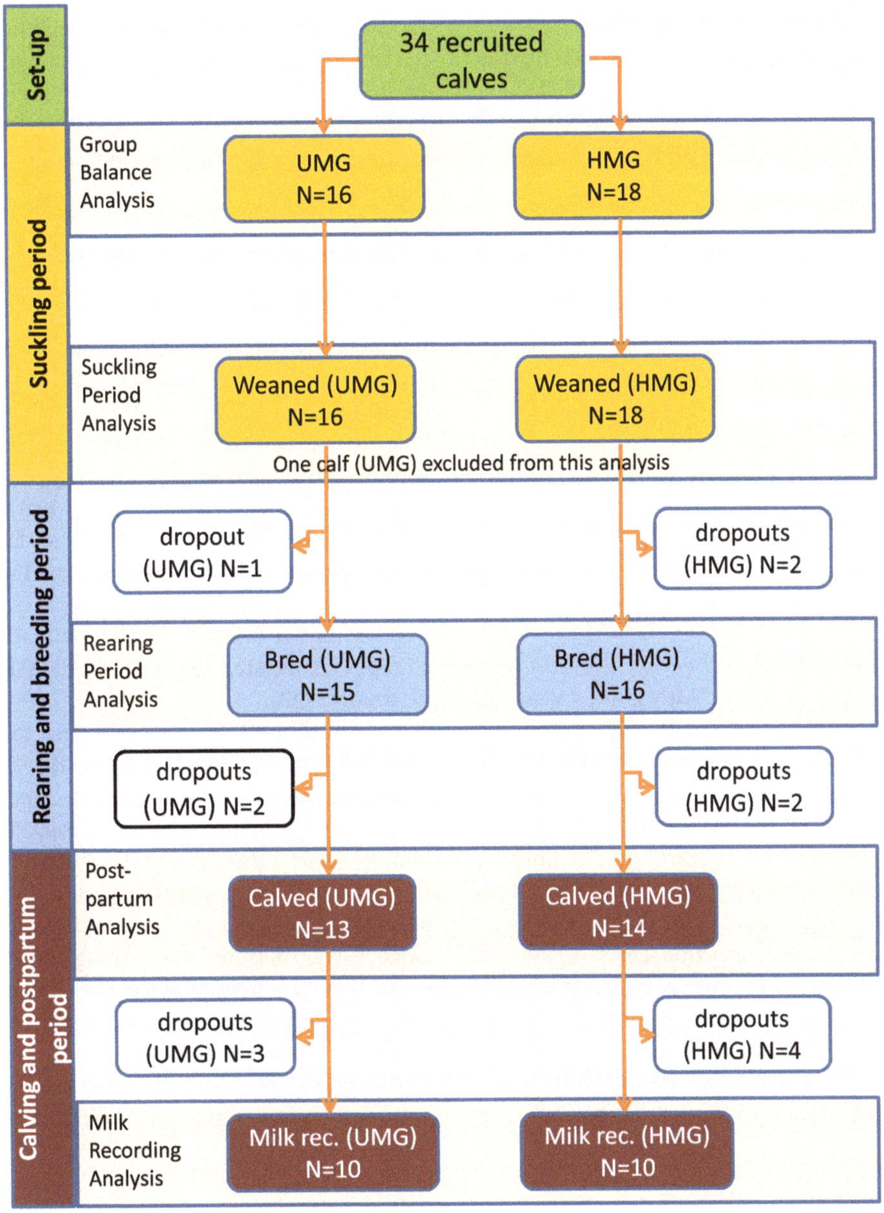

Figure 1 Study design and animals included in the trial. Number of animals in different periods and analyses in UMG and HMG, respectively; drop out reasons were death after illness and accidents, slaughtering because of infertility, late calving and lack of milk recording data; UMG = untreated milk group; HMG = heat-treated milk group.

if necessary fresh straw was substituted. Water was available ad libitum in 10 l buckets and replaced with fresh water twice a day. Hay provided from the farms of origin and concentrates (Kälber-Aufzuchtfutter Proflex Würfel, Alb. Lehmann, Birmenstorf, Switzerland) were also available ad libitum in hayracks and feeders, respectively.

For suckling, two automatic calf-feeders (U40, Urban, Hude-Wüsting, Germany) with two nipples each were installed. Calves were suckled for 12 weeks. The feeding plan of milk available at the automatic feeders is depicted in Table 1. Portions were limited to a maximum of two litres per meal. When calves entered the suckling station, milk was pumped from the storage churns to the feeder unit, heated to 40°C and offered at this temperature for a maximum of 20 minutes. After these 20 minutes, the milk not consumed was discarded. Daily consumable amounts of milk were available for 12 hours during the daytime and were recorded. No milk was available during the night. Calves were stepwise weaned starting in suckling week 7 (see Table 1). Suckling the juvenile

Table 1 Maximum amounts of milk (litres/calf/day or week, respectively) offered during the suckling period

Suckling week	Mean amount of milk (l) per calf and day	Milk (l) per calf and week
1	6	42
2	7	49
3	7	49
4	7	49
5	7	49
6	7	49
7	6	42
8	5.8	40.5
9	5.1	35.7
10	4.4	30.8
11	3.4	24.1
12	1.6	11.1
Maximum milk amount		471.2

Fed to 34 calves by the automatic feeders during the 12-week suckling period; weaning was performed stepwise starting from suckling week 7.

mammary glands of pen mates in both groups was documented as far as it was observed during the periods when observers were present (approximately 1.5 hours twice per day).

The internal surfaces of the feeders were cleaned by an automatic programme twice a day. For this purpose, a phosphoric acid-based cleaning agent (Unipred, Timac Agro Swiss, Sion, Switzerland) and hot water were pumped through the milk-conducting parts of the system. There were 12 hours in between the cleaning procedures. In the morning the hoses linking the feeders and the storage churns were also cleaned automatically. The stations were manually cleaned externally daily and the entire equipment twice a week.

Preparation of milk

The milk used for the pooled feeding milk (PFM) was provided by a farm in 10 km distance, with an evident herd health problem of *S. aureus* mastitis. Untreated cows that were positive for *S. aureus* at culture were considered as PFM contributors for the project; PFM was daily transported to the experimental unit in milk churns immediately after milking. For the UMG, untreated PFM was cooled down to 5°C with an immersion cooler (UK200, Alfa Laval, Lund, Sweden). For the HMG, heat-treated milk was prepared in a commercial 30 l auto-preserving cooker (WAT14, Weck, Wehr, Germany). For this purpose, PFM was heated to 61°C for one minute, filled into 40 l churns and cooled down to 5°C with another immersion cooler. Milk was stored at 5°C in the two 40 l milk churns for a maximum of 24 hours until consumed or discarded, respectively.

Clinical examination

General health examination and weighing (animal balance, Meier-Brakenberg, Extertal, Germany) were performed upon entering the experimental unit. During the suckling period, the two groups were observed daily, and general clinical examinations were conducted once a week by the first author, including evaluation of the general condition, body temperature (obtained as rectal temperature), the frequency of heartbeat and breath and auscultation of the respiratory tract. Furthermore, the gastrointestinal tract, focussing on appearance and consistency of feces, the central nervous and locomotor systems and the umbilicus were examined. Fever was diagnosed at body temperatures exceeding 39.5°C. Respiratory disease was defined as combination of breath frequency >45, pathological lung sounds and coughing present. Omphalitis was diagnosed in calves with a swollen, painful, warm umbilicus with discharge. Diarrhoea was defined as faeces of runny or watery consistency. During the rearing period, clinical examination was performed once a month, and weight was estimated based on heart girth measurement with a tape (Animeter, Albert Kerbl GmbH, Buchbach, Germany) [27].

If calves showed disease symptoms, the affected animals were individually treated. Treatments were primarily non-antimicrobial following published recommendations so as not to influence the microbial intake [28]. The homeopathic remedy corresponding to the symptoms was searched in the handbook, and 5 globuli of the C30 potency were given to the respective diseased animal twice a day (morning and evening) until the calves recovered. Remedies were adapted after a change of symptoms or if no amelioration of the calves condition was observed within 12 hours after treatment. Only in cases of therapy resistance or life-threatening diseases, conventional therapies with antimicrobials and NSAIDs were undertaken.

Study design

Both groups were observed during three stages of life: 1) the suckling period; 2) the rearing and breeding period; and 3) the calving and postpartum period. Figure 1 graphically depicts these periods and the involved animals. For the suckling period, calves were kept at the experimental unit. They were housed together with the penmates of their group and fed with the defined milk of their respective group throughout the suckling period. For the analysis of the suckling period, a disease was registered as such only after an individual calf had passed a disease free period of 5 days at the experimental unit, in order to exclude effects from the farms of origin.

After weaning, calves were housed in groups together with calves not participating in the project at their farms of origin. These farms all cooperated with specialised rearing farms located in the mountain regions of Switzerland

(1200 to 2300 m above s. l.). Calves were brought there at the age of 4 to 6 months. The first artificial inseminations were performed in winter 2010/2011. The timing of the first insemination depended on individual body-weight, but also according to the alpine pasturing seasons and the farmers' personal decisions.

Heifers returned for calving to their farms of origin between 5 months and one month prior to the calculated calving dates. Farmers were instructed to contact the first author immediately at the onset of calving symptoms to ensure that the very first milk prior to suckling or milking could be sampled.

Collection and analysis of milk samples

Of the PFM, daily milk samples were taken immediately after delivery at the experimental unit. Afterwards samples of the heat-treated PFM were taken from the auto-preserving cooker. Both samples were frozen at -18°C for later analysis. Individual composite samples of the PFM-providing cows were weekly taken prior to milking and analysed in a certified routine diagnostic laboratory (Idexx-Diavet, Bäch, Switzerland) to confirm the *S. aureus* shedding in the milk.

PFM samples were analysed at Agroscope research station (Agroscope, Berne, Switzerland) in summer 2012. All PFM samples were investigated in two steps: first the pooled samples were qualitatively analysed using classical culturing methods. In a second step, the number of CFU of *S. aureus* in milk was determined with plate counts. In preparation for the analyses, samples were thawed and pre-warmed in a water bath at 37°C for 10 minutes. Daily samples were merged to weekly pools and plated on 5% sheep blood agar (Biomérieux Suisse, Geneva, Switzerland) and CHROMagar Staph. aureus plates (CHROMagar, Paris, France), specific for *S. aureus* [29]. From heat-treated milk, 100 µl from each pool was plated. From untreated milk samples, only 30 µl was plated on blood agar after the first 10 processed pools grew too densely. All plates were incubated at 37°C. Results were obtained after 24 and 48 hours.

Colonies grown on blood agar were identified according to the guidelines of the National Mastitis Council (NMC) which include morphology, biochemical properties, and detection of haemolysis [30]. *S. aureus*, *S. uberis*, *S. dysgalactiae* and coliforms were referred to as major pathogens, whereas Corynebacterium sp. and CNS as minor pathogens. Concerning the CHROMagar plates, pink coloured colonies were considered to be *S. aureus* and were additionally identified by PCR for the *nuc*-gene [31]. This gene codes for the *S. aureus* specific enzyme thermonuclease. If colonies were *nuc*-positive, they were genotyped according to Fournier et al. [32]. For each PFM sample, one *S. aureus* colony was isolated. In cases, where colonies visually differed from each other, the according number of colonies was isolated.

The quantification of native *S. aureus* in the PFM was performed with serial dilutions. After thawing the pools in a water bath at 37°C for 10 minutes, two 1:10 dilutions with 0.9% sodium chloride solution were produced. Two 50 µl samples of pure milk and one 50 µl sample of each of the two 1:10 dilutions were plated on CHROMagar Staph. aureus plates with a spiral plater (EddyJet, iUL Instruments, Barcelona, Spain). The plates were incubated for 24 hours at 37°C and the colonies were manually counted. If the plates with pure milk were not countable, 1:100 dilutions were cultured and counted again.

Sterile quarter milk samples were taken from the heifers immediately after calving, following the guidelines of NMC [33]. A pre-milking teat disinfectant based on chloraminum (0.5%; Desinficin CL, DeLaval, Lund, Sweden) was applied before sampling. Of each quarter, two 10 ml sterile plastic tubes were filled. Immediately after transport (10 km), the milk samples were frozen at -18°C and stored until analysed. The samples were analysed at Agroscope according to the procedure for the PFM samples, as described above. For each quarter, one blood agar and one CHROMagar Staph. aureus plate was used. Milk was hand plated at 30 µl per plate on blood agar and 100 µl per plate on CHROMagar Staph. aureus by a sterile glass triangle. The plates were incubated at 37°C. Results were obtained after 24 and 48 hours and different pathogens were semi-quantitatively documented. In case of *S. aureus*-suspicious colonies, *nuc*-PCR and genotyping was performed as described above.

SCC and milk yield of the heifers were obtained from the first three routine monthly milk recordings provided by the Swiss breeding associations. Seven heifers (three of the UMG and four of the HMG) were excluded from this data collection, because the respective farm did not participate in the milk recordings.

Statistical methods

For all periods, variables of both groups were compared using univariate, univariable analysis methods with the exception of the general linear regression modelling that was univariate and multivariable. Group balance was checked in respect to the season of birth, breed, age, weight at arrival and clinical symptoms at arrival at the experimental unit using the Welch Two sample *t*-Test and Fisher's exact test. Because of unequal variances within both treatment groups, the Welch Two sample *t*-Test was chosen rather than Standard *t*-Test. The disease incidence rates (cases per 100 calf days at risk) and the specific disease incidence ratios with their 95% CIs for diarrhoea and respiratory disease during the suckling period were calculated and comparison between groups was performed using Chi-square-Test. In order to control for carry over effects from the farms of origin, the disease registration period for the analysis started for each calf individually after 5 disease-free days after arriving at the experimental unit. Survival analysis was conducted considering the time from the start of the disease registration period at the experimental unit to the first occurrence of the respective disease (diarrhoea and respiratory disease). Censoring happened at the day of the last feeding at the experimental unit. Survival differences were tested for statistical significance using a log-rank test. The weekly recorded growth rates were compared group-wise by calculating the mean daily growth rate and using the Welch Two Sample *t*-Test.

One calf from the UMG was excluded from the suckling period analysis, but not for further analysis of the post-weaning periods. With the requested five disease free days at the experimental unit before registering diseases, the start day for registering of this calf exceeded the mean start day of registering of the other calves (20th day of life) by 34 additional days. Moreover, this calf was treated with antibiotics.

For the growth and health evaluation during the rearing period, all remaining animals were considered, even if they failed calving afterwards and could not be analysed for calving and udder health parameters (rearing period analysis population in Figure 1). Body weights were only compared to day 440 in the life of every heifer, because that was the earliest age at first breeding. Linear regression modelling was used to model the effects on heifer growth. The mean weight gain per day at the rearing farms was set as the dependent variable. Independent variables were farm of origin, breed, birth season and treatment group as factors and disease days during the suckling period as continuous variable.

Somatic cell count values were transformed logarithmically to logSCC values in order to achieve normal distribution [34]. Comparison of logSCC and milk yield were performed by using the Welch Two Sample *t*-Test.

Significance level in all statistical calculations was set at $\alpha = 0.05$, statistical trends were indicated in case of $\alpha < 0.10$. The statistical analyses were calculated with the programme R ver. 2.15.2 base packages and the package epiR for incidence rate comparison [35]. For the post-hoc power analysis the package pwr was used in order to assess the ability of the study to detect prevalence differences in both groups. The function pwr.2p2n.test is based on Cohen's statistical power analysis methods [26].

Results

Effect of heating on bacterial load in milk

In the PFM samples of the untreated milk (41 pools), in all but one of the pooled samples SAGTB was identified. In this single sample *S. aureus* genotype S was detected instead. These results confirmed the cultural findings of the weekly composite samples of the milk-donating cows that were constantly positive for *S. aureus*. Apart from *S. aureus*, growth of CNS, *Streptococcus sp.*, *Corynebacterium sp.* and *Bacillus sp.* was discovered on the plates of the untreated milk pools.

In the heat-treated pools, bacterial growth was seen in all but one pool. Ten of 46 pools (22%) were positive for *S. aureus*. In 7 pools (15%), traces of SAGTB were found. In 3 pools (7%), *S. aureus* genotype S was found. *Corynebacterium sp.* was the most frequently discovered genus in the heat-treated milk; in some pools also CNS, *Bacillus sp.* and *Streptococcus sp.* were detected.

Quantification of *S. aureus* showed a maximum content of 68,050 CFU/ml and a minimum content of 1,330 CFU/ml in untreated milk. The median was 2,468 CFU/ml. In the heat-treated milk, 37 pools were negative for *S. aureus*, the maximum was 1,450 CFU/ml (median 0 CFU/ml). Compared to the corresponding untreated milk samples, there was a one log reduction (90% reduction of *S. aureus* CFU) in the heat-treated milk samples containing *S. aureus*.

Group balance

Considering the birth season, breed, age at arrival and body-weight at arrival, the two groups were not significantly different from each other at the beginning of the experiment. In contrast, with regard to the health status at arrival at the experimental unit, HMG calves showed clinical symptoms of disease in 9 out of 18 animals (50%), whereas, in the UMG 3 out of 16 calves (19%) showed clinical symptoms, although the difference was not significant ($p = 0.08$). Diarrhoea was the disease most often recorded (7 HMG calves, 2 UMG calves). With the requested five disease free days before registering diseases after arrival for the suckling period analysis, both groups entered the analysis period with a mean calf age of 20 days (SD ±5 days).

Suckling period

The median (quartile) of the total amount of milk consumed per calf in the trial was 436.3 (422.3-481.8) kg. Figure 2 shows the mean weekly bacterial intake of *S. aureus* CFU during the 12-week suckling period.

Disease incidence rate for diarrhoea in the UMG was 1.09 per 100 calf days at risk; in the HMG, the rate was 0.26 per 100 calf days at risk ($p < 0.05$). The incidence rate ratio was 4.22 [95% Confidence Interval (CI): 1.01; 24.68]. The survival analysis confirmed the significant differences between the groups concerning diarrhoea cases per day at risk ($p < 0.05$). While in the HMG the first analysed diarrhoea episodes ($n = 3$) lasted 2.3 days in mean (SD ±0.6 days), in the UMG ($n = 9$) they lasted 6.3 days in mean (SD ±3 days). The difference in duration was significant ($p = 0.004$). In the HMG, there were no second

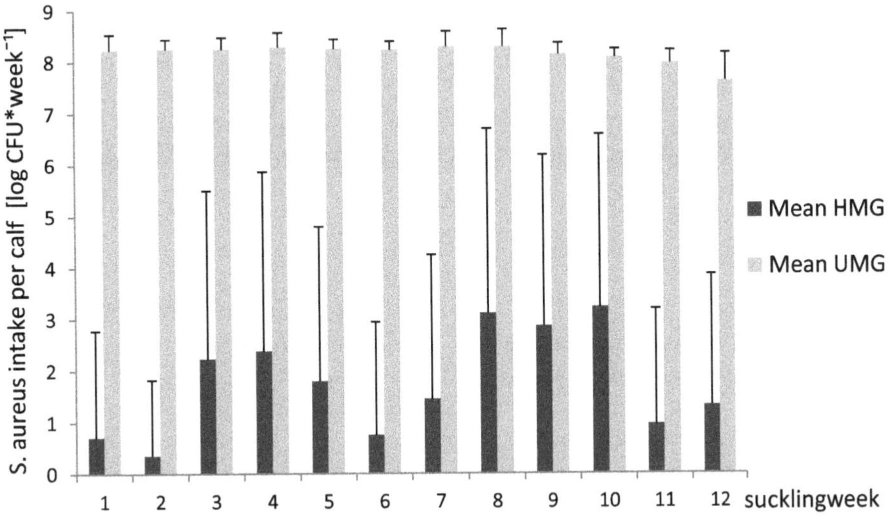

Figure 2 *S. aureus* intake during the suckling period. Mean and SD of total *S. aureus* CFU intake per calf and week during the 12-week suckling period of 34 calves fed untreated (UMG) and heat-treated milk (HMG), respectively; UMG = untreated milk group; HMG = heat-treated milk group; log = log10.

episodes during the suckling period, while in the UMG, there were three relapses, lasting 3.3 days in mean (SD ±1.52 days). The incidence rate for respiratory diseases did not differ significantly between the two groups. The mean daily weight gain during the suckling period was 715 g/day (SD ±9 g/day) in the UMG and 775 g/day (SD ±9 g/day) in the HMG ($p = 0.06$).

Conventional therapies were administered to two calves of each group. One calf of the UMG had a buccal abscess and severe lesions on the oral mucosa and the tongue. It was treated with Oxytetracyclin (Engemycin®, MSD Animal Health GmbH) 8 mg/kg once a day for four days administered by the intramuscular route and Tolfenamin Acid (Tolfendine®, Vetoquinol AG) 4 mg/kg once per day for two days administered by the subcutaneous route. Another calf of the UMG had a painful bronchitis or/and pleuritis and was treated with Benzylpenicillin (Norocillin®, Arovet AG) 12000 I.U. Penicillin/kg administered by the intramuscular route and Tolfenamid Acid (Tolfedine®, Vetoquinol AG) 2 mg/kg administered by the subcutaneous route both once per day for three days. One calf of the HMG was suspected to have osteomyelitis. It was treated with Benzylpenicillin (Norocillin®, Arovet AG) 12000 I.U. Penicillin/kg for four days administered by the intramuscular route. And the second calf of the HMG suffered from a severe omphalitis and was treated with Amoxicillin (Betamox®, Arovet AG) 15 mg/kg once a day for five days administered by the intramuscular route.

Suckling of pen mates and being suckled by pen mates was observed in all calves of both groups. Calves were observed being suckled on an average of 22% of the observation periods (SD ±8% of the observation periods). They were observed suckling pen mates on an average of 21% of the observation periods (SD ±7% of observation periods) each.

At the end of the suckling period, there were 16 calves in the UMG and 18 calves in the HMG (Figure 1).

Rearing and breeding period

During the rearing and breeding period, no significant differences in health and performance parameters for both groups were found.

Linear regression modelling showed that differences in weight gain depended on the farm of origin. The results are depicted in Table 2.

In this period, 7 heifers were lost. Three of these dropped out prior to breeding, one of the UMG and two of the HMG (Figure 1). The reasons for drop out were: Two heifers were killed in accidents (one of the UMG and one of the HMG) and two after illness (both HMG). Necropsy was performed in the latter two animals. The

Table 2 Results of the linear model performed for the rearing and breeding period

Variable	Level	Coefficient	Std. Error	T value	p
Intercept		711.3	36.6	19.4	<0.001
Treatment	HMG	Reference			
	UMG	−6.5	41.2	−0.2	0.87
Calf origin	Farm 1	Reference			
	Farm 2	−238.7	55.6	−4.3	<0.001
	Farm 3	33.1	48.4	0.7	0.50
	Farm 4	256.0	85.6	3.0	0.006

Final linear model after removing not significant variables by stepwise backward method (breed, birth season, and disease days during suckling time) for the significant influence factor "farm of origin" on weight gain during the rearing and breeding period; UMG was treated as fixed model factor of investigation.

first one died after a severe episode of diarrhoea and the resulting metabolic disorders. The other heifer had to be euthanised because of a progressive paralysis of the hind limbs, caused by a Waller degeneration of the neurons. Two heifers were infertile and therefore slaughtered (both UMG). One heifer of the UMG was calving late after all the others. It was therefore excluded from the postpartum and milk recording analysis.

Calving and postpartum period

Out of the initial 34 calves, 13 animals remained in the UMG and 14 in the HMG for the postpartum period analysis (Figure 1).

Examination of the heifer quarter milk samples revealed the growth of *S. aureus* in two animals, one from each group from two different farms. Both heifers showed three affected quarters and both of them carried *S. aureus* genotype C in all of these quarters. Both dams of the two heifers were never tested positive for *S. aureus* mastitis in the lactations before and after giving birth to these calves. Given the sample size considered, significance level of 0.05 and power set to 0.8 and assuming that no infections occurred in the HMG, power analysis indicates a total of 4 calves infected (31%) that would have been necessary to detect a statistically significant effect of feeding SAGTB contaminated milk on homologous SAGTB infection after calving.

Thirteen out of 14 heifers (20/56 quarters, 36%) of the HMG and 10 out of 13 heifers (18/52 quarters, 35%) of the UMG were infected with udder pathogens. The types of pathogens involved and their prevalences are depicted in Figure 3. Clinical symptoms of mastitis were not observed.

Results of 10 calves of each group were available for the analysis of the milk recording data (Figure 1). Comparison concerning SCC measurements and milk yield did not reveal any significant difference between the two groups. At first recording, there was a trend ($p = 0.09$) for higher logSCC values in the HMG (mean 2.5; SD ±1.8) as compared to the UMG (mean 1.2; SD ±1.5).

Discussion

The udder health data from this trial did not reveal any significant difference between calves fed with heat-treated and untreated mastitis milk. Results from the suckling period, however, indicated that calves fed with unheated mastitis milk suffered more diarrhoea days than calves receiving milk after heat treatment.

Blinding was not possible in this study as the same person prepared the milk and took care of the animals. Housing at the experimental unit was designed to separate the two groups from each other in order to prohibit microbial carryover on calves of the other group.

As we collected the milk sample from each heifer immediately after calving, it seems reasonable to rule out that the pathogens ingested during the suckling period might cause mastitis. Intramammary *S. aureus* infection discovered later might much more likely have been acquired via alternative infection pathways, for example during the milking process or by cow to cow transmission [36]. We intended to exclude these two factors with the chosen study design. In addition, there is evidence from the literature that the majority of intramammary infections occur at the beginning of the lactation [13]. Even subclinical cases should have been

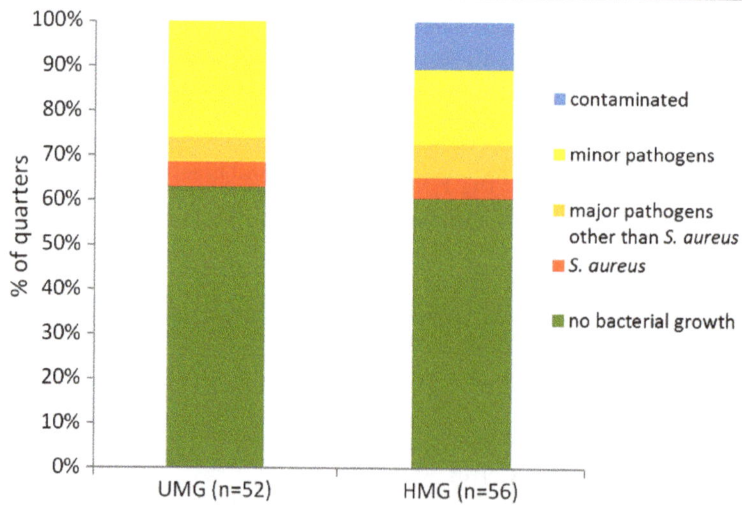

Figure 3 Results of the bacteriological culturing of heifers' quarter milk samples. Samples collected from 27 heifers (108 quarters) immediately after first calving; contaminated means three or more different pathogens per plate.

detected during laboratory analysis thanks to the high sensitivity of the agar plates [29]. Additionally, 100 µl of milk was plated on a separate plate per quarter to increase sensitivity [37].

Although under the conditions of this study, there was no influence of the quality of the feeding milk on udder health, this finding cannot be generalised, as the focus of the study was exclusively on the transmission of *S. aureus*. Other pathogens present in waste milk, such as *S. agalactiae* might have other transmission pathways and may infect suckling calves [17]. And the milk used for this trial came from one specific farm and contained mainly SAGTB.

The mean number of CFU of *S. aureus* in the feeding milk changed from week to week, according to the individual shedding patterns of the milk-donating cows. Nevertheless, we decided not to use inoculated milk, because we aimed at mimicking field conditions. Other dosages as well as combinations with other microorganisms might have caused other effects. Apart from the study of Barto et al. [20], precise quantities of *S. aureus* fed in milk can hardly be found in the literature.

Because of labour and material intensity the PFM samples were analysed after the analysis of the heifers' milk samples. Despite the long storage time the analysed quantities should reflect the original *S. aureus* content as it is recorded that freezing does not affect the viability of *S. aureus* in milk samples [38].

Investigations that indicate feeding mastitis milk to calves as being a risk factor for heifer mastitis might reflect an indirect correlation and have to be interpreted carefully. A possible interpretation is that farmers feed mastitis milk to calves due to a serious mastitis problem present in their herd, which represents another risk factor for heifer mastitis [16]. Feeding waste milk to calves could also be an indicator of lower hygiene standards on a respective farm so that the transmission of *S. aureus* is facilitated by an alternative infection pathway as, for example, not wearing gloves at milking or not cleaning milking equipment properly. Such effects were minimized by the particular design chosen in our study, as calves were not raised in herds with an evident mastitis problem.

Some epidemiological studies include the whole range of heifer mastitis pathogens, whereas we put the emphasis on one particular pathogen. Due to the lack of information about heifer SAGTB prevalence in general and, particularly, after exposition with SAGTB during the suckling period, sample size was primarily calculated based on the Cohen's effect size mentioned by Ivemeyer et al. [15] regarding postpartum SCC difference. Given our setting (considering 17 calves in each group, significance level 0.05, power 80%), we calculated that provided missing SAGTB cases in the HMG, a number of 4 cases (31%) of SAGTB infection in the UMG would detect a statistically significant effect of feeding contaminated milk to suckling calves on SAGTB prevalence in heifers. Results of the presented study suggest that expectable effects of a comparable setting have to be rather small, and with respect to Cohen [26] likely less than 0.2. Given the sample size calculation performed by Cohen's algorithms, this would require a sample size of nearly 1000 calves for the entire study to detect a difference of prevalences in both groups. Hence, the study results indicate that further research on this topic would have to consider an immense study setting upscaling. Considering the fact that in the current study not a single case of SAGTB was detected after exposition with SAGTB contaminated milk, it has to be evaluated carefully whether this effort is justified at all.

Analysis of the milk samples taken immediately after calving revealed that only two out of 27 heifers had intramammary infection with *S. aureus*. The genotypes, however, were different from those found in the milk during the suckling period. Therefore, it can be ruled out that bacteria from the milk fed in the experiment during the first months of life caused the mastitis of these two heifers in this later period of life. As the dams of these two heifers had never been tested positive for intramammary infections with *S. aureus* in the lactation before and after the birth of both calves, it is very unlikely that they had transmitted *S. aureus* via milk to their calves during the first days after birth.

Heat treatment of the feeding milk was thermisation. In the EU-definition [23] the phosphatase test is recommended to distinguish thermisation from pasteurisation. In this trial, we rather standardised the time-temperature combination to obtain a more exact and repeatable method. The phosphatase test, therefore, was not performed during the trial. By thermisation, inactivation of *S. aureus* was achieved in 78% of the samples. In the other 22% of samples, *S. aureus* CFU were reduced by one log compared to untreated milk. Also, the shift of the bacterial spectrum should be pointed out: in untreated milk, *Staphylococcus sp.* comprised the main part of the bacterial load. After heating, *Corynebacterium sp.* was most frequently found. The reduction of bacteria and the switch of the bacterial spectrum that disadvantaged the major pathogen fauna seem to have had an effect on calf health. On the other hand, the bacterial load was not reduced as effectively as after pasteurisation. Further investigation is needed to measure more accurately the cost effectiveness of thermisation compared to pasteurisation. The obtained reduction of the bacterial load, however, is expected not to be sufficient for some specific bacteria such as *Mycobacterium paratuberculosis* or *Mycoplasma sp.* [39,40]. We did not offer the prepared milk to the calves for longer than 12 hours per day. This was in accordance with the traditional Swiss feeding practice, where calves obtain their milk within this period. The same milk was used for both groups, but heat-treated for one group. Accordingly, the same amount of nutrients was provided to all calves but without the high bacterial load for the group obtaining the thermised milk. Side effects of heat-treating milk, such as the change of protein structure or loss of vitamins can be reduced to a minimum with the method advocated in our study [24].

The lower risk for diarrhoea in calves after ingesting heat-treated milk coincides with Jamaluddin's findings [41]. Episodes of diarrhoea in the group obtaining pasteurised milk were shorter and thereby less severe. *Staphylococcus aureus* is not a typical diarrhoea-causing pathogen in calves, but its enterotoxins are well known as contributing to immunosuppression which can explain a higher susceptibility to disease [42]. The presence of enterotoxins in milk was not analysed. But the minimal bacterial count of 10^5 CFU/g, necessary for enterotoxin formation, was reached in the pool milk. And high counts of viable bacteria are considered to play a role in adverse effects in calf health [5]. Some diarrhoea episodes, especially in the UMG, are overlapping with the episodes of other calves in the group (see Additional file 1). As causative organisms were not differentiated, a transmission of enteric pathogens from one calf to another cannot be excluded. Nevertheless, calves of the UMG seem to have been more prone to diarrhoea. During the suckling period, a trend to higher weight gains was observed in the group fed with heat-treated milk, which was also stated in earlier studies [2,41]. An explanation might be that diseases such as diarrhoea can adversely affect growth [43].

Conclusions

Feeding female calves with milk containing a high load of SAGTB under the conditions of this study did not affect udder health of these animals at first calving. As the power analysis revealed, generalisation of this result however, is not justified. The heat treatment of milk originating from cows suffering from *S. aureus* mastitis is a reasonable measure to reduce the risk of calf diseases, particularly diarrhoea. For small-scale farms without paratuberculosis problems, where a professional pasteuriser is not available, thermisation for one minute to 61°C might be a promising alternative to reduce calf morbidity. Furthermore, this technique is sustainable, as discarding the milk from cows with *S. aureus* mastitis can be avoided.

Abbreviations

CFU: Colony-forming units; CI: Confidence interval; CNS: Coagulase-negative staphylococci; *E. coli*: *Escherichia coli*; SAGTB: Genotype B of *S. aureus*; HMG: Heat-treated milk group; logSCC: Logarithmically transformed SCC by the use of log10; NMC: National Mastitis Council; NSAID: Non-steroidal anti-inflammatory drug; PFM: Pooled feeding milk; SCC: Somatic cell count; SD: Standard deviation; *S. aureus*: *Staphylococcus aureus*; *S. agalactiae*: *Streptococcus agalactiae*; *S. dysgalactiae*: *Streptococcus dysgalactiae*; *S. uberis*: *Streptococcus uberis*; UMG: Untreated milk group.

Competing interests

The authors declare that they have no competing interests.

Authors' contributions

KA carried out the practical work with the study animals, reviewed the literature, managed the data, performed the statistical analyses and drafted the manuscript. AM participated in the design of the study, helped with the practical work, gave practical advice and revised the manuscript. RB analysed the milk samples, helped to draft the manuscript and revised it carefully. HG participated in designing the study, coordinated and assisted the milk analyses and gave input to the manuscript draft. AS contributed to the study design, helped to coordinate the project partners, gave practical advice in all more complicated questions and was greatly involved in drafting the manuscript. PK was the project leader, coordinated the project, gave practical advice for all tasks, strongly supported the statistical analyses and substantially helped to draft the manuscript. All authors read and approved the final manuscript.

Acknowledgements

This study was mainly funded by the Swiss Federal Veterinary Office (BVET; Project number 1.08.22) and in parts by FiBL. The cooperating project partners at the University of Berne, Agroscope and FiBL and the farmers supporting our work are gratefully acknowledged. Thanks to Marcus Doherr for his input as statistical expert.

Author details

[1]Research Institute of Organic Agriculture, Ackerstrasse 21, 5070 Frick, Switzerland. [2]Agroscope Research Station, Schwarzenburgstrasse 161, 3003 Berne, Switzerland. [3]Department of Clinical Veterinary Medicine, Clinic for Farm Animals, Vetsuisse-Faculty, University of Berne, Bremgartenstrasse 109a, 3012 Berne, Switzerland. [4]bovicare GmbH, Hermannswerder Haus 14, 14473 Potsdam, Germany.

References

1. Biosuisse: Richtlinien für die Erzeugung, Verarbeitung und den Handel von Knospe-Produkten. [http://www.bio-suisse.ch/media/de/pdf2012/Regelwerk/rl_2012_d.pdf]
2. Godden SM, Fetrow JP, Feirtag JM, Green LR, Wells SJ: **Economic analysis of feeding pasteurized nonsaleable milk versus conventional milk replacer to dairy calves.** *J Am Vet Med Assoc* 2005, **226**(9):1547–1554.
3. Davis CL, Drackley JK: *The Development, Nutrition, and Management of The Young Calf.* Iowa City: Iowa State University Press; 1998.
4. Vasseur E, Borderas F, Cue RI, Lefebvre D, Pellerin D, Rushen J, Wade KM, De Passillé AM: **A survey of dairy calf management practices in Canada that affect animal welfare.** *J Dairy Sci* 2010, **93**(3):1307–1315.
5. Selim SA, Cullor JS: **Number of viable bacteria and presumptive antibiotic residues in milk fed to calves on commercial dairies.** *J Am Vet Medl Assoc* 1997, **211**(8):1029–1035.
6. Fox LK: **Prevalence, incidence and risk factors of heifer mastitis.** *Vet Microbiol* 2009, **134**(1–2):82–88.
7. Kalmus P, Viltrop A, Aasmäe B, Kask K: **Occurence of clinical mastitis in primiparous Estonian dairy cows in different housing conditions.** *Acta Vet Scand* 2006, **48**:21.
8. Compton CWR, Heuer C, Parker K, McDougall S: **Epidemiology of mastitis in pasture-grazed peripartum dairy heifers and its effects on productivity.** *J Dairy Sci* 2007, **90**:4157–4170.
9. Myllys V, Rautala H: **Characterization of clinical mastitis in primiparous heifers.** *J Dairy Sci* 1995, **78**:538–545.

10. Piepers S, Keeters K, Opsomer G, Barkema HW, Frankena K, De Vliegher S: **Pathogen group specific risk factors at herd, heifer and quarter levels for intramammary infections in early lactating dairy heifers.** *Prev Vet Med* 2011, **99:**91–101.
11. Reyher KK, Dohoo IR, Scholl DT, Keefe GF: **Evaluation of minor pathogen intramammary infection, susceptibility parameters, and somatic cell counts on the development of new intramammary infections with major mastitis pathogens.** *J Dairy Sci* 2012, **95:**3766–3780.
12. Waage S, Mørk T, Røros A, Aasland D, Hunshamar A, Ødegaard SA: **Bacteria associated with clinical mastitis in dairy heifers.** *J Dairy Sci* 1999, **82**(4):712–719.
13. Persson Waller K, Bengtsson B, Lindberg A, Nyman A, Ericsson Unnerstad H: **Incidence of mastitis and bacterial findings at clinical mastitis in Swedish primiparous cows - influence of breed and stage of lactation.** *Vet Microbiol* 2009, **134**(1–2):89–94.
14. Graber HU, Naskova J, Studer E, Kaufmann T, Kirchhofer M, Brechbühl M, Schaeren W, Steiner A, Fournier C: **Mastitis-related subtypes of bovine *Staphylococcus aureus* are characterized by different clinical properties.** *J Dairy Sci* 2009, **92**(4):1442–1451.
15. Ivemeyer S, Walkenhorst M, Heil F, Notz C, Maeschli A, Butler G, Klocke P: **Management factors affecting udder health and effects of a one year extension program in organic dairy herds.** *Animal* 2009, **3**(11):1596–1604.
16. Parker KI, Compton CWR, Anniss FM, Weir AM, McDougall S: **Management of dairy heifers and its relationships with the incidence of clinical mastitis.** *N Z Vet J* 2007, **55**(5):208–216.
17. Schalm O: **Streptococcus agalactiae in the udders of heifers at parturition traced to sucking among calves.** *Cornell Vet* 1942, **32:**49–60.
18. Johnson SD: **Raising heifer calves on mastitic milk.** *J Am Vet Med Assoc* 1947, **110:**179–182.
19. Roberson JR, Fox LK, Hancock DD, Gay CC, Besser TE: **Coagulase-positive staphylococcus intramammary infections in primiparous dairy cows.** *J Dairy Sci* 1994, **77**(4):958–969.
20. Barto PB, Bush LJ, Adams GD: **Feeding milk containing *Staphylococcus aureus* to calves.** *J Dairy Sci* 1982, **65**(2):271–274.
21. Aust V, Knappstein K, Kunz H-J, Kaspar H, Wallmann J, Kaske M: **Feeding untreated and pasteurized waste milk and bulk milk to calves: effects on calf performance, health status and antibiotic resistance of faecal bacteria.** *J Anim Physiol Anim Nutr (Berl)* 2013, **97**(6):1091–1103.
22. Jamaluddin AA, Carpenter TE, Hird DW, Thurmond MC: **Economics of feeding pasteurized colostrum and pasteurized waste milk to dairy calves.** *J Am Vet Med Assoc* 1996, **209**(4):751–756.
23. Council of the European Communities: Directive 92/46/EEC. Council directive laying down the health rules for the production and placing on the market of raw milk heat-treated milk and milk-based products. [http://ec.europa.eu/food/fs/sfp/mr/mr03_en.pdf]
24. Rukke EO, Sørhaug T, Stepaniak L: **Heat Treatment of Milk.** In *Encyclopedia of Dairy Sciences*, Volume 4. 2nd edition. Edited by Fuquay J, Fox P, McSweeney P. London: Academic Press; 2011:693–698.
25. Samelis J, Lianou A, Kakouri A, Delbès C, Rogeli I, Bogovic-Matijasić B, Montel MC: **Changes in the microbial composition of raw milk induced by thermization treatments applied prior to traditional Greek hard cheese processing.** *J Food Prot* 2009, **72**(4):783–790.
26. Cohen J: *Statistical Power Analysis for the Behavioral Sciences.* Hillsdayle: Lawrence Erlbaum Associates; 1988.
27. Gilbert RP, Bailey DR, Shannon NH: **Body dimensions and carcass measurements of cattle selected for postweaning gain fed two different diets.** *J Anim Sci* 1993, **71**(7):1688–1698.
28. Flury S, Gisler W, Knüsel A, Krüger C, Luder O, Schmidt A: *Handbuch zur homöopathischen Stallapotheke.* 2nd edition. Kriens: Brunner AG; 2007.
29. Graber H, Pfister S, Burgener P, Boss R, Meylan M, Hummerjohann J: **Bovine Staphylococcus aureus: diagnostic properties of specific media.** *Res Vet Sci* 2013, **95**(1):38–44.
30. Hogan JS, Gonzalez RN, Harmon RJ, Nickerson SC, Oliver SP, Pankey JW, Smith KL: *Laboratory Handbook on Bovine Mastitis (revised edition).* Madison, WI: National Mastitis Council; 1999.
31. Graber HU, Casey MG, Naskova J, Steiner A, Schaeren W: **Development of a highly sensitive and specific assay to detect *Staphylococcus aureus* in Bovine mastitic milk.** *J Dairy Sci* 2007, **90**(10):4661–4669.
32. Fournier C, Kuhnert P, Frey J, Miserez R, Kirchhofer M, Kaufmann T, Steiner A, Graber HU: **Bovine *Staphylococcus aureus*: association of virulence genes, genotypes and clinical outcome.** *Res Vet Sci* 2008, **85**(3):439–448.
33. National Mastitis Council: Microbiological Procedures for the Diagnosis of Bovine Udder Infection and Determination of Milk Quality: Procedures for Collecting Milk Samples. [http://www.nmconline.org/sampling.htm]
34. Ali AKA, Shook GE: **An optimum transformation for somatic cell concentration in milk.** *J Dairy Sci* 1980, **63**(3):487–490.
35. R Core Team: **R: A Language and Environment for Statistical Computing.** Vienna, Austria: R Foundation for Statistical Computing; 2012 [http://www.R-project.org/]. ISBN 3-900051-07-0.
36. Keefe G: **Update on control of staphylococcus aureus and streptococcus agalactiae for management of mastitis.** *Vet Clin North Am Food Anim Pract* 2012, **28**(2):203–216.
37. Sears PM, Smith BS, English PB, Herer PS, Gonzalez RN: **Shedding pattern of *Staphylococcus aureus* from Bovine intramammary infections.** *J Dairy Sci* 1990, **73**(10):2785–2789.
38. Petti CA, Caroll KC, Reimer LG: **Procedures for the Storage of Microorganisms.** In *Manual of Clinical Microbiology. Volume 1.* 9th edition. Edited by Murray PR, Baron EJ, Jorgensen JH, Landry ML, Pfaller MA. Washington, DC: ASM Press; 2007:55–61.
39. Butler JA, Sickles SA, Johanns CJ, Rosenbusch RF: **Pasteurization of discard mycoplasma mastitic milk used to feed calves: thermal effects on various mycoplasma.** *J Dairy Sci* 2000, **83**(10):2285–2288.
40. Stabel JR, Hurd S, Calvente L, Rosenbusch RF: **Destruction of mycobacterium paratuberculosis, salmonella spp., and Mycoplasma spp. in raw milk by a commercial on-farm high-temperature, short-time pasteurizer.** *J Dairy Sci* 2004, **87**(7):2177–2183.
41. Jamaluddin A: **Effects of Feeding Pasteurized Colostrum and Pasteurized Waste Milk on Mortality, Morbidity, and Weight Gain of Dairy Calves: Field Trial And Economic Analysis.** In *PhD Thesis.* University of California Davis: Department of Medicine and Epidemiology; 1995.
42. Moore DA, Taylor J, Hartman ML, Sischo WM: **Quality assessments of waste milk at a calf ranch.** *J Dairy Sci* 2009, **92**(7):3503–3509.
43. Donovan GA, Dohoo IR, Montgomery DM, Bennett FL: **Calf and disease factors affecting growth in female holstein calves in Florida, USA.** *Prev Vet Med* 1998, **33**(1–4):1–10.

23

Lipopolysaccharide derived from the rumen down-regulates stearoyl-CoA desaturase 1 expression and alters fatty acid composition in the liver of dairy cows fed a high-concentrate diet

Tianle Xu†, Hui Tao†, Guangjun Chang, Kai Zhang, Lei Xu and Xiangzhen Shen*

Abstract

Background: Dairy cows are often fed a high-concentrate diet to meet lactating demands, yet long-term concentrate feeding induces subacute ruminal acidosis (SARA) and leads to a decrease in milk fat. Stearoyl-CoA desaturase1 (SCD1) participates in fatty acid biosynthesis in the liver of lactating ruminants. Here, we conducted this study to investigate the impact of lipopolysaccharide derived from the rumen on SCD1 expression and on fatty acid composition in the liver of dairy cows fed a high-concentrate diet. Eight multiparous mid-lactating Holstein cows (455 ± 28 kg) were randomly assigned into two groups in the experiment and were fed a low-concentrate diet (LC) or high-concentrate diet (HC) for 18 weeks.

Results: The results showed that the total volatile fatty acids and lactic acid accumulated in the rumen, leading to a decreased rumen pH and elevated lipopolysaccharides (LPSs) in the HC group. The long chain fatty acid profile in the rumen and hepatic vein was remarkably altered in the animals fed the HC diet. The triglyceride (TG), non-esterified fatty acid (NEFA) and total cholesterol (TCH) content in the plasma was significantly decreased, whereas plasma glucose and insulin levels were increased. The expression of SCD1 in the liver was significantly down-regulated in the HC group. In regards to transcriptional regulators, the expression of sterol regulatory element binding transcription factors (SREBF1c, SREBF2) and SREBP cleavage activating protein (SCAP) was down-regulated, while peroxisome proliferator-activated receptor α (PPARα) was up-regulated.

Conclusions: These data indicate that lipopolysaccharide derived from the rumen down-regulates stearoyl-CoA desaturase 1 expression and alters fatty acid composition in the liver of dairy cows fed a high-concentrate diet.

Keywords: Lipopolysaccharide, Stearoyl-CoA desaturase 1, Long chain fatty acid, Liver, High concentrate diet, Dairy cows

Background

Dairy cows are often fed a high-concentrate diet to meet lactating requirements for high milk performance [1]. However, long-term feeding with a high-concentrate diet causes a decline in the rumen pH if organic acids, such as volatile fatty acids (VFAs) and lactic acid, accumulate in the rumen [2,3], and a chronic digestive disorder known as subacute ruminal acidosis (SARA) may occur. A rumen pH of less than 5.6 for over 3 h per day is used as a parameter to determine the occurrence of SARA [1]. Decreased rumen pH results in the release of lipopolysaccharides (LPSs), which originate from the cell-wall component of gram-negative bacteria [4].

Previous studies demonstrated that LPSs stimulate the gene expression of fatty acid synthetase (FAS) and acetyl-CoA carboxylase (ACC) in the liver of mice [5] but depressed stearoyl-CoA desaturase (SCD) expression in bovine hepatocytes [6]. SCD is a rate-limiting enzyme that catalyzes the synthesis of the monounsaturated fatty acids oleate (18:1) and palmitoleate (16:1) and forms triglycerides

* Correspondence: xzshen@njau.edu.cn
†Equal contributors
College of Veterinary Medicine, Nanjing Agricultural University, Nanjing 210095, China

and cholesterol esters [7]. Microarray assays have indicated that the gene expression profile was altered in the liver of SCD knockout mice, and the most obvious pattern was down-regulation of the genes involved in lipogenesis and up-regulation of the genes associated with fatty acid β-oxidation [8]. It was reported that LPS modulates lipid metabolism by inhibiting the clearance of triacylglycerol in the livers of bovine [9]. Furthermore, in a LPS-induced liver injury model, SCD1 expression was inhibited in the liver of mice, suggesting the potential action of LPS on SCD1 inhibition [10].

The liver is responsible for lipid metabolism in ruminant animals, and SARA is associated with liver abscesses, fatty liver and a whole-body inflammatory response when a high-grain diet is offered [11]. Therefore, the repartition of energy from production to anti-inflammation may exist in the liver and consequently lead to a negative energy balance during long-term high-concentrate supply. Many studies have been carried out on milk fat depression (MFD) in diet-induced SARA [12]. Some studies have focused on the *trans* fatty acid (i.e., *trans-10* C18:1n) pathway [13], while others have paid attention to LPS, which initiates the inflammatory response and influences the fatty acid profile in the rumen and milk [14]. Currently, several studies have been performed to evaluate hepatic lipid metabolism in dairy cows via exogenous LPS infusion [15].However, less information is available in regards to the alterations in hepatic lipid metabolism during long-term diet-induced SARA in dairy cows. Therefore, the present study was conducted to investigate the effects of a high-concentrate diet on the fatty acid composition and SCD1 expression in the liver of dairy cows.

Methods

Animals, diets and experimental design

Eight multiparous mid-lactating Holstein cows (455 ± 28 kg) were randomly assigned into two groups. One group was fed with a high-concentrate diet (HC) composed of 40% forage and 60% concentrate as a treatment, and the other group was offered a low-concentrate diet (LC) composed of 60% forage and 40% concentrate as a control for the 18-week experimental period. The ingredients and nutritional composition of the diets are presented in Table 1. The cows were fitted with a rumen fistula and hepatic catheters two weeks before the experiment and were ensured that they recovered from the surgery. The animals were maintained in individual tie stalls, fed at 0400, 1200, and 2000 h, and had free access to fresh water throughout the experimental time period.

The animal experiment was reviewed and approved by the Institutional Animal Care and Use Committee of Nanjing Agricultural University. The experiment was performed in accordance with the "Guidelines for Experimental Animals" of the Ministry of Science and Technology (Beijing, China).

Table 1 The ingredients in the diets and the nutritional composition

Ingredients, % of DM	LC[1]	HC[1]
Corn silage	30	20
Alfalfa	30	20
Maize	22.78	33.6
Wheat bran	5.15	15
Soybean meal	9.81	9
Calcium phosphate dibasic	0.92	0.53
Powder	0	0.52
Salt	0.35	0.35
Premix[2]	1	1
Total	100	100
Nutritional Composition[3]		
NE MJ/kg	6.32	6.74
CP %	16	16.2
EE %	3.96	4.15
NDF %	37.71	31.92
ADF %	22.75	17.55
NFC %	33.43	40.31
Ca %	0.9	0.8
P %	0.45	0.45

[1] LC, low concentrate; HC, high concentrate.
[2]The premix contained VA,1,900ku/kg; VD, 250ku/kg; VE, 3,000 mg/kg; Niacin, 4,000 mg/kg; Cu, 1,200 mg/kg; Fe, 525 mg/kg; Zn, 13,000 mg/kg; Mn, 5,500 mg/kg; I, 170 mg/kg; Co, 50 mg/kg; Se, 27 mg/kg.
[3]The calculated nutritional composition values.

Sample collection and analysis

The cows were milked at 0500, 1300, and 2100 h, and the milk yield was recorded daily. A 50-mL milk sample was taken to determine the milk fat and milk protein concentrations once a week (MilkoScan^{TM} FT1, FOSS, Denmark). Samples of the ruminal fluid were taken via the rumen fistula for 3 consecutive days during the 18th week, at 2-h intervals starting at 0400 h (after the morning feeding) for 12 hours. The samples were filtered through 2 layers of cheesecloth and stored at −20°C for the LPS, VFA, lactic acid and long-chain fatty acid analyses. A blood sample was taken at the same time as the ruminal fluid collection via the hepatic vein catheter and from the jugular vein using 5-mL vacuum tubes containing sodium heparin as an anticoagulant. The plasma was isolated from the blood samples by centrifugation at 3000 × g at 4°C for 15 min and was stored at −20°C for the LPS, biochemical parameter, hormones and long chain fatty acid analyses. Liver tissue samples were taken using a punch biopsy with a local anesthesia, and the samples were frozen in liquid nitrogen and then stored at −70°C until the quantitative Real-Time PCR and western blotting analyses.

LPS and biochemical parameters in the plasma and ruminal fluid

The LPS concentration in the ruminal fluid and plasma were determined using a chromogenic endpoint assay (CE64406, Chinese Horseshoe Crab Reagent Manufactory Co., Ltd., Xiamen, China) with a minimum detection limit of 0.05 EU/mL. The procedures were performed according to the manufacturer's instructions.

The analyses for the triglyceride, NEFA, total cholesterol and glucose concentrations were performed using commercial kits (Glucose Assay Kit, Rongsheng, Shanghai, China; Nonesterified Free Fatty Acids Assay Kit, Jiancheng, Nanjing, China; Total Cholesterol Reagent Kit, Dongou, Zhejiang, China; Lactic Acid Assay Kit, Jiancheng, Nanjing, China; Triglyceride Reagent Kit, Jiancheng, Nanjing, China) that used an enzymatic colorimetric method read by a microplate reader (Epoch, BioTek, USA). Plasma insulin and glucagon concentration was determined using an Iodine (^{125}I) Insulin Radioimmunoassay (RIA) Kit and Iodine (^{125}I) Glucagon Radioimmunoassay (RIA) Kit (Beijing North Institute of Biological Technology, Beijing, China) with Gamma Radioimmunoassay Counter (SN-6105, Hesuo Rihuan Photoelectric Instrument Co., Ltd, Shanghai, China). All of the procedures were performed according to the manufacturer's instructions.

Fatty acid analysis via gas chromatography

The VFA concentration in the ruminal fluid was determined via gas chromatography (GC) using a FFAP 123–3233, 30-m × 0.32-mm × 0.5-μm, capillary column (Agilent J&W GC Columns, Netherlands) on an Agilent 7890A (Agilent Technologies, USA) as described before with some modifications [16]. Crotonate was used as the internal standard.

The total lipids were extracted from the ruminal fluid and plasma using a mixture of polar and non-polar solvents according to Folch *et al.* at room temperature [17]. The fatty acid methyl esters (FAMEs) were prepared via esterification using sodium methoxide, followed by 14% borontrifluoride in methanol [18]. Heptadecanoic acid methyl ester served as the internal standard and was added to the samples prior to extraction and methylation. The FAME extracts were used for the gas chromatographic analysis of the total fatty acids. The fatty acid composition was determined using GC with a CP 7489, 100-m × 0.25-mm × 0.25-μm, capillary column (Agilent J&W Advanced Capillary GC Columns, Netherlands) on an Agilent 7890A (Agilent Technologies, USA) with an autosampler, flame ionization detector and split injection. The temperature programming was optimal for the separation of the majority of the C18:1 *trans* isomers. The initial oven temperature was 150°C, held for 5 min, then increased to 200°C at a rate of 2°C/min, held for 10 min, then increased to 220°C at 5°C/min and held for 35 min. Helium was used as carrier gas at a flow rate of 1 mL/min. The injector was set at 260°C and the detector at 280°C. The FAMEs were identified by comparing with the retention times of the standard.

RNA extraction, cDNA synthesis and quantitative real time PCR

The total RNA was extracted from 50 mg of liver tissue using the RNA iso Plus™ reagent (Takara Co., Otsu, Japan) via homogenization on ice. The purity and concentration of the RNA were measured using an Eppendorf BioPhotometer Plus (Eppendorf AG, Hamburg, Germany). The first-strand cDNA was synthesized using 250 ng of the total RNA template using the PrimeScript RT Master Mix Perfect Real Time kit (Takara Co., Otsu, Japan). The primers were designed using Premier 6.0 (Premier Biosoft International, USA) and were based on known cattle sequences or those cited in the published literature [19-21] (Table 2), and the primer efficiencies were evaluated prior to use. The qPCR was performed using the SYBR Premix Ex Taq™kit (Takara Co., Otsu, Japan) on an ABI 7300 Real-Time PCR System (Applied Biosystems, Foster City, CA, USA) according to the recommendations in the instruction manual. The standard PCR protocol was described in the manual: denaturing at 95°C for 15 s, then 40 cycles at 95°C for 5 s, and 60°C for 31 s. Glyceraldehyde phosphate dehydrogenase (GAPDH) served as the housekeeping gene for normalization, and the $2^{-\Delta\Delta Ct}$ method was used for the relative quantification.

Western blotting analysis

The liver samples were homogenized in RIPA lysis buffer (Beyotime, Shanghai, China) using 0.1 M PMSF using a Dounce homogenizer, and the lysate was centrifuged at 15,000 × g at 4°C for 20 min. The protein concentration of the supernatant was determined using bicinchoninic acid (BCA) and bovine serum albumin as standards (Pierce, Rockford, IL, USA). Equal amounts of protein were separated using 10% SDS-polyacrylamide gel electrophoresis (PAGE) and transferred onto a nitrocellulose membrane (Millipore, Danvers, MA) at 4°C. After blocking with 10% nonfat dry milk in tris-buffered saline at 4°C overnight, the membrane was washed and incubated with a primary antibody directed against SCD (Polyclonal antibodies raised in goat;sc-23016,Santa Cruz Biotechnology, diluted to 1:200) and HRP affinipure rabbit anti-goat IgG as the secondary antibody (E030130-01, Earth Ox, CA, diluted to 1:10,000). Visualization of the SCD protein was performed using the ECL western blot detection system (ECL plus, Beyotime, Shanghai China). The same membrane was then stripped with striping buffer (AR0153, Boster, Wuhan, China)and was normalized against β-tubulin (Polyclonal antibodies raised in goat,sc-9935, Santa Cruz Biotechnology, diluted

Table 2 The gene name, GeneBank accession number, sequence and product size of the primers used for the qRT-PCR

Gene	Accession #[1]	Forward Primer (5'-3')	Reverse Primer (5'-3')	Product Size
GAPDH[§]	NM_001034034	GGGTCATCATCTCTGCACCT	GGTCATAAGTCCCTCCACGA	176
ACC-α[§]	NM_174224	AGCTGAATTTTCGCAGCAAT	GGTTTTCTCCCCAGGAAAAG	117
FASN[§]	AF285607	GCATCGCTGGCTACTCCTAC	GTGTAGGCCATCACGAAGGT	136
LPL[§]	NM_001075120	GGGTTTTGAGCAAGGGTACA	GCCACAATGACCTTTCCAGT	193
FABP1&	FJ415874.1	GTTCATCATCACCGCTGGCT	CCACTGCCTTGATCTTCTCCC	101
PLIN2&	NM_173980.2	TTTATGGCCTCATGCTTTTGC	CTCAGAGCAGACCCCAATTCA	100
ACOX1&	NM_001035289.3	ACCCAGACTTCCAGCATGAGA	TTCCTCATCTTCTGCACCATGA	100
CPT1α&	FJ415874.1	TCGCGATGGACTTGCTGTATA	CGGTCCAGTTTGCGTCTGTA	100
SCD[§]	NM_173959.4	TTATTCCGTTATGCCCTTGG	GGTAGTTGTGGAAGCCCTCA	151
DGAT1&	FJ415874.1	CCACTGGGACCTGAGGTGTC	GCATCACCACACACCAATTCA	101
DGAT2&	FJ415874.1	CATGTACACATTCTGCACCGATT	TGACCTCCTGCCACCTTTCT	100
SREBF1c	FJ415874.1	CACTCGTCTTCCTCTGTCTC	GAGTGACTGGTTCTCCATAG	243
SREBF2&	NM_001205600.1	AGAGCAAACTCCTGAAGGGC	GGAGGCGACATCAGAAGGAC	103
SCAP	NM_001101889.1	CATCAAGCTCTACTCCATCC	CAATGGCAGCGTTGTCCAGCA	206
LXRα&	NM_001014861.1	CCCCATGACCGACTGATGTT	TGTCCTTCATCTGGCTCCACC	241
PPARα&	FJ415874.1	CATAACGCGATTCGTTTTGGA	CGCGGTTTCGGAATCTTCT	102

[§]published in [19,20]. &published in [21].
[1]Entrez Gene, National Center for Biotechnology Information (NCBI).

to 1:200). The procedures for the secondary antibody and visualization were the same as that used for SCD. The ECL signals were recorded using an imaging system (LAS4000, USA) and analyzed using Quantity One (Bio-Rad, USA).

Statistical analysis
All of the data were expressed as the mean ± SEM. The statistical data analysis was conducted via unpaired or paired Student's t-tests using IBM SPSS 20.0 Statistics for mac (IBM Inc., New York, USA). A difference was considered to be significant when $p < 0.05$.

Results
Rumen pH, LPS content in the rumen and plasma, and the milk composition
The pH value in the ruminal fluid is shown in Figure 1. The dynamic pH curve in the HC group was lower than that in the LC group during the long-term experiment. It showed that a pH value under 5.6 lasted for 223 minutes in the HC group, which indicated that SARA was successfully induced. The pH value of the HC group was significantly lower than that of the LC group after the morning feeding for 8 h ($p < 0.05$).

The LPS concentration in the ruminal fluid of the HC group was significantly increased, from 47.17×10^3 EU/mL to 79.04×10^3 EU/mL, compared to the LC group ($p < 0.01$). In the peripheral plasma, the LPS level in the HC group was 0.86 EU/mL, while it was 0.47 EU/mL in the LC group ($p < 0.001$).

Milk samples were collected from the 1st week to the 18th week to determine the change in the milk composition. The results showed that the milk yield, the percentage of milk fat and the milk fat yield decreased significantly in the HC group compared to the LC group ($p < 0.01$). However, the percentage of milk protein increased remarkably in the HC group ($p < 0.05$; Table 3).

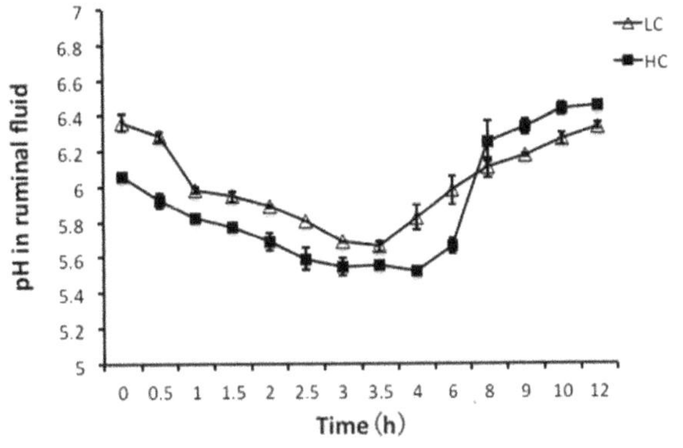

Figure 1 Comparison of the pH values in the ruminal fluid between the low-concentrate (LC) and high-concentrate (HC) groups. The data were measured using ruminal fluid samples collected at times ranging from 0 to 12 hours (shown on the x-axis) for three days during the 18th week. Significant differences were observed across all of the sampling times ($p < 0.05$), with the exception of 8 hours after the morning feeding. The error bars indicate the standard error of the mean. The data were compared using Student's *t*-test.

Table 3 The LPS, milk yield and milk composition in the dairy cows fed the low-concentrate (LC) and high-concentrate (HC) diets

Item	Treatment			
	LC	HC	SEM[1]	p-Value[2]
Rumen LPS, EU/mL[1] (×10³)	47.17	79.04[b]	7.94	<0.01
Plasma LPS, EU/mL[2]	0.47	0.86[ab]	0.08	<0.001
Milk[3]				
Milk yield, kg	28.05	26.92[b]	0.30	<0.01
Fat, %	3.40	3.09[b]	0.03	<0.01
Fat yield, kg/d	0.95	0.84[b]	0.02	<0.01
Protein, %	3.04	3.16[a]	0.03	<0.05
Protein yield, kg/d	0.85	0.85	0.02	0.98

[1]SEM = Standard error of the mean between the two groups.
[2]The LPS data were compared using Student's *t*-test. The milk samples were compared using a paired *t*-test.
[3]The milk samples were obtained from the 1st week to the 18th week.
[ab]indicates $p < 0.001$; [a]indicates $p < 0.05$; [b]indicates $p < 0.01$.

Table 4 The ruminal fluid composition, blood metabolites and hormone level in the dairy cows fed the low-concentrate (LC) diet and high-concentrate (HC) diet

	Treatment			
VFA profile[1]	**LC**	**HC**	**SEM**[2]	**p-value**[3]
Total VFA(mmol/L)	92.91	107.64[a]	3.66	<0.05
Molar proportion, mmol/mol				
Acetate	454.51	440.55	3.10	0.47
Propionate	247.68	281.83[b]	2.41	<0.01
Isobutyrate	25.19	24.68	0.38	0.83
Butyrate	210.11	195.99	1.77	0.17
Isovalerate	36.31	34.52	0.67	0.67
Valerate	21.25	19.93	0.24	0.38
Caproate	4.95	2.49[b]	0.17	<0.01
Acetate:Propionate	1.84	1.57[a]	0.02	<0.05
Propionate:Butyrate	1.18	1.44[b]	0.05	<0.01
Lactic acid	0.96	1.55[b]	0.20	<0.01
Plasma biochemical parameter[4]				
TG (mmol/L)	0.28	0.21[a]	0.05	<0.05
NEFA (mmol/L)	1.16	0.48[ab]	0.07	<0.001
TCH (mmol/L)	2.21	1.69[b]	0.31	<0.01
GLU (mg/dL)	45.23	56.02[b]	4.54	<0.01
Hormone level[4]				
Insulin (μIU/mL)	16.95	21.57[a]	1.01	<0.05
Glucagon (pg/mL)	191.23	161.02	18.28	0.28
Insulin:Glucacon	0.10	0.16[c]	0.01	0.07

[1]The volatile fatty acid and lactic acid concentrations and the mean proportion across the sampling times during the 18th week.
[2]SEM = Standard error of the mean between the two treatments.
[3]The data were compared using a paired *t*-test.
[4]The mean metabolite concentration in the jugular plasma across the sampling times. TG, triglyceride; NEFA: non-esterified fatty acid; TCH, total cholesterol; GLU, glucose;
[a]indicates $p < 0.05$; [b]indicates $p < 0.01$; [ab]indicates <0.001.

The VFA profile in the ruminal fluid and the plasma biochemical parameters

The VFA profiles in the ruminal fluid and the biochemical parameters in the peripheral plasma between the LC and HC groups are presented in Table 4. When compared to the LC group, the total VFA and lactic acid concentration in the rumen was significantly elevated in the HC group (105.95 *vs.* 92.91, $p < 0.05$ and 1.55 *vs.* 0.96 mmol/L, $p < 0.01$, respectively). The molar proportion (mmol/mol) of propionate was increased in the dairy cows fed the high-concentrate diet (281.83 *vs.* 247.68, $p < 0.01$), but the other proportional concentrations of the individual VFAs were unchanged. The ratio of propionate to butyrate (1.18 *vs.* 1.44, $p < 0.05$) was significantly increased, while the ratio of acetate to propionate (1.84 *vs.* 1.57, $p < 0.05$) was significantly reduced.

When compared to the LC group, the triacylglycerol ($p < 0.05$), NEFA ($p < 0.01$) and total cholesterol ($p < 0.01$) concentrations were significantly decreased in the peripheral plasma of the HC group. However, the glucose and insulin concentration were significantly enhanced ($p < 0.01$, $p < 0.05$ respectively) in the plasma of the high-concentrate diet group.

The long-chain fatty acid profiles in the ruminal fluid and the hepatic vein plasma

The LCFA profiles in the ruminal fluid and the hepatic vein plasma are shown in Tables 5 and 6, respectively. The LCFA concentration in the ruminal fluid and the hepatic vein of the HC group was lower than that in the LC group, specifically for palmitate C16:0 ($p < 0.05$) and palmitoleate C18:0 ($p < 0.05$). The desaturation index that was determined by calculating the plasma C16:1n-9/C16:0 ratio was decreased in the HC group ($p = 0.087$). Additionally, a decrease in the Δ^9 monounsaturated oleic acid (C18:1n-9) concentration was observed in the HC group. Compared with the LC group, the concentration of *cis9,trans 11* CLA was similar in the both rumen and plasma. The concentration of *trans11* C18:1 was increased in rumen of the HC group ($p = 0.067$), while in the plasma, the *trans11* C18:1 content was similar. In regards to α-linolenic acid (C18:3n-3), its content in the HC group was four-fold lower than that of the LC group. However, the content of both C22:0 and C22:1n-9 was significantly increased ($p < 0.05$) in the HC group. The desaturation index of C18:1n-9/C18:0 and *cis9, trans11* C18:2n/*trans11* C18:1n was decreased in the HC group, but no statistical significance was observed. Meanwhile, the presence of longer-chain saturated FAs (C20:0, $p < 0.01$; C21:0, $p < 0.01$), which are produced via ruminal microbial biohydrogenation, was decreased in the HC group.

Table 5 The fatty acid composition in the ruminal fluid of the dairy cows

μg/mL	Treatment			
	LC	HC	SEM[1]	p-Value[2]
C12:0	2.67	2.05	0.27	0.28
C13:0	3.31	2.92	0.30	0.56
C14:0	254.80	74.26	55.68	0.11
C15:0	6.25	3.47	1.05	0.21
C16:0	147.09	96.84^{b}	10.48	<0.01
*cis9*C16:1	3.17	2.32	0.11	0.27
C18:0	473.48	281.65^{ab}	38.02	<0.001
*trans11*C18:1n	4.53	11.40^{c}	1.93	0.07
C18:1n-9	11.36	8.93	0.65	0.05
C18:2n-6	10.15	7.18^{c}	0.89	0.09
C20:0	6.28	4.40^{b}	0.40	<0.01
6C18:3n	0.55	0.75	0.09	0.27
C20:1	0.52	0.35^{b}	0.04	<0.01
C18:3n-3	2.65	2.04	0.27	0.30
*cis9, trans11*C18:2n	6.61	5.27	0.45	0.15
C21:0	0.76	0.53^{b}	0.05	<0.01
C22:0	3.53	3.10	0.17	0.23
C23:0	1.21	1.12	0.06	0.51
C24:0	4.26	3.83	0.19	0.29
C24:1	0.69	0.60	0.04	0.29
C22:6n-3	2.21	1.83	0.13	0.17

[1]SEM = Standard error of the mean between the two treatments.
[2]The data were compared using Student's *t*-test.
[a]indicates $p < 0.05$; [b]indicates $p < 0.01$; [ab]indicates $p < 0.001$; [c]indicates $0.05 < p < 0.1$.

Table 6 The fatty acid composition in the hepatic vein plasma of the dairy cows

μg /mL	Treatment			
	LC	HC	SEM[1]	P-Value[2]
C12:0	3.57	4.04	0.20	0.26
C13:0	5.92	6.99	0.50	0.32
C14:0	66.06	74.32	13.66	0.79
C14:1	6.47	4.39^{c}	0.57	0.06
C15:0	8.81	6.43	0.84	0.17
C15:1	1.39	1.25	0.09	0.49
C16:0	92.47	71.10^{a}	5.14	<0.05
*cis9*C16:1	5.32	2.84^{c}	0.64	0.05
C18:0	121.33	85.5^{a}	8.81	<0.05
*trans11*C18:1n	2.92	2.93	0.09	0.96
C18:1n-9	62.24	40.28^{a}	5.69	<0.05
C18:2n-6	277.99	71.13^{a}	38.72	<0.05
C20:0	0.89	1.02^{a}	0.03	<0.05
C20:1	0.45	0.61^{c}	0.05	0.10
C18:3n-3	25.86	6.6^{b}	3.74	<0.01
*cis9, trans11*C18:2n	4.56	4.57	0.07	0.98
C21:0	1.27	1.00	0.19	0.51
C22:0	0.97	1.31^{a}	0.08	<0.05
C20:3n-6	27.01	10.80^{c}	4.41	0.07
C22:1n-9	1.24	1.57^{a}	0.08	<0.05
C20:4n-6	33.46	16.13^{b}	3.84	<0.01
C22:6n-3	10.57	6.20^{ab}	0.83	<0.001
Desaturation index				
cis9 C16:1/C16:0	0.05	0.04^{c}	0.01	0.09
cis9 C18:1/C18:0	0.51	0.47	0.02	0.37
cis9,trans11 C18:2/tran11 C18:1	1.66	1.55	0.06	0.37

[1]SEM = Standard error of the mean between the two treatments.
[2]The data were compared using Student's *t*-test.
[a]indicates $p < 0.05$; [b]indicates $p < 0.01$; [ab]indicates $p < 0.001$; [c]indicates $0.05 < p < 0.1$.

mRNA expression of the genes involved in lipid metabolism in the liver

The liver mRNA expression levels of the genes involved in lipid metabolism are presented in Figure 2. The expression levels of the genes associated with fatty acid uptake/transport, lipid formation, fatty acid oxidation and transcriptional regulators of lipogenic enzymes were remarkably altered between the HC and LC groups. There was a decrease in fatty acid binding protein 1 (FABP1) expression ($p = 0.09$) in the HC group compared to the LC group, and LPL was significantly down-regulated in the HC group. Compared with the LC group, the expression of perilipin 2 (PLIN2) ($p < 0.05$) was significantly decreased in the HC group, and there was 2-fold down regulation of SCD1 expression in the HC group. The expression of diacylglycerol acyltransferase (DGAT1 and DGAT2) was similar between the two groups, and the expression of ACCα and FAS, which are involved in *de novo* fatty acid synthesis, showed no significant difference between the HC and LC groups. However, the expression of carnitine palmitoyltransferase 1α (CPT1α) was up-regulated in the HC group ($p < 0.05$), whereas the expression of acyl-CoA oxidase 1 (ACOX1) was increased in the HC group ($p = 0.10$).

With respect to transcriptional regulators, the mRNA level of SCAP was down-regulated in the HC group ($p < 0.05$). Meanwhile, both the SREBF1c ($p = 0.09$) and SREBF2 ($p = 0.08$) mRNA expression levels were decreased in the HC group. However, the mRNA expression of PPARα was significantly increased in the HC group ($p < 0.05$). The expression of liver X receptor α (LXRα) showed no significant difference.

The protein expression of SCD1 in the liver

The protein expression of SCD in the liver is shown in Figure 3. The results demonstrated that the expression

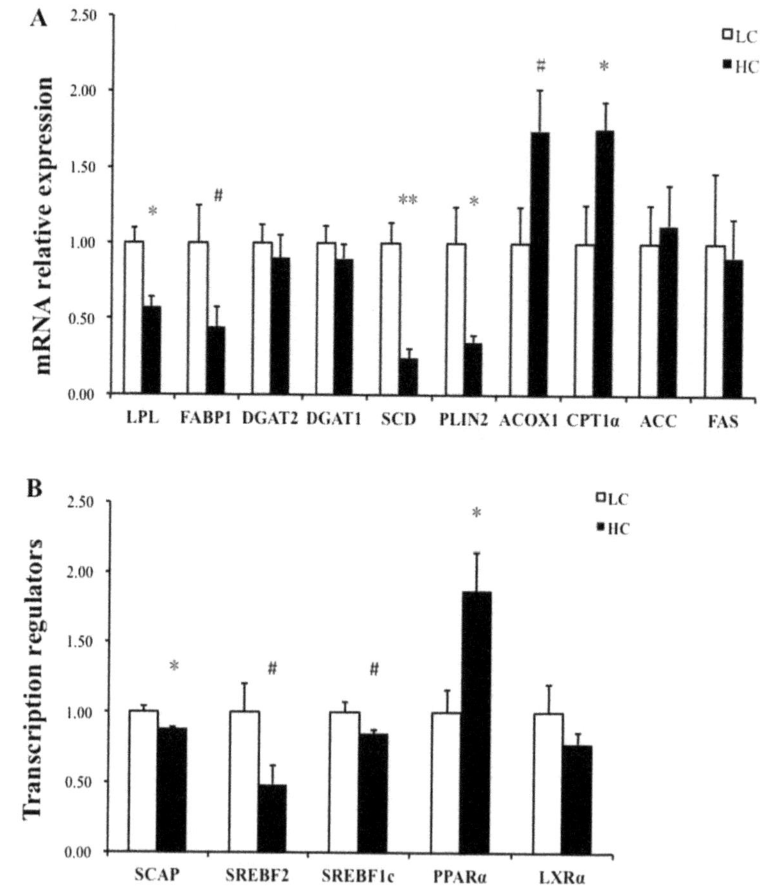

Figure 2 The hepatic gene expression profile analyzed via real-time PCR. Each value was normalized to the expression of GAPDH, and data were compared using Student's *t*-test between LC (n = 4) and HC (n = 4). **A.** The genes involved in desaturation, lipogenesis, fatty acid oxidation, TG synthesis and lipid droplet formation were measured in the liver tissue. The error bars indicate the standard error of the mean. * indicates $p < 0.05$; ** indicates $p < 0.01$; # indicates significance values between $0.05 < p < 0.1$. **B.** The genes involved in transcriptional regulation were measured in the liver tissue. The error bars indicate the standard error of the mean. * indicates $p < 0.05$; ** indicates $p < 0.01$; # indicates significance values between $0.05 < p < 0.1$.

of SCD in the liver was significantly down-regulated in the HC group compared to the LC group ($p < 0.05$).

Discussion

In this study, we showed that an altered fatty acid composition is induced by a HC diet. The reduced oleate and palmitoleate content may be associated with the down-regulated expression of SCD1 in the liver, which primarily resulted from the release of LPS during long-term HC feeding. These findings provide insights into the role of endogenous LPS on hepatic SCD1 expression in dairy cows and its relationship with fatty acid composition.

Previous studies have reported that SARA is characterized by declined feed intake, inflammation and depressed milk fat [12]. In our experiment, the duration of a rumen pH less than 5.6 lasted for 223 min/day in the cows fed a high-concentrate diet, meanwhile, a decrease in the milk yield (kg/d), milk fat (%) and milk fat yield (kg/d) was observed in the HC group. Therefore, our results are consistent with other studies.

Our data demonstrated that the total VFA and lactic acid levels in the ruminal fluid were significantly increased in the HC group. The ratio of acetate to propionate was decreased in the HC group due to an increase in propionate. Early experiments have also presented a low ratio of acetate to propionate in dairy cows fed with easily fermentable carbohydrates [22]. A previous study showed an increased ratio of ruminal propionate to butyrate in repartitioned milk from fat to lactose and protein [23]. In our study, the increased proportion of propionate may be related to glycogenesis [24,25]. Because most volatile fatty acids emerge in the portal vein after absorption from the digestive tract [26], alterations in volatile fatty acid concentrations may influence the metabolism in the liver.

When gram-negative bacteria in the rumen are lysed at low pH values, LPS is released and translocated into

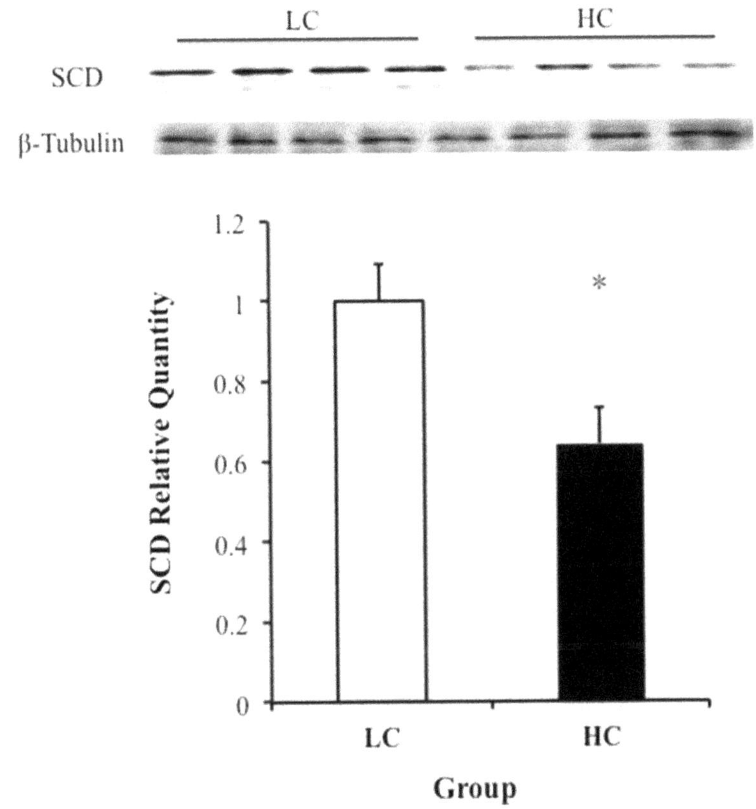

Figure 3 The western blotting analysis of SCD in the liver. The SCD content was assessed via western blotting of the livers of the LC (n = 4) and HC (n = 4) cows. The protein was quantified via band density measurements of the western blot. The band densities were normalized to the β-tubulin content within each sample. The data are expressed as the relative amounts of the two groups. * indicates p <0.05.The data were compared using Student's *t*-test.

the bloodstream, initiating an inflammatory response. In our experiment, the high endogenous LPS content may have triggered metabolic disorders in the digestive tract and liver. It has been documented that the liver has a strong ability to clear LPS [27-29]. A recent study showed that the LPS gene expression profile was altered in the liver of lactating goats fed a long-term high-concentrate diet, and the overall metabolism was shifted towards energy supply, in order to meet the higher energy expenditure demands for tissue anti-inflammation [30].

Our results indicated that SARA also influences plasma metabolites. Cholesterol in the plasma is negatively associated with the presence of LPS in the rumen [31,32], which explains our results of lower cholesterol concentrations in the plasma of the HC group. Among the hormones in peripheral plasma, insulin plays a crucial role in lipid metabolism, particularly in case of feeding cows with a high-concentrate diet. Decreased milk fat yield caused by SARA might have been because of an increased plasma insulin concentration and the ratio of insulin to glucagon [31,33,34], which has shown that high-concentrate diet resulted in greater plasma insulin concentration in our experiment. Due to the higher ruminal propionate and plasma glucose, increased plasma insulin might promote energy expenditure in hepatic through lipolysis and glycolysis, rather than fatty acid synthesis and gluconeogenesis [35], which in turn repressed the expression of lipogenic enzymes, such as SCD1. NEFAs are primarily mobilized from stored TGs in the adipose tissue [36]. In our experiment, the increased propionate and glucose resulted in a decrease in the NEFA concentration in the plasma, which is attributed to their inhibitory effect on adipose tissue lipolysis [35,37,38]. A lower TG concentration was observed in the HC group, which could be explained by the reduced adipose lipolysis and the restriction of biohydrogenation at low rumen pHs [39]. The restricted biohydrogenation led to a decreased saturated FA content, particularly of C16:0 and C18:0, in the rumen.

The fatty acid profiles in the ruminal fluid and the hepatic vein plasma were altered in this study. Decreased saturated FA concentrations in the ruminal fluid reflects the inhibition of biohydrogenation at lower pH values in the rumen of cows fed a long-term high-concentrate diet.

A previous study demonstrated that the conversion of long-chain fatty acids to TGs and phospholipids (PLs) in the livers of dairy cows is dependent on adipose tissue lipolysis [40]. The decrease in the palmitic (C16:0), stearic (C18:0) and oleic (C18:1) content in the liver may be explained by the decreased NEFA levels released from the adipose tissue and the lower production in the rumen. Similarly, a decrease in C18:2n-6 is associated with decreased lipolysis in the adipose tissue [41]. A decrease of both C18:2n-6 and C20:4n-6 in the hepatic vein plasma was observed in the HC group. It was reported that arachidonic acid (C20:4n-6),which comes from the cell cytoplasm, could be synthesized by Δ^5desaturase from C20:3n-6, and the latter could be desaturated and elongated from C18:2n-6 in the endoplasmic reticulum [42]. In addition, α-linoleic acid (C18:3n-3) could be desaturated and elongated to C22:6n-3, which could explain the decrease in C22:6n-3 in the HC group. Decreased C18:3n-3 may cause an accumulation of TGs [43], due to its function of enhancing the stability of apolipoprotein B:100 [44]. Furthermore, *trans11* C18:1n is considered to bethe precursor of *trans10* C:18:1n, which is a known inhibitor of milk fat in dairy cows [37]. Therefore, its increased levels in the rumen are associated with milk fat depression.

Fatty acid binding protein 1 (FABP1) is related to fatty acid uptake, transport, and metabolism and the activation of PPARα via NEFAs [45]. In our study, the down-regulated expression of FABP1 was likely observed because of the low NEFA concentrations in the peripheral plasma. As increased levels of *trans11* C18:1n in the rumen of the HC group emerged in the portal vein, it may have activated the PPARα pathway, as was demonstrated in a previous in vitro study [46].

In the liver, the activation of carnitine palmitoyltransferase 1α (CPT1α) and acyl-CoA oxidase 1 (ACOX1) is regulated by PPARα [8]. In dairy cows, the above enzymes are responsible for regulating the entry of LCFAs into the mitochondria for oxidation [47,48]. Because of the increased energy demand to resist inflammation during induced SARA, the gene expression of the above enzymes was increased in the HC group, which is consistent with the expression of the transcription factor PPARα. Additionally, shorter-chain fatty acyl-CoA, which is produced via peroxisomal fatty acid β-oxidation, is subsequently channeled to be oxidized completely in the mitochondria [45]. Therefore, the up-regulated expression of CPT1α suggests that hepatic energy export is necessary during induced SARA.

It was reported that the expression of SREBF1c and FAS, which are involved in lipid synthesis, was down-regulated in SCD1 knockout mice [8]. In our study, the expression of the transcription factor SREBF1c was decreased in the HC group, similar to its activation-dependent ligand SCAP, which could further explain the downregulation of SREBF2. In the liver, SREBF1c regulates the genes involved in fatty acid synthesis, while SREBF2 modulates the genes associated with cholesterol biosynthesis [49]. Therefore, the decreased cholesterol in the plasma could be regulated by SREBF2. SCD is a key lipogenic enzyme that regulates the synthesis of monounsaturated fatty acids, particularly oleate (C18:1) and palmitoleate (C16:1) [50]. The transcription of SCD is co-regulated by SREBF1c and PPARα [51,52]. The decreased C16:1 and C18:1n-9 content in the hepatic vein plasma indicated the downregulation of SCD in the dairy cows fed a high-concentrate diet. However, the *cis9,trans11*CLA content was unchanged because the process of *trans11*C18:1n desaturation to *cis9,trans11*CLA via SCD exists in most lipogenic and adipogenic tissues, except in the livers of rats and bovine [53,54]. Moreover, it is considered that the Δ^9 desaturation index poorly predicts the activity and/or expression of SCD [55]. Similar to SCD, the expression of perilipin 2 (PLIN2), which is involved in the intracellular accumulation of TGs and lipid droplet (LD) formation, was decreased in the HC group [56], which led to the attenuation of LD formation in the liver. Diacylglycerol acyltransferases (DGATs) plays key roles in the synthesis of TGs and very low-density lipoprotein (VLDL) secretion in the liver [57]. In our present study, the similar expression of DGAT1 and DGAT2 between the HC and LC group may be attributed to compensation of the down regulated SCD, which is partly due to a positive correlation between inflammation and fatty liver [58].

A previous study showed that the AMP-activated protein kinase (AMPK) signaling pathway is associated with the inhibition of SCD1 [59]. To some extent, this may contribute to the down regulation of SCD at the mRNA and protein levels. The unchanged mRNA expression of ACCα and FAS may not reflect the phosphorylation status of the enzymes. Therefore, further research is needed to elucidate the underlying mechanism.

Conclusions

In summary, lipid metabolism in the livers of dairy cows is influenced by long-term high-concentrate diet feeding. Lipopolysaccharide derived from the rumen down-regulates stearoyl-CoA desaturase 1 expression and alters fatty acid composition in the liver of dairy cows fed a high-concentrate diet. Our findings may shed light on the regulation of fatty acid metabolism and reprogramming in the livers of dairy cows fed high-concentrate diets.

Abbreviations

ACOX1: Acyl-CoA oxidase 1; CPT1: Carnitine palmitoyltransferase 1; DGAT: Diacylglycerol acyltransferase; FABP1: Fatty acid binding protein 1; FAME: Fatty acid methyl esters; LCFA: Long chain fatty acid; LPS: Lipopolysaccharide; LXR: Liver X receptor; PLIN2: Perilipin 2; PPARα: peroxisome proliferator-activated receptor α; SCAP: SREBP cleavage activating protein; SCD: Stearoyl-CoA desaturase; SREBF: Sterol regulatory element binding transcription factor; VFA: Volatile fatty acid.

Competing interests
The authors declare that they have no financial, personal or professional interests that would have influenced the content of the paper or interfered with their objective assessment of the manuscript.

Authors' contributions
XS and TX: conceived and designed the experiments. TX, HT, GC, KZ and LX: performed the experiments. TX and XS: analyzed the data. TX and XS: drafted the manuscript. All authors read and approved the final manuscript.

Acknowledgements
This study was supported by the National Basic Research Program of China (2011CB100802), National Natural Science Foundation of China (31172371; 30371040), the Federal Ministry of Food, Agriculture and Consumer Protection of Germany and the Ministry of Agriculture of China (Grant No. 30/2008-2009), and the Priority Academic Program Development of Jinagsu Higher Education Institutions (PAPD).
The authors are grateful to Dr. Dirk Dannenberger and Dr. Karin Nuernberg, from the Leibniz Institute for Farm Animal Biology, Germany, for their valuable advice on fatty acid analysis via gas chromatography (GC).

References
1. Gozho GN, Plaizier JC, Krause DO, Kennedy AD, Wittenberg KM. Subacute ruminal acidosis induces ruminal lipopolysaccharide endotoxin release and triggers an inflammatory response. J Dairy Sci. 2005;88:1399–403.
2. Chen Y, Oba M, Guan LL. Variation of bacterial communities and expression of Toll-like receptor genes in the rumen of steers differing in susceptibility to subacute ruminal acidosis. Vet Microbiol. 2012;159(3–4):451–9.
3. Plaizier JC, Krause DO, Gozho GN, McBride BW. Subacute ruminal acidosis in dairy cows: the physiological causes, incidence and consequences. Vet J. 2008;176(1):21–31.
4. Emmanuel DG, Dunn SM, Ametaj BN. Feeding high proportions of barley grain stimulates an inflammatory response in dairy cows. J Dairy Sci. 2008;91(2):606–14.
5. Arisqueta L, Nunez-Garcia M, Ogando J, Garcia-Arcos I, Ochoa B, Aspichueta P, et al. Involvement of lipid droplets in hepatic responses to lipopolysaccharide treatment in mice. Biochem Biophys Acta. 2013;1831(8):1357–67.
6. Jiang QD, Li HP, Liu FJ, Wang XJ, Guo YJ, Wang LF, et al. Effects of lipopolysaccharide on the stearoyl-coenzyme A desaturase mRNA level in bovine primary hepatic cells. Genet Mol Res. 2014;13(2):2548–54.
7. Ntambi JM. Regulation of stearoyl-CoA desaturase by polyunsaturated fatty acids and cholesterol. J Lipid Res. 1999;40:1549–58.
8. Ntambi JM, Miyazaki M, Stoehr JP, Lan H, Kendziorski CM, Yandell BS, et al. Loss of stearoyl-CoA desaturase-1 function protects mice against adiposity. Proc Natl Acad Sci U S A. 2002;99(17):11482–6.
9. Feingold KR, Staprans I, Memon RA, Moser AH, Shigenaga JK, Doerrler W, et al. Endotoxin rapidly induces changes in lipid metabolism that produce hypertriglyceridemia: low doses stimulate hepatic triglyceride production while high doses inhibit clearance. J Lipid Res. 1992;33:1765–76.
10. Chen C, Shah YM, Morimura K, Krausz KW, Miyazaki M, Richardson TA, et al. Metabolomics reveals that hepatic stearoyl-CoA desaturase 1 downregulation exacerbates inflammation and acute colitis. Cell Metab. 2008;7(2):135–47.
11. Kleen JL, Hooijer GA, Rehage J, Noordhuizen JPTM. Subacute Ruminal Acidosis (SARA): a review. J Vet Med. 2003;50:406–14.
12. Khafipour E, Krause DO, Plaizier JC. A grain-based subacute ruminal acidosis challenge causes translocation of lipopolysaccharide and triggers inflammation. J Dairy Sci. 2009;92(3):1060–70.
13. Gaynor PJ, Waldo DR, Capuco AV, Erdman RA, Douglass LW, Teters BB. Milk fat depression, the glucogenic theory, and trans-C18:1 fatty acids. J Dairy Sci. 1995;78:2008–15.
14. Zebeli Q, Ametaj BN. Relationships between rumen lipopolysaccharide and mediators of inflammatory response with milk fat production and efficiency in dairy cows. J Dairy Sci. 2009;92(8):3800–9.
15. Graugnard DE, Moyes KM, Trevisi E, Khan MJ, Keisler D, Drackley JK, et al. Liver lipid content and inflammometabolic indices in peripartal dairy cows are altered in response to prepartal energy intake and postpartal intramammary inflammatory challenge. J Dairy Sci. 2013;96(2):918–35.
16. Hamada T, Omori S, Kameoka K, Horii S, Morimoto H. Direct determination of rumen volatile fatty acids by gas chromatography. J Dairy Sci. 1968;51:228–9.
17. Folch J, Lees M, Sloane Stanley GH. A simple method for the isolation and purification of total lipides from animal tissues. J Biol Chem. 1957;226:497–509.
18. Chouinard PY, Corneau L, Barbano DM, Metzger LE, Bauman DE. Conjugated linoleic acids alter milk fatty acid composition and inhibit milk fat secretion in dairy cows. J Nutr. 1999;129:1579–84.
19. Akbar H, Schmitt E, Ballou MA, Correa MN, Depeters EJ, Loor JJ. Dietary lipid during late-pregnancy and early-lactation to manipulate metabolic and inflammatory gene network expression in dairy cattle liver with a focus on PPARs. Gene Regul Syst Bio. 2013;7:103–23.
20. Bionaz M, Loor JJ. Gene networks driving bovine milk fat synthesis during the lactation cycle. BMC Genomics. 2008;9:366.
21. Joseph SJ, Robbins KR, Pavan E, Pratt SL, Duckett SK, Rekaya R. Effect of diet supplementation on the expression of bovine genes associated with fatty acid synthesis and metabolism. Bioinf Biol Insights. 2010;4:19–31.
22. Fairfield AM, Plaizier JC, Duffield TF, Lindinger MI, Bagg R, Dick P, et al. Effects of prepartum administration of a monensin controlled release capsule on rumen pH, feed intake, and milk production of transition dairy cows. J Dairy Sci. 2007;90:937–45.
23. Harri M, Pekka H. Effects of the ratio of ruminal propionate to butyrate on milk yield and blood metabolites in dairy cows. J Dairy Sci. 1996;79:851–61.
24. Blanch M, Calsamiglia S, DiLorenzo N, DiCostanzo A, Muetzel S, Wallace RJ. Physiological changes in rumen fermentation during acidosis induction and its control using a multivalent polyclonal antibody preparation in heifers. J Anim Sci. 2009;87(5):1722–30.
25. Penner GB, Taniguchi M, Guan LL, Beauchemin KA, Oba M. Effect of dietary forage to concentrate ratio on volatile fatty acid absorption and the expression of genes related to volatile fatty acid absorption and metabolism in ruminal tissue. J Dairy Sci. 2009;92(6):2767–81.
26. Bergman EN. Energy contributions of volatile fatty acids from the gastrointestinal tract in various species. Physiol Rev. 1990;70:567–90.
27. Buttenschoen K, Radermacher P, Bracht H. Endotoxin elimination in sepsis: physiology and therapeutic application. Langenbecks Arch Surg. 2010;395(6):597–605.
28. Deng M, Scott MJ, Loughran P, Gibson G, Sodhi C, Watkins S, et al. Lipopolysaccharide clearance, bacterial clearance, and systemic inflammatory responses are regulated by cell type-specific functions of TLR4 during sepsis. J Immunol. 2013;190(10):5152–60.
29. Vels L, Rontved CM, Bjerring M, Ingvartsen KL. Cytokine and acute phase protein gene expression in repeated liver biopsies of dairy cows with a lipopolysaccharide-induced mastitis. J Dairy Sci. 2009;92(3):922–34.
30. Dong H, Wang S, Jia Y, Ni Y, Zhang Y, Zhuang S, et al. Long-term effects of subacute ruminal acidosis (SARA) on milk quality and hepatic gene expression in lactating goats fed a high-concentrate diet. Plos One. 2013;8(12):e82850.
31. Zebeli Q, Dunn SM, Ametaj BN. Perturbations of plasma metabolites correlated with the rise of rumen endotoxin in dairy cows fed diets rich in easily degradable carbohydrates. J Dairy Sci. 2011;94(5):2374–82.
32. Bertok L. Bile acids in physico-chemical host defence. Pathophysiol. 2004;11(3):139–45.
33. Piccioli-Cappelli F, Loor JJ, Seal CJ, Minuti A, Trevisi E. Effect of dietary starch level and high rumen-undegradable protein on endocrine-metabolic status, milk yield, and milk composition in dairy cows during early and late lactation. J Dairy Sci. 2014;97(12):7788–803.
34. Waggoner JW, Löest CA, Turner JL, Mathis CP, Hallford DM. Effects of dietary protein and bacterial lipopolysaccharide infusion on nitrogen metabolism and hormonal responses of growing beef steers. J Anim Sci. 2009;87:3656–68.
35. Guo Y, Xu X, Zou Y, Yang Z, Li S, Cao Z. Changes in feed intake, nutrient digestion, plasma metabolites, and oxidative stress parameters in dairy cows with subacute ruminal acidosis and its regulation with pelleted beet pulp. J Anim Sci Biotechnol. 2013;4:31.

36. Knegsel ATM, Brand H, Graat EAM, Dijkstra J, Jorritsma R, Decuypere E, et al. Dietary energy source in dairy cows in early lactation: metabolites and metabolic hormones. J Dairy Sci. 2007;90:1477–85.
37. Bauman DE, Griinari JM. Regulation and nutritional manipulation of milk fat: low-fat milk syndrome. Livest Sci. 2001;70:15–29.
38. Ametaj B, Bradford B, Bobe G, Nafikov R, Lu Y, Young J, et al. Strong relationships between mediators of the acute phase response and fatty liver in dairy cows. Can J Anim Sci. 2005;85:165–75.
39. Van Nevel CJ, Demeyer DI. Influence of pH on lipolysis and biohydrogenation of soybean oil by rumen contents in vitro. Reprod Nutr Dev. 1996;36:53–63.
40. Douglas GN, Rehage J, Beaulieu AD, Bahaa AO, Drackley JK. Prepartum nutrition alters fatty acid composition in plasma, adipose tissue, and liver lipids of periparturient dairy cows. J Dairy Sci. 2007;90:2941–59.
41. Rukkwamsuk T, Geelen MJH, Kruip TAM, Wensing T. Interrelation of fatty acid composition in adipose tissue, serum, and liver of dairy cows during the development of fatty liver postpartum. J Dairy Sci. 2000;83:52–9.
42. Ves-Losada A, Maté SM, Brenner RR. Incorporation and distribution of saturated and unsaturated fatty acids into nuclear lipids of hepatic cells. Lipids. 2001;36:273–82.
43. Mashek DG, Bertics SJ, Grummer RR. Metabolic fate of long-chain unsaturated fatty acids and their effects on palmitic acid metabolism and gluconeogenesis in bovine hepatocytes. J Dairy Sci. 2002;85:2283–9.
44. Wu X, Shang A, Jiang H, Ginsberg HN. Demonstration of biphasic effects of docosahexaenoic acid on apolipoprotein B secretion in HepG2 cells. Arterioscl Throm Vas. 1997;17:3347–55.
45. Loor JJ. Genomics of metabolic adaptations in the peripartal cow. Animal. 2010;4(07):1110–39.
46. Pawar A, Jump DB. Unsaturated fatty acid regulation of peroxisome proliferator-activated receptor alpha activity in rat primary hepatocytes. J Biol Chem. 2003;278(38):35931–9.
47. Dann HM, Drackley JK. Carnitine palmitoyltransferase I in liver of periparturient dairy cows: effects of prepartum intake, postpartum induction of ketosis, and periparturient disorders. J Dairy Sci. 2005;88:3851–9.
48. Vluggens A, Andreoletti P, Viswakarma N, Jia Y, Matsumoto K, Kulik W, et al. Reversal of mouse acyl-CoA oxidase 1 (ACOX1) null phenotype by human ACOX1b isoform. Lab Invest. 2010;90(5):696–708.
49. Bommer GT, MacDougald OA. Regulation of lipid homeostasis by the bifunctional SREBF2-miR33a locus. Cell Metab. 2011;13(3):241–7.
50. Hofacer R, Magrisso IJ, Jandacek R, Rider T, Tso P, Benoit SC, et al. Omega-3 fatty acid deficiency increases stearoyl-CoA desaturase expression and activity indices in rat liver: positive association with non-fasting plasma triglyceride levels. Prostag Leukotr Ess. 2012;86(1–2):71–7.
51. Biddinger SB, Almind K, Miyazaki M, Kokkotou E, Ntambi JM, Kahn CR. Effects of diet and genetic background on sterol regulatory element-binding protein-1c, stearoyl-CoA desaturase 1, and the development of the metabolic syndrome. Diabetes. 2004;54:1314–23.
52. Miyazaki M, Dobrzyn A, Sampath H, Lee SH, Man WC, Chu K, et al. Reduced adiposity and liver steatosis by stearoyl-CoA desaturase deficiency are independent of peroxisome proliferator-activated receptor-alpha. J Biol Chem. 2004;279(33):35017–24.
53. Shen X, Nuernberg K, Nuernberg G, Zhao R, Scollan N, Ender K, et al. Vaccenic acid and cis-9, trans-11 CLA in the rumen and different tissues of pasture- and concentrate-fed beef cattle. Lipids. 2007;42(12):1093–103.
54. Gruffat D, Torre ADL, Chardigny J-M, Durand D, Loreau O, Bauchart D. Vaccenic acid metabolism in the liver of rat and bovine. Lipids. 2005;40:295–301.
55. Invernizzi G, Thering BJ, McGuire MA, Savoini G, Loor JJ. Sustained upregulation of stearoyl-CoA desaturase in bovine mammary tissue with contrasting changes in milk fat synthesis and lipogenic gene networks caused by lipid supplements. Funct Integr Genomics. 2010;10(4):561–75.
56. Chang BH, Chan L. Regulation of triglyceride metabolism. III emerging role of lipid droplet protein ADFP in health and disease. Am J Physiol Gastrointest Liver Physiol. 2007;292(6):G1465–8.
57. Yamazaki T, Sasaki E, Kakinuma C, Yano T, Miura S, Ezaki O. Increased very low density lipoprotein secretion and gonadal fat mass in mice overexpressing liver DGAT1. J Biol Chem. 2005;280(22):21506–14.
58. Bradford BJ, Mamedova LK, Minton JE, Drouillard JS, Johnson BJ. Daily injection of tumor necrosis factor-{alpha} increases hepatic triglycerides and alters transcript abundance of metabolic genes in lactating dairy cattle. J Nutr. 2009;139(8):1451–6.
59. Dobrzyn P, Dobrzyn A, Miyazaki M, Cohen P, Asilmaz E, Hardie DG, et al. Stearoyl-CoA desaturase 1 deficiency increases fatty acid oxidation by activating AMP-activated protein kinase in liver. Natl Acad Sci USA. 2004;101(17):6409–14.

Productive and reproductive performances of dairy cattle herds in Treviso province, Italy (2009–2012): an assessment of the potential impact of Schmallenberg virus epidemic

Marica Toson[1†], Lapo Mughini-Gras[1,2*†], Katia Capello[1], Laura Gagliazzo[1], Laura Bortolotti[1], Matteo Mazzucato[1], Stefano Marangon[1] and Lebana Bonfanti[1]

Abstract

Background: Schmallenberg virus (SBV) has spread across Europe since mid-2011, causing unspecific and transitory symptoms in ruminants and congenital malformations in their offspring. Evidence for the impact of SBV on cattle (re)productive performance is limited. Using a comprehensive data set from a SBV-affected province in North-East Italy, this study aimed at assessing the potential impact of SBV emergence on 11 productive and reproductive performance indicators of dairy cattle herds, accounting for weather conditions and other herd-level factors that could also influence these indicators.

Results: A total of 127 farms with an average of 71 cows per farm (range 29–496) were monitored monthly from January 2009 to June 2012. Mixed-effects linear models for longitudinal data were used to assess the average variation in herds' performance indicators over semesters (Jan-Jun 2009, Jul-Dec 2009, Jan-Jun 2010, Jul-Dec 2010, Jan-Jun 2011, Jul-Dec 2011, Jan-Jun 2012) and trimesters therein. Taking the second semester of 2011 as reference, significant decreases in the average lactation length (−6 days, on average) and calving-to-conception interval (−4 days, on average) were observed relative to the same semesters of the years 2010 and 2009, respectively. Similarly, during the last trimester of 2011, which is most likely to cover the SBV infection period in the study area, there was an average decrease of −4 days (lactation length) and −7 days (calving-to-conception interval) compared to the same trimesters of the years 2010 and 2009, respectively. However, the observed decreases actually represent a positive outcome that is not as such imputable to SBV emergence, but rather reflects other beneficial changes in farm management. None of the other indicators showed significant variations, confirming the relatively mild expression of SBV infection in cattle.

Conclusions: Although the emergence of SBV might have significantly affected the (re)productive performance of some individual farms, we concluded that overall at the province level there were no significant variations attributable to SBV, at least not in a way that would lead to negative effects on farm profitability.

Keywords: SBV, Orthobunyavirus, Dairy cattle, Italy

* Correspondence: lapo.mughini.gras@rivm.nl
†Equal contributors
[1]Istituto Zooprofilattico Sperimentale delle Venezie (IZSVe), Viale dell'Università 10, 35020 Legnaro, (Padua), Italy
[2]National Institute for Public Health and the Environment (RIVM), Centre for Infectious Disease Control (Clb), PO Box 1-3720, BA, Bilthoven, The Netherlands

Background

Schmallenberg virus (SBV), a newly discovered *Orthobunyavirus* (family *Bunyaviridae*), was first identified in November 2011 from dairy cattle with unspecific and transitory symptoms, including fever, decreased milk production and diarrhoea, in North Rhine-Westphalia, Germany [1]. Schmallenberg virus shows high homology to viruses of the Simbu serogroup, the members of which are typically transmitted through *Culicoides* biting midges [2, 3] and can cause congenital defects in domestic animals [4, 5].

The first clinical evidences of SBV appeared in August 2011, not only in Germany, but also in Belgium and the Netherlands. Because examination of archived samples did not indicate prior SBV circulation, the most accredited hypothesis is that the virus was introduced in Europe between spring and summer 2011, and that the excess of congenital malformations, abortions, and stillborns that have been reported in domestic ruminants since December 2011 are the result of SBV infections in pregnant animals between summer and autumn 2011 [6]. By May 2013, a total of 8730 bovine and ovicaprine herds in 22 European countries, including Italy, were reported as being SBV-positive based on laboratory confirmation of suspected cases [7].

In Italy, the first laboratory-confirmed SBV case was reported at the beginning of February 2012 from a small multi-species farm in the province of Treviso, Veneto region (Fig. 1), where a dystocic goat died because of the

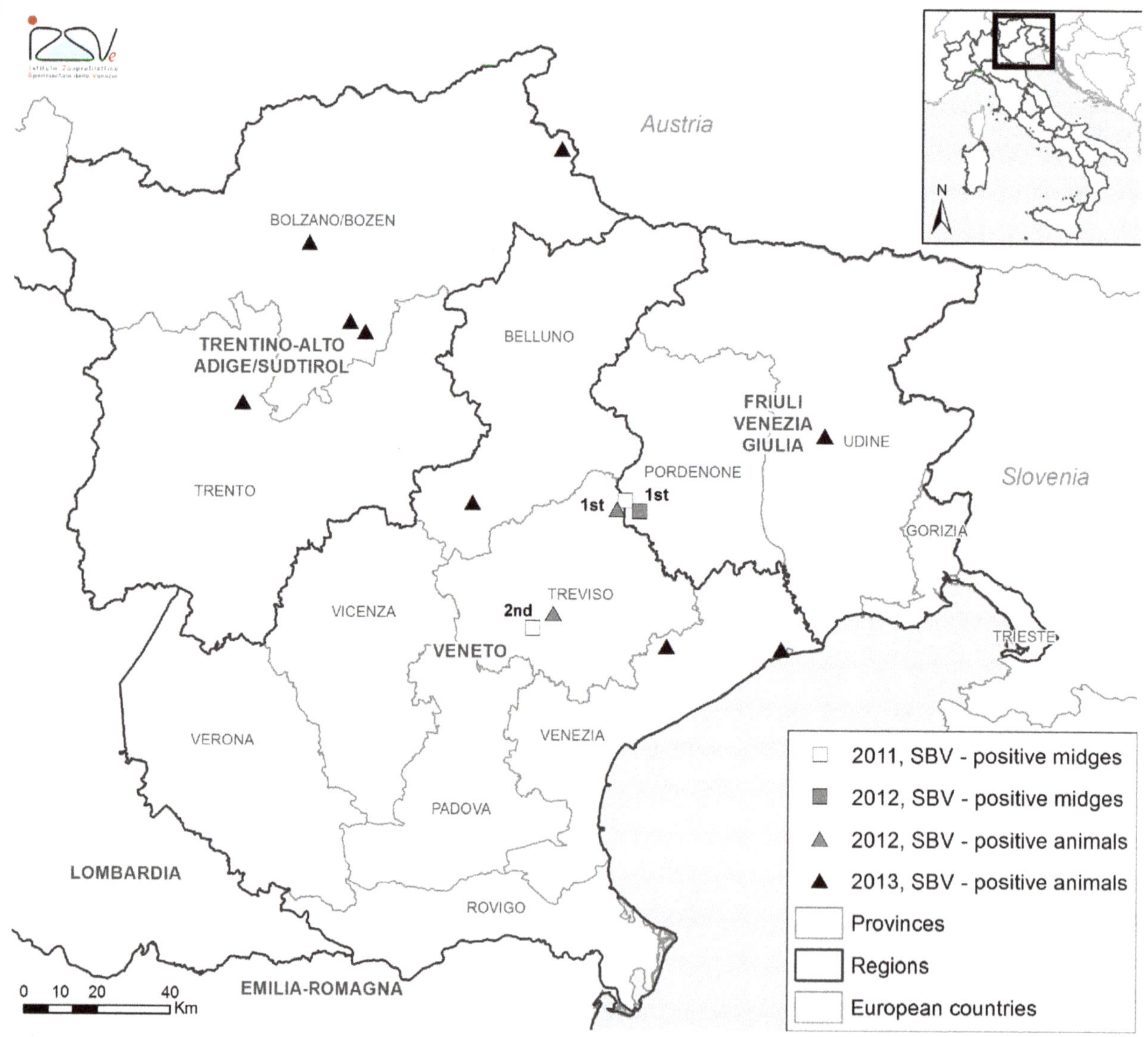

Fig. 1 Map of North-East Italy, including the study area (Treviso province), with indication of positivities for Schmallenberg virus in *Culicoides* biting midges and susceptible animals (affected farms) from 2011 to 2013. The first SBV-positive animal was detected in Treviso province in a dystocic goat on the first semester of 2012 (1st). Midges were found positive in Treviso province on the second semester of 2011 (2nd) and in Pordenone province on the first semester of 2012 (1st)

retention of a malformed foetus [8]. Epidemiological investigation around this farm excluded any prior introduction of animals from other European countries, suggesting that the virus had circulated locally. Moreover, six pools of *Culicoides* midges collected as part of the national surveillance programme for bluetongue virus (BTV) between September and November 2011 within a 50 km radius from the first SBV case tested positive for SBV [2, 8]. This provided evidence that SBV had circulated in the local vector population almost five months before detecting the first case, but that any circulation before September 2011 was unlikely [2, 8].

Evidence for the impact of SBV on animal (re)productive performance is scarce. A few studies have assessed the impact of SBV infection as: 1) direct impact on adult animals, including their clinical manifestations and herd- or area-level variation in key performance parameters [9–11]; 2) direct impact on newborns, including congenital malformations, abortions, and stillbirths [7, 9, 12]; and 3) indirect impacts, including trade restrictions and treatment costs [9, 13]. Although SBV may cause clinical signs of pyrexia and reduced milk production, cases are typically identified only when animals appear with congenital malformations among offspring whose mothers acquired infection during a particular period of pregnancy [9]. The lack of a clear clinically overt presentation is one of the main reasons as to why there is a large degree of under-ascertainment for SBV [7], especially in cattle, for which it is known that SBV infection is relatively mild, with a low incidence of birth malformations [7, 9–11]. Accordingly, the impact of SBV on adverse pregnancy outcomes (abortions, stillbirths, and congenital malformations) of dairy cattle is believed to be limited and lower than that on milk production or return to service [6]. A recent Dutch study reported that calf mortality differed very little between herds with low and high SBV seroprevalence during the second half of 2011, a finding that mirrors the relatively mild expression of SBV infection in adult cows [10]. Conversely, SBV seemed to have had a relatively more pronounced impact on milk production. In a predefined 4-week period during which most SBV infections were expected to have occurred in the Netherlands, the overall drop in milk production per cow was quantified as ~30 kg (in herds with high SBV seroprevalence) and ~50 kg (in herds with reported SBV-associated symptoms) [10]. However, potential confounders such as weather conditions, particularly temperature and relative humidity, which are known to affect milk production levels [14], were not accounted for.

This study aimed at assessing variations imputable to SBV emergence in 11 productive and reproductive performance indicators of dairy cattle herds in the province of Treviso, i.e. Italy's area where SBV was at first detected [8] and post-epidemic modelling studies have predicted it as one of the areas with the highest risk of SBV spread in Europe [15]. The analysis accounted for weather conditions and other herd-level factors that could also influence these indicators. Yet, it could only assess the presumptive/circumstantial effect of SBV on dairy cattle herd performances since no exact SBV data were available for the study area.

Results

The productive and reproductive performance indicators of dairy cattle herds in the study area are summarized by semester in Table 1. Data were collected monthly from January 2009 to June 2012 at the herd level ($n = 127$ herds). The average number of herd-month observations by semester per each indicator varied from 591 (Jan-Jun 2012) to 726 (Jan-Jun 2012), depending on the number of missing values. Some indicators (i.e. average daily milk production and total number of registered newborns) showed a clear seasonal pattern, with a tendency towards decreased milk production in the second semester (Jul-Dec) of each year (which comprises the summer) and increased number of newborns in the first semester (Jan-Jun) of each year (which comprises the spring). Accordingly, the average number of inseminations per pregnancy leading to a healthy newborn increased in the second semester of each year. The indicator average lactation length showed a drop in Jul-Dec 2011, which was preceded in first half of 2011 by another drop as compared to the previous semesters. The same was observed for the average calving-to-conception interval.

The mixed-effects linear models showed that only two (out of the 11) performance indicators of interest varied significantly with respect to the reference semester (Jul-Dec 2011). These indicators were the average lactation length and the average calving-to-conception interval (Table 2). Except for the second semester of 2009, all other semesters showed a significantly longer lactation length – also known as days in milk (DIM) – than that of the reference semester (Table 2). Looking specifically at the same semesters over the years, herds' lactation length was, on average, 6.3 days longer in the second semesters of 2010 compared to the reference semester (Table 2). Concerning the average calving-to-conception interval, a significant difference of approximately 4 days was observed between the second semester of 2009 only (LS-mean = 163.7 days) and the reference semester (LS-mean = 159.63 days) (Table 2).

The effects of the control variables included in the final models for DIM and calving-to-conception interval are reported in Table 3. Except for the herd size and cows' average age, all the other covariates in the final model for DIM were significantly associated with the outcome. An increase in the average monthly temperature, calving-to-conception interval, proportion of pregnant cows, and proportion of

Table 1 Mean values (±standard deviation) of the productive and reproductive performance indicators of dairy cattle herds of Treviso province, North-East Italy. Data were collected monthly at the herd level ($n = 127$ herds), from January 2009 to June 2012

	Herd-level performance indicators	Semesters						
		Jan-Jun 2009 ($n = 710$)	Jul-Dec 2009 ($n = 620$)	Jan-Jun 2010 ($n = 726$)	Jul-Dec 2010 ($n = 636$)	Jan-Jun 2011 ($n = 747$)	Jul-Dec 2011 ($n = 632$)	Jan-Jun 2012 ($n = 591$)
1	Average daily milk production per cow (L/day/cow)	29.08 (±5.46)	27.44 (±4.54)	29.23 (±4.52)	27.05 (±4.57)	29.16 (±4.69)	27.33 (±4.60)	29.32 (±4.69)
2	Average milk protein content (%)	3.43 (±0.19)	3.43 (±0.16)	3.39 (±0.14)	3.4 (±0.28)	3.36 (±0.14)	3.43 (±0.18)	3.42 (±0.15)
3	Average milk fat content (%)	3.65 (±0.33)	3.73 (±0.32)	3.67 (±0.31)	3.71 (±0.43)	3.61 (±0.32)	3.68 (±0.32)	3.72 (±0.30)
4	Average number of lactations per cow (no.)	2.37 (±0.30)	2.36 (±0.30)	2.34 (±0.28)	2.34 (±0.27)	2.33 (±0.27)	2.33 (±0.28)	2.33 (±0.28)
5	Average lactation length (days)	200.75 (±34.37)	203.41 (±37.20)	203.42 (±38.30)	203.83 (±38.73)	198.8 (±33.28)	195.03 (±32.89)	198.89 (±31.81)
6	Average calving-to-conception interval (days)	163.65 (±34.65)	167.28 (±38.08)	164.03 (±38.36)	166.25 (±38.95)	159.34 (±34.71)	158.44 (±34.60)	157.5 (±34.46)
7	Average number of inseminations per pregnancy (no.)	2.38 (±0.63)	2.36 (±0.66)	2.43 (±0.74)	2.47 (±0.76)	2.39 (±0.78)	2.34 (±0.71)	2.36 (±0.69)
8	Average number of inseminations per pregnancy in primiparous cows (no.)	2.36 (±0.83)	2.36 (±0.89)	2.44 (±0.88)	2.48 (±0.98)	2.42 (±0.95)	2.32 (±0.87)	2.39 (±0.84)
9	Average number of inseminations per pregnancy in multiparous cows (no.)	2.39 (±0.71)	2.35 (±0.67)	2.42 (±0.81)	2.48 (±0.84)	2.37 (±0.86)	2.35 (±0.77)	2.33 (±0.77)
10	Average number of inseminations per pregnancy leading to a registered newborn (no.)[a]	0.66 (±0.58)	0.81 (±0.73)	0.67 (±0.64)	0.79 (±0.75)	0.67 (±0.63)	0.76 (±0.74)	0.65 (±0.61)
11	Total number of registered newborns (no.)[b]	5.04 (±4.18)	4.47 (±3.69)	5.34(±4.37)	4.74 (±3.80)	5.34 (±4.38)	4.71 (±4.22)	4.98 (±3.69)
12	Number of cows per farm (no.)	70.27 (±45.49)	70.54 (±48.23)	70.67 (±48.17)	71.52 (±50.00)	72.72 (±49.51)	73.73 (±52.68)	64.65 (±22.36)
13	Cows' average age (months)	46.19 (±5.98)	46.66 (±6.67)	46.61 (±6.38)	46.73 (±5.98)	46.69 (±5.16)	46.77 (±5.18)	46.85 (±5.85)
14	Average length of dry period (days)	76.78 (±59.67)	76.03 (±66.95)	74.16 (±41.76)	74.15 (±38.01)	70.1 (±20.19)	67.96 (±13.57)	67.09 (±14.23)
15	Proportion of pregnant primiparous cows (%)	44.63 (±14.18)	45.74 (±15.34)	43.79 (±15.59)	45.96 (±15.99)	46.53 (±15.82)	46.48 (±16.22)	44.84 (±16.07)
16	Proportion of pregnant multiparous cows (%)	41.66 (±13.53)	42.21 (±14.07)	40.4 (±14.93)	43.13 (±14.41)	42.62 (±13.76)	42.42 (±14.75)	40.72 (±14.92)
17	Proportion of artificial inseminations (%)	91.67 (±20.39)	91.48 (±21.35)	91.61 (±22.42)	89.95 (±25.04)	89.02 (±26.32)	89.62 (±25.56)	89.65 (±25.39)
18	Proportion of culled cows (%)	2.54 (±3.09)	3.04 (±3.18)	2.27 (±2.29)	2.83 (±2.55)	2.37 (±2.40)	2.85 (±2.74)	2.31 (±2.35)
19	Cows' average age at first delivery (months)	27.15 (±4.04)	27.87 (±6.67)	28.17 (±6.71)	31.32 (±39.57)	21.07 (±12.97)	20.9 (±12.00)	19.25 (±12.88)

[a]Calculated as the ratio between the average number of inseminations per pregnancy (indicator 7) and the total number of registered newborns (indicator 11) of the following 9th month

[b]Includes only healthy calves that survived for at least 21 days after birth and the farmers were then obliged to register to the local health authority

artificial (over natural) inseminations performed in the farm was significantly associated with an increase in DIM. Conversely, an increase in the length of the dry period and proportion of culled cows was associated with a decrease in DIM. Except for the herd size, cows' average age, average monthly temperature and proportion of pregnant cows, all the other covariates in the final model for the calving-to-conception interval were significantly associated with the outcome and had associations similar to those found for DIM.

The trimester-based analysis (Table 4) revealed that DIM and calving-to-conception interval were again the

Table 2 Adjusted estimates of the effects of semester on the average lactation length and average calving-to-conception interval of 127 dairy cattle herds of Treviso province, Italy, obtained from the final mixed-effects linear models with first order autoregressive covariance structure fitted to monthly measurements (January 2009 to June 2012)

Semester	Average lactation length (days)[a]				Average calving-to-conception interval (days)[b]			
	Least square mean†	Standard error	Coefficient	*p*-value	Least square mean†	Standard error	Coefficient	*p*-value
Jan-Jun 2009	203.50	2.13	7.65	0.0002	161.20	2.63	1.57	0.4448
Jul-Dec 2009	199.27	2.11	3.41	0.0671	163.72	2.62	4.09	0.0295
Jan-Jun 2010	202.21	2.03	6.36	0.0003	160.79	2.55	1.16	0.4923
Jul-Dec 2010	202.11	2.07	6.26	<0.0001	161.09	2.59	1.46	0.3077
Jan-Jun 2011	200.87	2.04	5.02	<0.0001	159.49	2.57	−0.14	0.8899
Jul-Dec 2011	195.85	2.12	Ref.	-	159.63	2.64	Ref.	-
Jan-Jun 2012	199.97	2.25	4.12	0.0042	159.98	2.72	0.35	0.7870

[a]Estimates are adjusted for variables 6 and 12–18 and for mean temperature (see Tables 1 and 3), all included as fixed effects. Random effects were set at the farm level and used sinusoidal terms for seasonal control

[b]Estimates are adjusted for variables 12–18, and for mean temperature (see Tables 1 and 3), all included as fixed effects. Random effects were set at the farm level and used sinusoidal terms for seasonal control

†Also referred to as marginal means, are the means after controlling for the other covariates included in the models

only indicators with significant variations as compared with the reference trimester (Oct-Dec 2011). The lactation length was, on average, 4.1 days significantly longer ($p = 0.035$) in Oct-Dec 2010 (LS-mean = 202.14 days) than in Oct-Dec 2011 (LS-mean = 198.00 days), but this was not the case ($p = 0.193$) in Oct-Dec 2009 (LS-mean = 200.96). Regarding the calving-to-conception interval, a difference of 7.7 days was observed between the last trimester of 2009 (LS-mean = 164.34 days) and the reference trimester (LS-mean = 156.63 days). The associations of the control covariates (Table 5) did not differ from those of the semester-based analysis.

Discussion

This study assessed the potential impact of SBV emergence on a number of productive and reproductive performance indicators of dairy cattle herds in a SBV-affected province of North-East Italy, while accounting for a range of potential confounders, including weather conditions. Results indicated a statistically significant drop in the average lactation length and in the average calving-to-conception interval during the second semester of 2011, but none of the other indicators showed significant variations. These results were reproduced using a narrower temporal window based on trimesters.

A reduced number of DIM is strictly related to a decrease in the number of days between calving and conception, also known as days open, as cows that are re-impregnated early in lactation will enter earlier the dry period. Shortening the lactation period by means of early start of re-insemination (6–8 weeks postpartum) is a desirable target in modern dairy farm management [16–18]. Extended lactation periods are in fact an indicator of reproductive problems in the herd (e.g. decreased fertility, abortion, foetal resorption, non-return, etc.). The advantage of short lactation cycles is that

Table 3 Associations of the control variables included in the 'semester-based' mixed-effects linear models for the average lactation length and average calving-to-conception interval of 127 dairy cattle herds of Treviso province, Italy, with first order autoregressive variance structure fitted to monthly measurements (January 2009 to June 2012)

Control variable	Average lactation length (days)[a]			Average calving-to-conception interval (days)[a]		
	Coefficient	F-value	p-value	Coefficient	*F*-value	*p*-value
Herd size	−4.69	1.63	0.2047	−1.32	0.07	0.7858
Cows' average age (months)	−0.09	0.51	0.4737	−0.19	2.21	0.1369
Average calving-to-conception interval (days)	0.15	85.63	<0.0001	-	-	-
Average monthly temperature	0.82	167.86	<0.0001	−0.05	1.25	0.2643
Proportion of pregnant primiparous cows (%)	0.33	213.63	<0.0001	−0.01	0.47	0.4928
Proportion of pregnant multiparous cows (%)	0.46	275.41	<0.0001	−0.07	9.21	0.0024
Average length of dry period (days)	−0.03	4.16	0.0414	−0.07	21.11	<0.0001
Proportion of artificial inseminations (%)	0.11	7.27	0.0071	0.43	120.17	<0.0001
Proportion of culled cows (%)	−0.45	61.78	<0.0001	−0.28	32.04	<0.0001

[a]Estimates are adjusted for semester (see Tables 2), included as fixed effect. Random effects were set at the farm level and used sinusoidal terms for seasonal control

Table 4 Adjusted estimates of the effects of trimester on the average lactation length and average calving-to-conception interval of 127 dairy cattle herds of Treviso province, Italy, obtained from the final mixed-effects linear models with first order autoregressive co-variance structure fitted to monthly measurements (January 2009 to June 2012).

Trimester	Average lactation length (days)[a]				Average calving-to-conception interval (days)[b]			
	Least square mean†	Standard error	Coefficient	*p*-value	Least square mean[c]	Standard error	Coefficient	*p*-value
Jan-March 2009	200.27	2.33	2.27	0.3540	165.97	2.81	9.34	0.0004
Apr-Jun 2009	203.17	2.25	5.17	0.0303	160.03	2.75	3.40	0.1815
Jul-Sep 2009	197.72	2.28	−0.28	0.9052	162.91	2.77	6.28	0.0113
Oct-Dec 2009	200.96	2.27	2.96	0.1929	164.34	2.76	7.71	0.0012
Jan-March 2010	203.15	2.21	5.15	0.0212	162.01	2.70	5.38	0.0185
Apr-Jun 2010	201.95	2.18	3.95	0.0703	161.71	2.68	5.08	0.0199
Jul-Sep 2010	200.74	2.24	2.74	0.1935	162.40	2.73	5.77	0.0054
Oct-Dec 2010	202.14	2.24	4.14	0.0349	159.11	2.73	2.48	0.1921
Jan-March 2011	199.91	2.20	1.91	0.2916	158.10	2.71	1.47	0.3853
Apr-Jun 2011	199.82	2.22	1.82	0.2688	156.94	2.73	0.31	0.8338
Jul-Sep 2011	193.57	2.30	−4.43	0.0006	157.47	2.79	0.84	0.4645
Oct-Dec 2011	198.00	2.30	Ref.	-	156.63	2.79	Ref.	-
Jan-March 2012	200.70	2.38	2.70	0.0639	157.42	2.85	0.79	0.5395
Apr-Jun 2012	206.23	2.50	8.23	<.0001	158.22	2.91	1.59	0.3545

[a]Estimates are adjusted for variables 6 and 12–18 and for mean temperature (see Tables 1 and 5), all included as fixed effects. Random effects were set at the farm level and used sinusoidal terms for seasonal control

[b]Estimates are adjusted for variables 12–18, and for mean temperature (see Tables 1 and 5), all included as fixed effects. Random effects were set at the farm level and used sinusoidal terms for seasonal control

[c]Also referred to as marginal means, are the means after controlling for the other covariates included in the models

more peaks of lactation will occur in a given time span, resulting in more milk yield and more calves delivered. Moreover, cows that are re-inseminated early in lactation perform better in terms of pregnancy rates [19]. Conversely, increased days open affect farm profitability via increased breeding costs coupled with reduced milk production due to the post-peak flattening of the lactation curve. Besides, increased days open in a previous parity have been associated with a higher risk of death and live culling around the subsequent calving, resulting in increased replacement costs [20]. It follows, therefore, that the significant decrease in DIM and days open observed here actually represents a positive outcome, which is not as such imputable to SBV emergence. If SBV would have had (detrimental) effects on these two indicators via, for instance, the onset of clinical disease, non-return and adverse pregnancy outcomes, an increase, rather than a decrease, in DIM and days open had to be expected. As it is highly unlikely that SBV improved the performance of dairy cattle herds, other factors than SBV are likely to be

Table 5 Associations of the control variables included in the 'trimester-based' mixed-effects linear models for the average lactation length and average calving-to-conception interval of 127 dairy cattle herds of Treviso province, Italy, with first order autoregressive variance structure fitted to monthly measurements (January 2009 to June 2012)

Control variable	Average lactation length (days)[a]			Average calving-to-conception interval (days)[a]		
	Coefficient	*F*-value	*p*-value	Coefficient	*F*-value	*p*-value
Herd size	−4.53	1.51	0.2211	−1.34	0.08	0.7822
Cows' average age (months)	−0.12	0.82	0.3643	−0.12	0.92	0.3381
Average calving-to-conception interval (days)	0.16	91.00	<.0001	-	-	-
Average monthly temperature	0.94	168.66	<.0001	−0.01	0.02	0.8754
Proportion of pregnant primiparous cows (%)	0.33	216.91	<.0001	−0.01	0.44	0.5086
Proportion of pregnant multiparous cows (%)	0.45	261.98	<.0001	−0.07	8.31	0.0040
Average length of dry period (days)	−0.03	3.25	0.0717	−0.07	23.67	<.0001
Proportion of artificial inseminations (%)	0.11	7.70	0.0056	0.42	114.97	<.0001
Proportion of culled cows (%)	−0.42	52.11	<.0001	−0.29	35.20	<.0001

[a]Estimates are adjusted for trimester (see Tables 4), included as fixed effect. Random effects were set at the farm level and used sinusoidal terms for seasonal control

involved, e.g. the voluntary waiting period, artificial insemination technique, calving season (including changes in estrous synchronization in response to market needs), farm management policy, herd size, production level and parity [21], but also mere yearly variation. As some of these factors, such as the herd size, season, production level and parity were actually controlled for in our models, other (hitherto unknown) changes in farm management practices would probably explain the observed differences. Besides this, co-morbidities in the farms might have had an effect on the performance indicators. However, only three farms notified problems of mastitis caused by *Staphylococcus aureus* and *Streptococcus uberis*, but no other diseases were reported during the second semester of 2011, making concurrent health problems unlikely to have contributed significantly to the observed decrease in DIM and days open.

The main limitation of this study is that it was not possible to distinguish between "SBV-infected" and "SBV-free" herds, nor between different gradients of within-herd SBV prevalence. Rather, we assessed the overall impact of SBV based on a pre- and post-emergence design, assuming a homogeneous spread of infection across the study area. This assumption has some plausibility given the limited extension of the study area and the evidenced ability of SBV to spread rapidly and extensively [22]. Moreover, in the study area the prevalence of SBV in biting midges of the Obsoletus complex has been reported to be much higher (~10 times) than that of BTV [2]. Similar findings have been reported from other European countries [23, 24], suggesting a higher vector competence for SBV, which might counterbalance its shorter viraemia compared to BTV [1], a feature that draws further attention to the SBV potential for spreading. Accordingly, a model for between-farm SBV spread has shown that non-midge-borne transmission routes are unnecessary to explain the rapid spread of SBV [15]. Given that all herds in the study area were assumed to be equally exposed to SBV and effects at the herd level might be detectable at a finer scale depending on the within-farm prevalence, our results might be biased towards no significant effect because of insufficient insights into the spatial heterogeneity of SBV spread within the study area. Similarly, the use of relatively large temporal windows (semesters and trimesters), because of the difficulties in pinpointing the exact 1–2 week period in which animals were clinically ill, could have masked an actual effect. However, it can also be argued that any significant effect of SBV, including repercussions on farm profitability, would be better appreciable in a six to three month period interval, as finer temporal subdivisions would have complicated the interpretation of the results given the occurrence of small-scale punctual fluctuations in the (re)productive indicators. After all, a study in the Netherlands [11] was able to detect significance variations in herds' performances attributable to SBV using a 3-month reference period.

As mentioned in the introduction, Veldhuis et al. [10] were able to demonstrate a significant, yet limited, reduction in milk production and calf mortality in the Netherlands during the hypothesized SBV transmission period by distinguishing between herds with high and low seroprevalence and between farms with and without reported clinical cases. The same approach allowed to highlight an increased rate of abortions, malformed lambs, dystocia and lamb mortality in SBV-positive versus SBV-negative sheep flocks in Belgium [12]. The lack of prevalence or clinical data for the herds under study was the main reason as to why we could not apply such an analytic approach, which was certainly likely to provide more insights in the impact of SBV. However, Veldhuis et al. [11] did not distinguish between herds based on their SBV infection levels, but performed separate analyses for the Dutch national population of dairy cattle herds and for the subgroup of herds reporting malformations in newborn calves, and found small but significant variations in several (re)productive parameters. Therefore, the present study can only make the point that, irrespective of the SBV infection levels and rates of clinical illness, overall at the province level the (re)productive performance of dairy cattle herds in the study area did not seem to have been significantly affected by the emergence of SBV, at least not in a way that there were detrimental effects on farm profitability. This is in line with the results of Wernike et al. [25] who monitored a farm near the city of Schmallenberg, Germany, between May 2011 and January 2012. All tested animals were SBV-positive after week 41 of 2011. Yet, no decrease in milk yield nor diarrhoea was observed, only transient fever. Moreover, no abortions, stillbirths or malformed calves were observed despite that at the end of September 2011 most cows were between 75 and 175 days of pregnancy, i.e. the critical gestation period. Similarly, the relatively low number of SBV-associated congenital malformations compared to the high level of SBV infection as indicated by seroprevalence studies suggest that these occur only rarely in calves [9]. Furthermore, calves infected *in utero* can clear SBV infection [26], and farms adopting calving patterns with the critical period falling mostly into periods of low vector activity (usually Dec-Mar) can experience lower impact [9]. Therefore, although it cannot be excluded that in some instances the impact of SBV may have been considerable at the individual farm level, it was not unexpected to find here that overall at the province level this impact did not appear to be significant. Our results agree, to a major extent, with those of a recent study on the impact of SBV on milk production, reproductive performance, and calf mortality in dairy cattle in Kleve district, Germany [11]. These authors could only find a small increase in the number of inseminations in the assumed SBV period, but no significant changes in calf mortality or milk production were found, indicating that SBV had a very limited impact, but also that there is considerable variation in the extent of such impact between countries.

Underreporting and under-ascertainment of SBV cases is known to be considerable [7]. This was especially true at the beginning of the epidemic, when the mild and transitory nature of the clinical disease, combined with the (initial) lack of diagnostic insights, led many cases to pass unnoticed. Even after the discovery of the virus and the launch of the international alert, many farmers might have not reported suspected cases because of the fear of bearing financial consequences. Thus, even if the reported cases could merely depict the enforcement of enhanced monitoring/control activities by local authorities, development of diagnostic tools and recognition of the lack of SBV zoonotic potential, a possible explanation for the absence of an evidenced negative effect of SBV in the study area may be due, to some extent, to the absence of reported clinical cases in 2011 [7, 9, 27].

The major impact of SBV originated from restrictions on international trade of susceptible animals, including semen and embryos [6]. Of course the rapid spread of the virus in a completely naïve host population resulted in some direct impacts. However, whether this scenario will happen again is a matter of discussion, especially since SBV is likely to remain endemically in Europe, allowing for herd immunity to persist [9]. New SBV outbreaks will therefore occur if the level of herd immunity declines or introductions of novel SBV strains occur. Nonetheless, based on our study, it seems that the future impact of SBV on dairy cattle will depend more on the trade restrictions applied than on the worsening of herds' performance. It should be reminded, however, that our results are not based on empirical data on the exact SBV status of the herds, so we are likely to have underestimated the effect of SBV because the proportion of SBV-infected animals within a herd is variable and probably never 100 %. Moreover, the detrimental effects of SBV infection are only present during a limited period of gestation, and a large part of the animals in the herds may have been infected outside this critical period, having no detectable effect on performance.

Conclusions

We assessed the potential impact of SBV emergence on several productive and reproductive performance indicators of dairy cattle herds in a SBV-affected province of North-East Italy, accounting for weather conditions and other herd-level factors that could also influence these indicators. Significant decreases in the average lactation length and calving-to-conception interval were observed during the likely SBV infection period. However, such decreases actually represent a positive outcome which is not imputable to SBV emergence *per se*, but rather mirrors other beneficial changes in farm management. While it is highly unlikely that SBV would have improved the performance of dairy cattle herds, the lack of significant variations in dairy herds' (re)productive performance imputable to SBV is in line with the relatively mild expression of acute SBV infection in adult cattle. However, our results also contradict previous reports on reproductive problems, and may have underestimated the effect of SBV emergence since no data on the SBV status of the herds and animals therein were available to inform the analysis on the infection levels of the cattle population under study. Although the emergence of SBV might have significantly challenged some individual farms provided that most SBV infections had occurred in the critical gestation period, overall at the province level their (re)productive performances did not seem to have been significantly affected, at least not in a way that there were detrimental effects on farm profitability.

Methods

Study area and herd-level performance indicators

The study area consisted of the province of Treviso (2477 km^2) in North-East Italy (Fig. 1). All dairy cattle farms subscribed to the Treviso's Provincial Cattle Breeders Association (Italian acronym: APA) were included in the study. These were 127 intensive farms, with an average of 71 cows per farm (range 29–496) and a closed production system. These farms are under continuous monitoring of a number of productive and reproductive performance indicators by the APA, the membership of which is voluntary. APA members have access to several services aimed at improving livestock productivity and welfare. Such services include, among others, the monitoring of how farms perform, both in relation to their own historical records and for benchmarking against other farms. Herd-level monthly measurements of 19 (re)-productive performance indicators of the abovementioned 127 farms were provided by the APA for the period between January 2009 and June 2012. These indicators, together with a summary of their values by semester, are reported in Table 1.

Indicators represent monthly averages (indicators 1–10, 13–15 and 19, see Table 1) or total monthly counts (indicators 11 and 12) based on the whole farm and are recorded automatically by a herd management software dedicated to the individual follow-up of every animal in the farm. Averages of milk production per cow and fat/protein milk content are determined in bulk milk or recorded individually at milking of each cow through their pedometers. The reproductive status of every animal at any point in time, including their inseminations, pregnancy diagnoses, deliveries, start and end dates of lactation and dry periods, is also individually recorded. Indicators are then automatically calculated by the APA based on the recorded data. Sporadically, because of various technical or logistical problems, some indicators are incalculable for specific farms at a given point in time; these missed observations are then treated as missing values. Abortions in these farms are notifiable to local veterinary services and are handled according to the regional control programme in force for bovine abortion [28].

Weather data

Monthly mean temperature (°C) and relative humidity (%) data for the period between January 2009 and June 2012 were collected through 21 official weather stations distributed all over the province of Treviso. These data were provided by the Environmental Protection Agency of Veneto region (Italian acronym: ARPAV); a governmental agency with multiple missions related to environmental policy and research and is in charge, among other tasks, for the management of all weather stations of Veneto region. This includes collecting, validating, analyzing and disseminating meteorological data to external users. Each of the 127 farms included in the study was anchored to its closest weather station.

Study periods

The overall study period spanned between January 2009 and June 2012. Significant variations in herds' performance imputable to SBV were expected to be detectable during the second semester of 2011, which covers the most likely period of primary exposure to SBV in continental Europe [6, 9, 10, 15, 27]. Particularly, most ruminants in the study area were likely to have come into contact with the virus between October and December 2011. This was indicated by the detection in the study area of the first case of SBV-induced foetal malformations in the full-term dystocic goat at the beginning of February 2012 [8], as the most susceptible gestation period in ovicaprines is between 45 and 60 days of pregnancy [9]. Moreover, virus detection in midges has demonstrated that SBV had been circulating in the study area since September 2011, yet any circulation before then was deemed unlikely [2]. Furthermore, a convenience sample of 487 sera collected in January and in March 2012 at 14 cattle farms in the study area within the framework of statutory surveillance activities for brucellosis and bovine leukosis, were tested for SBV at the Istituto Zooprofilattico Sperimentale delle Venezie using the commercial ID Screen® Schmallenberg virus competition multi-species ELISA test (ID.vet Innovative Diagnostics, France), according to manufacturer's instructions. Serological testing revealed an overall positivity rate for SBV of 76 % (range 30-100 %, unpublished data). Although no sera collected in 2011 were available for SBV testing, this, together with the above-mentioned viral detections in midges and animals in 2011, provided evidence that exposure to SBV was already serologically detectable at the earliest beginning of 2012, supporting the notion that SBV had shortly before spread over the study area, i.e. in the last semester of 2011, and most likely in Oct-Dec 2011. Therefore, the second semester of 2011 and particularly the last trimester therein were taken as the reference periods (i.e. the SBV infection periods) for assessing the possible impact of SBV (see below).

A putative effect of the clinical disease on the productive parameters (i.e. milk production and milk protein and fat contents) and on the reproductive parameters regarding immediate fertility of the conception frame (i.e. number of lactations, lactation length, calving-to-conception interval, number of inseminations per pregnancy - since pregnancies would already have occurred during the acute phase of the infection) was indeed expected to occur in the SBV infection periods. Only the reproductive indicators related to newborns (e.g. number of inseminations per pregnancy leading to a healthy newborn and total number of healthy newborns) were lagged 9 months later in time, as they would have been affected by the intrauterine infection rather than the acute (symptomatic) phase.

Given the limited extension of the Treviso province, all cattle herds in the study area were assumed to have been exposed to SBV. This was supported by the extraordinary ability of SBV to spread rapidly over large geographical areas, as demonstrated by several seroprevalence studies indicating that virtually every cow housed in SBV-affected areas had come into contact with the virus [22, 29–31]. As neither prevalence nor clinical data on SBV were available for the herds in question, this study could only assess whether there were significant variations in province-level (re)productive performances of dairy cattle herds imputable to the emergence of SBV, regardless of their within-herd prevalence and rate of clinical illness.

Data analysis

Each of the first 11 performance indicators listed in Table 1 was modelled, as dependent variable, using mixed-effects linear models for longitudinal data, given that repeated measurements were made on the same farms over time; a random effect at the farm level was included in all models. The other eight indicators listed in Table 1, together with the mean temperature and relative humidity, were considered for inclusion as control variables in the models. Except for herd size, which is a dichotomous variable of large *vs.* small farms, all other control variables were treated as continuous covariates. Analyses were performed using the procedure PROC MIXED in SAS v.9.3, with the RANDOM and REPEATED statements to model the random effect at the farm level and the multiple observations made over time on the same farms. Variables were selected based on previous studies, biological plausibility of being influenceable by SBV, and scientific interest of the research team. Collinearities between variables were checked by looking at their correlation matrix (Additional file 1), and selection between collinear variables was based on the improvement in model fit as revealed by the Akaike information criterion (AIC) values.

The independent variable of interest to describe the average variation in herds' performance over the study period was the semester, which was included as categorical fixed effect (Jan-Jun 2009, Jul-Dec 2009, Jan-Jun 2010, Jul-Dec 2010, Jan-Jun 2011, Jul-Dec 2011, Jan-Jun 2012), with the semester Jul-Dec 2011 (SBV infection period) being the reference category. Further models were built narrowing the SBV infection period by splitting the above semesters in trimesters as to assess variations in herds' performance relative to the reference trimester Oct-Dec 2011. All comparisons focused on the same periods of the years 2010 and 2009.

To control for unobserved covariates with a systematic behaviour in time, sinusoidal terms were included in the models as random effects. The structure for the random effect was selected using restricted maximum likelihood (REML) estimation, by comparing the −2 REML log-likelihood for the reference model with the −2 REML log-likelihood of the nested model [32]; all random effects were significant. Several covariance structures (autoregressive, compound symmetry, and unstructured) were assessed to model the residuals associated with multiple observations originating from the same farm, and the first order autoregressive covariance structure was identified as the best one to capture the within-herd correlation.

Non-significant ($p > 0.05$) fixed effect terms were dropped from the models in stepwise backward fashion, after having evaluated the results of Type III F-test. Additionally, the effect of removing variables on the other covariates included in the models was also monitored, and variables causing a significant change in the other covariates when removed were retained in the model to control for their effect regardless of significance. Normality of residuals was checked using normal probability plots to confirm absence of any remaining structure not accounted for by the models. For the sake of simplicity, only the final models of those performance indicators showing significant variations compared to the reference semester were presented. Results were expressed as adjusted means (LS-means) and standard errors. The basic equation of the regression models with n candidate independent variables ($n = 1, 2...21$) used in this study was as follows:

$$E\left(y_{ji}\right) = \beta_0 + \beta_1 X_{1j} + \beta_2 X_{2t} + \beta_3 X_{3ij} \ldots + \beta_n X_{nij} + u_{oj} + u_{1i}\sin(2\pi \times \text{month}_i/12) + u_{2i}\cos(2\pi \times \text{month}_i/12) + \varepsilon_{ji}$$

where $E(y_{ji})$ is the expected value of the dependent variable in herd j ($j = 1, 2...127$) and month i ($i = 1, 2...42$). The parameter β_1 is the fixed effect regression coefficient of the independent variable X_{1j} denoting the size of the jth herd; β_2 is the fixed effect regression coefficient of the independent variable X_{2t} specifying the tth semester ($t = 1, 2...7$) or the tth trimester ($t = 1, 2...14$) of observation; $\beta_3 \ldots \beta_n$ are the fixed effect regression coefficients of the n herd- and month-varying independent variables $X_{3ij} \ldots X_{nij}$ as reported in Table 1, including the weather variables; u_{0j} is the random effect associated with the intercept for herd j, and u_{1i} and u_{2i} are the sine and cosine random effects for month i, respectively; ε_{ji} is the random error term for herd j in month i.

Ethics statement

Data on cattle performance were collected by the Treviso's Provincial Cattle Breeders Association (APA) among its member farms as part of their regular monitoring activities of productive and reproductive performance indicators to assess how the farms perform in relation to their own historical records and for comparison to other farms. Data were automatically obtained from bulk milk, at milking via cows' pedometers, and/or by veterinarians during routine reproduction management practices in cattle following standards of good veterinary practice in Italy and in the EU. No ethical approval was required for this study.

Competing interests

The author(s) declare that they have no competing interests.

Authors' contributions

LB and SM conceived and designed the study. MM and LB collected the data. MT, LMG, KC, and LG performed the analyses. LMG, MT, and KC drafted the manuscript. All authors contributed to data interpretation and manuscript writing/editing. All authors read and approved the final manuscript.

Acknowledgements

This study was supported by the Regional Government of Veneto (Giunta Regionale del Veneto) with the project "Piano di sorveglianza delle zoonosi e di attuazione di una adeguata rete di sorveglianza epidemiologica" (DGRV

1986/14). The funder had no role in study design, data collection, analysis, interpretation, or writing of this paper. The authors are grateful to the Provincial Cattle Breeders Association of Treviso for providing the data, and to V.U. Santucci, M. Morini, and S. Nardelli for support in the serological analyses and review of the article.

References

1. Hoffmann B, Scheuch M, Höper D, Jungblut R, Holsteg M, Schirrmeier H, et al. Novel orthobunyavirus in Cattle, Europe, 2011. Emerg Infect Dis. 2012;18:469–72.
2. Goffredo M, Monaco F, Capelli G, Quaglia M, Federici V, Catalani M, et al. Schmallenberg virus in Italy: a retrospective survey in *Culicoides* stored during the bluetongue Italian surveillance program. Prev Vet Med. 2013;111:230–6.
3. Balenghien T, Pagès N, Goffredo M, Carpenter S, Augot D, Jacquier E, et al. The emergence of Schmallenberg virus across *Culicoides* communities and ecosystems in Europe. Prev Vet Med. 2014. doi:10.1016/j.prevetmed.2014.03.007.
4. Parsonson IM, McPhee DA. Bunyavirus pathogenesis. Adv Virus Res. 1985;30:279–316.
5. Shope RE. Bunyaviruses. In: Baron S, editor. Medical microbiology. Galveston, TX, USA: University of Texas Medical Branch at Galveston; 1996.
6. Beer M, Conraths FJ, van der Poel WHM. "Schmallenberg virus"–a novel orthobunyavirus emerging in Europe. Epidemiol Infect. 2013;141:1–8.
7. Afonso A, Abrahantes JC, Conraths F, Veldhuis A, Elbers A, Roberts H, et al. The Schmallenberg virus epidemic in Europe-2011-2013. Prev Vet Med. 2014. doi:10.1016/j.prevetmed.2014.02.012.
8. Monaco F, Goffredo M, Federici V, Carvelli A, Capobianco Dondona A, Polci A, et al. First cases of Schmallenberg virus in Italy: surveillance strategies. Vet Ital. 2013;49:269–75.
9. EFSA – European Food Safety Agency, 2013. "Schmallenberg" virus: analysis of the epidemiological data. http://www.efsa.europa.eu/it/supporting/pub/429e.htm (accessed 3 May 2014).
10. Veldhuis AMB, Carp-van Dijken S, van Wuijckhuise L, Witteveen G, van Schaik G. Schmallenberg virus in Dutch dairy herds: potential risk factors for high within-herd seroprevalence and malformations in calves, and its impact on productivity. Vet Microbiol. 2014;168:281–93.
11. Veldhuis AM, Santman-Berends IM, Gethmann JM, Mars MH, van Wuyckhuise L, Vellema P, et al. Schmallenberg virus epidemic: impact on milk production, reproductive performance and mortality in dairy cattle in the Netherlands and Kleve district Germany. Prev Vet Med. 2014;116:412–22.
12. Saegerman C, Martinelle L, Dal Pozzo F, Kirschvink N. Preliminary survey on the impact of Schmallenberg virus on sheep flocks in South of Belgium. Transbound Emerg Dis. 2013. doi:10.1111/tbed.12047.
13. Martinelle L, Dal Pozzo F, Gauthier B, Kirschvink N, Saegerman C. Field veterinary survey on clinical and economic impact of schmallenberg virus in belgium. Transbound Emerg Dis. 2014;61:285–8.
14. West JW, Mullinix BG, Bernard JK. Effects of hot, humid weather on milk temperature, dry matter intake, and milk yield of lactating dairy cows. J Dairy Sci. 2003;86:232–42.
15. Gubbins S, Richardson J, Baylis M, Wilson AJ, Abrahantes JC. Modelling the continental-scale spread of Schmallenberg virus in. Europe: Approaches and challenges. Prev. Vet. Med; 2014. doi:10.1016/j.prevetmed.2014.02.004.
16. Sørensen TJ, Østergaard S. Economic consequences of postponed first insemination of cows in a dairy cattle herd. Livest Prod Sci. 2003;79:145–53.
17. Inchaisri C, Jorritsma R, Vos PLAM, van der Weijden GC, Hogeveen H. Economic consequences of reproductive performance in dairy cattle. Theriogenology. 2010;74:835–46.
18. Inchaisri C, Jorritsma R, Vos PLAM, van der Weijden GC, Hogeveen H. Analysis of the economically optimal voluntary waiting period for first insemination. J Dairy Sci. 2011;94:3811–23.
19. Yusuf M, Nakao T, Yoshida C, Long ST, Gautam G, Ranasinghe RMSBK, et al. Days in milk at first AI in dairy cows; its effect on subsequent reproductive performance and some factors influencing it. J Reprod Develop. 2011;57:643–9.
20. Pinedo PJ, De Vries A. Effect of days to conception in the previous lactation on the risk of death and live culling around calving. J Dairy Sci. 2010;93:968–77.
21. Oseni S, Misztal I, Tsuruta S, Rekaya R. Seasonality of days open in US Holsteins. J Dairy Sci. 2003;86:3718–25.
22. Chenais E, Ståhl K, Frössling J, Blomqvist G, Näslund K, Svensson L, et al. Schmallenberg virus beyond latitude 65°N. Transbound Emerg Dis. 2013. doi:10.1111/tbed.12195.
23. De Regge N, Deblauwe I, De Deken R, Vantieghem P, Madder M, Geysen D, et al. Detection of Schmallenberg virus in different *Culicoides* spp. by real-time RT-PCR. Transbound Emerg Dis. 2012;59:471–5.
24. Elbers ARW, Meiswinkel R, van Weezep E, Sloet van Oldruitenborgh-Oosterbaan MM, Kooi EA. Schmallenberg virus in *Culicoides* spp. biting midges, the Netherlands, 2011. Emerg Infect Dis. 2013;19:106–9.
25. Wernike K, Silaghi C, Nieder M, Pfeffer M, Beer M. Dynamics of Schmallenberg virus infection within a cattle herd in Germany, 2011. Epidemiol Infect. 2014;142:1501–4.
26. De Regge N, van den Berg T, Georges L, Cay B. Diagnosis of Schmallenberg virus infection in malformed lambs and calves and first indications for virus clearance in the fetus. Vet Microbiol. 2013;162:595–600.
27. EFSA – European Food Safety Agency, 2012. "Schmallenberg" virus: Analysis of the epidemiological data and Impact assessment. EFSA J. 10, 2768. http://www.efsa.europa.eu/en/efsajournal/pub/2768.htm (accessed 29 May 2014).
28. IZSVe – Istituto Zooprofilattico Sperimentale delle Venezie. The diagnosis of infectious abortion in cattle – Guidelines [in Italian]. Padua, Italy: Istituto Zooprofilattico Sperimentale delle Venezie, 2011. Available at: http://www.izsvenezie.it/index.php?option=com_content&view=article&id=1535:Gallery&catid=67&Itemid=179&pdf=749798-La-diagnosi-di-aborto-infettivo-nella-specie-bovina-LG-in-VET1/.
29. Méroc E, Poskin A, Van Loo H, Quinet C, Van Driessche E, Delooz L, et al. Large-scale cross-sectional serological survey of Schmallenberg virus in Belgian cattle at the end of the first vector season. Transbound Emerg Dis. 2013;60:4–8.
30. Veldhuis AMB, van Schaik G, Vellema P, Elbers ARW, Bouwstra R, van der Heijden HMJF, et al. Schmallenberg virus epidemic in the Netherlands: spatiotemporal introduction in 2011 and seroprevalence in ruminants. Prev Vet Med. 2013;112:35–47.
31. Steinrigl A, Schiefer P, Schleicher C, Peinhopf W, Wodak E, Bagó Z, et al. Rapid spread and association of Schmallenberg virus with ruminant abortions and foetal death in Austria in 2012/2013. Prev Vet Med. 2014;116(4):350–9. doi:10.1016/j.prevetmed.2014.03.006.
32. West BT, Welch KB, Galecki AT. Linear mixed models: a practical guide using statistical software. Boca Raton, FL, USA: Chapman & Hall/CRC; 2006.

PERMISSIONS

All chapters in this book were first published in VR, by BioMed Central; hereby published with permission under the Creative Commons Attribution License or equivalent. Every chapter published in this book has been scrutinized by our experts. Their significance has been extensively debated. The topics covered herein carry significant findings which will fuel the growth of the discipline. They may even be implemented as practical applications or may be referred to as a beginning point for another development.

The contributors of this book come from diverse backgrounds, making this book a truly international effort. This book will bring forth new frontiers with its revolutionizing research information and detailed analysis of the nascent developments around the world.

We would like to thank all the contributing authors for lending their expertise to make the book truly unique. They have played a crucial role in the development of this book. Without their invaluable contributions this book wouldn't have been possible. They have made vital efforts to compile up to date information on the varied aspects of this subject to make this book a valuable addition to the collection of many professionals and students.

This book was conceptualized with the vision of imparting up-to-date information and advanced data in this field. To ensure the same, a matchless editorial board was set up. Every individual on the board went through rigorous rounds of assessment to prove their worth. After which they invested a large part of their time researching and compiling the most relevant data for our readers.

The editorial board has been involved in producing this book since its inception. They have spent rigorous hours researching and exploring the diverse topics which have resulted in the successful publishing of this book. They have passed on their knowledge of decades through this book. To expedite this challenging task, the publisher supported the team at every step. A small team of assistant editors was also appointed to further simplify the editing procedure and attain best results for the readers.

Apart from the editorial board, the designing team has also invested a significant amount of their time in understanding the subject and creating the most relevant covers. They scrutinized every image to scout for the most suitable representation of the subject and create an appropriate cover for the book.

The publishing team has been an ardent support to the editorial, designing and production team. Their endless efforts to recruit the best for this project, has resulted in the accomplishment of this book. They are a veteran in the field of academics and their pool of knowledge is as vast as their experience in printing. Their expertise and guidance has proved useful at every step. Their uncompromising quality standards have made this book an exceptional effort. Their encouragement from time to time has been an inspiration for everyone.

The publisher and the editorial board hope that this book will prove to be a valuable piece of knowledge for researchers, students, practitioners and scholars across the globe.

LIST OF CONTRIBUTORS

Luigi Calamari, Annarita Ferrari, Andrea Minuti and Erminio Trevisi
Istituto di Zootecnica, Facoltà di Scienze Agrarie, Alimentari e Ambientali, Università Cattolica del Sacro Cuore, Piacenza 29122, Italy

Andrew Bates
Vetlife, Centre for Dairy Excellence, Wilson Street, Geraldine, New Zealand

Rory O'Brien and Simon Liggett
Disease Research Limited, Invermay Agricultural Centre, Mosgiel, New Zealand

Frank Griffin
Department of Microbiology & Immunology, University of Otago, Dunedin, New Zealand

Marco Tolone, Maria Teresa Sardina and Baldassare Portolano
Dipartimento Scienze Agrarie e Forestali, Università degli Studi di Palermo, Viale delle Scienze, Palermo 90128, Italy

Cristian Larrondo and José M. Yáñez
Faculty of Veterinary and Animal Sciences, University of Chile, Av. Santa Rosa, La Pintana, Santiago 11735, Chile

Scott Newman
Genus plc, Hendersonville, TN 37075, USA

Morteza H. Ghaffari, Jayden A. R. MacPherson and Michael A. Steele
Department of Agricultural, Food and Nutritional Science, University of Alberta, Edmonton T6G 2P5, Canada

Harma Berends
Trouw Nutrition R&D, 5830 AE Boxmeer, The Netherlands

Junsheng Dong, Jianji Li, Luying Cui, Yefan Wang, Jiaqi Lin, Yang Qu and Heng Wang
College of Veterinary Medicine, Yangzhou University, Yangzhou, Jiangsu 225009, China
Jiangsu Co-innovation Center for the Prevention and Control of Important Animal Infectious Diseases and Zoonoses, Yangzhou, Jiangsu 225009, China

Bradley N. White
Natural Resources DNA Profiling and Forensics Centre, DNA Building, Trent University, 2140 East Bank Drive, Peterborough, ON, Canada

Vythegi Srithayakumar
Natural Resources DNA Profiling and Forensics Centre, DNA Building, Trent University, 2140 East Bank Drive, Peterborough, ON, Canada
Livestock Gentec, Department of Agricultural, Food and Nutritional Science, University of Alberta, Edmonton, AB T6G2P5, Canada

Gordon B. Mitchell
National and OIE Reference Laboratory for Scrapie and CWD, Ottawa Laboratory Fallowfield, Canadian Food Inspection Agency, 3851 Fallowfield Road, Ottawa, ON, Canada

Guangjun Chang, Lailai Wang, Nana Ma, Wenwen Zhang, Huanmin Zhang, Hongyu Dai and Xiangzhen Shen
Guangjun Chang and Lailai Wang contributed equally to this work. College of Veterinary Medicine, Nanjing Agricultural University, Nanjing, Jiangsu, China

Kaushi Kanankege, Scott Wells, Andres Perez and Julio Alvarez
Department of Veterinary Population Medicine, University of Minnesota, St. Paul, MN, USA

Gustavo Machado
Department of Veterinary Population Medicine, University of Minnesota, St. Paul, MN, USA
Department of Population Health and Pathobiology, College of Veterinary Medicine, North Carolina State University, 1060 William Moore Drive, Raleigh, NC, USA
Raleigh, USA

Val Schumann
Minnesota DHIA, Minnesota Dairy Herd Improvement Association, Buffalo, MN, USA

N. Wagmann and P. Zanolari
Clinic for Ruminants, Vetsuisse Faculty, University of Bern, Bremgartenstrasse 109a, 3012 Bern, Switzerland

C. Spadavecchia and U. Morath-Huss
Department of Veterinary Clinical Science, Anaesthesiology and Pain Therapy Division, Vetsuisse Faculty, University of Bern, Länggassstrasse 124, 3012 Bern, Switzerland

G. Schüpbach-Regula
Veterinary Public Health Institute, Vetsuisse Faculty, University of Bern, Schwarzenburgstrasse 155, 3097 Liebefeld, Bern, Switzerland

Alessandra Crisà and Bianca Moioli
Consiglio per la ricerca in agricoltura e l'analisi dell'economia agraria (CREA) Animal production research centre, Via Salaria 31, 00015 Monterotondo, Rome, Italy

Fabrizio Ferrè
Department of Pharmacy and Biotechnology (FaBiT), University of Bologna Alma Mater, Via Belmeloro 6, 40126 Bologna, Italy

Giovanni Chillemi
Applications and Innovation Department, CINECA, SCAI Super Computing, Via dei Tizii 6, 00185 Rome, Italy

B. Mair, M. Drillich, D. Klein-Jöbstl, P. Kanz, L. Meyer and M. Iwersen
Department for Farm Animals and Veterinary Public Health, Clinical Unit for Herd Health Management in Ruminants, University Clinic for Ruminants, University of Veterinary Medicine Vienna, 1210 Vienna, Austria

S. Borchardt
FirstFarms Slovakia, 900 68 Plavecký Štvrtok, Slovakia
Clinic for Animal Reproduction, Faculty of Veterinary Medicine, Freie Universität Berlin, Berlin, Germany

I. Schwendenwein
Department for Pathobiology, Central Clinical Pathology Unit, University of Veterinary Medicine Vienna, 1210 Vienna, Austria

Mette Bisgaard Petersen, Matthew Denwood and Liza Rosenbaum Nielsen
Department of Veterinary and Animal Sciences, Faculty of Health and Medical Sciences, University of Copenhagen, Grønnegårdsvej 8, 1870 Frederiksberg, Denmark

Nadeeka K. Wawegama, Philip F. Markham and Glenn F. Browning
Asia-Pacific Centre for Animal Health, Melbourne Veterinary School, Faculty of Veterinary and Agricultural Sciences, The University of Melbourne, Parkville, VIC 3010, Australia

C. Gerspach and R. Klingler
Department of Farm Animals, Vetsuisse Faculty, University of Zurich, Winterthurerstrasse 260, Zurich, Switzerland

S. Imhasly
Institute of Pharmacology and Toxicology, Vetsuisse Faculty, University of Zurich, Winterthurerstrasse 260, 8057 Zurich, Switzerland

M. Hilbeb and M. Ruetten
Institute of Veterinary Pathology, Vetsuisse Faculty, University of Zurich, Winterthurerstrasse 268, Zurich, Switzerland

S. Hartnack
Section of Epidemiology, Vetsuisse Faculty, University of Zurich, Winterthurerstrasse 270, Zurich, Switzerland

Alessandro Foddai and Anette Boklund
Section of Epidemiology, National Veterinary Institute, Technical University of Denmark, Bülowsvej 27, DK-1870 Frederiksberg C, Denmark

Anders Stockmarr
Data Analysis Section, Department of Applied Mathematics and Computer Science, Technical University of Denmark, Matematiktorvet, DK-2800 Lyngby, Denmark

Longping Li
College of Animal Science and Technology, Northwest A&F University, YangLing, Shaanxi, People's Republic of China
Life Science Research Center, Yulin University, Yulin 719000, People's Republic of China
Shaanxi Provincial Engineering and Technology Research Center of Cashmere Goat, Yulin University, Yulin 719000, People's Republic of China

Xin Zhao
Department of Veterinary Population Medicine, University of Minnesota, St. Paul, MN, USA
Department of Animal Science, McGill University, 21,111 Lakeshore, Ste. Anne de Bellevue, Quebec H9X 3V9, Canada

Di Jin, Guangjun Chang, Kai Zhang, Junfei Guo, Tianle Xu and Xiangzhen Shen
College of Veterinary Medicine, Nanjing Agricultural University, Nanjing, People's Republic of China

Hui Dong, Yao Yao Lu, Rui Jing Su, Meng Yao Wang, Yi Bao Jiang and Yu Rong Yang
College of Animal Science and Veterinary Medicine, Henan Agricultural University, Zhengzhou, People's Republic of China

Ying Hua Wang
Center for Animal Disease Control and Prevention of Henan Province, Zhengzhou, China

James L. N. Wood
Disease Dynamics Unit, Department of Veterinary Medicine, University of Cambridge, Madingley Road, Cambridge CB3 0ES, UK

Bedaso Mammo Edao
Disease Dynamics Unit, Department of Veterinary Medicine, University of Cambridge, Madingley Road, Cambridge CB3 0ES, UK
College of Veterinary Medicine, Addis Ababa University, Bishoftu, Ethiopia

Teshale Sori
College of Veterinary Medicine, Addis Ababa University, Bishoftu, Ethiopia

Gizachew Hailegebreal
South Agricultural Research Institute, Worabe Agricultural Research Center, Worabe, Ethiopia

Adrian M. Whatmore and Stefan Berg
Animal and Plant Health Agency, Woodham Lane, Addlestone KT15 3NB, UK

Gizat Almaw
National Animal Health Diagnostic and Investigation Centre, Sebeta, Ethiopia

Aboma Zewude, Yemiserach Zeleke and Gobena Ameni
Aklilu Lemma Institute of Pathobiology, Addis Ababa University, Addis Ababa, Ethiopia

Alexander Tavella, Astrid Bettini, Monia Cocchi, Ilda Idrizi, Stefano Colorio, Laura Viel and Claudia Zanardello
Istituto Zooprofilattico Sperimentale delle Venezie, Viale dell'Università 10, 35020 Legnaro, Italy

Patrik Zanolari
Clinic for Ruminants, Vetsuisse-Faculty, University of Bern, Bremgartenstrasse 109A, 3012 Bern, Switzerland

Index

CPSIA information can be obtained
at www.ICGtesting.com
Printed in the USA
LVHW060855030122
707669LV00002B/164

9 781639 874484